J. R. (James Richard) Ainsworth Davis

An Elementary Text-book of Biology

Comprising Vegetable and Animal Morphology and Physiology

J. R. (James Richard) Ainsworth Davis

An Elementary Text-book of Biology
Comprising Vegetable and Animal Morphology and Physiology

ISBN/EAN: 9783744751094

Printed in Europe, USA, Canada, Australia, Japan

Cover: Foto ©berggeist007 / pixelio.de

More available books at **www.hansebooks.com**

AN ELEMENTARY
TEXT-BOOK OF BIOLOGY:

COMPRISING

VEGETABLE AND ANIMAL MORPHOLOGY AND PHYSIOLOGY.

BY

J. R. AINSWORTH DAVIS, B.A.,

TRINITY COLLEGE, CAMBRIDGE;
PROFESSOR OF BIOLOGY AND GEOLOGY IN THE UNIVERSITY COLLEGE OF WALES, ABERYSTWYTH.

With Numerous Illustrations and Glossary.

SECOND EDITION, REVISED AND ENLARGED.

PART II.—ANIMAL MORPHOLOGY AND PHYSIOLOGY.

LONDON:
CHARLES GRIFFIN & COMPANY, LIMITED;
EXETER STREET, STRAND.
1893.
[*All Rights Reserved.*]

TABLE OF CONTENTS.

	PAGE
INTRODUCTION,	1

Chapter I.—Protozoa.

§ 1. Amœba (Proteus Animalcule),	7
§ 2. Vorticella (Bell Animalcule),	11
§ 3. Gregarina,	17

Chapter II.—Cœlenterata.

§ 4. Hydra (Fresh Water Polype),	19
Further remarks on Hydrozoa,	27

Chapter III.—Platyhelmia (Flat-worms).

§ 5. Distoma (Liver-fluke),	28
Further remarks on Flukes,	39
§ 6. Tænia (Tapeworm),	39
Further remarks on Tapeworms,	46

Chapter IV.—Nemathelmia (Thread-worms).

§ 7. Ascaris (Round-worm),	47
Other Thread-worms,	53

Chapter V.—Annelida (Segmented Worms).

§ 8. Lumbricus (Earthworm),	54
Sexual reproduction,	71
§ 9. Hirudo (Leech),	75

CHAPTER VI.—ARTHROPODA.

	PAGE
§ 10. Astacus (Crayfish),	83
Other Crustacea,	107

CHAPTER VII.—MOLLUSCA.

§ 11. Anodonta and Unio (Fresh Water Mussels),	109
§ 12. Helix (Snail),	122

CHAPTER VIII.—VERTEBRATA ACRANIA.

§ 13. Amphioxus (Lancelet),	134

CHAPTER IX.—PISCES (Fishes).

§ 14. Scyllium (Dogfish),	153

CHAPTER X.—AMPHIBIA.

§ 15. Rana (Frog),	174

CHAPTER XI.—AVES (Birds).

§ 16. Columba livia and Gallus bankiva (Pigeon and Fowl),	229

CHAPTER XII.—MAMMALIA.

§ 17. Lepus cuniculus (Rabbit),	268

CHAPTER XIII.—COMPARATIVE ANIMAL MORPHOLOGY AND PHYSIOLOGY,	310
CHAPTER XIV.—MAN,	321
CHAPTER XV.—CLASSIFICATION AND DISTRIBUTION OF ANIMALS,	339

LIST OF ILLUSTRATIONS.

(Names in italics indicate the Sources whence derived.)

FIGURE		PAGE
1. Amœba and Vorticella (*Haddon*, after *Howes* and *Stein*),		8
2. Group of Vorticellæ (*Marshall* and *Hurst*),		12
3. Gregarines (after *Stein* and *Bütschli*),		17
4. Hydra (Original),		21
5. Distoma—Gut (from *Claus*, after *Leuckart*),		29
6. ,, Excretory, Nervous, and Reproductive Organs (after *Sommer*),		30
7. ,, Reproductive Organs (*Claus*, after *Sommer*),		32
8. ,, Development (*Claus*, partly after *Leuckart*),		37
9. Tænia (*Landois* and *Stirling*),		40
10. ,, Reproductive Organs (after *Sommer*),		42
11. ,, Ripe Egg (*Landois* and *Stirling*),		44
12. ,, Cysticerci (,, ,,),		44
13. ,, ,, everted (*Landois* and *Stirling*),		45
14. Ascaris (*von Jaksch*),		47
15. ,, (Original, and after *Leuckart*, *Vogt* and *Yung*, *Bütschli*, and *Van Beneden*),		50
16. Oxyuris (*von Jaksch*),		53
17. Trichina (,,),		53
18. Lumbricus—Gut, Circulatory Organs, Nervous System (Original),		56
19. ,, Reproductive Organs (*Marshall* and *Hurst*),		63
20. ,, Development (after *Wilson*),		69
21. Asterias—Polar Cells (after *Fol* and *Hertwig*),		73
22. ,, Fertilization (after *Fol*),		74
23. Hirudo—General Dissections (after *Leuckart*, and *Vogt* and *Yung*),		77
24. ,, Nephridium and Eye (after *Vogt* and *Yung*, and *Whitman*),		77
25 and 26. Astacus—External Characters (Original),		85
27. Astacus—Appendages (Original),		88
28. ,, Gut and Kidney (Original),		94
29. ,, Reproductive Organs (Original).		99
30. ,, Eye (after *Carrière*),		104
31. ,, Blastula and Gastrula (after *Reichenbach* and *Huxley*),		105
32. Unio—General Dissection (*Claus*, after *Grobben*),		110
33. Anodonta—Nervous System (*Claus*, after *Keber*),		112

LIST OF ILLUSTRATIONS.

FIGURE		PAGE
34. Anodonta—Transverse Sections and Gill (Original),		116
35. Helix—Digestive Organs and Nervous System (Original),		124
36. ,, Buccal Mass (*Claus*, after *Keferstein*),		126
37. ,, Reproductive Organs (Original),		130
38. Amphioxus (*Marshall* and *Hurst*),		137
39. ,, Sections (Original),		140
40. ,, Blastula and Gastrula (*Claus*, after *Hatschek*),		148
41. ,, Sections of Embryos (*Haddon*, after *Hatschek*),		149
42. ,, Larval Stages (*Claus*, after *Hatschek*),		150
43. Scyllium—Skull, &c. (Original),		156
44. ,, Skeleton of Pectoral Fins (Original),		159
45. ,, ,, Pelvic Fins (Original),		159
46. ,, General Dissection (Original),		163
47. ,, Nerves (after *Wiedersheim*),		170
48. Diagrams of Membranous Labyrinth (*Bell*, after *Waldeyer*),		173
49. Rana—Skin (after *Wiedersheim*),		177
50. ,, Endoskeleton (Original, and after *Ecker*),		181
51. ,, General Dissection (Original),		190
52. Unstriated Muscle-Fibres (*Landois* and *Stirling*),		192
53. Rana—Heart (after *Ecker*),		195
54. ,, Arteries (,,),		196
55. ,, Veins (,,),		199
56. ,, Posterior Lymph-Hearts (*Ecker*),		200
57. ,, Muscle-Fibres from Heart (*Landois* and *Stirling*),		201
58. Small Artery showing the Coats (,,),		201
59. Capillaries (*Landois* and *Stirling*),		202
60. Rana—Female Reproductive Organs (after *Ecker* and *Wiedersheim*),		207
61. Histology of Nerve (*Landois* and *Stirling*),		214
62. Rana—Sense Organs (after *Ecker* and *Wiedersheim*),		219
63. Diagrammatic Horizontal Section of Eye (*Landois* and *Stirling*),		220
64. Diagram of Layers of Retina (*Landois* and *Stirling*),		221
65. Rana—Cleavage of Oösperm (*Haddon*, after *Ecker*),		223
66. ,, Early Stages (*Haddon* after *Götte*),		223
67. ,, Diagrammatic Longitudinal Section through Embryo (*Haddon*, after *Götte*),		225
68. Columba—Endoskeleton (Original),		235
69. ,, ,, (,,),		240
70. ,, General Dissection of Male (Original),		243
71. ,, Urinogenital Organs (Original),		250
72. ,, Diagrams of the Membranous Labyrinth (*Bell*, after *Waldeyer*),		255

LIST OF ILLUSTRATIONS.

FIGURE			PAGE
73.	Gallus—Diagrammatic Longitudinal Section through Unincubated Egg (*Claus*, after *Balfour* and *Allen Thomson*),		257
74.	,, Surface Views to show Cleavage in the oösperm (*Haddon*, after *Coste*),		258
75.	,, Section through part of Unincubated Blastoderm (*Haddon*, after *Klein*),		259
76.	,, Transverse Section through Front End of Primitive Streak in First Day Chick (*Haddon*, after *Balfour*),		259
77.	,, Surface View of First Day (20 hrs.) Chick (*Kölliker*),		260
78.	,, Surface View of Chick, rather later than Fig. 77 (*Kölliker*),		261
79.	,, Surface View of Second Day Chick (*Kölliker*),		262
80.	,, Development of the Eye (*Haddon*, partly after *Marshall*),		263
81.	Transverse Section through Embryo Duct at Third Day (from *Haddon*, after *Balfour*),		264
82.	Gallus—Diagram to Illustrate the Embryonic Appendages (*Haddon*, after *Foster* and *Balfour*),		266
83.	Six Stages in the Development of Hair (*Haddon*, after *Wiedersheim*),		270
84.	Lepus—Endoskeleton (Original),		277
85.	,, General Dissection of Head and Thorax (Original),		283
86.	,, Histology of Liver (*Landois* and *Stirling*, after *Hering*),		289
87.	Muscle-Fibres from Mammalian Heart (*Landois* and *Stirling*),		292
88.	Structure of Mammalian Kidney (*Landois* and *Stirling*),		296
89.	Lepus—Urinogenital Organs (Original),		297
90.	Histology of Striated Muscle (*Landois and Stirling*),		300
91.	Diagrams of Membranous Labyrinth (*Bell*, after *Waldeyer*),		304
92.	Lepus—Formation of Blastocyst (*Haddon*, after *E. Van Beneden*),		305
93.	,, Blastocyst (*Haddon*, after *Kölliker*),		307
94.	,, Head of Ten Day Embryo (*Haddon*, after *Kölliker*),		307
95.	,, Embryonic Appendages (*Haddon*, after *Bischoff*),		309
96.	Man—Skull (*Macalister*),		324
97.	,, Sacrum (,,),		325
98.	,, Coccyx (,,),		325
99.	,, Bones of Hand (,,),		327
100.	,, Pelvis, (,,),		328
101.	,, Bones of Foot (,,),		328
102.	,, Cæcum, &c. (,,),		329
103.	,, Aortic Arches (,,),		330
104.	,, Development of Veins (*Macalister*),		331
105.	,, ,, Urinogenital Organs (*Macalister*),		332
106.	,, Brain-exterior (*Macalister*),		335
107.	,, ,, section (,,),		336
108.	,, Fœtal Membranes (*Macalister*, after *Longet*),		337

AN ELEMENTARY
TEXT-BOOK OF BIOLOGY.

PART II.—ANIMAL MORPHOLOGY AND PHYSIOLOGY.

INTRODUCTION.

1. **ZOOLOGY**, the branch of Biology which deals with animals, is such an immense subject that, like Botany, it is conveniently split up into a number of subdivisions. The most important of these are—(1) **Animal Morphology**, dealing with the structure and form of animals; (2) **Animal Physiology**, which treats of their actions or functions; (3) **Development**, the application of morphology and physiology to the study of immature forms; (4) **Classification** or arrangement; (5) **Distribution**, in space and time; and (6) **Phylogeny**, the province of which is to make out the past history of animal groups (phyla). We are mainly concerned in this volume with the first three branches, and, to a much smaller degree, with the questions of classification.

The range of these subdivisions will be comprehended more clearly if we consider what questions they seek to answer in regard to some particular animal, say, for example, the common Frog.

(1) **Morphology** takes note of the external characters, such as shape, colour, divisions of the body, &c., and by means of dissection determines that various systems of organs (digestive, &c., &c.) are present, which have a definite arrangement, and which can

be more or less divided up into smaller parts visible with the eye or with the aid of a lens. The stomach, for instance, is found to have a wall consisting of certain layers or coats. So far we have had to do with *External Morphology* and *Anatomy*. But, with the assistance of the compound microscope, a further analysis is possible, and the special aim of *Histology* is to consider minute structure. Such an analysis demonstrates that the body of a Frog is composed of an exceedingly large number of miscroscopic units, **cells**, differing very much in size, shape, and form, like cells being aggregated into masses known as **tissues**, such as muscular, nervous, &c. There are, besides, other elements formed by or from cells. It is usual to regard the cell as the morphological unit, but the higher powers of the microscope prove that the cell itself is most wonderfully complex.

Morphology, however, in the modern sense, is not a mere descriptive study, but endeavours to discover not only *how* things are but also *why* they are. In the Frog, for example, there is a slender tube, the pineal body, running from the brain to the roof of the skull. In the tadpole it stretches right up to the skin, but the formation of the skull-roof pinches off its end, which remains in the adult frog as a "brow-spot." What, then, is the meaning of the pineal body and brow-spot? No light is thrown upon this question by the study of the Frog only, but by employing the comparative method a definite answer can be given. It is known that in certain lizards the pineal body has the structure which is characteristic of *eyes*, and we, therefore, conclude that the Frog's remote ancestors possessed an unpaired eye in the top of the head, which has since disappeared, leaving an insignificant rudiment. In numberless other instances *Comparative Morphology* (including Comparative Anatomy, &c.) helps to clear up otherwise unintelligible matters.

(2) From the **physiological** standpoint, a Frog is a machine capable of performing various kinds of work. Physiology is concerned, in fact, with the uses of the different parts of the body. It investigates, for example, the processes by which food is digested, absorbed, and built up into living tissue; the circulation of the blood; and the functions of the brain. The problems of physiology, like those of morphology, often require comparative treatment, and they are commonly more difficult to solve.

(3) **Development** (Embryology, Ontogeny) as applied to the

Frog traces its life-history from the egg to the adult condition, and is separated from the two preceding branches merely for the sake of convenience. In the solution of many biological problems it is often of the greatest possible use. An animal in the course of its development passes through a series of stages which to some extent indicate its pedigree, or, as it has been put, "climbs up its own genealogical tree." Thus the Frog starts active life as a gill-breathing aquatic tadpole which is practically a fish in structure, and points to descent from fish-like ancestors.

(4) **Classification** refers the Frog to a definite place in a system of grouping, in which animals are best arranged according to those characters which denote blood-relationship. It is found possible to divide the animal kingdom into several large branches or phyla, which are again subdivided, and the process continued until the individual is reached. The relative status of the various groups may be indicated as follows:—

PHYLUM
    SUB-PHYLUM
        Class
            ORDER
                Family
                    Genus
                        Species
                            Variety.

The Frog is placed in the important phylum **CHORDATA**, which includes all animals that possess, temporarily or permanently, an elastic supporting rod, the notochord, below the central nervous system, as well as clefts which place the cavity of the throat in communication with the exterior. Mammals, Birds, Reptiles, Frogs, Fishes, and certain somewhat lower forms are here included. The Frog's sub-phylum, that of the VERTEBRATA, embraces animals among the common characters of which are possession of a more or less complete brain-case and spinal column. It includes the above-mentioned animals except the forms lower than fishes. Vertebrates are subdivided into several classes, one of which, the AMPHIBIA, comprises frogs, toads, newts, salamanders, and the like, all of which pass through a tadpole-stage.

Those Amphibia which, like frogs and toads, have four well-developed limbs, but are tailless, constitute the order *ANURA*, and this includes a number of families, of which one, **Ranidæ**, comprehends frogs with long hind limbs, webbed hind feet, teeth in the upper jaw, and tongue fixed to the front of the mouth-floor. The most important genus in this family is Rana, to which the common and edible frogs belong, and which is characterized by the rudimentary nature of the thumb, by the presence of teeth on the roof of the mouth, and the deeply forked tongue. Some forty "kinds" or *species* of frog are included in the genus Rana, and this is a convenient place to explain the principle upon which scientific names are given to animals. On the binominal system, which is universally adopted, each animal receives two names, that placed first being the generic, while the other is the specific name. The common or Grass Frog is known as Rana *temporaria*, and the Edible Frog as Rana *esculenta*. The species *temporaria* is marked by a number of distinctive features, among which may be placed the presence of a dark blotch on each side of the head, and the absence of croaking sacs in the male. It is scarcely possible to say what is actually meant by the term "species," and there is much difference of opinion in certain cases. The following definition, given by De Candolle, will serve as well as any:—
"A species is a collection of all the individuals which resemble each other more than they resemble anything else, which can by mutual fecundation produce fertile individuals, and which reproduce themselves by generation, in such a manner that we may from analogy suppose them all to have sprung from one single individual." It is generally found impossible to obtain crosses between different species, or if such crosses (hybrids) are produced these are, as a rule, infertile.

The Frog is not a very good illustration of varieties, to explain which other animals may be taken. The Clouded Yellow Butterfly (Colias *edusa*), for instance, is usually orange and black, but a small proportion of yellow and black individuals are found, and these constitute a variety* of the species *edusa*. Another good example is the Field Snail (Helix *hortensis*), in which the colour and striping of the shell vary in the most remarkable way, so that the species *hortensis* has been split up into a large number of varieties. In such a case, however, the varieties are perfectly

* Hyale.

fertile among themselves, and the crosses (mongrels) between them are also fertile, though there are exceptions to this among plants.

The old view as to the "origin of species" was that they were all separately created, but it is now almost universally held by scientists that living species have been evolved from pre-existing ones, and that varieties are to be regarded as species in process of evolution. A brief sketch of the evolution theory will be given in another place.

(5) **Distribution.**—Rana temporaria is a very widespread species, ranging over the greater part of Europe, N. Africa, N. Asia to Japan, and N. America. It is absent, however, from Iceland and N. Scandinavia.

We have less definite knowledge as to the range in time of this species, but in this country the bones of a frog closely resembling it have been found in a deposit* accumulated at a time when the British fauna included species of Elephant, Hippopotamus, Rhinoceros, and Hyæna.

(6) **Phylogeny** (Ætiology) makes use of data supplied by all the other departments to work out the evolution of groups. The Amphibia must be regarded as one branch of a huge genealogical tree, its orders corresponding to subdivisions of this branch, its families and genera to still smaller ramifications, and its species to the ultimate twigs.

2. Differences between Animals and Plants.—No difficulty is experienced in distinguishing a higher animal from a higher plant. A Frog, for instance, differs in the compactness of its form from, say, a tree, and this difference is correlated with the nature of the food. A tree spreads out leaves for the absorption of carbon dioxide, while it takes up by its roots water charged with inorganic salts. Its food is of simple kind and gaseous or liquid in nature; hence the diffused branching form which offers a large external absorptive surface. A Frog, on the other hand, requires much more complex food, part of which must be albuminous in nature, and it can only get this by devouring other organisms, in this case insects; while tadpoles chiefly live on plants, upon which, indeed, all animals directly or indirectly depend. This complex organic food being solid the Frog requires an internal digestive cavity for its reception, and there is an obvious

* The Cromer Forest-Bed.

necessity for powers of locomotion, which are unnecessary to higher plants. A compact form is obviously advantageous under the circumstances, and the possession of a well-developed nervous system and sense-organs bringing the various activities of the animal into touch with one another and the outer world becomes readily intelligible.

Plants and animals, however, must be regarded as of common descent, forming as it were the two diverging limbs of a V, the higher forms being situated at the ends of those limbs, and the lower forms near the point of union. Hence, as we pass down in the scale the differences between the two kingdoms become less and less marked, until, in the simplest cases, it is often scarcely possible to say whether a given organism be plant or animal.

The distinction which perhaps holds most generally is found in the nature of the food, and plants are also usually characterized by the presence of protective membranes composed of cellulose $(C_5 H_{10} O_5)n$, a substance closely allied to starch. It is, indeed, the presence of these membranes which renders it necessary that the food should be in a gaseous or liquid condition.

3. Distinctions between Living and Non-Living Matter.—An organism is bounded by curved surfaces, while masses of non-living matter are either shapeless (amorphous) or else of crystalline form, in which case they are usually limited by flat faces meeting in straight edges. An organism, too, is of excessively complicated physical and chemical structure, and its living part is always composed of a substance known as **protoplasm**, of which more will be said in the sequel. This complexity is related to a process of constant chemical change (metabolism), involving continual loss of substance, which must be compensated by the intaking of food. This is built up into fresh protoplasmic molecules, which are intercalated between those already existing. By this process of intussusception growth may be effected, up to a certain limit in the case of each organism. A mineral mass, say, for example, a crystal of copper sulphate, is not of such exceedingly complex nature, nor are its molecules subject to constant down-breaking and up-building. It may be kept in an unaltered form for an indefinite period, and since it does not eliminate waste products does not require food. If placed in a saturated solution of copper sulphate it exhibits a kind of growth, by addition of new layers to its outside (accretion), and such increase may go on indefinitely so long as fresh solution is available.

An organism, again, has a life-history, passing through a cycle of changes, which either terminate in death or else in loss of individuality (*cf.* Section on Amœba). This is not the case with a mineral.

4. Biogenesis and Abiogenesis.—All existing organisms, so far as we know, have been derived from pre-existing individuals by processes of reproduction (biogenesis). It was, however, formerly held that some organisms could spring directly from non-living matter (abiogenesis, spontaneous or equivocal generation). This belief was gradually limited to the lowest forms of life, and even for them has now been disproved. A flask partly filled with broth or hay-infusion soon swarms with such organisms if left freely exposed to the air; but if all germs are killed by continued boiling, and the entry of fresh ones prevented by plugging the neck of the flask with sterilized cotton-wool, none of them make their appearance. Such experiments, however, do not prove that abiogenesis has not occurred during some former period of the earth's history.

CHAPTER I.—PROTOZOA.

§ 1. AMŒBA (Proteus Animalcule).

THIS is a microscopic animal, varying much in size. It is found on the surface of the mud in fresh-water pools, on damp earth, in organic infusions, and elsewhere.

MORPHOLOGY.

1. External Characters.—The form of an active Amœba is constantly changing, but it is always irregular. The semi-fluid, transparent protoplasm of which the body is wholly composed is thrust out into bluntish lobes (pseudopodia), which vary continually in number and shape. In most Amœbæ these can be emitted from any part of the body.

8   AN ELEMENTARY TEXT-BOOK OF BIOLOGY.

2. Structure (Fig 1, A).—An outer perfectly transparent *exoplasm* can be distinguished from a more granular internal *endoplasm*. The exoplasm is apparently stiffer than the endoplasm, but there is no clear line of demarcation, and as part of the body may be outside one moment, and inside next, the apparent difference is probably caused by a tendency of the granules to collect

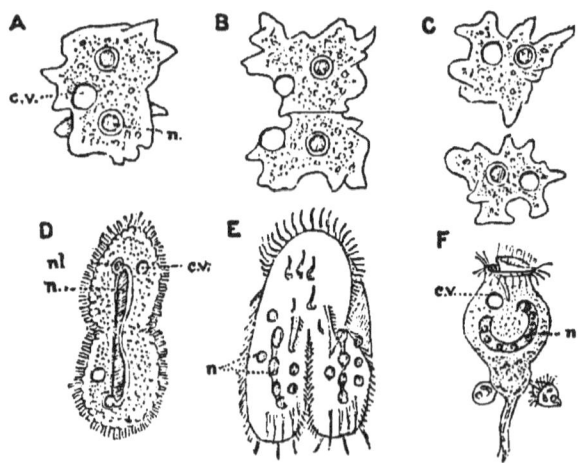

Fig. 1.—AMŒBA AND VORTICELLA (from *Haddon*, after *Howes*, and *Stein*), enlarged.—A-C, Stages in fission of *Amœba;* D and E, fission of other Protozoa; F, conjugation of *Vorticella.* *c.v.*, Contractile vacuole; *n*, nucleus.

in the centre. Some amœbæ, however, possess a firmer exoplasm in which a sort of fibrillation can be made out.

Within the granular portion two important structures are present:—(1) The **nucleus** (*n*), a rounded or ovoid mass, consisting of a modification of protoplasm, and denser than the rest of the body. In the living animal it is inconspicuous, but becomes very obvious on treatment with weak acid or a staining solution. In some cases, at any rate, the nucleus is invested by a delicate membrane, and consists of two substances, one, *chromatin*, staining readily—the other, *achromatin*, staining with difficulty. The arrangement of the chromatin differs with the species, but part of it is frequently aggregated into a central particle, the *nucleolus*. (2) The **contractile or pulsating vacuole** (*c.v.*), a spherical space with fluid contents, which alternately increases and diminishes in size, in a rhythmic manner. There may be more than one.

More or less food, surrounded by fluid, is usually present in the endoplasm, occupying spaces known as **food-vacuoles**. Gas-containing vacuoles have also been observed in some specimens.

PHYSIOLOGY.

The functions carried on by Amœba and, indeed, by all other organisms may be conveniently considered under five headings.

1. Nutrition.—The food of Amœba consists of small organisms and organic particles, which its pseudopodia first come into contact with, and then flow around. All parts of the body alike serve for the reception of food. The complex chemical composition of the substances utilized, and their *ingestion* in the solid state are worthy of notice, as typical *animal* characteristics. Each mass enters the body surrounded by a small quantity of water, and a food-vacuole is thus constituted, the temporary breach in the protoplasm being at once closed up. Within the body, mechanical and chemical influences are brought to bear upon the food, it being subjected, on the one hand, to slow rotation in the endoplasm, while on the other, either the protoplasm itself or a digestive juice formed (secreted) by it reduces the digestible portions to a state of solution, or fine division. The food thus digested is then *assimilated*, that is, built up into living protoplasm, while the indigestible and undigested remnants are thrown out at any convenient point.

Encystment.—If conditions of temperature or food-supply are unfavourable, Amœba possesses the power of assuming a spherical form, and then secreting a firm structureless coating or *cyst*, which is possibly of a horny nature, and is a very considerable protection.

2. Katabolism.—The complex protoplasmic molecules are constantly breaking down into simpler substances, and potential energy is thus transformed into kinetic. This process is, as a rule, far more rapid in animals than in plants, and greater activity is consequently displayed by the former.

The products of Katabolism (Katastates) are:—

  a. **Secretions**, which are utilized in some way before passing out of the body. It is probable, for example, that Amœba *secretes* something which acts chemically upon the food, and corresponds to the digestive juices of other forms.

b. **Excretions** (waste-products), which pass out of the body at once. These are—water, carbon dioxide (CO_2), and compounds of ammonia.

Katabolism is, practically, a process of oxidation, the result being that animals, like most plants, require a supply of free oxygen, and the term **Respiration**, or breathing, is applied to the taking in of oxygen with concomitant excretion of carbon dioxide (and water).

The *contractile vacuole*, which is alternately enlarged to a certain maximum size by the gradual aggregation of liquid, and obliterated by the contraction of the surrounding protoplasm, appears to communicate with the exterior. It probably serves as a receptacle in which can collect, at one time, oxygenated water from the exterior, at another time waste-products. The general surface of the body is also doubtless respiratory, oxygen diffusing in and carbon dioxide diffusing out.

3. **Reproduction** (Fig. 1, A-C).—This is asexual, and usually effected by *binary fission*, the animal splitting into two equal parts. The nucleus elongates, becomes dumb-bell-shaped, and separates into two, each half being surrounded by a moiety of the protoplasm, which has meanwhile been gradually constricted. The two new individuals grow to the adult size, divide again, and so on, there being no known limit to the process.

An Amœba, in fact, instead of dying, divides into two new and vigorous individuals. Hence it has been said to be, in a sense, immortal.

4. **Contractility** is one of the primary properties of protoplasm, in virtue of which a change of form is effected. The process does not involve diminution in volume, since reduced breadth in one or more directions is made up for by equivalent increase in other directions. In Amœba there are irregular contractions of the protoplasm leading to the formation of pseudopodia. Where one of these is about to be thrust out, the exoplasm (which is probably specially contractile) is protruded as a clear knob, into which, as it increases in size, the endoplasm enters. An anterior *progressing region*, actively flowing into pseudopodia, and a more passive *following region*, can be made out in some Amœbæ. These regions are in ordinary cases determined by the direction of movement,

and simply correspond to the anterior and posterior ends for the time being. There are, however, some species which can only thrust out pseudopodia from part of the body, and which therefore possess *permanent* progressing and following regions.

It appears probable that a delicate film of firmer nature rapidly forms on the surface of naked protoplasm. The direction of pseudopodia may perhaps, in some Amœbæ, be determined by the rupture of such a film at definite points. The movements of Amœba are so characteristic that the special epithet, "amœboid," is given to all similar movements, wherever occurring.

5. Irritability and Spontaneity.—That Amœba is affected by external agents or stimuli may readily be noticed. Pseudopodia which come in contact with food-particles flow around them, but inedible substances are usually rejected. Increase of temperature enhances activity, but as $35°C.$ is neared, this is diminished, and arrested when that point is reached. Death is caused by a rise of $5°$ or $10°$ more. On the other hand, activity is diminished by cooling, and arrested at the freezing-point. Weak electric shocks cause the animal to assume the spherical form, and become quiescent.

Amœba is constantly moving, and it cannot be doubted that some of its movements are *spontaneous*—*i.e.*, not directly due to external stimuli, but resulting from internal causes, such, perhaps, as chemical decomposition.

§ 2. VORTICELLA (Bell Animalcule).

Vorticella is an animal of microscopic dimensions, which is found attached to marine and fresh-water plants, &c., and may be distinguished by the aid of a lens.

MORPHOLOGY.

1. External Characters (Fig. 2).—The body is shaped something like a pear or rounded cone, and, in most individuals, is attached to some object by an elongated slender stalk, into which its apex is drawn out. It is convenient to speak of the attached end of the body as *proximal*, the free end as *distal*. The animal possesses considerable powers of movement, and may be found fully expanded, fully retracted, or in an intermediate condition.

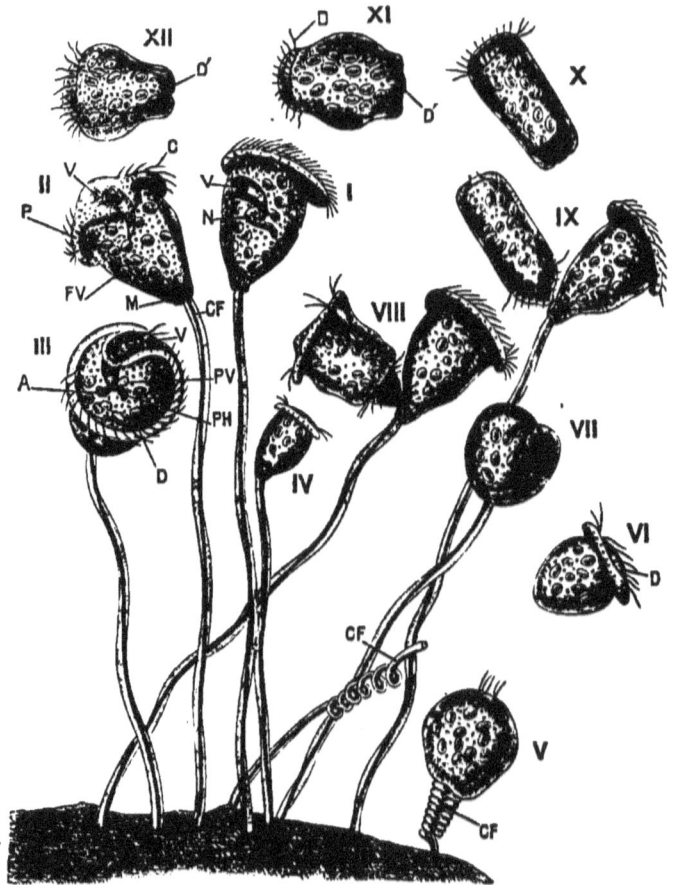

Fig. 2.—GROUP OF VORTICELLÆ (from *Marshall* and *Hurst*). × 220.— I., II., and III. show the animal in various positions; IV. is a much smaller specimen drawn to the same scale; V. shows a specimen made to contract by action of alcohol; VI. is detached from its stalk, and swimming away freely, disc forwards; VII., VIII., and IX. show three stages of fission; X., XI., and XII. show the separated individual swimming by means of the aboral circle of cilia; XI. is slightly contracted; XII. strongly contracted after the cover-glass has been tapped. A, Food-vacuole discharging contents at anus; C, cilia of the disc; C F, contractile fibre of stalk; D, disc; D', disc contracted; F V, food-vacuole; P H, gullet; P V, pulsating vacuole; P, peristome; M, myophan striation; N, nucleus; V, vestibule.

When it is completely expanded the stalk is straight, or slightly curved, while the distal end, or *disc*, forms a flattish projection bordered by a thickened rim or *peristome*. Within the peristome is a groove which surrounds the disc and forms on one side a deep depression, the *vestibule*. In the groove is a row of delicate, hair-like, protoplasmic processes (cilia). The retracted state is initiated by the withdrawal of the disc; then follows the folding-in of the peristome, so that the body becomes rounded, and, at the same time, the stalk is thrown into a tightly-coiled spiral.

Smaller free-swimming, stalkless individuals are also met with, simpler in structure and possessing a circlet of cilia near the arboral end of the body.

2. **Structure.**—The body is transparent as in Amœba, and its internal parts can therefore be readily studied. Covering the whole external surface, cilia excepted, is a thin elastic membrane, the *cuticle*, which may exhibit transverse striations. It is thinnest on the disc and peristome, and thickest on the stalk, of which it forms the sheath. The cuticle is secreted by the underlying protoplasm, and owing to its presence the body possesses a constant form, while the protrusion of lobe-like pseudopodia is rendered impossible. The cilia, which may be compared to permanent thread-shaped pseudopodia, pass through holes in the cuticle. The protoplasm is divided into a firm, finely granular, external *ectosarc* or *cortical layer*, and a semi-fluid *endosarc*. The deeper part of the ectosarc exhibits more or less distinct longitudinal striations (myophan striations), probably due to a fluting of the internal surface. This deeper part is produced into a contractile filament, which is slightly twisted in a spiral manner, and traverses the cavity of the hollow stalk, attached here and there to the firm sheath.

As in Amœba, a nucleus and contractile vacuole are present. The former is a horse-shoe-shaped band placed in the broad end of the body, just within the ectosarc; the latter is spherical, and situated nearly the vestibule. There is a small round *paranucleus* near the nucleus.

A rudimentary **digestive apparatus** can here be distinguished for the first time. The vestibule, which performs the function of a mouth, is continued into a short tube, the *pharynx*, which is lined by a continuation of the cuticle, and is provided with numerous short cilia. It passes down into the body with a somewhat curved course, and ends abruptly within the endosarc. A small area in the vestibule, beneath the disc, is devoid of cuticle, and serving, as it does, for the extrusion of remnants of the food, may be termed the *anus*. Within the endosarc numerous *food-vacuoles* can usually be distinguished.

PHYSIOLOGY.

Vorticella, like Amœba, is an unicellular animal, but the single cell of which its body is composed is very much specialized or differentiated, since different parts of it are modified for the performance of different functions.

1. **Nutrition.**—The combined action of the cilia produces a current by which minute organisms and organic particles are carried down to the end of the pharynx, and, together with a small amount of water, forced into the endosarc. There, in *food-vacuoles*, they slowly pass down one side of the body and up the other, the water being gradually absorbed, and the nutritious parts digested as in Amœba. The undigested remnants are ejected from the anus, which is only visible at the moment of extrusion, and are carried out of the vestibule by the ciliary current.

Under unfavourable conditions Vorticella sometimes detaches itself from its stalk, and swims away by means of its cilia, becoming re-attached, and developing a new stalk, if a suitable spot is reached. As in Amœba, unfavourable conditions may also lead to *encystment*, the body, either whilst attached to its stalk or after separation, becoming rounded, and secreting a protective horny cyst.

In one species of Vorticella, *V. viridis*, *chlorophyll* is diffused throughout its body, which no doubt enables it to live, in part, like a green plant, utilizing carbon dioxide as a source of carbon.

2. **Katabolism.**—As in Amœba, something akin to a *digestive secretion* is probably formed by the protoplasm. The same waste-products, namely, carbon dioxide, water, and ammonia compounds, result from the breaking-down of the body, and, as before, the *contractile vacuole* is probably **excretory** and **respiratory**, waste-products passing from it into the vestibule, with which it has been observed to communicate, while, on the other hand, oxygenated water may be taken up into it from the exterior. The

ciliary current brings dissolved oxygen with it, and also carries away waste-products.

3. Reproduction (Fig. 2).—This is *asexual*, and usually takes place by equal *binary fission* in a longitudinal direction. The animal broadens, its nucleus elongating, and then a furrow appears, which rapidly deepens so that the animal is cleft down to its stalk, the nucleus and contractile vacuole being halved. Two equal-sized individuals result, one of which remains attached to the stalk, while the other develops a circlet of cilia near its proximal end, becomes detached, and swims away, later on becoming fixed by its proximal end and developing a stalk. Thus Vorticella, although a fixed form, can readily spread from place to place, and it is, therefore, not to be wondered at that it has a very wide distribution, especially if it be remembered that the process of fission only takes an hour or two for its completion.

Small free-swimming individuals (microzooids), similar in structure to the large free forms, may be produced by unequal fission, in which case an ordinary zooid divides into two parts, one large, the other small,—or else continued fission may take place in which equal bipartition is immediately followed by rapid division of the half to be detached into eight microzooids. Using large letters for the larger zooids (macrozooids) and small ones for the microzooids, the different kinds of fission may be expressed as follows:—

Equal Fission. *Unequal Fission.*

$$A \div \begin{cases} B \text{ (fixed)} \\ B' \text{ (free)} \end{cases} \qquad A \div \begin{cases} B \text{ (fixed)} \\ b \end{cases}$$

Continued Fission.

$$A \div \begin{cases} B \text{ (fixed)} \\ B' \div \begin{cases} c\ c\ c\ c \\ c\ c\ c\ c \end{cases} \end{cases}$$

A process, known as **conjugation**, which has an important bearing on the origin of sexual reproduction, also frequently occurs in Vorticella. A microzooid comes into contact, by its ciliated end, with a large fixed macrozooid, at a point near the junction of the stalk. The two gradually fuse together, parts of the paranuclei uniting. Increased vigour (rejuvenescence) appears to be imparted by this process, which generally shows itself in greater reproductive energy, as displayed in fission. More rarely, encystment may follow, when the nucleus becomes larger and much longer, ultimately breaking up into a number of *spores*, oval bodies, each provided with a circlet of cilia at one end. These are liberated by the rupture of the cyst, and becoming attached by their ciliated ends, grow up into adult forms. Before this they may, however, increase by a process of fission.

4. **Contractility.**—The presence of a firm cuticle prevents the formation of pseudopodia, though protoplasmic currents are observable in the endosarc, by which the food-vacuoles are carried round the body, and such streamings are also seen in the firmer ectosarc, though to a much less extent. Locomotion is effected in the free forms by cilia, while the currents that bring food and oxygen, and carry away waste, are also due to ciliary action. Each cilium is a delicate thread of protoplasm, protruded from the ectosarc through a pore in the cuticle. By means of the alternate contraction of its longitudinal halves, bending and straightening are produced in turn.

The retraction of the body is due to the ectosarc, and in this case contraction takes place is the direction of the myophan striation, whence it appears probable that the deeper layer is the part mostly concerned in the process. The thread which traverses the stalk contracts in such a definite way, getting shorter and broader, that it deserves to be called a specially contractile or muscle fibre. The spiral direction taken by the fibre causes the stalk to be readily thrown into spiral folds. The much slower process of expansion appears to be largely due to the elasticity of the cuticle, and this is especially true of the stalk.

5. **Irritability and Spontaneity.**—These phenomena are much more definitely exhibited than in the case of Amœba. The animals appear sensitive to the slightest touch, which causes them to contract rapidly. The same effect may be produced by irritant solutions, such as weak acetic acid, and by other stimuli. **Spontaneity** is shown by the way in which the cilia work together to a common end, instead of acting irregularly. Spontaneity is further exemplified by free-swimming individuals in fixing and conjugation.

§ 3. GREGARINA.

Gregarines are Protozoa, often of worm-like form, in which a relatively large size may be attained (up to $\frac{3}{4}$ of an inch long). They are parasitic, living at the expense of other animals, and as they are found within the bodies of these "hosts," may be termed *endoparasites*. Since they are surrounded by abundance of nutritious food, there is no necessity for locomotor or current-producing organs, and, in fact, there are neither pseudopodia nor cilia, and digestive organs are absent as well.

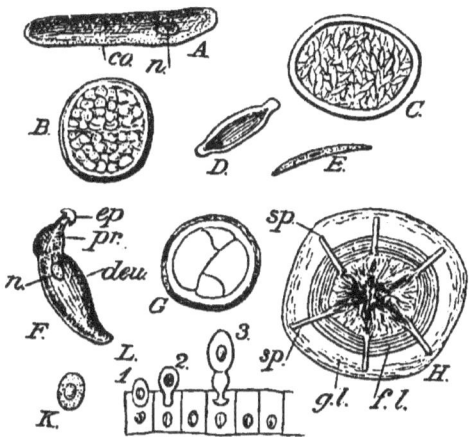

Fig. 3.—GREGARINES (after *Stein* and *Bütschli*). Enlarged to various scales.—A-E, Stages in the life-history of *Monocystis agilis*; A, adult; co, cortical layer (dotted); n, nucleus; B, cyst, with developing spores; C, cyst, with ripe spores; D, a spore with contained falciform young; E, one of the falciform young; F-K, stages in the life-history of *Gregarina blattarum*; F, young adult; ep, epimerite; pr, protomerite; deu, deutomerite containing nucleus, n; G, cyst containing two conjugated individuals; H, ripe cyst; g.l, gelatinous layer; f.l, firm stratified layer, within which is a network, still enclosing spores in the centre; sp, sporoduct; K, a spore; L, epithelial cells from intestine of cockroach, with attached stages (1, 2, 3) of developing spore-contents.

The vesiculæ seminales of the earthworm are infested with one kind of Gregarine, to which the name of **Monocystis** has been applied. This has a cylindrical body (Fig. 3), which in the largest species may be $\frac{1}{5}$ of an inch long. It is covered with a firm cuticle, within which is a well-marked ectosarc or cortical layer, in which an appearance of longitudinal striation may be observed, probably due to an internal fluting, as in Vorticella. Worm-like movements can be performed by the contraction of this layer. Within the ectosarc is a more fluid endosarc, and there is a large nucleus at one end. There is no contractile vacuole, which is not surprising when the sluggish nature and consequently slow katabolism of the animal is considered.

Monocystis undergoes encystment as a regular part of its life-history, and this is usually preceded by the conjugation of two individuals, which come into contact by their anterior ends, sometimes fuse, but in any case assume a spherical form and become surrounded by a firm cyst, said to be formed from the cuticle and an altered part of the underlying protoplasm. The contents of the cyst now break up into a large number of spindle-shaped spores (pseudonavicellæ) invested in firm envelopes, and then, by a process of division, eight nucleated sickle-shaped cells (falciform young) are formed within each spore, a small part of the protoplasm, however, being left unused (residual core). The falciform young are ultimately liberated by rupture of the cyst and the spore-envelopes. They are capable of movement by alternate bending and straightening, and make their way into those cells (spermatospores) from which the male sex-cells of the earthworm are developed, thus becoming intracellular parasites. After a time they quit these cells and gradually assume the adult form.

Another common kind of Gregarine is **Gregarina (Clepsidrina) blattarum,** found in the intestine of the cockroach. A mature individual, when young, possesses an elongated tapering body (Fig. 3), divided into three regions, (*a*) an anterior epimerite ("cap"), provided with hook-like cuticular processes, and passing into (*b*) a protomerite, separated by a transverse septum of ectosarc from (*c*) a much larger deutomerite, in which the nucleus is contained. These regions are probably not distinct cells, but merely parts of one cell. Sooner or later the epimerite is thrown off.

Two individuals unite, the anterior end of one being opposite the posterior end of the other, and a complicated cyst is secreted, consisting of a number of firm layers with an external gelatinous investment. Repeated nuclear division presumably takes place, and a considerable number of barrel-shaped spores are formed just within the wall of the cyst, migrating later on to the centre. Part of the protoplasm remains as a kind of network, and a number of tubes (sporoducts) are formed, which are at first directed inwards, but are later on turned inside out (everted), projecting from the cyst, and serving as channels through which the spores escape. This, of course, involves absorption of the firm cyst at points where sporoducts are formed. From each spore a single embryo escapes, which makes its way into one of the epithelial cells lining the intestine of the cockroach. The regions of the body are now gradually developed, the protomerite and deutomerite projecting into the cavity of the intestine, while the epimerite remains imbedded in the cell, and is soon thrown off. The young Gregarina absorbs the digested food with which it is surrounded, and quickly grows to the adult size.

CHAPTER II.—CŒLENTERATA.

§ 4. HYDRA (Fresh Water Polype).

THIS is a small animal common in ponds, ditches, and stagnant streams. Extended specimens measure from half an inch in length downwards. There are two common kinds, the Brown Hydra (*Hydra fusca*), and the rather smaller Green Hydra (*Hydra viridis*).

MORPHOLOGY.

1. **External Features** (Fig. 4).—The body of Hydra, when extended (A), is in the form of a hollow cylinder, usually attached to some object by its closed proximal end, which constitutes an adhesive disc (the "foot"). The distal end is terminated by a conical projection, the *hypostome*, in the centre of which is a rounded opening, the *mouth* (*m*). Several (5 to 8) slender prolongations of the body, known as *tentacles* (*tn*), project, at regular intervals, from the base of the hypostome. Both body and tentacles may

undergo various stages of contraction or shortening. In extreme contraction, the entire animal appears (A') like a small, rounded, gelatinous knob, brown or green, according to the species. The external appearance presents certain characteristic features in connection with reproduction (*see below*).

By looking down upon the distal end of Hydra, it is seen that a number of radiating lines, all passing through similar parts, can be drawn from the centre of the mouth along the tentacles. This sort of symmetry, common among lower animals, is of the kind termed *radial*.

2. **Body-wall.**—Amœba, Gregarina, and Vorticella are *unicellular*, each being made up of one *cell* or *morphological unit*. These animals therefore possess diverse parts in virtue of the specialization, *differentiation*, of the protoplasm of single cells. Hydra, on the other hand, is *multicellular*, being made up of very numerous cells, each of which is morphologically equivalent to an Amœba. These cells, however, are modified in various ways for the performance of diverse functions. This is the principle of **physiological division of labour**, and the accompanying diversity of form is termed **morphological differentiation**. Aggregates of cells, similar in form and mode of origin, and specially capable of carrying on a particular function or functions (instead of *all* functions, as in Amœba), are known as **tissues**, and these again are interwoven into **organs**, digestive, reproductive, &c. In Hydra this specialization is not very complete; but, as the animal scale is ascended, physiological division of labour and morphological differentiation become more and more marked.

Histology.—The body of Hydra contains a large digestive cavity, prolonged into the hollow tentacles. The wall of this cavity is the *body-wall*, and the cells composing it are divided into two distinct layers, an external *ectoderm* (*ec*), and an internal *endoderm* (*en*), about twice as thick. Between the two is a very thin structureless membrane, the *mesoglœa* (intermediate or supporting lamella) (*i.l*), not composed of cells.

The **ectoderm** (*ec*) is mainly made up of large somewhat conical cells (C) which are broadest externally, while the narrow internal end of each of them is drawn out into one or more contractile tail-like processes, which may branch. These "tails" take a longitudinal direction, and are closely attached to the mesoglœa. Each large ectoderm cell possesses a clear external border, and contains a large nucleus with nucleolus. Adjacent cells touch one another externally, whilst, internally, spaces are left between them which are occupied by *interstitial cells (i.c)*. These are small nucleated cells present in all parts of the body but the foot, and often crowded together so as to obscure their individual outlines.

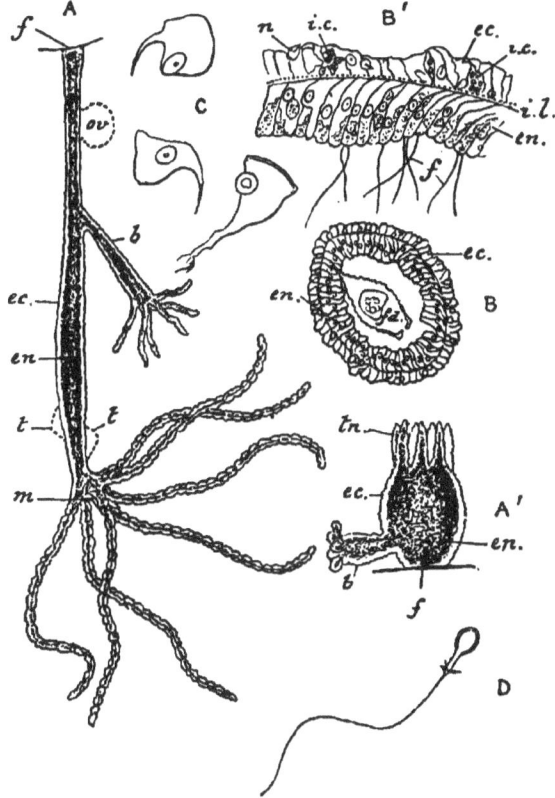

Fig. 4.—HYDRA, various scales.—A, Budding specimen of *H. fusca* extended; A', ditto, retracted; B, tranverse section of *H. fusca;* B', part of same on larger scale; C, isolated tailed cells; D, isolated thread-cell, from *H. viridis; f,* foot; *m,* mouth; *tn,* tentacles; *ec,* ectoderm; *en,* endoderm; *b,* bud; *ov,* typical position, &c., of ovary; *t,* do., of testes; *i.l,* intermediate lamella, the dots above this represent the cut ends of "tails;" *i.c,* interstitial cells; *n,* thread-cells; *f,* flagella; *fd,* organism (in B) in digestive cavity, as food.

Within some of the interstitial cells (cnidoblasts) oval vesicles (nematocysts) are developed, each of which contains a coiled-up filament. Nematocysts are chiefly found in the distal end of the body, especially in the tentacles, where they are aggregated into "batteries," giving rise to an irregular surface. If a little weak acid is added to the water which contains a Hydra, the threads will be rapidly shot out.

The cnidoblasts are at first situated deeply in the ectoderm, but as their nematocysts ("thread-cells") develop, force their way to the surface. Several kinds of nematocyst may be distinguished, of which the largest consist of a transparent vesicle containing fluid, and (when examined in the discharged condition) this it seen to be produced (D) into a long tapering neck, which towards its end bears a few large backwardly directed spines, and is continued into an extremely long and delicate hollow filament or thread. In an unused nematocyst, the neck is turned into the vesicle, within which the filament is coiled up. The remains of the cnidoblast persist as a layer of protoplasm surrounding the vesicle and produced externally into a slender sensitive process, the cnidocil, which slopes towards the free end of the body or tentacle, as the case may be. There are also other much smaller nematocysts with shorter, thicker threads, and without barbs. Small star-shaped *nerve-cells*, connected with the protoplasmic investment of the thread-cells, are said to exist in the ectoderm.

The **endoderm** (*en*) is made up of a single layer of large, irregular, granular cells, each containing a big flattened nucleus with nucleolus, and one or more vacuoles, often of great size. The endoderm cells abut at one end upon the supporting lamella (and appear to be here produced into *transrersely*-arranged tails), while the other end projects into the digestive cavity, and can be thrust out into pseudopodia. This end often bears one or more elongated protoplasmic threads (flagella*) (*f*), which are capable of being withdrawn, like pseudopodia. The external half of each cell contains in *Hydra viridis* numerous small globular bodies, coated with chlorophyll. *Hydra fusca* possesses small bodies of similar nature, but containing particles.

Reproductive Organs.—Male and female sex-organs are differentiated, and, as both occur in the same animal, Hydra is said to

---

* Flagella are longer than cilia, capable of more complex movement, and either occur singly or associated in small numbers.

be *hermaphrodite*. They are only found, as a rule, during the autumn months.

(1) **The Male Organs** consist of a varying number (1 to 20) of *spermaries* (testes) (t, t), usually placed near the distal end of the body. Each is a conical or rounded elevation, the *wall* of which is formed by large ectoderm cells, while within is an aggregate of interstitial cells. The spermaries are, in fact, projections caused by the increase of these cells at particular points of the body. These contained *germinal cells* develop into *sperms* (spermatozoa), minute tadpole-like bodies, with oval heads and long vibratile tails. The head of the sperm is mainly formed from the nucleus of the germinal cell, covered by a thin film of its protoplasm, the rest of which is drawn out into the tail.

(2) **The Female Organs** are known as *ovaries* (*ov*), and are typically developed at the proximal end in the same manner as the spermaries. *Hydra viridis* usually has but one ovary; *H. fusca* may possess from one to eight. Each is a rounded projection, much larger than a spermary, and, when mature, contains a single large egg-cell or *ovum*. The young ovary contains at first a large number of germinal cells. One of these, occupying a central position, grows more vigorously than the others, becomes amœboid, and uses them as food, amœba-fashion. This relatively large cell is the ovum. It assumes a spherical form on attaining its full size, and though resembling ordinary cells in structure, its parts receive special names. The protoplasm is termed *vitellus*, and contains numerous highly refractive bodies, known as *yolk-spherules*. The large spherical nucleus, or *germinal vesicle* contains a nucleolus, the *germinal spot*.

PHYSIOLOGY.

As might naturally be expected, the ectoderm is more especially concerned with the function of irritability and spontaneity, while the endoderm carries on digestion.

1. Nutrition.—The food consists of small animals, often of relatively high organization, such, for example, as the active little Crustacea known as water fleas. The tentacles are the agents by which these are conveyed to the mouth, and their nematocysts play an important part in this connection. The following description applies to the large forms. In the quiescent state, the neck and filament of the thread-cell are *inside* the vesicle, which contains a poisonous fluid. If now a small animal touches the cnidocil (trigger-hair), the neck is first turned inside out, *everted* (*d*), and then, by means of its spines, becomes fixed in the animal. The thread is now everted within the wound, and the poisonous fluid exerts a paralyzing influence. The mouth of the animal is very extensile, and capable of taking in morsels of considerable size (B *fd*). Within the digestive cavity movement of fluids, &c., is mainly effected by the action of the endodermal flagella. The endoderm-cells vary much in size at different times, and can extend themselves so as to almost obliterate the cavity. It is stated that some of these cells are glandular, secreting a fluid which diffuses into the digestive cavity, where it dissolves and breaks down the food; but digestion is also largely *intra-cellular*, the pseudopodia ingesting solid particles in the same way as in Amœba. Within these cells the food is reduced to solution, or fine division, and readily passes to the other parts of the body by diffusion. Undigested remnants are passed out through the mouth.

In *Hydra viridis* the presence of chlorophyll probably enables the nutrition to be partly effected as in green plants.

2. Katabolism.—Secretions are elaborated by glandular cells in various parts of the body. All parts of the body alike give rise to waste-products, which are secreted into the surrounding water, and in the process of Respiration oxygen is taken in (*cf.* p. 10).

3. Reproduction is of two kinds, asexual and sexual, the one irrespective of, the other dependent on, the reproductive organs.

(1) **Asexual Reproduction** (A and A') is effected by *gemmation*, and is not limited to any special time of year, but depends upon the food supply and temperature. A small knob grows out from the side of the body, and not only the ectoderm and endoderm but also the digestive cavity of the parent are continued into it. This bud (*b*) lengthens, develops a mouth and tentacles, and is finally pitched off at its proximal end, which becomes an attachment disc. Several buds in the same or different stages of development may be present on one individual at the same time, and, under favourable circumstances, these may themselves bear buds before becoming detached. *Fission* is not known to occur in nature, but a Hydra may be cut into several pieces, each of which can develop into a new and perfect individual.

(2) **Sexual Reproduction.**—The wall of the ripe ovary bursts, and part of the surface of the now spherical ovum is exposed. By the rupture of the spermary at its apex, the sperms are liberated, and by the action of their tails move rapidly head first through the water. *Impregnation* or *fertilization* of the ovum now takes place, that is, a *sperm* (derived from the same or a different individual) fuses with it to form an *oösperm*. This union of a small active male cell with a relatively large and passive female one, is the essential part of sexual reproduction.

4. **Contractility.**—Amœboid and ciliary movements are seen in the endoderm. Besides these, Hydra is capable of movements on a larger scale. The body and tentacles can be extended and contracted, their appearance varying very much in consequence, and curvings, very complex in the case of the tentacles, can also be effected. These movements may lead to *locomotion*, either slow by the gradual shifting of the foot, or more rapid by the alternate sucker-like attachment of the foot and the mouth. Hydra has also been seen creeping along on its tentacles, the body being held almost vertical. It can also float passively near the surface of the water, mouth downwards.

The retraction of the body subserves *protection*, whilst the tentacular movements are mainly concerned with the procuring of food. All these movements are the result of longitudinal and tranverse contractions, separate or variously combined. The tails of the large ectoderm cells are arranged longitudinally, and by their shortening or contraction furnish the retracting element. If the endoderm-cell really possesses traversely-arranged tails, their contraction would give the necessary extending element. If this is not the case, the conical parts of the tailed-cells could effect the same purpose, by contracting traversely, the result being a corresponding increase in length.

Discharge of a nematocyst is caused by the contraction of the cnidoblast. By this means the contained fluid is put under pressure, and everts first the neck, and then the hollow filament. There is no contrivance for effecting the reverse process, and nematocysts, when once used, are therefore discarded.

5. **Irritability and Spontaneity.**—Hydra is extremely sensitive to external stimuli, and this is especially well seen in the movements of retraction and in the discharge of nematocysts. The cnidocils are **sensory**, that is to say, specially modified for the

reception of external stimuli, in this case chemical. They are turned in those directions whence food is most likely to approach, *i.e.*, the oral end of body and tips of tentacles. The external surfaces of the large ectoderm-cells appear to receive those stimuli which lead to movements.

Spontaneity is best shown by the co-ordination of the cells for the performance of useful movements, such as those of locomotion, which cannot all be directly due to external influences. The co-ordinative power certainly resides in the ectoderm, either in the large cells or nerve-cells, or probably in both. The discharge of the nematocysts also seems to depend on the "will" of the animal. If a cnidocil is touched, an impulse of some sort appears to be transmitted to the connected nerve-cell, which, in the case of food, sends back an impulse leading to the eversion of the cell, while a particle of sand, say, causes no such disturbance.

DEVELOPMENT.

In each ovary, as a general rule, only one ovum is developed, which when almost mature projects freely to the exterior, as a result of the rupture of the cells which cover it. Before fertilization the ovum divides into two unequal parts, the smaller of which is known as a polar cell. A second polar cell is next formed by further subdivision, and the nucleus of the ovum is now known as the female pronucleus. The two polar cells perish without taking any share in the further development.

Fertilization now ensues, a single sperm fusing with the ovum, and its nucleus (male pronucleus) uniting with the female pronucleus to constitute the nucleus of the oösperm (segmentation nucleus). The oösperm at once commences to develop, and during the earlier stages remains attached to the parent. The first process is that of **segmentation** or **cleavage**, consisting of a series of cell-divisions by which the unicellular oösperm is converted into a cellular mass. It is, in fact, a case of continued fission, but the products of division, instead of separating and becoming distinct organisms, as in Vorticella and the like, remain connected together and constitute the rudiment of a multicellular animal. The cleavage in Hydra consists of a series of bipartitions, so that the embryo consists of 2, 4, 8, 16, 32, &c., cells successively. Cleavage is here said to be both *complete* (holoblastic) and nearly *regular*, for the whole of the oösperm divides, and the resultant cells (blastomeres) are of about the same size. At the end of cleavage the embryo is a hollow sphere or **blastula** (blastosphere), the cavity of which is known as the **blastocœle** (segmentation cavity). Cells are now budded off into the blastocœle from the attached side of the blastula, until the embryo becomes a solid mass, consisting of an external layer of cells, the **ectoderm** (epiblast) covering an internal cellular core, the **endoderm** (hypoblast). At the same time a double protective investment, the outer layer of which is firm and chitinous, is secreted by the ectoderm. The changes above described occupy about four days, and, after their completion, the embryo falls from the ovary into the mud, where its development is slowly completed during the winter months.

The further changes consist in the formation of a digestive cavity by absorption of some of the endoderm cells, while the remaining ones constitute a layer surrounding this cavity. The ectoderm at the same time becomes differentiated, and the mesoglœa makes its appearance. Later on the outer membrane is ruptured by increase in size and elongation of the embryo, the mouth is then formed as a perforation, tentacles grow out as hollow processes of the body-wall, and the inner membrane is cast off.

The development of Hydra has been investigated by several writers, whose accounts are conflicting. The brief outline given above is probably correct in the main.

Further remarks on Hydra:—

It must be remembered that although Hydra is a simple and readily obtainable type of the lower multicellular animals, it is by no means a typical example of the group, *i.e.*, the **Hydrozoa**, to which it belongs. These are for the most part colonial, the colonies being produced by budding. If, in Hydra, the buds, instead of becoming detached, remained united together, something resembling one of these "hydroid zoophytes" would be produced. These colonial forms give rise, in a large number of cases, to free-swimming sexual individuals, which are somewhat umbrella-shaped, and are popularly known as "jelly-fish," technically as "medusæ." This is a good example of "alternation of generations," where asexual and sexual stages alternate in the life history of the same form. Thus:— A, an asexual hydroid colony gives rise to S, sexual medusæ, which again produce A, and so on, indefinitely. By taking a selected series of forms it can be shown that the sexual organs of Hydra are probably morphologically equivalent (homologous) to medusæ. From (1) forms with free-swimming medusæ, we can pass to (2) forms with medusa-like buds, and thence to (3) forms with sexual buds obscurely medusa-like,—lastly to (4) Hydra, with sexual buds as mere knobs of ectoderm. If this reasoning be correct, Hydra is a degenerate form—*i.e*, derived from ancestors more highly differentiated than itself.

CHAPTER III.—PLATYHELMIA (Flat-worms).

§ 5. DISTOMA (The Liver-fluke).

THE best-known kind of fluke is *Distoma hepaticum*, the Liver-fluke. This is an internal parasite (endoparasite) when adult about an inch and a half long, which lives in the bile-ducts of several animals, more especially the sheep, and causes the disease known as liver-rot.

MORPHOLOGY AND PHYSIOLOGY.

1. External Characters.—The animal (Fig. 6) is *bilaterally symmetrical*, right and left halves, anterior and posterior ends, and upper (dorsal) and lower (ventral) surfaces being distinguishable. The animal can be divided into corresponding halves by one plane only, the median vertical.

The body is flattened and somewhat leaf-shaped, its broad dorsal and ventral surfaces passing into sharp edges at the sides and behind. The gently curved outline is interrupted at the anterior end by the *head-papilla*, a projection which tapers to a blunt point, and ends in an adhesive cup or sucker which looks somewhat ventralwards. Behind the head-papilla the sides of the body curve outwards and backwards till the maximum breadth is attained at about the junction of the anterior and middle thirds, while the hinder two-thirds narrow gradually to the posterior end of the body. There is a second sucker ($v.s$) on the ventral surface, a little way behind the head-papilla.

Four median apertures are present on the surface of the body. These are:—(1) The *mouth* (m), in the centre of the anterior sucker; the minute *excretory pore* ($ex.p$) at the posterior end; (3) the *genital aperture* ($g.o$) on the ventral surface between the suckers; and (4) the minute opening of the vagina (Laurer's canal) on the dorsal surface.

With the exception of the smooth suckers the body surface is rough, owing to the presence of an immense number of minute backwardly directed spines, which project from the *cuticle*, a firm membrane which invests the body and is secreted by an underlying *epidermis*, that consists of a single layer of large granular cells. The hard cuticle protects and supports the body, and its spines prevent the animal from slipping back as it makes its way along the bile-ducts of its host.

The digestive and other organs are imbedded in a mass of tissue, which, since it supports the various parts and connects them together, may be termed connective tissue. Polyhedral nucleated cells are the main constituent of this tissue in the Liver-fluke, but there is also a certain amount of non-cellular fibrous material between these.

2. Digestive Organs (Fig. 5).—The mouth leads into a short tube which quickly forks, the two limbs of the fork passing back and giving off numerous branching processes, all of which end blindly. There is no anus.

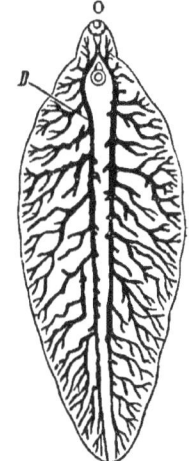

Fig. 5.—DISTOMA (from *Claus* after *Leuckart*) — Alimentary canal. O, Mouth, a short distance behind which is ventral sucker. D, right limb of intestine.

The oval *mouth* (O) is situated in the middle of the anterior sucker, and leads into a small *mouth-cavity*, which is followed by the oval, thick-walled *pharynx* that passes into a very short, straight, delicate tube, the *gullet* or *œsophagus*, this again opening into a bifurcated *intestine*. Each half of this is a thin-walled, fairly wide tube (D), running back to the end of the body, and situated near the middle line. A number of small pouches project from its inner side, whilst a large number of pouches, mostly much branched, extend from its outer side to the edge of the body.

Special *muscles*—i.e., bands of contractile fibres—are connected with the pharynx. A sheath of such fibres, the *protractor muscle*, closely surrounds this organ, and, on the other hand, is connected with the anterior sucker, while *retractor muscles* slant back from the pharynx to the dorsal wall of the body.

Histology.—The pharynx is lined by a continuation of the general cuticle. Its wall is very muscular, the fibres taking various directions. The gullet and intestine have very thin walls, consisting of an epithelial layer of cells, external to which is a structureless membrane. The term *epithelium* is applied to membranes, formed by one or more layers of cells, which cover external and line internal surfaces. The epithelium

30   AN ELEMENTARY TEXT-BOOK OF BIOLOGY.

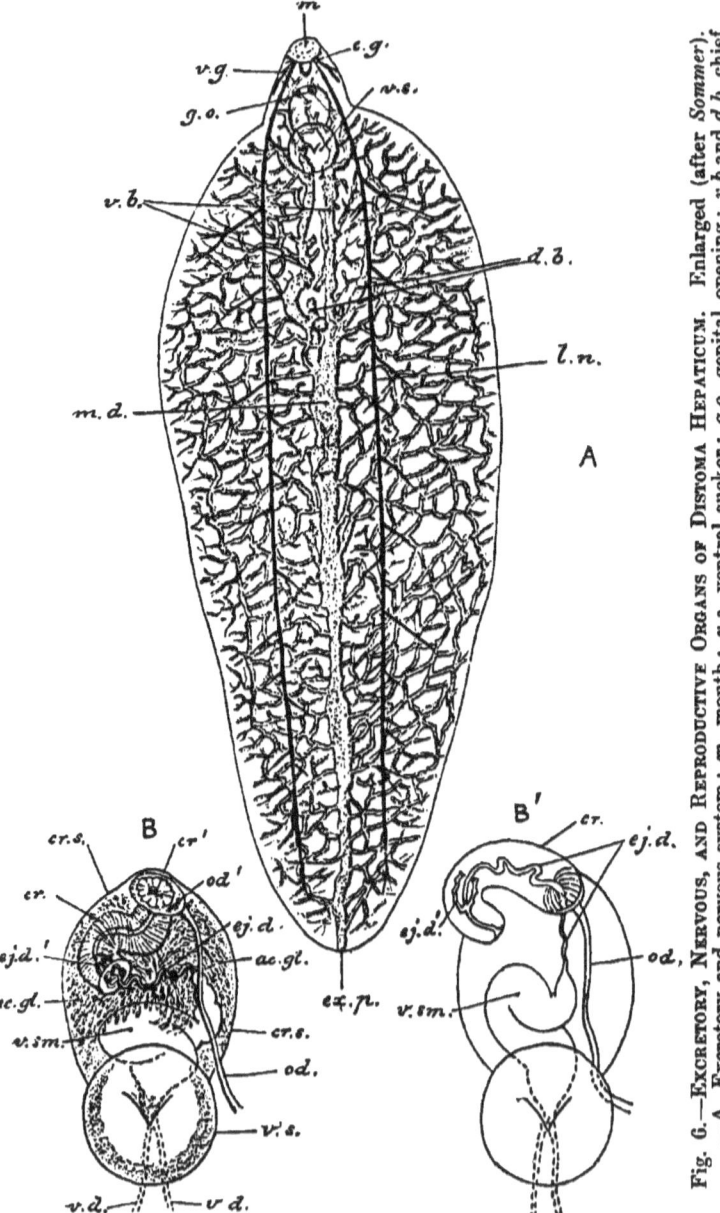

Fig. 6.—Excretory, Nervous, and Reproductive Organs of Distoma Hepaticum. Enlarged (after Sommer). —A, Excretory and nervous system; m, mouth; v.s, ventral sucker; g.o, genital opening; r.b ard d.b, chief ventral and dorsal branches of excretory duct (one of the latter only partly drawn); m.d, main duct; ex.p, excretory pore; c.g, cerebral ganglion; v.g, ventral ganglion; l.n, lateral nerves. B and B', Reproductive organs contained in cirrus-sac, cr.s. In B the penis is partly everted. v.s, Ventral sucker; od, oviduct; od', opening of ditto; v.d, vasa deferentia; v.sm, vesicula seminalis; ej.d, ejaculatory duct; ej.d', aperture of same on papilla; cr, penis; cr', opening of ditto; ac.gl, accessory gland (not shown in B').

in question is *simple*—only one cell thick—and *columnar*, the constituent cells being longer than broad. They are amœboid and contain large nuclei.

The food consists of the bile of the host, together with blood and disintegrated liver-substance, resulting from the "rot." This is taken in by the action of the pharynx, which acts somewhat like a piston. By the retractor muscles it is drawn back, when the food rushes into and fills the mouth-cavity. The mouth is now closed and the pharynx brought forwards by the protractor muscle. As, meanwhile, its cavity is enlarged by the contraction of radial fibres in its wall, food passes into it, and (the cavity being diminished by the contraction of other fibres), is then squeezed on into the gullet and intestine. These present a large surface which can at once absorb the fluid part of the food. The amœboid epithelial cells effect intra-cellular digestion, and perhaps also secrete a digestive fluid which, being poured into the gut, can bring solid food into solution or a fine state of division. Undigested matter is ejected from the mouth. The digested material diffuses through the intestinal wall to the other parts of the body.

3. Excretory Organs (Fig. 6, A).—Those organs are termed "excretory" (in the narrowest sense), the function of which is to get rid of nitrogenous waste. They here form a network of canals ramifying throughout the body. The smallest of these each terminate in a pear-shaped "flame" cell, containing a large vacuole, and named from the flickering appearance it presents in the living state, and which is probably due to the presence of a row of cilia projecting into the vacuole. The cavities of the flame-cells open into the canals where they terminate, and also, perhaps, into the minute spaces in the tissues, which collectively represent the continuous space, or *body-cavity*, found around the organs of most animals. The small tubes branch freely, and are united into a network. They open into larger tubes, which, in the anterior quarter of the body, are continuous with four still larger longitudinal trunks, two dorsal ($d.b$) and two ventral ($v.b$), and in the posterior three-quarters with a median longitudinal trunk, formed by the union of the four anterior trunks. This *main duct* ($m.d$), the largest excretory trunk, runs back, just beneath the dorsal wall of the body, to open by the minute *excretory pore* ($ex.p$).

Histology.—The delicate walls of the excretory tubes are made up of a single elastic, structureless layer. They contain a clear fluid, in which are numerous highly refracting granules. The presence of guanin ($C_5 H_5 N_5 O$) has been proved.

4. Reproductive Organs (Fig. 7).—Distoma is hermaphrodite, and its very complex reproductive organs lie, for the most part, ventral to the intestine. The *genital aperture* (Fig. 6, *g.o*) is a small oval pore situated in the middle line between the anterior and ventral suckers, and rather nearer the latter. It leads into a shallow pit, the *genital sinus* (Fig. 6, B), on the right of which is the larger *male aperture* (*cr'*), and on the left the *female aperture* (*od'*).

(1) **The Male Organs** consist of two *spermaries* (testes) (T), one behind the other, extending over a considerable area in the middle region of the body. Each is made up of several much-branched tubes, which unite to form a tube or duct of smaller calibre, the *spermiduct* (vas deferens). The spermiducts run forwards, one on each side of the middle line, to the level of the ventral sucker (S). In the latter part of their course they converge (Fig. 6, B, B'), and at the above level open into a spindle-shaped thick-walled tube, the *vesicula seminalis* (*v. sm*), which tapers into a very delicate convoluted tube, the *ejaculatory duct* (*ej.d*), that again opens into a shorter and stouter tube, the *penis* (cirrus) (*cr*), opening at the male aperture. Around the end of the ejaculatory duct a mass of nucleated cells is arranged, each of which opens by an excessively fine tube into the duct. They are the unicellular *accessory glands* (*ac.gl*).

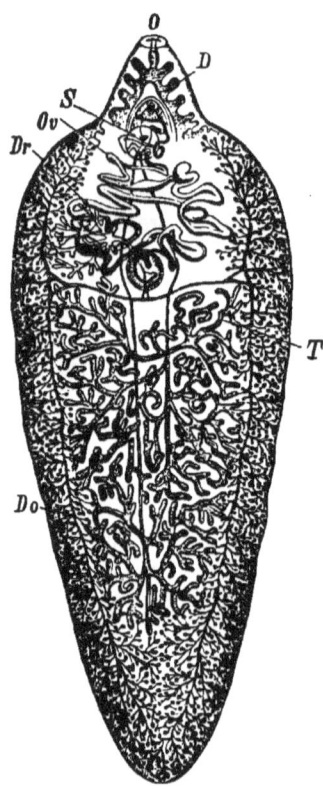

Fig. 7. — DISTOMA HEPATICUM (from *Claus* after *Sommer*).—O, Mouth; D, limb of intestine; S, ventral sucker, in front of which is genital opening; T, testis, in front of which is the rounded shell-gland; Do, yolk-gland; Ov, oviduct; Dr, ovary.

A *gland* is an organ essentially composed of one or more epithelial-cells, in which some special secretion or excretion is formed. The simple *unicellular*, i.e., one-celled, condition is seen in the accessory glands.

The vesicula seminalis, ejaculatory duct, accessory glands, and penis are contained in the *cirrus-sac* (*cr.s*), a hollow oval body with muscular walls, lying between the ventral sucker and genital opening.

(2) **The Female Organs** are made up of the following parts:— The unpaired *ovary* (Dr) lies on the right side, in front of the spermaries. It is composed of several branched tubes, which unite together to form the *oviduct* (Ov). This is at first very narrow, then gradually enlarges and becomes considerably convoluted, and finally narrowing, opens by the female aperture (Fig. 6, B, *od'*). The dilated part of the oviduct (uterus) is generally found full of eggs, in which embryos are beginning to develop. Two glands are connected with the female organs. (1) The *yolk-glands* (Do), which secrete a nutritive substance (yolk) for the use of the developing embryos, are paired, and each of them is made up of an immense number of minute rounded bodies (acini) of glandular nature, and extends over a lateral strip outside the other reproductive organs. Behind the spermaries the two meet in the middle line. From the acini fine ducts proceed, which unite into larger ones that finally open into a longitudinal duct which runs along the inner side of the gland. Just in front of the anterior spermary a transverse duct arises from this trunk, which unites with its fellow in the middle line to form a small *yolk-reservoir*, from which the short unpaired *yolk-duct* runs forwards to join the oviduct not far from its commencement. Just before their union a fine tube, the *vagina* (Laurer's canal), through which sperms are introduced into the oviduct, passes to the yolk-duct from the dorsal surface, where it opens by a minute aperture. (2) The *shell-gland*, by which the egg-shells are secreted, is similar in structure to the accessory gland. It is a rounded mass, the elements of which open into the oviduct near its junction with the yolk-duct. (*See Fig. 7, just below convoluted oviduct.*)

Histology.—The tubules of the spermary are supported by a structureless wall, on the outside of which filamentous contractile cells are arranged longitudinally. There appears to be a lining of germinal epithelial cells, sperm-mother-cells, some of which become free, and give rise to tufts of sperms (spermatozoa). Only part of the substance of the mother-cells is used up in this process. Each sperm possesses a small oval head, and long vibratile tail. The walls of the vesicula seminalis, cirrus-sac, and penis contain muscle-fibres variously arranged, and the last is lined by a continuation of the spiny cuticle covering the body.

The *ovarian tubules* possess supporting layers, and are lined by germinal epithelium, the cells of which become *ova*. Each of these has a small vitellus, and large germinal vesicle, with germinal spot. In the acini of the *yolk-glands* a formation of *yolk-cells* occurs. These contain a large amount of nutritious matter. A yellowish fluid, which can harden into shell substance, is secreted by the *shell-glands*.

5. **Muscular System.**—This is made up of bands and sheets of contractile muscle-fibres, by means of which definite movements are effected. When these fibres contract they become shorter and broader, their ends being thus brought closer together so that the parts to which they are attached tend to approach. In the case of fibres surrounding a cavity in a circular direction (pharynx, for example), narrowing of the cavity takes place, and the reverse is effected by longitudinal fibres. The movements of locomotion are mainly due to the *dermal musculature*, a sheet of muscle intimately connected with the skin, and together with it making up the body-wall. The sheet is composed of three layers, external, middle, and internal, the fibres of which take respectively transverse (circular), longitudinal, and oblique directions. Closely connected with these are the suckers, which are muscular cups of rather complex structure. There are also a great many muscular bands running through the connective-tissue which fills up the space between the organs. These bands take a dorso-ventral direction, and are also connected together so as to form a close network.

Locomotion is effected in the following way:—The ventral sucker having been fixed, the head-papilla is elongated by the contraction of its circular muscle-layer. The anterior sucker then attaches itself, the ventral sucker being at the same time loosened, and the body is dragged forward by the contraction of its longitudinal muscle-layer.

6. **Nervous System** (Fig. 7).—Under this term are included

those organs which have more especially to do with irritability and spontaneity.

Around the pharynx is placed a *nerve-ring*, the plane of which slopes downwards and backwards. In its dorsal part there is a swelling, the *cerebral ganglion* (*c.g*), on either side, and a similar unpaired *ventral ganglion* (*v.g*) below. From each cerebral ganglion two small filaments, *nerves*, are given off to the head-papilla, in which they branch, whilst a stouter *lateral nerve* (*l.n*) runs back almost to the end of the body, giving off branches as it does so. It lies just within the ventral body-wall, below the reproductive organs, and is separated from its fellow by about one-third the breadth of the body. Excessively fine filaments are given off from the ventral ganglion to surrounding parts.

Histology.—Two elements make up the essential part of the nervous system—(*a*) *Nerve-cells* (ganglion-cells), (*b*) *Nerve-fibres*. The nerve-cells are most numerous in the ganglia, but are also present in less abundance in parts of the nerve-ring. They are irregular in shape, with several projections or processes; their protoplasm is clear and they contain a very distinct refractive nucleus. The nerve-fibres are extremely delicate threads, which make up the nerves and most of the nerve-ring. They are continuous on the one hand with the processes of the nerve-cells, on the other with the various organs of the body.

The nerve-cells are connected with one another by means of their processes.

DEVELOPMENT (*Fig*. 8).

The life-history of the Liver-fluke, like that of many other parasites, exemplifies **alternation of generations**, a phenomenon far less general in the case of animals than in the case of plants.

From the egg a young fluke does not proceed direct, but a number of *asexual stages*—*i.e.*, forms capable of asexual multiplication, intervene between it and the sexually mature adult. The sequence of events is as follows:—(1) A *ciliated embryo* escapes from the egg and becomes parasitic within the lung-chamber of a water-snail, there degenerating into (2) a shapeless sac or *sporocyst*. (3) Within the sporocyst numerous cylindrical *rediæ* are produced by a process of internal budding, and these feed upon the liver of the snail. There are usually several generations of rediæ. (4) A number of tadpole-shaped *cercariæ* are formed in the redia by budding. The cercaria makes its way out of the snail, and after losing its tail, encysts upon grass. (5) A young fluke is formed within the cyst from the remains of the cercaria. (6) If the cyst is now swallowed by a sheep the young fluke emerges, makes its way up the bile-duct into the liver, and becomes sexually mature.

Cleavage (segmentation), which occurs within the oviduct, is regular and complete. The ovum, together with a number of yolk-cells, is enclosed in a horny shell to form the *egg*. This is oval, with a smooth surface. The ovum lies at one end, and a small circular area of the transparent shell is here marked off as the lid or *operculum*. The further development occurs outside the body of the sheep, the eggs passing to the exterior with the excrement. It leads to the formation of a **free embryo** (*a* and *b*), which gradually comes to occupy most of the cavity of the egg, as the yolk-cells are used up. When fully developed, a sudden elongation of its body causes the operculum to fly open, and it is thus liberated. The body is somewhat conical, and its thick anterior end possesses a short retractile *head-papilla*. The external layer of the body-wall is formed by flattened *ectoderm* cells which, except on the head-papilla, bear long locomotor cilia. The deeper layer of the body-wall is granular, with rudimentary muscle-fibres, a pair of excretory funnels, and near the anterior end a pair of *eye-spots*. These are two small refracting cells, placed close together, and each containing a crescentic mass of pigment. The two crescents placed back to back present a somewhat **X**-shaped appearance. The eye-spots are imbedded in an ectodermic thickening, the *cerebral ganglion*. The interior of the body is mostly filled with rounded *germinal-cells*, but there is also an oval mass of *endoderm* cells (b, D), representing a rudimentary and mouthless gut (digestive tube). The germinal and other cells which come between this and the ectoderm are collectively known as the *mesoderm* (mesoblast).

If the free embryo finds itself in water or among damp herbage, it moves actively about by means of its cilia, and should it meet with a small water-snail, *Limnaea truncatula*, within about eight hours development proceeds, but not otherwise. The head-papilla is lengthened, and the embryo, revolving rapidly, bores by its means into the snail. Within this host, generally in the lung-chamber, it becomes a **sporocyst**. The ciliated ectoderm-cells are lost, and the body gradually becomes an elongated and rather irregular sac. Its wall is covered externally with a cuticle, below which are feebly developed muscle-fibres, within which is a layer of epithelium lining the body-cavity. Several excretory funnels are present in the body-wall, and the eye-spots, separated from each other, can still be recognised. The sporocyst may multiply by transverse fission, and it produces asexually within it a number

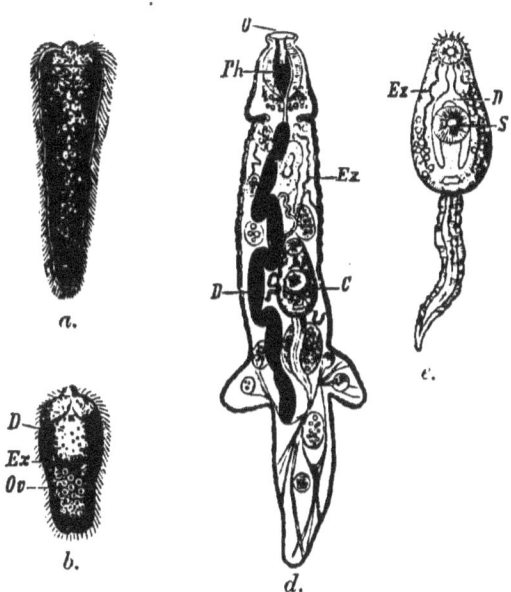

Fig. 8.—DEVELOPMENT OF DISTOMA (from *Claus*, partly after *Leuckart*).
—*a*, Free swimming ciliated embryo: *b*, the same contracted, with rudimentary gut, D, germinal cells, Ov, and excretory funnels, Ex: *d*, redia; O, mouth; Ph, pharynx; D, intestine; Ex, excretory tubule; C, contained cercariæ: *e*, free cercaria; S, ventral sucker; D, intestine; Ex, excretory tubule; the little circles in the body represent cystogenous cells.

of the next, or **redia** stage, in the following manner :—The body-cavity contains a large number of germinal cells, partly those of the free embryo, partly budded off from the lining epithelium, lying loosely within it. These become aggregated into small solid masses, *morulæ*, each of which becomes a *gastrula* (*i.e.*, a two-layered embryo with a mouth and rudimentary digestive cavity), by an inpushing or *invagination* of cells on one side. This becomes a redia, which breaks through the wall of the sporocyst, the ruptured place afterwards closing up.

Redia (*d*).—Body elongated and cylindrical. Near the anterior end is a thickened ring or "collar," while not far from the posterior end a pair of blunt processes project, one on each side. The body wall has a structure similar to that of the sporocyst, but the muscle is better developed. A *mouth* (O) is present at the anterior end of the body, which leads into a muscular *pharynx* (Ph). This is continued into a simple *digestive sac* (D). Branched *excretory trunks* (Ex) commencing in ciliated funnels are present.

The redia wanders over the body of the snail and feeds upon its tissues, especially the liver. The posterior processes act as foot-stumps, and the collar serves as a relatively fixed point upon which the anterior end can move when feeding. Within the body-cavity, *daughter-rediæ*, or the next, **cercaria** stage (*c*), may be developed, the former only in warm weather. In either case some of the epithelial cells lining the body-cavity enlarge and segment to produce morulæ, which become gastrulæ, and gradually assume the form of daughter-rediæ, or cercariæ, which escape by a special *birth-opening* behind the collar.

The **Free Cercaria** (*e*) resembles a minute tadpole. It has a rounded flattened body armed anteriorly with minute spines, and possessing anterior and ventral (S) suckers, and a muscular tail. The *body-wall* contains numerous large granular lime-secreting cells (cystogenous cells). A *mouth, pharynx, gullet*, and simple forked *intestine* (D) are present, as well as *excretory organs* (Ex). The cercaria makes its way out of the snail, loses its tail, and attaches itself to a grass-stem, the cystogenous cells pouring out a secretion which hardens into a bright white cyst, within which it develops into a young fluke. If this encysted stage is swallowed by a sheep, the gastric juice dissolves the cyst, and the young fluke makes its way up the bile duct into the liver, there becoming sexually mature in about six weeks. It is worthy of remark that the adult sexual stage lives in the higher vertebrate host, where it has a better chance of being preserved.

The following table summarizes the succession of stages:—

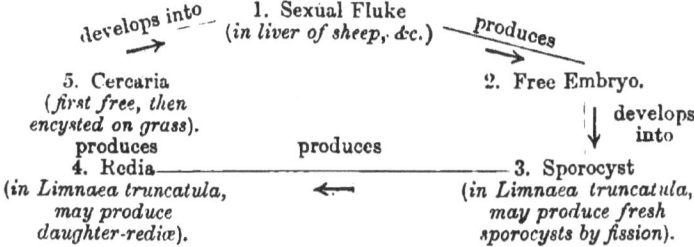

Further Remarks on Flukes.

"Alternation of Generations" includes a number of by no means equivalent cases. One of the simplest is that described (p. 27) for certain Hydrozoa, where an asexual hydroid (A) produces a sexual medusa (S) by budding or fission. The sequence is A S, A S, &c., and this simple alternation of asexual and sexual stages may be termed metagenesis. The "internal budding" by which, in the Liver-fluke, redia and cercaria are produced is not, strictly speaking, an asexual process, nor, on the other hand, is it a normal sexual process. The germinal cells appear to be precociously formed ova which can develop without fertilization (parthenogenetically). Such an alternation of normal sexual reproduction with parthenogenesis may be termed heterogamy. Writing S for the sexual fluke, s^1 for the sporocyst, and s^2 for the redia, this particular instance, reduced to its simplest form is s^1 s^2 S, s^1 s^2 S, &c. The ciliated embryo is included in the s^1, since it develops into the sporocyst, and similarly the cercaria is merged in the S.

D. hepaticum is but infrequently found parasitic in man, a far more important fluke, from a medical point of view, being D. hæmatobium (Bilharzia), which chiefly occurs in Egypt, Abyssinia, and S. Africa. It infests the portal vein and its branches, and is remarkable among flukes in having the sexes separate. The male is elongated and worm-like; upon its ventral surface is a groove in which the similarly shaped female is carried. Life-history unknown.

§ 6. TÆNIA (The tapeworm).

Tapeworms are endoparasites, nearly all of which, when sexually mature, inhabit the intestines of vertebrate animals. One of the best known types is Tænia solium, the Common Tapeworm, which during the early stages of its existence is found in the muscles of the pig (intermediate host), and, when mature, in the small intestine of man (final host).

MORPHOLOGY AND PHYSIOLOGY.

The tapeworm is bilaterally symmetrical and made up (cf. Fig. 9) of a minute scolex (head and neck), about $\frac{1}{25}$ of an inch long, and

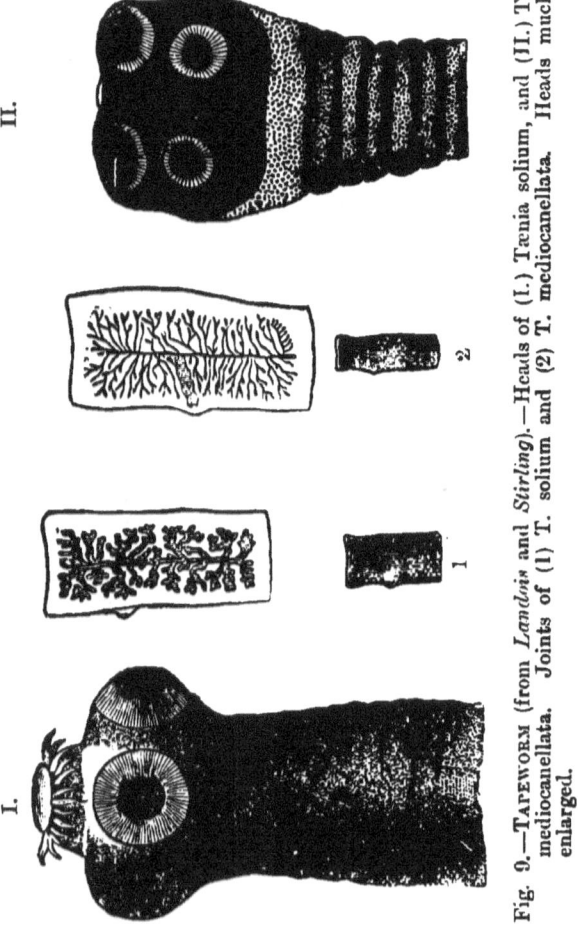

Fig. 9.—TAPEWORM (from *Landois* and *Stirling*).—Heads of (I.) Tænia solium, and (II.) T. mediocanellata. Joints of (1) T. solium and (2) T. mediocanellata. Heads much enlarged.

a body some six to ten feet in length, consisting of 800 to 900 joints (proglottides), which near the scolex are extremely small and narrow, while at the other extremity they attain the length of half an inch and about two-thirds that breadth. The scolex is slightly, the proglottides very much flattened dorso-ventrally, but there is no obvious difference between the dorsal and ventral surfaces. The somewhat pear-shaped **scolex** bears four muscular suckers on its broadest part, and at its free head-end there is a round projection (the rostellum), in which are imbedded some twenty-four chitinous hooks, arranged in a double circlet. By means of these hooks and suckers the tapeworm firmly fixes itself to the wall of the intestine. The difference in size between the proglottides is accounted for by the fact that they are not all of the same age, those next the scolex being youngest. In fact, during the life of the scolex, new joints are continually being developed by the alternate constriction and growth of its narrow neck-end. A proglottis taken from about the middle of the body will be found to possess a complicated set of hermaphrodite reproductive organs, while the oldest proglottides are full of eggs containing embryos. About the middle of one edge of a proglottis is seen a small elevation, the *genital papilla*. These papillae are alternately right and left in successive proglottides. As new joints are developed at the head-end old ones become detached and pass out of the body of the host. No sexual organs are found in the scolex, which is often regarded as an asexual individual from which, by means of budding, a chain of sexual individuals (proglottides) arise. It is, however, more likely that the entire worm is a single individual.

The tapeworm is entirely devoid of digestive organs, its food consisting of the highly nutritious ready digested material with which it is surrounded, and which can readily diffuse into its body. Although the animal is invested by a complex four-layered cuticle this does not prevent such diffusion, as the two outer layers are traversed by numerous pores. Below the cuticle there is an epidermis composed of a single layer of spindle-shaped cells. The hooks in the head are cuticular thickenings.

Excretory Organs are present similar to those found in the Liver-fluke. Like the other internal structures these are imbedded in a mass of parenchyma, which serves as a kind of packing-tissue and is made up of variously shaped cells and a granular matrix through which detached nuclei, fat-globules, and calcareous particles are scattered. A main excretory tube runs along each side of the body (Fig. 10), not far from its margin, and is connected with its fellow in the head and also by a transverse commissure running across each proglottis near its posterior side. In the young tapeworm, before any proglottides have been detached, the two excretory trunks converge posteriorly and pass to a vesicle which opens to the exterior by a terminal pore. Numerous delicate excretory tubules appear to be present in the parenchyma, and these communicate on the one hand with the main trunks, while on the other they terminate in flame-cells, which probably

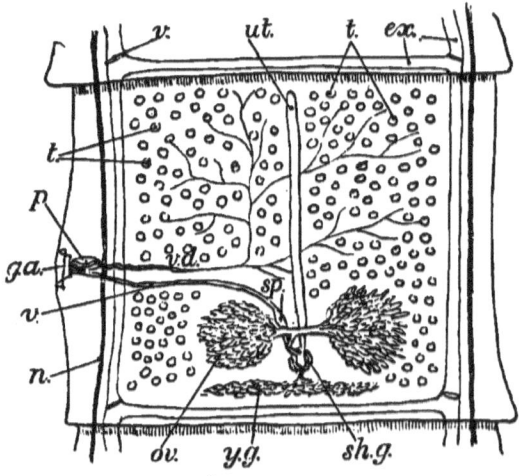

Fig. 10.—TAPEWORM (after *Sommer*). Enlarged. Proglottis of *Tænia mediocanellata*, with developed sexual organs.—*t*, Spermaries (testes); *v.d*, spermiduct (vas deferens); *p*, penis, contained in cirrus sac; *or*, ovary; *y.g*, yolk gland; *sh.g*, shell gland; *ut*, uterus; *sp*, spermotheca; *v*, vagina; *g.a*, genital atrium; *ex*, main excretory tube, with valves, *v*; *n*, lateral nerve-cord.

communicate with minute spaces in the tissues. These spaces collectively constitute a body-cavity. The main excretory trunks possess valves by which the fluid they contain is prevented from flowing along them towards the front end. Xanthin ($C_5H_4N_4O_2$) and guanin ($C_5H_5N_5O$) are said to occur in this fluid.

The complicated hermaphrodite reproductive system (Fig. 10) of the tapeworm bears a general resemblance to that of the Liverfluke. (1) The **male organs** consist of an immense number of minute rounded *spermaries* (*t*), from which proceed exceedingly delicate ducts that by their continued union form a *spermiduct* (*v.d*). This is continuous with a muscular, eversible *penis* (*p*) (cirrus), contained in a *cirrus-sac* (*c.s*) and opening into a depression, the genital atrium (*g.a*) which occupies the summit of the genital papilla.

Each spermary essentially consists of germinal cells (sperm-mother-cells) each of which produces a number of sperms. In this process the nucleus divides repeatedly to form the heads of the sperms, and their vibratile tails are then differentiated from the protoplasm. (2) The **female organs** consist of two ovaries, an oviduct, uterus, vagina with spermotheca, a yolk-gland, and a shell-gland.

The *ovaries* (*ov*) are collections of much branched tubules situated in the posterior part of the proglottis. These tubules unite to form an oviduct which communicates on the one hand with a blindly ending tube, the *uterus* (*ut*), and on the other with a *vagina* (*v*), dilated internally into a *spermotheca* (*sp*) and opening into the genital atrium. The *yolk-gland* (*yk*) is a tubular network situated behind the ovaries and opening into the oviduct, which also receives in this region the numerous minute ducts of the unicellular elements which make up the rounded *shell-gland* (*sh.g*).

Self-fertilization appears to be the rule, and takes place, after closure of the genital atrium, by contractions of the spermiduct and penis—the result being that sperms are forced through the vagina to the spermotheca, and thence to the oviduct, where they meet with and fertilize the ova. Each fertilized ovum, together with a number of yolk-cells, is surrounded by an egg-shell to constitute an egg, and the eggs pass on to the uterus.

The tapeworm possesses considerable power of movement, being able to elongate or shorten itself, raise or lower its hooks, and attach or detach its suckers. The **muscles** by which these various movements are made consist of—(1) a superficial longitudinal layer by which the body can be shortened, (2) transverse muscle-bands situated more deeply and serving to elongate the body, (3) dorso-ventral fibres by which the proglottides can be flattened, (4) radial and circular muscles in the suckers, and (5) minute elevator and depressor bands attached to the hooks.

The muscles are for the most part attached to the firm cuticle, and they are made up of exceedingly delicate muscle-cells which are of considerable length and taper at both ends.

The **nervous system** of a tapeworm consists of two longitudinal cords, one of which traverses each side of the body near its margin (Fig. 10), and of a transverse *brain-commissure* by which these two cords are connected in the head. Nerves run from the commissure to the hooks, suckers, and other parts of the scolex. Histologically the nervous system is essentially made up of nerve-cells and nerve-fibres, the former of which are not limited to the cerebral commissure.

DEVELOPMENT.

The very numerous eggs pass into the uterus, which, at first a simple tube, becomes much branched (Fig. 9), and fills up the greater part of the ripe proglottis, while at the same time the remaining sexual organs gradually abort.

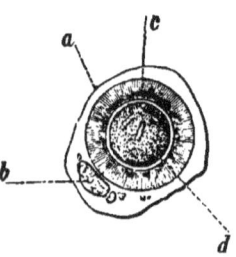

Fig. 11.—TAPEWORM (from *Landois* and *Stirling*).— Ripe egg from uterus of Tænia solium: *a*, albuminous envelope; *b*, remains of yolk; *c*, covering of embryo; *d*, embryo with hooklets.

The minute oösperm first undergoes complete cleavage, and then is gradually converted into a spherical six-hooked embryo (Fig. 11), in the development of which the yolk-cells enclosed in the egg are gradually used up.

The ripe proglottides pass out of the body of the host, and gradually decay, the innumerable embryos contained being thus liberated. If any of these are now swallowed by a pig (or other warm-blooded animal) the egg shells are dissolved, and the embryos, by means of their hooks, penetrate the walls of the intestine and pass into blood-vessels, being then carried in the circulation to the muscles. In these they pass into the *cysticercus* or *bladder worm* stage (Fig. 12), consisting of a fluid-filled vesicle, (proscolex) into which a tapeworm-head (scolex) projects as a hollow bud, the whole being surrounded by a firm cyst developed from the surrounding tissue as a result of irritation. Development proceeds no further in the pig, but if pork infested by these cysts ("measly" pork) is taken into the stomach of a human being the cyst is dissolved, the scolex is everted (Fig. 13), and the proscolex digested. The scolex then passes on into the in-

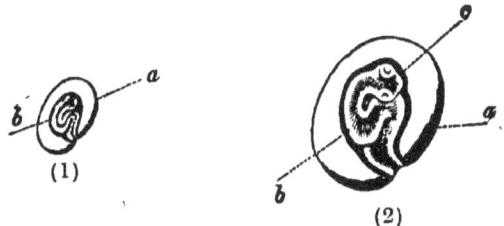

Fig. 12.—TAPEWORM (from *Landois* and *Stirling*).—Cysticerci of Tænia solium removed from their cysts, (1) natural size, (2) magnified; *a*, proscolex ; *b*, proscolex with hooks and suckers, *c*.

testine, attaches itself by means of its hooks and suckers to the mucous membrane, becomes solid, and develops a chain of proglottides. The bladder worm was not at first known to be a stage in the life-history of tapeworm, and hence received a special generic

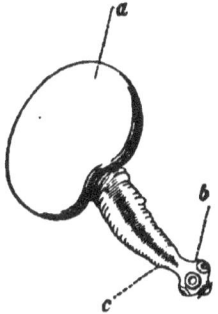

Fig. 13.—TAPEWORM (from *Landois* and *Stirling*).—Cysticercus of Tænia solium with everted scolex: *a*, proscolex; *b*, head with hooks and suckers ; *c*, neck.

name, Cysticercus, the particular one belonging to Tænia solium being known as *Cysticercus cellulosæ*.

Those who regard the tapeworm as an example of alternation of generations, recognise three stages in its life-history, two asexual (proscolex, A ; scolex, A′), and one sexual (proglottis, S), the cycle being A A′ S.

Further Remarks on Tapeworms.

The following tapeworms are, next to *T. solium*, of most importance:—

Name.	Intermediate Host (sheltering Cysticercus).	Final Host (sheltering adult).
Tænia mediocanellata (saginata). No hooks. Very numerous proglottides.	Ox.	Small intestine of Man.
T. echinococcus. Minute. Numerous very small hooks. Only three or four proglottides.	Cysticercus (=*Echinococcus veterinorum*). Very large, with numerous tapeworm heads; in man and domestic animals.	Small intestine of dog.
Bothriocephalus latus. Flat head; no hooks; two adhesive grooves. Very numerous broad proglottides, with ventral genital and uterine openings.	Embryo ciliated, passes into first int. host, and thence to muscles of second iut. host (pike, or burbot).	Intestine of Man; rarely cat or dog.
Tænia cœnurus. 24 to 32 small hooks. Numerous proglottides.	Cysticercus (= *Cœnurus cerebralis*) in brain of sheep, causing "staggers."	Intestine of sheep-dog.
T. serrata. 38 to 48 hooks of two sizes. Numerous proglottides, with prominent posterior angles.	*Cysticercus pisiformis.* Liver and mesentery of rabbit.	Intestine of dog.

Certain tapeworms are known (parasitic when mature in the intestines of fishes) which have no proglottides. The scolex in these cases possesses a set of hermaphrodite reproductive organs, and, in many respects, is not unlike a fluke.

CHAPTER IV.—NEMATHELMIA (Thread-worms).

§ 7. ASCARIS (The Round-worm).

THE large group of Nemathelmia or Thread-worms, includes both free and endoparasitic forms. The latter are by far the more numerous, and of these the most convenient type-genus is *Ascaris* (the round-worm), one species of which *A. lumbricoides* infests the small intestine of the human subject, while the much larger *A. megalocephala* commonly occurs in the intestines of the horse. The following description applies to both species, but when measurements are given they refer to A. lumbricoides.

MORPHOLOGY AND PHYSIOLOGY.

1. **External Characters** (Figs. 14 and 15).—The bilaterally symmetrical body is cylindrical and tapers to a blunt point at each end. The dorsal and ventral surfaces are much alike. Thread-worms are not, as a rule, hermaphrodite, and in Ascaris the two sexes are readily distinguishable, for the male is not only smaller than the female (♂ about 6 inches, ♀ about 8 inches long), but the posterior end of its body is sharply bent up on the ventral side (Fig. 15).

The colour is whitish, and in the fresh condition the worm is somewhat translucent, there being, however, four more opaque longitudinal streaks, named from their relative positions the dorsal, ventral, and lateral lines. The *mouth*, situated at the anterior end, is guarded by three projecting lips, one of which is dorsal, while the other two meet together in the mid-ventral line. The gut has also a posterior opening, readily seen as a ventral slit, not

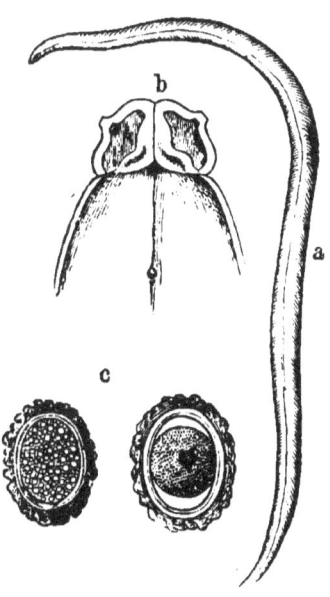

Fig. 14.—ASCARIS. *a*, Female specimen of A. lumbricoides; *b*, head; *c*, egg.—From *v. Jaksch* (Clin. Diagnosis).

far from the hind end of the body. This aperture is termed *anus* in the female, but *cloacal opening* in the male, because it is common to the digestive and reproductive organs. The female possesses a rounded genital aperture, situated in the mid-ventral line about one-third of the way back. In both sexes there is a minute *excretory pore*, placed mid-ventrally not far from the front end.

2. The **body-wall** (Fig. 15) consists of a firm, chitinous several-layered cuticle, covering an epidermis, internal to which is a layer of longitudinally directed muscle-cells.

In a young specimen the *cuticle* is thin and the underlying *epidermis* a distinctly cellular layer, but in the adult the formation of a thick complex cuticle (*c*) seems to have caused degeneration in the epidermis (*ep*), which is then a thin granular stratum through which numerous nuclei are scattered, and which is traversed by gelatinous fibres. It is, however, much thickened mid-dorsally, mid-ventrally, and, especially, laterally, to form the *dorsal* (*d.l*), *ventral* (*v.l*), and *lateral lines* (*l.l*) to which reference has already been made. Each of these is lined by a layer of cuticle which is folded in so as to nearly divide the line into two.

The *muscle-layer* is divided into four parts by the epidermal lines, and is made up of flat muscle-cells. Each of these is about $\frac{1}{7}$ of an inch long, and consists of a transversely striated contractile fibre, internal to which is a nucleated protoplasmic part produced into a number of threads. The cell is directed longitudinally, with its fibre closely attached to the epidermis, its protoplasm projecting into the body-cavity, and its flat sides facing adjacent muscle-cells. The thread-like internal processes run to the dorsal line in the upper half of the body, and in the lower half of the body to the ventral line.

3. The **Digestive Organs** (Fig. 15) consist of a narrow tube which runs straight from mouth to anus or cloacal aperture. This *alimentary canal* or *gut* is divided into three well-marked regions (1) the fore-gut (stomodæum), (2) mid-gut (mesenteron), and (3) hind-gut (proctodæum). The essential difference between these is their mode of origin. In the course of development (1) and (3) are formed as infoldings or in-pushings from the exterior, and are therefore lined by ectoderm, while (2) is lined by endoderm.

The **fore-gut** consists of a thick-walled *gullet* (œsophagus) into which the mouth opens. Its wall, starting from the inside, is made up of (a) a thick cuticle, and (b) a layer of epithelium, continuous respectively with cuticle and epidermis of body-wall. while outside these is (c) a muscular layer, the cells of which are radially arranged. The gullet is about $\frac{1}{4}$ of an inch long, and is sharply marked off by a constriction from the **mid-gut**. This, consists of an *intestine*, somewhat flattened dorso-ventrally, and with an exceedingly thin wall made up of a single layer of columnar epithelial cells, lined by a thin cuticle and covered externally by a structureless membrane. The intestine passes into a short **hind-gut** or *rectum*, the walls of which are thickened and resemble those of the gullet in structure. In the male, the terminal part of the rectum is a cloaca, since it receives the genital duct.

A definite *body-cavity* is found outside the wall of the gut, but it is for the most part reduced to a system of narrow spaces by the projecting parts of the muscle-cells. The body-cavity contains a clear albuminous perivisceral fluid.

Ascaris lives, for the most part, in the small intestine, and is consequently surrounded by food much of which is digested—*i.e.*, in a dissolved or finely-divided state. The gullet acts as a kind of suction-pump, its radial muscle-fibres causing enlargement of its cavity, when food rushes into it, after which the elasticity of the cuticular lining comes into play, and, the lips being approximated, drives the food into the intestine. Here the digested part of it diffuses into the perivisceral fluid. There is no special provision for setting up currents in this, but the general movements of the body effect an indefinite kind of circulation.

4. **Excretory Organs** are represented by a narrow tube, imbedded in each lateral line (Fig. 15), and ending blindly behind, while the two tubes unite anteriorly to form an unpaired portion which opens by the excretory pore.

5. **Reproductive Organs** (Fig. 15).—(1) The **male organs** chiefly consist of a much convoluted tube, some seven or eight times the length of the body, ending blindly at one end and opening by the other end into the ventral side of the cloaca. The greater part of this tube is very slender and constitutes the *spermary*, which merges into a much shorter thicker portion, the *vesicula seminalis*, and this again into an extremely short and narrow ejaculatory duct, the walls of which are very muscular.

50   AN ELEMENTARY TEXT-BOOK OF BIOLOGY.

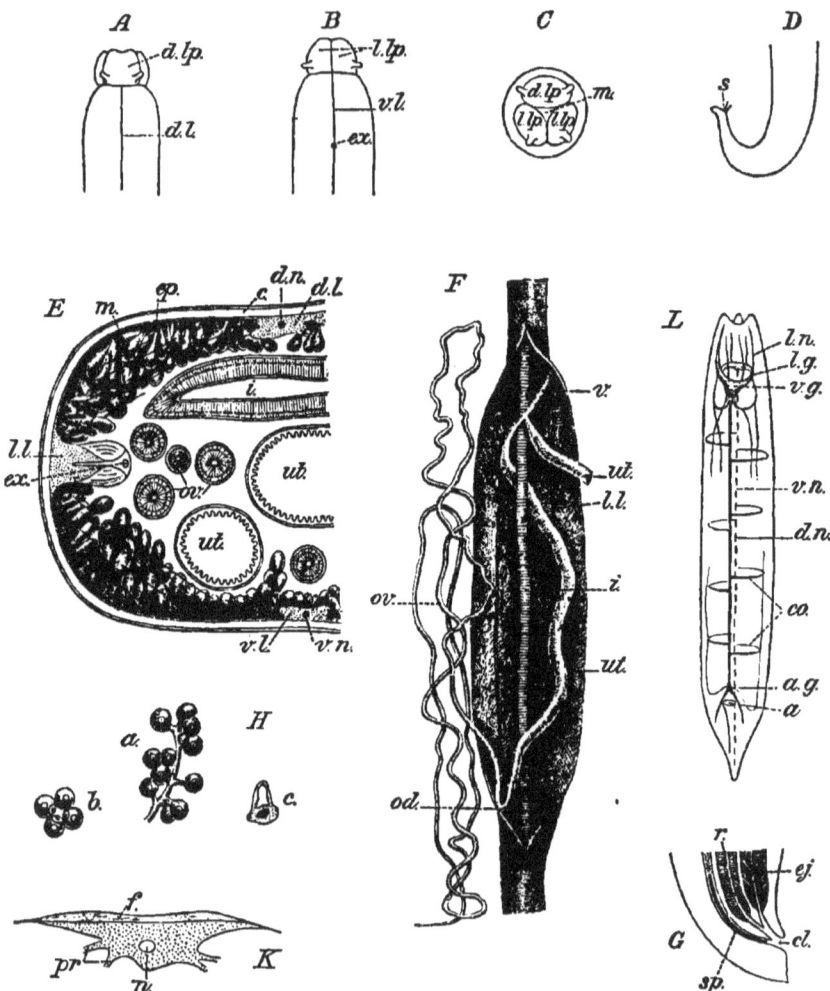

Fig. 15.—ASCARIS (A, B, C, D, H, a, c, K, G, after *Leuckart;* E, mainly after *Vogt* and *Yung;* H, b, after *Van Beneden;* L, after *Bütschli*).— A-D (enlarged), dorsal, ventral, and front views of anterior end, and side view of posterior end of ♂ ; *d.lp* and *l.lp,* dorsal and lateral lips, possessing tactile papillæ, and surrounding mouth, *m; d.l,* dorsal line ; *ex,* excretory pore ; *sp,* copulatory spicules. E, Part of transverse section of ♀, much enlarged ; *c,* cuticle ; *ep,* epidermis, thickened into

The spermary is lined by nucleated protoplasm, and its cavity is traversed by a cord or rachis from which *mother-sperm-cells* arise. Each of these divides into two, and each of the halves by further division produces a group of four cells (spermatocytes) which divide once more. Of the eight cells thus originated four are small and four large. The former are in the centre of the group and constitute a *sperm-blastophor*, the latter are external and become *sperms*. At first the sperms are spherical, but after transfer to the female become conoidal. They never assume the tadpole-shape seen in so many other cases. The basalpart of the conoidal sperm is protoplasmic and nucleated, while the narrow part is a highly refractive "cap" probably composed of nutritive material analogous to yolk, and covered by a layer of protoplasm external to which is a firm membrane. The consecutive stages in sperm-development (spermatogenesis) can be easily followed in a single spermary by examining its successive portions.

Accessory male organs are present in the form of two curved chitinous spicules which are contained in a pouch that opens dorsally into the cloaca. These spicules can be protruded and retracted by special muscles.

The vesicula seminalis serves for the storage of sperms, and, by the contraction of the muscular walls which it and the ejaculatory duct possess, sperms can be ejected into the vagina of the female, this process being aided by the copulatory spicules.

---

dorsal line ($d.l$) traversed by dorsal nerve ($d.n$), ventral line ($v.l$) traversed by ventral nerve ($v.n$), and lateral line ($l.l$) traversed by excretory tube (ex) and strengthened by internal cuticular thickening; m, layer of muscle-cells, with striated part external and protoplasmic part internal, the oblique lines represent processes of the latter running to median lines; i, intestine; ov, ovary, traversed by rachis bearing germ-cells; ut, ut, uteri. F, Female Ascaris opened from ventral side, and reproductive organs of left side cut short (semidiagrammatic); ov, right ovary; od, right oviduct; ut, ut, uteri; v, vagina; i, intestine; $l.l$, lateral line. G, Posterior part of ♂ reproductive organs, longitudinally bisected (enlarged); cl, cloacal opening; ej, ejaculatory duct; sp, spicule in sac; r, rectum. H, Stages in sperm-development (much enlarged); a, part of rachis, with mother-sperm-cells; b, group of four spermatocytes, surrounding a four-celled sperm-blastophor; c, a sperm, nucleus below and refractile "cap" above. K, A muscle-cell (much enlarged); f, contractile fibrous outer part, inner protoplasmic (dotted) part, with nucleus (n), and processes (pr). L, Diagram of nervous system; $v.g$, ventral ganglion; $l.g$, lateral ganglion; $a.g$, anal ganglion; $l.n$, lateral nerve; $d.n$, dorsal nerve; $v.n$, ventral nerve; co, commissural nerves.

(2) The **female organs** consist of two convoluted tubes, ten times the length of the body, which unite to form a short unpaired section that opens to the exterior.

Three regions are distinguishable in each tube, ovary, oviduct, and uterus, while the unpaired part is a vagina. The *ovary* is analogous in structure to the spermary, like it being traversed by a rachis, from which in this case ova arise. The mature *ovum* is of ovoid shape, invested by a delicate vitelline membrane except at one point (the micropyle), and consisting of vitellus (protoplasm) with numerous contained yolk particles, nucleus (germinal vesicle), and nucleolus (germinal spot). The wall of the *oviduct* contains a muscle layer, and so does that of the much wider *uterus*, which also possesses a lining raised into numerous longitudinal ridges. The wall of the narrower *vagina* contains internal circular and external longitudinal muscle-layers.

Ascaris megalocephala presents special facilities for the study of ovum-development (oögenesis) and fertilization, since a single female specimen furnishes all the various stages.

The sperms, ejected into the vagina, crawl by amœboid movements to the upper ends of the uteri, where they meet and unite with the mature ova. In their upward course the sperms pass along the grooves between the ridges into which the uterine lining is raised, and so escape, being swept back by the descending current of ova. The fertilized ova are surrounded by thick eggshells secreted by the glandular lining of the uteri, and also by firm membranes developed within the shell by the oösperms themselves.

6. The **Nervous System** (Fig. 15) is made up of a *ring*, which closely surrounds the anterior part of the gullet, and gives off ill-defined *longitudinal nerves*, six in front and six behind. Two of the former (the largest) run in the lateral lines, and the rest near the median lines,—one of the latter is dorsal, one ventral, while the others are sublateral and do not extend far back. The dorsal and ventral nerves are connected by a number of transverse commissures. The circumœsophageal ring is somewhat swollen at the origins of the lateral and ventral nerves to form two *lateral ganglia* and a *ventral ganglion*. There is also a small *anal ganglion* in front of the anus (or cloacal aperture). The ganglion-cells are most abundant in these ganglia, but are not limited to them.

7. Sense Organs are only represented by *tactile papillæ*, in which nerves terminate. There are two of these on the dorsal lip, one on each ventral lip, and others near the anus (or cloacal aperture).

LIFE-HISTORY.

In both *Ascaris lumbricoides* and *A. megalocephala* the eggs pass out of the intestine of the host, and embryos develop within them which may be liberated on damp soil, &c. Infection results from the swallowing of these eggs or embryos. *A. lumbricoides*, however, may perhaps pass into an intermediate host, *Julus guttulatus*, a small millipede.

Other Nemathelmia.

Oxyuris vermicularis (the thread-worm) is probably the commonest parasite of the kind found in the human subject. It occurs throughout the

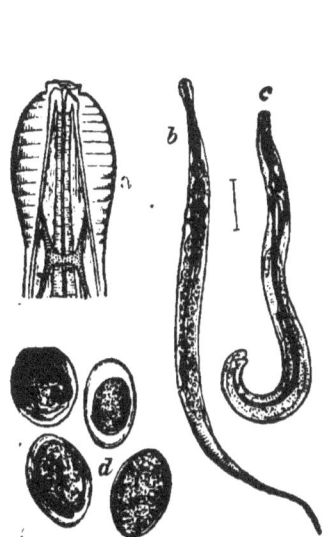
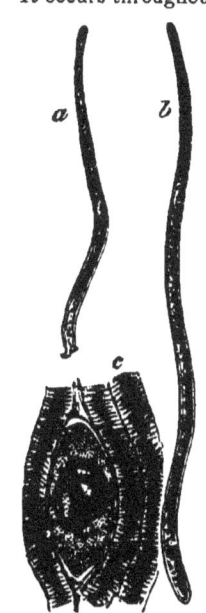

Fig. 16.—OXYURIS (from *von Jaksch*).—*a*, Head; *b*, female; *c*, male; *d*, eggs.

Fig. 17.—TRICHINA (from *von Jaksch*).—*a*, Male; *b*, female; *c*, encysted young.

large intestine, especially in children, its head-quarters being the cæcum. The female is rather less than ½ an inch long, and the male about half that length (Fig. 16). The eggs, which are laid in an advanced state of development, are passed out of the body of the host in vast numbers. There is no intermediate host, and infection is easy, while it is an unusually difficult parasite to get rid of, as its numbers are kept up by self-infection.

Trichina spiralis (Fig. 17) is a minute parasite (♂ $\frac{1}{15}$ inch, ♀ $\frac{1}{5}$ inch long), the adult sexual stage of which occurs in the intestines of vertebrates, especially rat, pig, and man. The young are born alive, bore into the wall of gut where they enter blood-vessels, and thus reach the muscles within which they encyst. Infection is caused in the human subject by the eating of "trichinized" pork (*i.e.*, pork containing encysted trichinæ) which has not been sufficiently cooked.

CHAPTER V.—ANNELIDA (Segmented Worms).

§ 8. LUMBRICUS (Earthworm).

IN this country there are several species of Earthworm, which differ from one another in comparatively minor details. The largest is *Lumbricus herculeus*. It and the other kinds are found in damp earth, &c., in which they burrow. The variations in size are very considerable.

MORPHOLOGY AND PHYSIOLOGY.

1. **External Characters.**—The body is much elongated, and sub-cylindrical. The hinder part is somewhat flattened from above downwards. As in crawling animals generally the symmetry is bilateral, and the *dorsal* surface is readily distinguishable by its dark-reddish colour from the paler *ventral* surface. The body is segmented, that is, made up of a series of transverse rings or **segments** (metameres). These are very numerous (as many as 150), and vary much in size. They are largest at the anterior end, where also the shallow grooves, of which one (or more) encircles each segment, are most evident. The front end of the body tapers to a blunt point, the apex of which is formed by the small *upper lip* (prostomium). This overhangs the transversely crescentic *mouth* which is mainly bounded by the first or *mouth-*

segment (peristomium). A small vertical oval opening, the *anus*, is seen at the posterior end of the body.

The eleventh, like all succeeding segments, has a minute median *dorsal pore*, which communicates with the body-cavity, and is placed in the groove separating the segment from the one in front of it. Every segment, except the first three and the last, possesses a pair of extremely small *excretory apertures* on its ventral surface.

The worm is hermaphrodite, and the openings of its reproductive organs are found on certain of the anterior segments. The dorsal and lateral regions of some of the segments between 29 and 36 inclusive (in *L. herculeus*, 32-37 inclusive) are thickened into a band, the *clitellum*, which varies in size according to the sexual condition.

Running along each side of the body are two double rows of minute, backwardly directed bristles, *setæ*, which can be readily felt by drawing a worm backwards between the fingers. One row is lateral and placed where the pigment of the dorsal surface shades off, the other ventral. Each segment, therefore, except the first few and sometimes the last few, possesses *eight* setæ.

2. **Skin.**—The body is covered by cuticular and epidermal layers, internal to which come the muscles of the body-wall. The *cuticle* is a very thin iridescent membrane, traversed by numerous pores and of chitinous nature. It is secreted by the underlying *epidermis*, which, except in the clitellum, is made up of a single layer of nucleated columnar cells, many of which are unicellular glands, of the type known as *goblet-cells*. These are oval, and filled with liquid secreted by the cell-protoplasm. The epidermis also lines deep narrow pouches (*setigerous sacs*), in which are secreted the hard curved setæ, which project at the surface for only about one-fifth of their length. Each sac contains one bristle, and when this falls out it is replaced by another developed in the same sac. Muscular bands pass from the sacs to the body-wall. The ventral setæ of the genital and clitellar regions are very slender. In the former region some of the ventral setigerous sacs are glandular and much enlarged (capsulogenous glands).

The epidermis is much thickened and very glandular in the clitellum, which also contains a network of blood-vessels.

3. **Digestive Organs** (Fig. 18, A).—The alimentary canal or gut

is a straight tube, of varying calibre, traversing the whole length of the body, and separated from the body-wall by a spacious

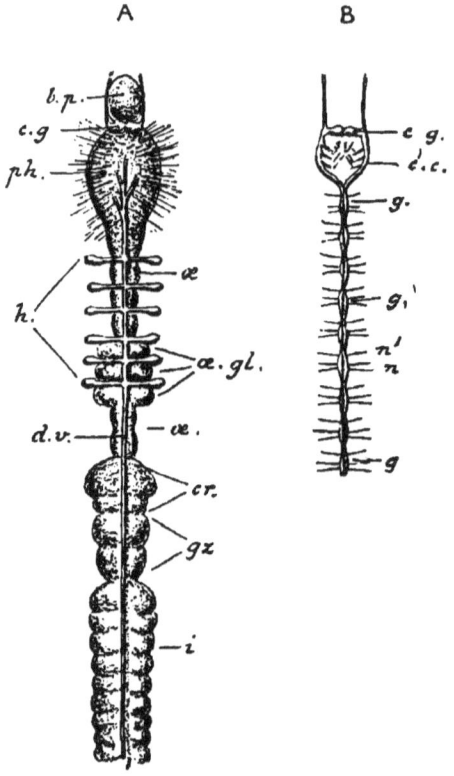

Fig. 18.—DIGESTIVE AND CIRCULATORY ORGANS AND NERVOUS SYSTEM OF EARTHWORM.—A, Digestive and circulatory organs; *b.p*, buccal pouch; *ph*, pharynx; *œ*, œsophagus; *œ.gl*, œsophageal pouches and calciferous glands; *cr*, crop; *gz*, gizzard; *i*, intestine; *c.g*, cerebral ganglia; *h*, hearts; *d.v*, dorsal vessel. B, Nervous system; *c.g*, cerebral ganglia; *c.c*, circumœsophageal commissure; *g.g*, ganglia of ventral cord; *n*, nerves, from ganglia; *n'*, nerves, from between ganglia; *sy*, sympathetic nerves.

cœlom or *body-cavity*, which is divided into compartments corresponding to the segments, and separated imperfectly from one another by thin transverse partitions, the *mesenteric septa*, which are perforated below. These septa are firmly united to the gut, which they hold in place. Fore-gut (buccal pouch, pharynx), mid-gut (gullet, crop, gizzard, intestine), and hind-gut (rectum) are distinguishable.

The mouth, for which the prostomium and peristomium form upper and lower lips, leads into a small, thick-walled eversible *buccal pouch* (*b.p*). This is connected by muscle-fibres with the wall of the body in front of it. The walls of the *mouth-cavity* contained in the pouch are raised into numerous folds. Next follows the oval *pharynx* (*ph*) which is placed in segments 3-6. It is very thick-walled, and connected by numerous retractor muscle-fibres with the body-wall. The posterior end of the pharynx is continued into the tubular *gullet* or *œsophagus* (*œ*) which occupies segments 7-14, and bulges out slightly in each of them. Three pairs of swellings (*œ.gl*) are to be seen on the sides of the gullet in segments 10, 11, and 12 respectively. The anterior pair (œsophageal pouches) are outgrowths from the gullet, while the others (œsophageal or calciferous glands) are glandular thickenings in which carbonate of lime is secreted. The calciferous glands contain a number of small cavities, which communicate with the œsophageal pouches. In 13 to 16, the œsophagus expands into a rounded chamber, the *crop* (*cr*), with rather thicker walls. Behind this comes the thick-walled *gizzard* (*gz*) of similar shape, extending back to about segment 19. The rest of the alimentary canal is almost entirely formed by the thin-walled *intestine* (*i*), which is dilated in each segment. Its dorsal wall is pushed in, so to speak, and forms a thick, longitudinal ridge, the *typhlosole*, which projects into the intestinal cavity from its dorsal side. Its surface is raised into transverse ridges. The outside of the intestine is covered by a thin layer of yellowish-brown cells (chloragogen cells), which also fill up the cavity of the typhlosole, and form the so-called *liver*. The gut is completed by a very short thin-walled *rectum* which occupies the last segment.

The alimentary canal is lined with simple columnar *epithelium*, the surface of which, except in the intestine, is covered by a cuticle that is especially thick and firm in the gizzard. The glandular cells of the calciferous glands are of epithelial nature, and so are a number of small glands which open into the pharynx. A *sub-epithelial* layer follows, composed of connective-tissue, in which is a network of blood-vessels. Externally is a *muscular layer*, composed of an internal sheet of circular, and external sheet of longitudinal fibres. The muscle layer is thicker, and has a more complicated arrangement in the pharynx, crop, and gizzard than elsewhere. Outside the muscle there is a layer of delicate flattened peritoneal epithelium. The "hepatic cells" are elongated and fusiform, with well-marked nuclei, and granular protoplasm often containing concretions. They are closely connected with the blood-vessels.

Earthworms live principally on vegetable food, but animal substances can be used as well. Much earth is swallowed by them, chiefly for the sake of the contained organic matter, but also as a means of excavating burrows. Darwin has shown "that in many parts of England a weight of more than ten tons of dry earth annually passes through their bodies, and is brought to the surface on each acre of land." Pieces of leaf, &c., are partially digested before being taken into the body by a fluid poured on to them from the mouth, and apparently secreted by the pharyngeal glands. Food is taken in by alternate dilatation and contraction of the mouth, aided by a sucking action of the pharynx, and organic acids contained in it are neutralized by the carbonate of lime secreted in the calciferous glands.*

Food accumulates in the crop, and is then passed on to the gizzard, where the contractions of that organ grind it up. This process is largely helped by small stones which have been swallowed and act as millstones. The contractions of the muscular wall of the intestine cause the substances which enter it to be passed slowly backwards. They are meanwhile subjected to the action of a digestive juice, similar to that poured from the mouth, and probably secreted by the intestinal epithelium. This apparently contains *ferments* † which bring the starch and proteids into a state of solution, the former being converted into grape-sugar, the latter into peptones. Any fat swallowed is brought into a very fine state of division—*i.e.*, emulsified. The particular ferments which bring about these changes can only work well in an alkaline solution, hence the use of the œsophageal glands.

* This lime may also be regarded as an excretion. A superabundance of it is taken in with the food, and is thus got rid of.
† **Ferments** are bodies which excite chemical changes in other bodies without themselves entering into the reactions. They are (1) *Living—e.g.*, Yeast and Bacteria ; (2) *Non-living—e.g.*, digestive ferments, complex nitrogenous bodies found in digestive fluids.

The undigested remnants are passed out from the anus as cylindrical "worm-castings." These form little heaps on the surface of the ground, and as they are composed of earth brought from below, a mixing of soil is gradually effected.

The intestine, partly owing to the presence of the typhlosole, offers a large absorbent surface to the digested materials. The products of digestion readily pass into the blood-vessels ramifying in the intestinal wall, and are thence distributed to the body at large.

4. Circulatory Organs (Fig. 18, A).—These place the different parts of the body in communication as regards oxygen and the products of digestion and katabolism. They may be divided into (1) Blood System, (2) Cœlom and Cœlomic Fluid.

(1) **Blood System.**—This is made up of a closed series of tubes containing bright red *blood*, in which are contained minute colourless nucleated cells, the *blood-corpuscles*. The red fluid in which these are suspended is known as *plasma*. The blood, after death, or on removal from the body, *coagulates*—i.e., sets into a jelly-like mass. There are five chief longitudinal vessels, a large number of transverse vessels, and delicate capillary networks which connect up the various branches.

Running in the median line on the dorsal side of the alimentary canal, and visible through the skin in a fresh specimen, is the *dorsal vessel* (*d.v*). This breaks up in front into a network of small vessels on the pharynx. A similar *ventral vessel* runs longitudinally below the alimentary canal. A longitudinal *sub-neural vessel* runs below the ventral nerve-cord, and a small *lateral neural vessel* on each side of it. These longitudinal trunks are connected by transverse *lateral vessels*, of which the most important are the *hearts*. These are six pairs of large lateral vessels in segments 6-11, which connect the dorsal and ventral trunks, and are closely attached to the anterior faces of the septa. The typical arrangement of the transverse trunks in a segment of the intestinal region is as follows:—(*a*) A pair of *parietal* (commissural) *vessels*, receiving numerous twigs from the body-wall, connect the dorsal and sub-neural trunks; (*b*) The intestine is supplied by numerous afferent branches from the ventral, which break up into a network, from which the blood returns by two efferent branches to the dorsal; (*c*) Each nephridium possesses a very rich plexus of vessels, supplied by a branch from the ventral and sending one to the corresponding parietal.

A large *lateral œsophageal vessel* runs along each side of the gullet, opening into the dorsal vessel in segment 10, and returning blood from the gullet and pharynx.

Course of the Circulation.—The blood is kept circulating by means of the hearts, which act as force pumps. Their walls are muscular, and they contract rhythmically from above downwards, the hinder ones contracting first. The large vessels also have contractile walls by which the circulation is assisted. The blood flows from back to front in the dorsal vessel, and the contrary way in the ventral one. The parietal vessels return blood purified in the skin and nephridia to the dorsal vessels, which also receives blood from the gut by means of the lateral œsophageal and efferent intestinal vessels. This blood passes forwards in the dorsal vessel and through the hearts to the ventral vessels, in which the blood flows backwards and passes by afferent trunks to the gut, nephridia, and body-wall.

The delicate networks of capillary vessels which complete the closed blood-system are of great physiological importance, since their extremely thin walls allow of diffusion. In this way, for example, the tissues are nourished.

There are no special organs of respiration and this function is performed by the general surface of the body, the carbon dioxide diffusing out of the blood-vessels underlying and penetrating the skin, while oxygen diffuses into them. The red colour of the blood is due to *hæmoglobin*, a complex compound of carbon, hydrogen, oxygen, nitrogen, sulphur, and iron, which enters into loose combination with the oxygen taken into the body, readily parting with it again to the tissues.

(2) **Cœlom and Cœlomic Fluid.**—The *cœlom* or *body-cavity* is imperfectly separated into compartments by the septa, and communicates with the exterior by the dorsal pores, excretory tubes (nephridia), and oviducts. It contains a milky coagulable cœlomic (perivisceral) fluid, in which are suspended numerous irregular nucleated cells (cœlomic corpuscles), with granular protoplasm, capable of amœboid movements. The body-cavity is lined by cœlomic epithelium composed of a single layer of flattened nucleated cells, from which these corpuscles are constantly being budded off.

5. Excretory Organs.—Each segment, with the exception of the first three and the last, contains a pair of *nephridia* (segmental organs). Each nephridium is a very long convoluted tube, of varying calibre in its different regions, attached by a membrane to the back of the septum bounding the front of the segment, and placing the cœlom in communication with the exterior. It is thrown into three main folds running parallel to the septum in a vertical direction, and forming an internal *short loop*, a middle *long loop*, and a wide external *end loop*. The nephridium commences with a *ciliated funnel* which lies in the *preceding* segment. The margin of the funnel is formed by a row of elongated ciliated cells, and its back by a very large crescentic cell. The slit-like internal opening (nephrostome) of the nephridium is situated within the funnel. The rest of the tube is divided into four sections : (1) An extremely long and delicate *narrow section* which runs back from the funnel through the septum, round the short loop, up and down one side of the long loop, and back round the short loop, after which it passes into (2) a much wider *middle section*, contained entirely in the long loop, and followed by (3) a *wide section*, which begins with a dilated part, runs along one side of the long loop, round the short loop, and across to the end loop, where it is succeeded by (4) a much larger *muscular section* ("bladder") which opens to the exterior by a small ventral aperture, the *nephridiopore*. The greater part of the nephridium (*i.e.*, from the funnel to (4)) is made up of tubular "drain-pipe cells," placed end to end, and the cavity of this part is therefore intracellular—*i.e.*, *within* and not *between* cells. The tubular cells differ in size and character in the different regions of the nephridium.

Cilia are found in parts of the narrow section and throughout the middle section, while the cells of the middle and wide sections are very large and full of excretory granules. The muscular section is lined by cells (its cavity being therefore intercellular), and its thick wall contains a network of muscle-fibres.

The nephridium possesses a very rich network of capillary blood-vessels, supplied from the ventral vessel and returning its blood into one of the parietals.

The nephridia are specially concerned with the excretion of water and nitrogenous waste, the former entering them by the ciliated funnels, and the latter being probably excreted by the glandular drain-pipe cells of the middle and wide sections. It is not unlikely that the chloragogenous cells have to do with excretion, for they are closely connected with the blood-vessels, resemble excretory cells in their granular character, and are known to break down into the cœlom, the materials thus formed being very probably got rid of by the nephridia.

The *urine*, as the excreted substances may be collectively termed, passes down the nephridia by the action of the cilia, and collecting in the dilated muscular section is expelled by its contraction.

6. Reproductive Organs (Fig. 19).—The Earthworm is hermaphrodite, and its sexual organs vary much in size, according to the time of year, being largest in the summer, which is the breeding-season. The segments in which they are placed are anterior to the clitellum. Their colour is white, and they lie below and at the sides of the gut, in its œsophageal region.

(1) **Male Organs.**—There are two pairs of minute flattened *spermaries* (T), produced into finger-like processes behind. These are attached, near the nerve-cord, to the back of the septa forming the anterior boundaries of segments 10 and 11. In the sexually mature state they project into the cavities of the seminal reservoirs. The most conspicuous parts of the male apparatus are the *vesiculæ seminales*, three pairs of large white pouches which may completely overlap the œsophagus, and are situated in segments 9-12. The anterior pair (A) lie in segment 9, and are united into a median seminal reservoir (B) in segment 10. This also receives the middle (C) pair of vesiculæ situated in segment 11, while the posterior vesiculæ are situated in segment 12, and unite into a similar reservoir in segment 11. Projecting into the floor of each reservoir is a pair of large plicated *seminal funnels* (S F) turned towards the testes. From each funnel a delicate tube, the *vas efferens*, proceeds, which, after forming a coil, takes a backward course. The two vasa efferentia on each side unite together in segment 12 into a straight tube, the *spermiduct* (V D), which passes back through segments 13 and 14 to open by the male pore on segment 15, just external to the ventral setæ.

The *spermary* may be regarded as a thickened and specialized part of the epithelium lining the body-cavity. It is made up of numerous *germinal-cells* which, as mother-sperm-cells, pass into the vesiculæ seminales, the cavities of which are traversed by numerous strands of connective tissue. In the interstices between these the mother-sperm-cells undergo repeated division or seg-

ANNELIDA. 63

mentation to produce *sperm-morula*, each of which contains a central sperm-blastophor covered by numerous small cells, *spermatocytes*, which become *sperms*. These possess a long cylindrical

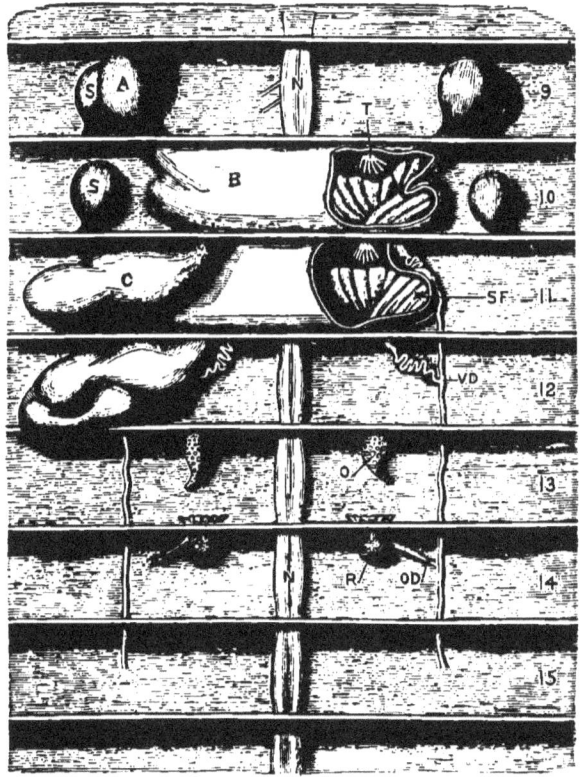

Fig. 19.—EARTHWORM (from *Marshall* and *Hurst*). Plan of the reproductive organs as seen from above after removal of the alimentary canal. The body-wall is pinned down flat.—A, Anterior vesicula seminalis, uniting with its fellow to form the anterior seminal reservoir (B) which is cut open on one side to display an anterior spermary (testis) (T) and an anterior seminal funnel. C, Middle vesicula opening into B. In segment 12 is seen one of the posterior vesiculæ, which unite to form the posterior reservoir in 11. This is cut open to display a posterior spermary (testis) and posterior seminal funnel (S F). VD, Vas deferens or spermiduct; O, ovary; O D, oviduct; R, oötheca; S, spermotheca; N, nerve-chain; 9-15, ninth to fifteenth segments.

head and vibratile tail. They remain for some time attached by their heads to the blastophor, but are finally liberated.*

(2) **Female Organs.**—The *ovaries* (O) are two pear-shaped bodies, rather larger than the spermaries, and having a similar position in segment 13. Their narrow ends are directed backwards. The *oviducts* (OD) are short, wide tubes, lying mainly in segment 14, on the ventral side of which they open by the *female pores*, much smaller than the male pores, but occupying a similar position. Each oviduct ends internally in a funnel which pierces the septum between segments 13 and 14, projecting into the cavity of the former. The funnel communicates with a small pouch-like *oötheca* (R) (receptaculum ovorum), in which a few ripe ova temporarily accumulate. The *spermothecæ* (receptacula seminis) are two pairs of rounded sacs, lying in segments 9 and 10 (S). They open laterally between segments 9 and 10, and 10 and 11, just on a level with the lateral setæ.

The *ovaries* consist of germinal cells which develop into ova. Each ovary contains many ova in different stages of development, the ripest being at the free end. The *ovum* is spherical in shape, and covered externally by a delicate *vitelline membrane*. The granular protoplasm contains a clear *germinal vesicle*, with *germinal spot*. The ripe ova burst out of the ovary into the body-cavity, and are taken up by the ciliated funnels of the oviducts.

Although the Earthworm is hermaphrodite, self-fertilization does not occur. Two individuals copulate and fertilize each other. This takes place on the surface of the ground during the warm spring and summer months. The two apply themselves by their ventral surfaces, their heads being in opposite directions. Adhesion is effected by the slender genital setæ, the prominent ventral edges of the swollen clitella, and a firm secretion of the clitella and capsulogenous glands. The sperms pass back from the male pores of either worm to the spermothecæ of the other, along grooves visible at the time. They are aggregated within the spermothecæ into thread-like packets (spermatophores), by means of a glutinous secretion.

The ova are laid in firm capsules, probably formed by the hardening of a fluid secreted by the clitellum. In *L. herculeus* many ova are passed into each capsule, together with a few

* The contents of the vesiculæ commonly contain various stages in the life-history of the Earthworm Gregarine, *Monocystis lumbrici* (see p. 17).

spermatophores (composed of sperms derived from another individual) from the spermothecæ. The egg-capsule or cocoon is passed forwards over the front end of the worm, and cylindrical to begin with immediately becomes spindle-shaped by closure of its ends. The spermatophores break down within the capsule, and the liberated sperms fertilize the ova, polar cells having previously been formed.

7. **Muscular System.**—As in the Fluke, most of the muscle is found in the body-wall. Underlying the skin is a continuous *circular layer* in which the pigment that gives the dorsal surface its characteristic tint is imbedded. Delicate bands, the *protractor muscles* of the setæ, pass from this layer to the inner ends of the setigerous sacs. Beneath the circular layer is a *longitudinal layer* of about the same thickness, subdivided into longitudinal bands, a broad *dorsal* band between the two rows of lateral setæ, a much narrower *ventral* band between the ventral setæ, a *lateral* band of about the same width between the lateral and ventral setæ on either side, and finally four excessively thin strips (two *dorso-lateral* and two *ventro-lateral*), one between the setigerous sacs in each double row. Each band is made up of plates, perpendicular to the circular layer, in which the fibres are arranged so as to give a feather-like appearance in cross-section. Connected with the longitudinal layer are delicate *retractor muscles* of the setæ, attached to the setigerous sacs where they reach the middle of the circular layer.

The septa contain muscular fibres, imbedded in much connective-tissue, and taking various directions. They are covered by the peritoneal epithelium of the body-cavity, with which also the longitudinal muscle layer is lined.

The muscle-fibres are very slender cells, united together by their tapering ends.

In locomotion the circular layer contracts so as to advance the front part of the body, and the hinder part is then dragged up by contraction of the longitudinal layer. The reverse movement can also be effected, and complex curves described. The setæ act as hold-fasts, their direction and amount of protrusion being regulated by their retractor and protractor muscles.

8. The **Nervous System** (Fig. 20, B) consists of a ring encircling the gut at the junction of the buccal pouch and pharynx, a ventral cord, and numerous nerves. The dorsal side of the ring is thick-

ened into two *cerebral ganglia* (*c.g*, A and B), while its sides, the *connectives* (*cc*), run downwards and backwards and unite together to form the ventral cord. This dilates slightly into a ganglion (*g*) in each segment, the ganglia being best marked in the posterior part of the body. The double nature of the cord is indicated by a longitudinal furrow running along its upper surface.

A nerve passes from each cerebral ganglion to the prostomium and peristomium, numerous nerves run out from the first ventral ganglion, while behind this there are typically three pairs of nerves in each segment, two arising from its ganglion (*n*), and one (septal nerves) from the cord in front of this (*n'*).

A number of delicate nerves pass from the inner sides of the connectives to the pharynx, in the walls of which they form a delicate ganglionated network. This arrangement is called the sympathetic nervous system (*sy*).

The nervous system is essentially made up of nerve-cells and nerve-fibres. In the nerve-ring and cord there is a very delicate connective-tissue framework (neuroglia) in which these elements are imbedded. The nerve-cells occupy the anterior part of the cerebral ganglia and the under side of the ventral cord both in and between the ganglia. The ventral cord is surrounded by a firm muscular sheath in which the neural vessels run, and imbedded in the dorsal part of the sheath there are three *giant-fibres*, which perhaps have a supporting function, but in any case are modified nerve-fibres.

The central organs certainly constitute a correlating apparatus, but scarcely anything is known of their mode of action.

9. **Special Sense Organs** appear to be represented only by "goblet-bodies," consisting of aggregates of slender epidermal cells connected with nerve-fibres, and occurring on the prostomium, peristomium, and, less abundantly, on the anterior segments. The Earthworm is, however, by no means devoid of special senses. The entire surface of its body is endowed with a delicate sense of **touch**, while certain kinds of food (*e.g.*, bits of onion) are specially preferred and readily discovered, which seems to prove possession of both **smell** and **taste**. The structure and position of the prostomial and peristomial goblet-bodies indicate that these are the special organs of taste and smell. Although earthworms possess nothing that can be described as a definite sense of **sight**, they are sensitive to intense light, which they shun, and thus escape

many enemies. The anterior part of the body is specially sensitive to light, which perhaps, as Darwin suggested, passes through the skin and acts directly on the cerebral ganglia. It has, however, been stated that all the pigmented regions of the body are affected by blue, violet, and ultra-violet rays. The sense of **hearing** is entirely absent.

DEVELOPMENT (Fig. 20).

The egg-capsules of the Earthworm are deposited during spring and summer in damp earth, a few inches from the surface. They are olive-green in colour, spindle-shaped, and about $\frac{1}{3}$ of an inch long. Each capsule contains a mass of slimy albumen and several eggs, of which, however, only one usually comes to maturity.

Fertilization takes place within the capsule, after which the oösperm undergoes continued cell-division (cleavage, segmentation) to form a hollow sphere (blastula, blastosphere) with a cellular wall. The sphere is then converted into a double-walled bag or gastrula by inpushing of its wall. The inner layer of the gastrula is endoderm, the outer ectoderm. At the same time the third germinal layer, mesoderm, begins to develop by division of two large cells (mesoblasts) found near the posterior end of the body. The embryo gradually becomes more and more elongated, the various organs being at the same time differentiated from the three germinal layers. The surrounding albumen is used as food, and one embryo develops more rapidly than the others which it utilizes in the same way. In the case of capsules which are kept in the laboratory, hatching takes place in from two to three weeks.

1. **Cleavage** (Segmentation).—This is complete (holoblastic) and irregular, the entire oösperm dividing, but the resultant cells being of unequal size. The divisions take place in a regular way (B, C) until a 7- or 8-celled embryo is produced, consisting of 2 large, 2 medium-sized, and 3 or 4 (probably ectodermic) small cells. Beyond this there seems to be no definite order. The completely segmented oösperm is a *blastula* (blastosphere) containing a large *segmentation-cavity* (*s.c*) (blastocœle), bounded on one side by small and on the other by large cells, which respectively become ectoderm and endoderm (D, E, F). Two large cells (mesoblasts) are also distinguishable (E, M) from which the whole of the

mesoderm arises later on, and even during the blastula stage they divide to produce mesodermic cells which partly block up the blastocœle (F, $m.s$).

2. **Gastrulation.**—This is effected on the third day by a modification of what is known in other more typical cases as *emboly* or *embolic invagination*. This is a sort of inpushing, the nature of which may be realised by taking one of the perforated india-rubber balls used by small boys as squirts, and collapsing it so as to form a double-walled cup. The earthworm blastula first of all flattens (F), and becomes an oval plate, with an upper layer of small cells (*ec*) constituting the *ectoderm* (epiblast) and a lower layer of clear columnar cells (*en*) constituting the *endoderm* (hypoblast). *Mesoderm* (mesoblast) is also present (*ms*). The plate next becomes concave below, and its edges approach. The embryo is now a *gastrula* (G, H, K) with an internal digestive cavity, the *archenteron* (*ar*), opening by a longitudinal ventral slit, the *blastopore*. The embryo is elongated in the direction which correspond to the long axis of the future worm, and the two mesoblasts (M) are at its posterior end. In fact, bilateral symmetry is well established.

Fig. 20.—EARTHWORM DEVELOPMENT (after *Wilson*). Enlarged to various scales.—A, Unsegmented ovum, surrounded by the vitelline membrane, with two cell-groups formed by division of polar cells. B, First cleavage (in the plane of the polar cells). C, Six-celled stage. D, Optical section of young blastula. E, Surface view of young blastula, showing the two mesoblasts. F, Optical section of a flattening blastula, seen from the side. G, Ventral view of early gastrula (anterior end directed upwards), with wide blastopore. H, Similar view of rather older gastrula, with slit-like blastopore. K, Left lateral view (in optical section) of established gastrula, in which mesoderm bands have met above mouth. L, Right lateral view (partly in optical section) of embryo with established germinal bands and fore-gut. M, Ventral view of same embryo. N, Transverse (slightly oblique) section of rather younger embryo, cutting through germinal bands, the left neuroblast, and one of the right nephroblasts. O, Right lateral view (in optical section) of older living embryo, showing fore-gut, mid-gut, septa, sections of body-cavity, and cerebral ganglia. P, Left lateral view of embryo in middle of development, showing segments, prostomium, head-cavity, nerve-ring and cord, setæ, nephridia, and lateral vessel. *ar*, Archenteron; *c.g*, cerebral ganglia; *cœ*, cœlom; *d*, septum; *ec*, ectoderm; *en*, endoderm; *l.v*, lateral vessel; M, primary mesoblast; *m*, mouth; *ms*, mesoderm (upper *ms* in O = migratory mesoderm budded off from main band); N, nephroblast; Nb, neuroblast; *n.c*, neural cord (nerve-cord in P); *np*, nephridia; *np.c*, nephric cords; *ps*, prostomium; *S.c*, segmentation cavity; *st*, fore-gut (stomodæum).

ANNELIDA.

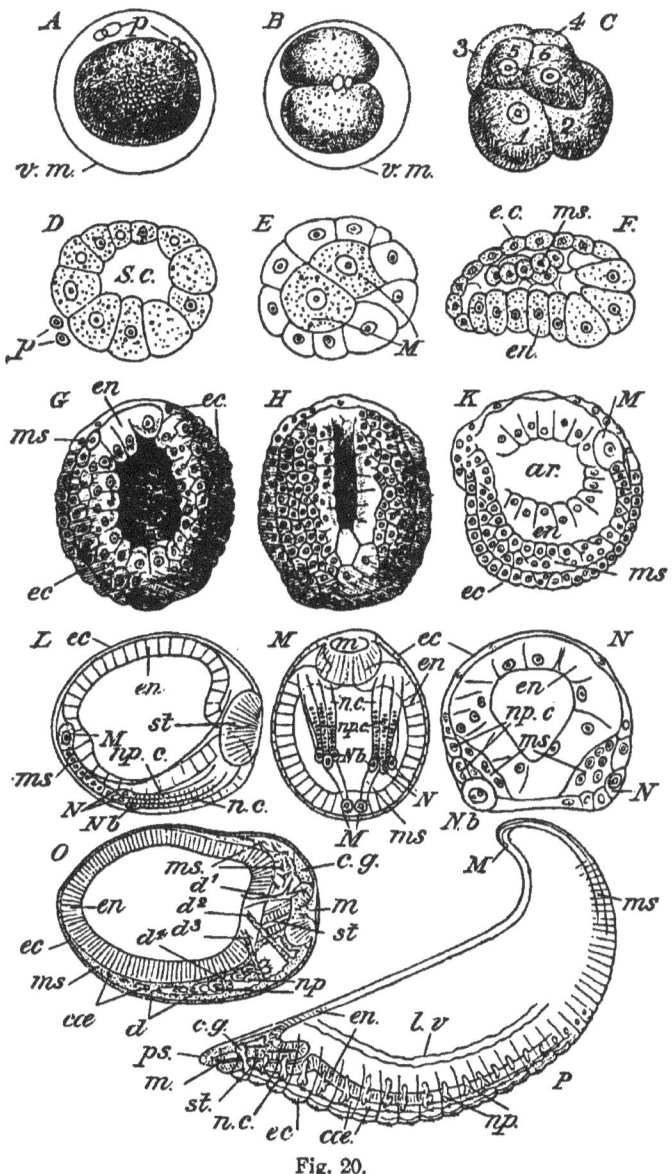

Fig. 20.

The blastopore immediately closes up except in front, where it acts as a mouth, and at this stage the swallowing of albumen begins, and the embryo becomes distended and thin-walled as a result (L, O).

3. Fate of the Germinal Layers.—The *ectoderm* produces the epidermis with its setigerous sacs, the nervous system, the nephridia (except their funnels), and the epithelium lining fore- and hind-guts. The *mesoderm* gives rise to nearly everything between the epidermis and the epithelium of the gut, including circulatory and reproductive organs, the funnels of the nephridia, and all the muscular and connective tissue. The *endoderm* originates the epithelium of the mid-gut.

Ectoderm.—The fore-gut is formed at a comparatively early stage as an ingrowth of ectoderm (*st*) at the unclosed mouth-end of the blastopore (L, M). Soon after gastrulation cilia are developed along a median ventral strip of the ectoderm, and by their action cause the embryo to rotate in the surrounding albumen. The fore-gut is also ciliated. Not long before hatching the hindgut is formed as a small ectodermic involution which soon meets and fuses with the mid-gut.

At the time when the fore-gut is being formed a thickened band can be made out running along each side of the body, and known as a **germinal band**. It is partly constituted by three longitudinal rows of ectodermal cells, each of which is derived by constant subdivision from a cell (teloblast) at its posterior end. These rows and their teloblasts are at first superficial, but later on sink in and are covered by other ectodermal cells. The middle region of a fully developed germ-band (L, M, N) consists of three layers—

(*a*) *Outer layer*, of ordinary ectodermal cells.
(*b*) *Middle layers*, of cells derived from ectodermal teloblasts.
(*c*) *Inner layer* of mesoderm.

The internal teloblasts are known as neuroblasts (Nb), and the neural cords (*n.c*) which they produce thicken and become the right and left halves of the central nervous system, shifting down and uniting ventrally to form the cord and uniting above the mouth to form the nerve-ring (see P). The two external teloblasts on each side are termed nephroblasts (N), because the nephridia (except funnels) grow out from the nephric cord (*np.c*) which they originate. The two nephric cords on each side soon unite together, and the setigerous sacs as well as the nephridia grow out from the band thus formed.

Mesoderm.—The whole of this is developed from the primary mesoblasts (M). These lie behind the other teloblasts and give rise to a mesodermal band (*ms* in K to O) on each side which forms the middle layer of the germinal bands. From the front ends of these bands stellate cells bud off in front and form a mass above the mouth (see O) by which the two bands are connected together. Similar cells are budded off in the trunk-region, but these do not develop into any adult structures.

The mesodermic bands gradually broaden and thicken, their posterior ends being smallest because they have only recently been derived from the teloblasts. In each band a series of cavities (*cœ*) appear from before backwards. These become the sections of the adult cœlom and indicate the position of the future segments. Where the bands unite an unpaired "head-cavity" (P) is developed. The mesoderm which forms the outer boundary of each cavity gives rise to part of the body-wall, that making up the inner boundary originates part of the gut-wall, while the partitions between adjacent cavities become septa.

The mesodermal bands gradually get broader by the division of their cells and unite together from before backwards, first ventrally and then dorsally, making their way between the other layers, and ultimately forming a continuous sheet of mesoderm. The corresponding cœlomic cavities of the two bands fuse completely together dorsally, but remain separated ventrally by a thin mesentery. The blood-vessels are developed as spaces in the mesoderm which at first (except in the case of the hearts) have no proper walls of their own. The dorsal vessel has a double origin, being formed from two lateral vessels (*l.v* in P) which gradually shift upwards and fuse together. Each nephridial funnel is formed by the subdivision of a large mesodermal cell which soon becomes connected with an ingrowth from the nephric band.

The **endoderm** becomes the lining of the mid-gut, including the epithelial part of the œsophageal pouches and calciferous glands.

Sexual Reproduction.

There is some difficulty in comparing the reproduction of unicellular animals (Protozoa) and multicellular animals (Metazoa), because in the former case all the functions are performed by the same cell, while in the latter there are special reproductive cells. The permanent conjugation of Vorticella, however, where a small active microzoid unites with a larger more passive macrozoid, appears to be equivalent to an act of sexual reproduction—*i.e.*, to the union of a male cell (sperm) with a female one (ovum). The process in Vorticella is in reality a very complicated one, and the paranuclei (micronuclei) are alone concerned in it. The ordinary nuclei (macronuclei) appear to have as their function the regulation of nutrition. In many free-swimming ciliated Protozoa "temporary" conjugation takes place between two equal-sized individuals. They apply themselves closely by their oral surfaces and exchange parts of their paranuclei, after which separation takes place, the nuclear apparatus is reconstructed, and fission is resumed. It has been demonstrated that, in ciliated Protozoa, this process is necessary for the continuance of the species, as without it asexual reproduction cannot take place beyond a certain number of generations. In one of the simplest cases (*Colpidium*) the changes which take place in the paranucleus are as follows:—The division is always of the "indirect" kind (mitosis, karyokinesis), and involves more than simple elongation, constriction, and division; (1) Paranucleus divides into two, n^1, n^1, and (2) these subdivide again n^2, n^2, n^2, n^2. (3) Of these four parts three are absorbed by the protoplasm, while the fourth again divides into two, n^3, n^3. (4) One of these migrates into the other individual and fuses with the corresponding part. (5) Each individual now contains a conjugation nucleus $[n^3 + n_3]$ formed by fusion of ¼th of one paranucleus with ¼th of the other. This compound structure divides twice, two of the quarters passing to one end and two to the other. Meanwhile the original nuclei have been absorbed and the individuals have separated. Transverse fission now takes place, the quarters above-mentioned becoming the nuclei and paranuclei of the daughter forms. The complete process may be graphically represented as follows, where N = nucleus, n = paranucleus, and the indices are written above (*e.g.*, n^3) for one individual and below (*e.g.*, n_3) for the other. Elements enclosed in parentheses are absorbed. v = subdivisions of conjugation nucleus.

	INDIVIDUAL 1.	INDIVIDUAL 2.	
STAGE 1.	$n \begin{cases} n^1 \begin{cases} (n^2) \\ (n^2) \end{cases} \\ n^1 \begin{cases} (n^2) \\ n^2 \end{cases} \end{cases}$ $\begin{Bmatrix} n^3 \\ n^3 \end{Bmatrix} \longrightarrow\\ \longleftarrow$ $\begin{Bmatrix} -n_3 \\ -n_3 \end{Bmatrix}$ $\begin{Bmatrix} (n_2) \\ (n_2) \\ (n_2) \\ n_2 \end{Bmatrix} n_1 \Bigg\} n$		
	(N)		(N
STAGE 2.	$[n^3 + n_3] \begin{cases} v \begin{Bmatrix} v^1 \\ v^1 \end{Bmatrix} = N \text{ and } n \\ \cdots\cdots\cdots\cdots\cdots\cdots\cdots \\ v \begin{Bmatrix} v^1 \\ v^1 \end{Bmatrix} = \text{new} \\ \quad N \text{ and } n \end{cases}$	new line of fission. new N and n	$\begin{matrix} N \text{ and } n = \begin{Bmatrix} v^1 \\ v^1 \end{Bmatrix} v \\ \\ \text{new} = \begin{Bmatrix} v^1 \\ v^1 \end{Bmatrix} v \end{matrix} \Bigg\} [n_3 + n^3]$

The processes of conjugation just described involve complete reconstruction of the nuclear elements by (*a*) rejection and absorption of parts of the old constituents, (b) addition of new constituents from another individual. In Vorticella there is in addition *dimorphism*, the conjugating elements differing in size and activity. This is a form of physiological division of labour, for a small active cell is best fitted for the mere transfer of nuclear material, and a large cell is a convenient storehouse of nutriment for developmental needs. This is the meaning of the striking differences between sperm and ovum in the sexual reproduction of the Metazoa. Again, in the maturation of both ova and sperms the elimination of nuclear material just described for ciliate Protozoa is paralleled. This is the import of the polar cells or bodies which are separated off from ripening ova. An ovum is commonly invested by a cell-wall, the *vitelline membrane*, within which are the protoplasm and nucleus (germinal vesicle). The protoplasm (vitellus) consists of a delicate living network, the meshes of which enclose different substances, especially nutritive granules or food-yolk, the amount of which varies with the size of the ovum. The nucleus is covered by a delicate nuclear membrane, and it consists of a modification of protoplasm roughly divisible into (a) *achromatin*, staining with difficulty, and traversed by (b) an easily stained convoluted thread or network of *chromatin*, part or all of which may be condensed into a rounded nucleolus (germinal spot). Fig. 21 represents the stages in the formation of polar

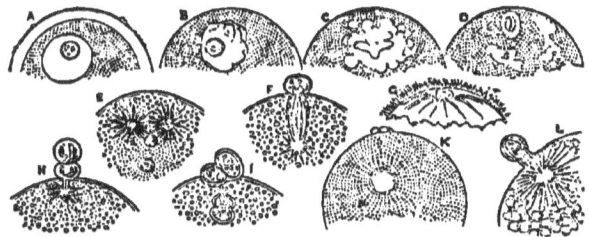

Fig. 21.—FORMATION OF POLAR CELLS IN A STARFISH (*Asterias glacialis*). A.K, after *Fol*; L, after *O. Hertwig*.—A, Ripe ovum, with excentric germinal vesicle and spot. B-D, Gradual metamorphosis of germinal vesicle and spot, as seen in the living egg, into two asters. F, Formation of first polar cell and withdrawal of remaining part of nuclear spindle within the ovum. G, Surface view of living ovum, with first polar cell. H, Completion of second polar cell. I, A later stage, showing the remaining internal half of the spindle in the form of two clear vesicles. K, Ovum, with two polar cells and radial striæ round female pronucleus, as seen in the living egg. [E, F, H, I], From picric acid preparations. L, Expulsion of first polar cell.

cells in the case of a starfish (*Asterias glacialis*). In A are seen the excentric germinal vesicle and spot, which become less distinct in B, C, and D. A "nuclear spindle" is then formed, consisting of threads along which half the chromatin passes to one end and half to the other. At each end of the spindle a small "central corpuscle" or *centrosoma* appears, which is the centre of a radiating sun-like figure, and appears to be the seat of a special sort of protoplasm (archoplasm) that determines the division and movement of the chromatin. The spindle now places itself vertically to the surface of the ovum, and part of it, together with a minute quantity of protoplasm, is pinched off as the first polar cell (F, L, G). A second division (H, I, K) leads to the production of a second polar cell, and the nucleus, now reduced by three-fourths, travels back to the centre of the ovum. It is now termed the *female pronucleus*, and does not contain enough chromatin to enable the ovum to develop further. In certain ova which can develop without fertilization (*i.e.*, are *parthenogenetic*) only one polar cell is formed. Weismann supposes that the germinal vesicle contains two kinds of protoplasm, (1) ovogenetic substance, which presides over the growth of the ovum, and (2) germ-plasma, which enables it to develop into an embryo. He believes the first polar cell (or only one in cases of parthenogenesis) to consist of (1), the second of half (2). A certain amount of germ-plasma is imagined to be requisite for development, and when two polar cells are formed too little is left for the purpose. Fertilization according to this would seem to mean the importation of germ-plasma in sufficient quantity to make up the deficit. In Ascaris the polar cells are not formed till after the entry of the sperm.

In *spermatogenesis* something akin to the formation of polar cells is to be observed. The whole of a mother-sperm-cell does not become converted into sperms; there is always a residue forming a sperm-blastophor or an equivalent to it. A typical sperm is somewhat tadpole-shaped. The head contains a nucleus rich in chromatin, and covered by a thin layer of protoplasm drawn out into the vibratile tail, by means of which the sperm is propelled head first.

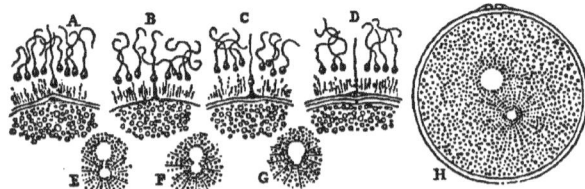

Fig. 22.—FERTILIZATION OF OVUM OF A STARFISH (*Asterias glacialis*)— (after *Fol*).—In A-D the sperms are represented as imbedded in the mucilaginous coat of the ovum. In A a small prominence is rising from the surface of the ovum towards the nearest sperm. In B they have nearly met, and in C they have met. D, The sperm has penetrated the ovum, and a vitelline membrane which prevents the entry of other sperms has been formed. H, Ovum showing polar cells and approach of the ♂ and ♀ pronuclei; the protoplasm is radially striated round the former. E, F, G, Later stages in the coalescence of the two pronuclei.

Fig. 22 illustrates fertilization (impregnation) as it occurs in a starfish (*Asterias glacialis*). A single sperm penetrates the ovum, with the protoplasm of which its protoplasm fuses, while its nucleus, now known as the *male pronucleus*, unites with the female pronucleus to form the *segmentation-nucleus*. The oösperm constituted by the union of the male and female cells can now develop into an embryo. The phenomena of fertilization have been studied with great care in Ascaris, and it has been shown that the chromatin elements of the pronuclei become intimately united, and that when the oösperm undergoes cleavage, the chromatin is halved at each cell-division in a very exact manner. Centrosomata in the protoplasm play an important part in cleavage.

§ 9. HIRUDO (The Leech).

The medicinal Leech (*Hirudo medicinalis*) is a slimy, flattened worm-like animal some 2 to 6 inches long, found abundantly in the freshwater pools and marshes of the Continent. It also occurs in this country. There is a sucker at each end of the body, and by the alternate attachment of these locomotion is usually effected. The animal can also swim by means of wavelike contractions of the body. It is a matter of ordinary knowledge that the Leech is a blood-sucking ectoparasite.

MORPHOLOGY AND PHYSIOLOGY.

1. **External Characters.**—The bilaterally symmetrical body, when fully extended, is somewhat strap-shaped, broadest about the middle, and with the dorsal surface rather more convex than the ventral. A large amount of contraction can, however, take place, by which the body is much shortened and its shape altered. A number of transverse grooves divide the body into about 95 rings or *annuli*, these, however, being superficial and not corresponding to the true segments, which are 26 in number, as indicated by the structure of the nervous and excretory systems. Except near the ends each segment is made up of five annuli, of which the last is distinguished by special colour-markings, and the first by a transverse series of small white papillæ. There is a considerable difference in appearance between the dorsal and ventral surfaces, for the former is much darker and marked on each side by three longitudinal bands, of which the two outer are diversified by dark dots, specially conspicuous in the last annulus of each segment. The segments making up the anterior and posterior ends are shorter than the others and contain fewer annuli.

An oval ventral sucker (Fig. 23) terminates the front end, and it appears to be formed by fusion of a lip-like *prostomium* with the first two annuli of the most anterior or *peristomial segment*. The *mouth* is placed within this sucker, and round its dorsal margin ten minute *eyes* can be recognized as black spots. A large round sucker (Fig. 23), also facing ventrally, occupies the posterior end of the body, and the small dorsal *anus* opens just in front of it. There are two other unpaired openings, both in the median ventral line, one, the *male aperture*, in the second annulus of the 6th segment, the other being the *female aperture* in the second annulus of the 7th segment. There are also 17 pairs of minute *excretory pores* by which the nephridia open in the hindmost annuli of segments 2-18. The setæ of the earthworm are quite unrepresented in the leech.

2. The **Skin** consists of cuticle, epidermis, and dermis. The *cuticle* is thin, elastic, and frequently cast and renewed. It is secreted by an underlying *epidermis*, which is mainly made up of a single layer of cells shaped like mallets with the handles turned inward. There are also numerous *epidermal glands*, each of which is unicellular and opens on the surface by a very narrow tube or duct. These glands are of two kinds—(1) *mucous glands*, scattered all over the body and secreting the slime which makes the body of a leech so slippery, and (2) *clitellar glands*, occurring in segments 5-7, and secreting the materials from which the cocoons are made. There is no swollen clitellum as in the earthworm, and a further difference is that the reproductive organs open in, not in front of, the clitellar region. The *dermis* is a moderately thick gelatinous layer containing a number of branched cells, and traversed by pigmented fibres and a network of capillary vessels, both of which penetrate between the internal "handles" of the epidermal cells. There are also muscle-fibres in the dermis.

Respiration is effected by the skin (*i.e.*, is cutaneous), and absorption of oxygen from the exterior with corresponding elimination of carbon dioxide is rendered possible by the superficial position of the capillary network, as described above.

3. The **Digestive Organs** (Fig. 23) consist of a tube running straight from mouth to anus, and divisible into fore-gut (buccal cavity and pharynx), mid-gut (gullet, crop, and stomach), and hind-gut (rectum).

The **fore-gut** (stomodæum) commences with a conical buccal cavity, situated at the bottom of the anterior sucker with three flat muscular *jaws* projecting into it. One of these is dorsal, the others ventro-lateral, and they are arranged in a three-rayed

ANNELIDA. 77

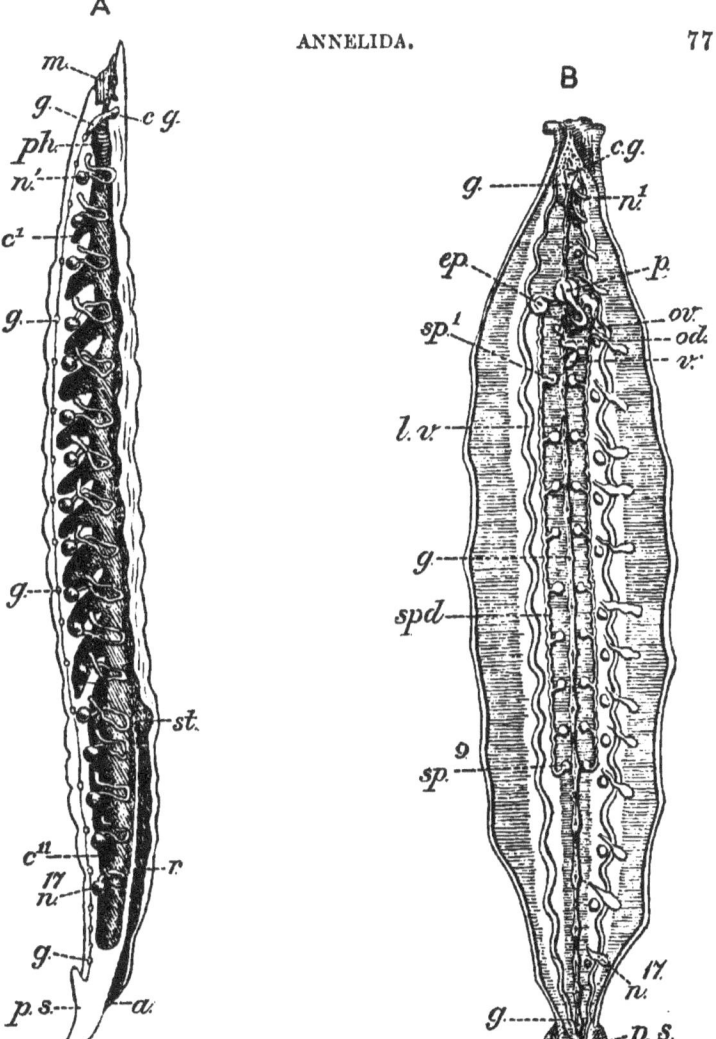

Fig. 23.—LEECH. A (slightly altered after *Leuckart*), Diagram of side-dissection, showing general relation of organs. B (slightly altered after *Vogt and Yung*), Diagram of dorsal dissection, showing relation of excretory, reproductive, and nervous systems.—*p.s.*, Posterior sucker; *m*, mouth; *ph*, pharynx; c^1, first pouch of crop; c^{11}, eleventh pouch of crop; *st*, stomach; *r*, rectum; *a*, anus; *l.v.*, lateral vessel; n^1 and n^{17}, first and seventeenth (last) nephridia; sp^1 and sp^9, first and ninth spermaries (testes); *spd*, spermiduct (vas deferens); *ep*, epididymis; *p*, penis; *ov*, ovary in sac; *od*, oviduct; *v*, vagina; *c.g.*, cerebral ganglia; *g, g, g, g*, ganglia of ventral chain.

manner, thus λ. The cuticle covering each of them is thickened along its margin into a large number of sharp calcified teeth. The remainder of the fore-gut is constituted by an ovoid muscular *pharynx*, with which the buccal cavity communicates by an extremely small aperture. Its wall is thick, and numerous dilator muscles radiate from it to the body-wall. The pharynx is surrounded by a large number of *salivary glands*, each of which is a large cell from which a slender duct runs forwards to open on one of the jaws.

The **mid-gut** (mesenteron) commences with a short and narrow *gullet* (œsophagus) leading from the pharynx to a very large *crop* which constitutes the greater part of the gut. The crop is a thin-walled tube which gives off 11 pairs of lateral pouches, corresponding but not limited to segments 4-14, and increasing in size from before backwards. The last pair are much the largest, and extend backwards by the sides of the intestine. A small rounded *stomach* terminates the mid-gut.

The **hind-gut** (proctodæum) consists of a narrow tube, the *rectum* or *intestine*, running from the stomach to the anus. A spiral fold projects into its interior.

The food consists of blood sucked from a higher animal, attachment being effected by the anterior sucker, and a three-rayed cut made by a sawing action of the jaws. Meanwhile a fluid is poured out from the salivary gland by which the blood of the victim is prevented from coagulating. The pharynx acts as a suction-pump, its cavity being alternately enlarged by contraction of the dilator muscles and diminished by contraction of its muscular wall. A very large amount of blood can be sucked at one time, and this is stored up in the capacious crop. Meals are often infrequent and they take a corresponding time to digest, even as much as nine months. This is accounted for by the fact that digestion and absorption take place only in the relatively small stomach.

4. The **Circulatory Organs** consist of a blood-system which freely communicates with a reduced cœlomic system, both alike containing blood coloured red by hæmoglobin. There are numerous colourless corpuscles.

The blood-vessels are distinguished from the cœlomic spaces by the possession of definite muscular walls lined by epithelium. A large *lateral vessel* runs along each side and is connected with its fellow by transverse ventral branches. Numerous branches are given off to the gut, excretory organs, and reproductive organs.

The *cœlom* or *body-cavity* is largely filled up with connective tissue, which unites the internal organs firmly together. The chief cœlomic spaces are, a *dorsal sinus*, running above the gut, and a *ventral sinus* surrounding the nerve-cord and connected with spaces which surround the internal ends of the nephridia. These sinuses have no definite walls and are simply spaces in the connective tissue. The vessels and sinuses communicate together by means of capillary networks, the chief of which are found in

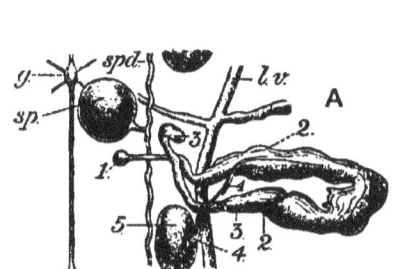
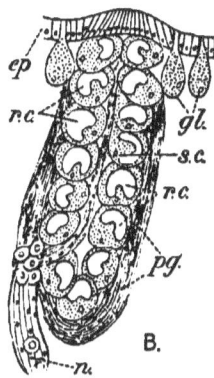

Fig. 24.—LEECH.—A (slightly altered after *Vogt and Yung*—enlarged), a nephridium with adjacent organs, seen from above (*1*) testes lobe, followed by *2, 2*, principal lobe, which again passes into the apical lobe, *3, 3*; the vesicle duct, *4*, arises from junction of *1* and *2*, and runs to vesicle *5*, which opens to exterior; *sp*, one of right spermaries (testes); *spd*, right spermiduct (vas deferens); *l.v*, lateral vessel of right side; *g*, one of ventral ganglia. B (after *Whitman*), longitudinal section of eye, much enlarged; *ep*, epidermis; *gl*, glandular cells of epidermis; *pg*, pigmented sheath; *r.c*, refracting cells; *s.c*, sense cells; *n*, optic nerve.

the skin and surrounding the crop. The latter plexus is made up of irregular tubules, the walls of which are composed of large brown granular cells, and which constitute what is known as the *botryoidal tissue*.

There are no hearts as in the earthworm. The course of the circulation is not accurately known. The general movements of the body largely aid in the circulation.

80   AN ELEMENTARY TEXT-BOOK OF BIOLOGY.

5. **Excretory Organs.**—Seventeen pairs of complicated nephridia succeed one another from the second to the eighteenth segment. Their internal ends lie within, but do not open into, special sections of the body-cavity communicating with the ventral sinus, and their external ends open by ventral pores.

A nephridium taken from the middle of the series is an upwardly projecting ∩-shaped loop, the limbs of which are anterior and posterior. The walls of the loop are traversed by a complicated system of intracellular ducts. The anterior limb commences with a blind cauliflower-shaped end which lies in the special sinus, and then runs upwards to be succeeded by the posterior limb which ends blindly below. A narrow *vesicle-duct* runs from the anterior limb to a *vesicle* which opens by a short tube to the exterior. The vesicle is muscular and lined by ciliated epithelium, which also extends into the tube leading from it. As in earthworm, therefore, the nephridial cavity is mainly intracellular, but its last section is intercellular. The former part is closely associated with an elaborate network of capillary vessels.

Nitrogenous waste has been detected in the vesicles.

6. **Reproductive Organs** (Fig. 23).—The unpaired ventral apertures of the hermaphrodite system have already been described.

The **male organs** consist of spermaries, ducts, and a protrusible penis.

There are nine pairs of spherical *spermaries* (testes) in the ventral region of segments 8 to 16. The testes on each side communicate by means of wavy *vasa efferentia* with a sinuous *spermiduct* (vas deferens) which enlarges in the sixth segment into a convoluted *epididymis*. From this a short duct runs to the base of the protrusible *penis* which lies in the same segment.

The **female organs** consist of ovaries, oviducts, and vagina, all of which are placed in the seventh segment.

The *ovaries* are two small thread-like structures enclosed in rounded sacs, from which two narrow *oviducts* proceed. These unite to form an unpaired convoluted tube which is continuous with a muscular sac-like *vagina*.

Numerous *sperms* with rounded heads and vibratile tails are produced in the testes by the division of mother-sperm-cells, and passing down the spermiducts are bound into cylindrical packets (sperm-ropes or spermatophores) by a secretion of the epididymes. The penis is a copulatory organ by which these packets are introduced into the vagina of another individual. In this process two individuals mutually fertilize each other, their ventral surfaces being applied and their anterior ends turned in opposite directions.

7. **Muscular System.**—The body-wall, within the skin, consists of three muscle-layers, an external, a middle, and an internal. The thin external layer is composed of fibres taking a *circular* direction, and within this comes the thin middle layer in which the fibres are *oblique*, while the very much thicker internal layer is made up of numerous bundles of *longitudinal* fibres. There are also numerous dorso-ventral muscle-bands, the ends of which spread out in a fan-like way within the epidermis. The suckers are made up of radial and circular muscle-fibres.

The muscular tissue is composed of elongated spindle-shaped muscle-cells, each of which consists of a longitudinally striated external (cortical) part and a granular nucleated internal (medullary) part. The former is specially contractile, the latter protoplasmic.

Locomotion is chiefly effected by a looping movement or by swimming. In the first method the posterior sucker is attached to some object and the body stretched forwards by contraction of the circular muscle-layer. Then the anterior sucker is attached and the body is dragged forwards by contraction of the longitudinal layer. Swimming is a much more complex affair; in it the body is moved in a wave-like fashion.

8. The **nervous system** consists of a very narrow nerve-ring round the pharynx, a ganglionated ventral cord, a sympathetic system, and nerves connected with these.

The nerve-ring is thickened above into *cerebral ganglia*, which supply the jaws, sense-organs of the head, and other anterior structures. The sides of the ring are *connectives* which unite these ganglia with a pair of *infra-œsophageal ganglia*, from which five pairs of nerves are given off. The *ventral cord* is made up of two closely united longitudinal halves, which enlarge to form a pair of ganglia in the first annulus of each segment. Of the 23 pairs thus constituted the first and largest are the infra-œsophageal ganglia. Then follow 21 much smaller pairs, from each of which two dorsal and two ventral nerves are given off to the corresponding segment, and the cord is terminated by the somewhat larger 23rd pair, which supply the posterior sucker and give origin to a number of nerves.

The *sympathetic* or *visceral system* is constituted by a slender nerve running on the under surface of the crop, which it supplies with numerous branches. This nerve is probably connected in front with the ventral cord; behind it bifurcates to supply the last pair of crop-pouches.

The large pear-shaped nerve cells make up the external part of the ganglia.

9. The most important sense-organs are the ten pairs of *eyes* situated in the dorsal margin of the anterior sucker. Each of these (Fig. 24) is a cylindrical cup, covered by elongated transparent epidermal cells and perforated by a nerve near its internal end. The *axis* of the cup is occupied by slender cells in continuity with nerve fibres, and it is lined with large cells containing refractive masses. A pigment-layer surrounds the eye except on its outer side.

A large number of tactile organs are found on the upper side of the front part of the body, and there is a ring of tactile papillæ on the first annulus of every segment. These all closely resemble the eyes in structure, but are devoid of pigment. It is probable that the eyes in this case are modified tactile organs. The leech possesses a very delicate sense of touch.

LIFE-HISTORY.

The breeding-season is spring. The fertilized ova are laid in ellipsoidal cocoons, and these are deposited in the damp earth at the margin of the water. Each cocoon is about an inch long, and has a thick wall composed of a firm inner layer and a spongy outer layer which is supposed to prevent desiccation. The cocoon contains as many as twenty eggs floating freely in an albuminous liquid. The cocoon is made from the hardened secretion of the clitellar glands, and in the first stage of its formation is a cylindrical band surrounding the clitellar region. The leech draws its body back out of this band, the ends of which then close up.

The course of development is in many respects similar to that of the earthworm, and germinal bands of the same nature are present. The embryos are soon liberated from the eggs and swim about in the surrounding albumen by which they are nourished. The young leeches escape from the cocoon after several weeks have elapsed.

CHAPTER VI.—ARTHROPODA.

§ 10. ASTACUS (The Crayfish).

THE Crayfish is a small greenish lobster-like animal, living in streams and canals, under the banks of which it burrows. An average mature specimen is about five inches long. There is only one British species, *Astacus fluviatilis*, a large variety of which, *A. fluviatilis* var. *nobilis*, is common on the Continent, where it is largely used as food. As the points of difference are but small, the following description will apply to either:—

Crayfishes are mainly carnivorous, devouring water-snails, tadpoles, insect larvæ, worms or other animals, dead or alive, but they also feed on vegetable matter. They either walk by means of their four pair of jointed legs, or else swim swiftly by alternately bending and straightening their powerful tails, which propel them backwards; they can also employ their large nippers for climbing.

The sexes are distinct, but there is not much difference externally between the male and female. The most obvious distinction is that the tail of the latter is somewhat broader, and in specimens examined during winter or spring a considerable number of rather large eggs will be found attached to its under side. These are hatched in May or June.

MORPHOLOGY AND PHYSIOLOGY.

1. External Characters (Figs. 25, 26, and 27).—The bilaterally symmetrical body is *segmented*, as in the Earthworm, but the segments are definite in number, and, instead of being nearly all alike, vary much in character, and many of them are fused together. Owing to these differences the body is marked out into regions from before backwards, *head*, *thorax*, and *abdomen* or tail. The two first are united to form the *cephalothorax*, which is covered by a firm continuous shell, the *carapace*, dorsally and laterally. The segments of all the regions bear paired, jointed *appendages* of various kinds, which differ greatly in form and function. Since the body is covered by a firm calcified cuticle, segmentation and jointed appendages are a mechanical necessity to an animal of such active habits.

(1) The **abdomen**, as the least modified part, affords a convenient point of departure for study. It is made up of seven segments, movably joined together, the first six of which bear appendages. The abdominal region, like all the rest of the body, is covered by a firm chitinous *exoskeleton* which is largely calcified, but parts of it remain soft and joints are 'thus formed. The 3rd, 4th, and 5th abdominal segments are most typical, and similar in the two sexes. The exoskeleton of each of these segments consists of a calcified ring connected by uncalcified parts with the segments in front and behind. The broad, strongly convex dorsal and lateral part of the segment is the *tergum* (*t*). The front part of this is overlapped by the preceding tergum, and its hinder part overlaps the one behind it. Between the two extends an uncalcified *intertergal membrane*, which is folded when the tail is straight. The sides of the terga of adjacent segments are united by peg and socket joints, allowing of upward and downward movements. The tergum passes down on each side into a small pointed projection (*pl*), the *pleuron* (larger in the female), which is V-shaped in transverse section, being made up of an outer and an inner limb. On the under side of the segment is a slender transverse bar, the *sternum* (*st*), and broad uncalcified intervals, the *intersternal membranes*, separate adjacent sterna. The *sternum* passes on each side into a very short but broader *epimeron* (*epn*), the junction between the two being marked by the attachment of the appendage. The foregoing parts form a perfectly continuous ring with no sharp lines of demarcation.

The appendages are *swimmerets* (Figs. 26 and 27, M), and each of them is a small elongated ⋏-shaped limb, consisting of a proximal* stalk, the *protopodite* (*pr*), and two distal* branches, an internal *endopodite* and external *exopodite* (*ex*).

The *protopodite* is made up of a very short proximal and a larger distal joint. The *endopodite* and *exopodite* both possess a relatively long proximal joint, and a rather longer distal part imperfectly divided into rings. All parts of the swimmeret are more or less beset with stiff bristles or *setæ*.

* The *proximal* and *distal* ends of an appendage, &c., are the ends respectively further from and nearer to its free end.

ARTHROPODA. 85

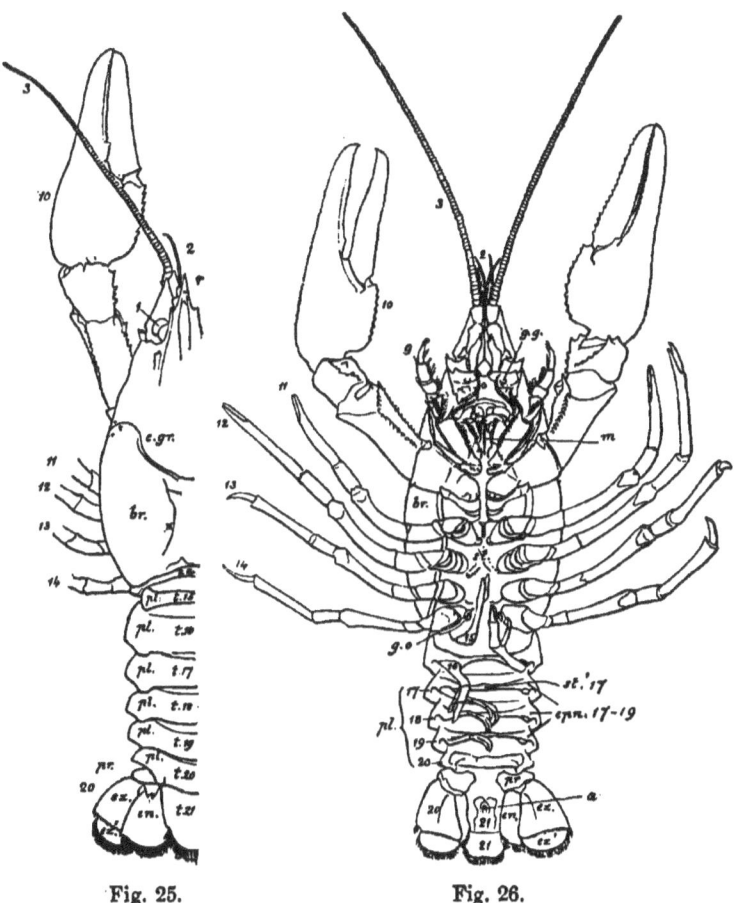

Fig. 25.   Fig. 26.

Fig. 25 DORSAL and Fig. 26 VENTRAL VIEW OF MALE CRAYFISH.—r, Rostrum; $c.gr$, cervical groove; ×, placed on branchio-cardiac groove; br, branchiostegite; t, tergum; st, sternum; pl, pleuron; ep, epimeron; pr, protopodite; ex, exopodite; en, endopodite; the appendages and parts of certain segments are numbered 1 to 21, but probably the eyestalks (1) are not appendages, therefore subtract one from each number to make it correct; $g.g$, excretory opening; m, mouth; a, anus; $g.o$, genital opening.

86 AN ELEMENTARY TEXT-BOOK OF BIOLOGY.

The 6th abdominal segment (20 in Figs. 25 and 26) differs from the preceding, mainly in the great size of the swimmerets. The protopodite is broad, thick, and undivided, while two broad oval plates fringed with setæ represent endopodite and expodite, the latter being two-jointed. The last abdominal segment or *telson* (21) possesses neither pleura nor appendages. It is divided into two by an imperfect transverse joint, and upon its little-calcified sternal surface the *anus* (*a*) opens as a longitudinal slit with thickened edges. The telson and large 6th pair of swimmerets make up together the *tail-fin*. The 1st and 2nd abdominal segments mainly differ in the character of their appendages, but in addition to this the former is small, possesses no pleura, and hinges on to the cephalothorax in a somewhat complicated manner. In the *female*, the 2nd pair of abdominal appendages are swimmerets, the 1st pair, when present, rudimentary swimmerets. The appendages of these segments are modified in the male into *copulatory organs* (Fig. 26, 15 and 16; Fig. 27, K and L).

In the *cephalothorax* (Figs. 25 and 26) the tergal and pleural regions are covered by a continuous carapace, produced in front into a sharp spine, the *rostrum* (*r*). On the upper surface of the carapace is a transverse groove, which slopes downwards and forwards to its front edge. This *cervical groove* (*c.gr*) marks the boundary between the head and thorax.

(2) The **thorax** is composed of eight fused segments, the last of which is slightly movable. In many of the lower Crustacea the thoracic segments are freely movable upon one another like the abdominal segments of the crayfish, and particular interest attaches to the genus *Nebalia*, which is in many ways transitional between the lower and higher forms. This is a small shrimp-like animal with a thorax composed of eight segments and covered by a large fold growing back from the head. The part of the crayfish's carapace behind the cervical groove probably corresponds to such a fold, which in this case, however, has fused closely with the thorax. The *tergal region* of this fold is rather narrow, and its lateral boundaries are indicated by a *branchio-cardiac groove* (×) running back from the cervical groove. The pleural region of the fold on each side is a large curved plate, the *branchiostegite* (*br*), which bends outwards and downwards to the bases of the thoracic appendages and covers a large *gill-chamber*. It is united with the head along the cervical groove. The *epimera* are very large and

thin, forming on each side a plate sloping steeply downwards, which constitutes the inner wall of the *gill-chamber*. Grooves converging upwards mark the boundaries of the segments. The thoracic sterna (*st*) are narrow elongated plates, lying between the bases of the appendages, and, except the last, closely united together. The last four of these appendages are elongated *walking-legs* (11-14), the two first of which are *chelate*, *i.e.*, with pincers, and possess no exopodite. The two-jointed protopodite is continuous with a five-jointed endopodite, the last joint but one of which, in the two first, is produced distally alongside the last joint, which works against it to form a claw. The broad proximal joint of the protopodite bears in the first three walking-legs a membranous plate, the *epipodite* (*ep*., I., Fig. 27), which projects into the gill-chamber, and on which gill-filaments are arranged in a plume-like way. The *male genital pore* (*g.o*) is placed on the inner side of the proximal joint of the protopodite of the last walking-leg. The *female genital pore* is found in a similar position on the second walking-leg.

The fourth pair of thoracic appendages are the large *forceps* (10). These are similar in structure to the chelate walking-legs, but their claws are very much larger. The first three pairs of thoracic appendages are relatively small and flattened, and are termed *foot-jaws* (maxillipedes). They are directed forwards and work against one another from side to side, their inner margins being provided with stout setæ. By turning a crayfish on its back the last pair (27, H) will readily be seen, and upon removing or turning back these the smaller second pair (27, G) will come into sight. These similarly conceal the still smaller first pair (27, F). They possess all the typical regions, but there is considerable variation in detail as may be gathered from Fig. 27.

(3) The **head** (Figs. 25 and 28, B) is probably made up of five segments closely united, and bears five pairs of appendages. That part of the carapace in front of the cervical groove represents the terga. It is produced in front into the rostrum, on either side of which is a deep notch, in which the stalked eye is placed. This region is much broader than the corresponding thoracic one, and bounds the front of each gill-chamber. The pleural region is rudimentary, and represented by the edge of the carapace, while the epimera are narrow and the sterna represented by several small median pieces. On the sternal surface, between the 3rd

88   AN ELEMENTARY TEXT-BOOK OF BIOLOGY.

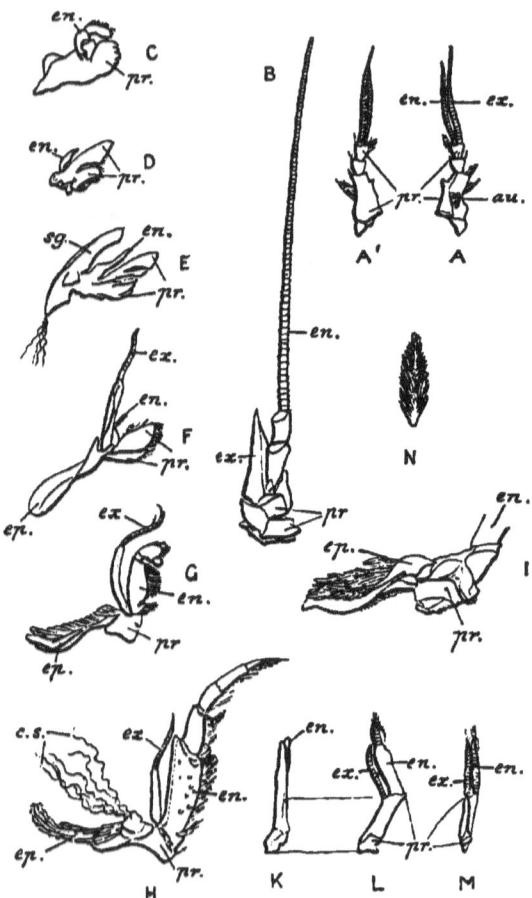

Fig. 27.—APPENDAGES OF ASTACUS, Right side.—A and A', Upper and lower views of antennule; B, antenna; C, mandible; D, mx. 1; E, mx. 2; F, mxpd. 1; G, mxpd. 2; H, mxpd. 3; I, base of third walking-leg; K and L, 1st and 2nd abdominal appendages of male; M, typical swimt.; N, arthrobranchia; *pr*, protopodite; *en*, endopodite; *ex*, exopodite; *ep*, epipodite; *c.s*, coxopoditic setæ; *sg*, scaphognathite; *au*, auditory opening.

and 4th segments, is the elongated and slit-like mouth (*m*), with a posterior lip, the *metastoma* (28 A, *mt*), from which a pointed lobe grows out on each side, and an overhanging pointed anterior lip, the *labrum* (28 A, *lb*), both of which are supported by calcareous plates. In front of the mouth the sternal region is sharply bent up, and, owing to this *cephalic flexure*, the anterior appendages are directed forwards (28, B). The last three pairs of headappendages are small jaws, covered over by the foot-jaws. The hindmost are the *second maxillæ* (27, E), which resemble the first foot-jaws, but are more delicate. The exopodite and epipodite are fused into a relatively large elongated plate, which lies in the front of the gill-chamber, and has received the special name of *scaphognathite* (*sg*). These appendages overlap the still weaker *first maxillæ* (D), which are devoid of endopodite. Lying at the sides of the mouth are the *mandibles* (C), in which exopodite and epipodite are absent. The protopodite is a firm, transversely elongated structure, the inner end of which possesses two strongly toothed ridges. The endopodite is a small three-jointed palp. The two segments in front of the mouth bear slender forwardly directed tactile appendages. Those of the 2nd segment are the large feelers or *antennæ* (B). Each of these consists of a protopodite with two stout cylindrical joints, the proximal of which has a small tubercle on its under surface, upon which is the *excretory* or *renal aperture* (28, B). To the protopodite a scale-like exopodite, the "squame," is attached externally, and a very long slender endopodite internally. This is composed of two stout proximal joints and a terminal part made up of a great many small rings, imperfectly jointed together. The small feelers or *antennules* (A and A') attached to the 1st segment are of much less size. They lie rather internal to the antennæ, and are directed forwards. The protopodite is three-jointed, and its proximal joint, which is much the largest, abuts against its fellow in the middle line. It is somewhat trihedral, and on its upper surface is a small longitudinal slit (*au*), the *auditory aperture*, over which a horizontal brush of setæ attached to its outer edge projects. The endopodite and exopodite are two short, slender, imperfectly-jointed filaments, of which the latter is rather the longer.

Between the sterna of the thorax and posterior part of the head a number of cuticular folds project into the body. These *apodemes* are calcified and make up a kind of open framework, the *endophragmal system*, which imperfectly roofs over a cavity known as the *sternal canal*, which is traversed by the ventral nerve-cord.

It will be seen from the foregoing that the body of the crayfish may be regarded as made up of twenty segments or metameres all *serially homologous*—that is, reducible to the same type. The differences between the several regions are chiefly differences of proportion, while certain parts are suppressed in some segments, and the anterior segments have fused together. These various differences have been brought about by division of physiological labour. The head bears the chief sense-organs and contains the ganglia which supply them. Its three posterior pairs of appendages are jaws, the last of which help to renew the water in the gill-chambers. The anterior thoracic appendages are jaws, then follow forceps which seize food, serve as weapons, and enable the animal to climb, lastly come walking-legs. The thorax also undertakes the work of respiration, and affords firm points of origin to the powerful muscles which move the tail. This is the swimming organ, and it also has secondary functions in relation to reproduction.

2. **Skin.**—The true skin is formed by an *epidermis* with a thin underlying *dermis*. The former secretes the exoskeleton, a many-layered pigmented cuticle, differing in degree and not in kind from that of a worm, and largely impregnated with salts of lime (seven-eighths carbonate and one-eighth phosphate). The numerous setæ are cuticular structures.

The epidermis is a single layer of columnar cells, the outlines of which are indistinct. Slender prolongations of the epidermis traverse the cuticle and end at the bases of the setæ. Numerous tubular *cement-glands*, lined by glandular epidermal cells, open on the ventral surface of the abdomen in the female. The thin dermis is composed of connective tissue, in which delicate blood-vessels and nerves run. The exoskeleton is made up of numerous closely-united layers, which are traversed by an immense number of delicate vertical "pore-canals." Most of the setæ are two-jointed pinnate bristles slightly sunk in small pits of the outer surface. Each of these setæ is hollow and contains a granular core, which is only separated by a thin transverse cuticular layer from the epidermic process which occupies the underlying pore-canal.

ARTHROPODA. 91

3. The **muscular system** is conveniently considered here, since its arrangement is dependent upon the structure of the exoskeleton, to the inside of which the muscles are attached.

Locomotion consists in swimming, walking, or climbing. The first results from alternate *flexion* (bending ventralwards), and *extension* (straightening), of the abdomen. In accordance with this, two great *flexor muscles* arise from the roof of the sternal canal, and are inserted into the abdominal sterna, and two much smaller *extensor muscles* have their origin and insertion in the sidewalls of the thorax and abdominal terga respectively. During flexion the intertergal membranes are stretched and the intersternal ones folded, *vice versâ* during extension. The lateral peg-and-socket joints only permit upward and downward movements, while excessive extension is prevented by the overlapping terga. Movements of the walking-legs and forceps effect the other two kinds of locomotion, and the successive segments of these appendages are connected by hinge-joints, the axes of which all take different directions, so that the limb can be bent in various planes and execute complex movements. Some of the limb-muscles are *extrinsic*, taking origin from the exoskeleton of the body; others are *intrinsic*, having both origin and insertion within the limb.

The jaws are moved to and from the middle line by *adductor* and *abductor* muscles. The mandibular adductors are easily seen in dissection. The claws of the first three pairs of thoracic limbs are opened and shut by abductor and adductor muscles which work the terminal joint against the produced part of the last joint but one. In many cases it is advantageous for a large muscle to be attached to a small and definite area. This is managed by means of a firm *tendon*, formed as a cuticular infolding to which the muscle-fibres are connected.

The muscles are made up of numerous slender *muscle-fibres* resulting from the modification of cells. Each of these fibres is invested by a delicate membranous sheath (sarcolemma), and, owing to the regular alternation of dark and clear bands, is *transversely striated*. Hence the term "striped" or "striated" muscle. Longitudinal striations also occur, and along these the fibre can be split, after death, into *primitive fibrillæ*. Beneath the sarcolemma a number of longitudinal rows of *muscle-corpuscles* are arranged, each of which consists of a nucleus surrounded by a small quantity of protoplasm.

The *motor nerve-fibres* which supply muscle come into very close relation with its fibres. The primitive-sheath becomes continuous with the sarcolemma, and the axis-cylinder passes into a granular, nucleated *end-plate*, resting upon the muscle substance.

4. Digestive Organs (Fig. 28, A).—The forceps and clawed walking-legs are accessory to digestion, and the six pairs of jaws overlapping one another outside the mouth are still more closely connected with the same function. But all these are more or less modified appendages, and outside the alimentary canal. This is a tube which runs straight from mouth to anus and consists of a capacious fore-gut (gullet and stomach), a very small mid-gut into which a large digestive gland opens, and an elongated tubular hind-gut (intestine).

The ventral *mouth* (*m*), which possesses anterior and posterior lips (*l* and *mt*), leads into the **fore-gut** (stomodæum). This consists of a short, wide, upwardly-directed gullet (œsophagus) (*œ*), and a large sac, the *stomach*, into which this opens. The stomach fills most of the cavity in the head, and is divided into a larger *cardiac* (*ca*) part in front and a smaller *pyloric* (*py*) part behind, the two being separated by a deep constriction. The fore-gut is an ingrowth from the exterior, and it is not, therefore, surprising to find it lined by a firm cuticle, continuous at the mouth with the exoskeleton. The cuticular lining of the stomach is locally thickened and calcified into hard plates and bars (sclerites) which form a chewing-apparatus or *gastric mill*, situated for the most part in the posterior part of the cardiac division. Numerous projecting setæ constitute a *strainer* in the pyloric part of the stomach. The sclerites of the gastric mill are united together into an elastic hexagonal framework with anterior and posterior sides connected by a jointed rod. These can be diagrammatically represented as in annexed plan, the arrow pointing to the front.

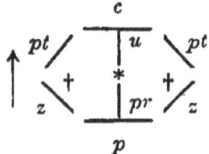

A broad *cardiac sclerite* (*c*) lies transversely in the dorsal wall of the cardiac division. Attached to the middle of this is a small *uro-cardiac sclerite* (*u*), directed downwards and backwards in the front wall of the constriction separating the two regions of the stomach. These two sclerites form a ⊤-shaped arrangement. A second and somewhat similar ⊥-shaped combination is formed by a *pyloric sclerite* (*p*) placed across the dorsal wall of the pyloric part of the stomach. To this is jointed a *pre-pyloric sclerite* (*pr*), forming the stem of the ⊥, which takes a downward and backward course in the back wall of the constriction above-mentioned, and joins the end of the uro-cardiac sclerite. At the junction of the two there is a red-coloured *median tooth* (∗), which projects into the cardiac chamber. The diagram here shown indicates the relation of these five sclerites as seen from the side.

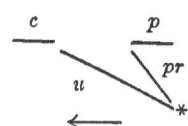

From the extremity of the cardiac sclerite on either side a small *ptero-cardiac sclerite* (*pt*) runs back, and unites with a *zygo-cardiac sclerite* (*z*) that runs forwards from the end of the pyloric sclerite. The inner side of each zygo-cardiac sclerite is thickened into a red elongated *lateral tooth* (†), which projects into the cardiac cavity, and is marked by numerous transverse ridges.

Two *anterior gastric muscles* (28, *a.g.m*) take their origin on the inner side of the carapace, and run backwards and downwards to be inserted into the cardiac sclerite, while two *posterior gastric muscles* (*p.g.m*) run downwards and forwards from a similar origin to be inserted into the pyloric sclerite.

By the contraction of the anterior and posterior gastric muscles the cardiac and pyloric sclerites are pulled away from each other, which involves—(1) the conversion of the hexagon into a rectangle, causing the lateral teeth to approach; (2) the pulling out of the sharp fold made by the urocardiac and prepyloric sclerites, so that the median tooth passes downwards and forwards. Hence the three teeth meet together in the centre, and effectually chew anything that comes between them. The elasticity of the framework then comes into play and separates the teeth.

In addition to the sclerites forming the gastric mill, others of less importance are found in the cardiac wall; and in summer a round button-like calcareous *gastrolith* is often seen projecting into the cardiac cavity on either side.

The communication between the cardiac and pyloric cavities is very narrow, partly owing to the external constriction and partly to hair-fringed projections. This is the commencement of the *strainer*, the rest of which is formed by hairy cushions projecting into the pyloric cavity and reducing it to a fissure, anchor-shaped in cross-section. A valve composed of five flaps separates the stomach from the mid-gut.

The extremely short **mid-gut** ($m.g$) is not lined by cuticle. Its dorsal wall is produced upwards and forwards into a short blind tube, the *cæcum* (*cæ*). A large brownish three-lobed *digestive gland* ($h.p$) (liver, hepato-pancreas) lies on each side within the cephalo-thorax, and opens by a short, wide duct (bile duct) ($b.d$) into the mid-gut, of which it was originally an outgrowth.

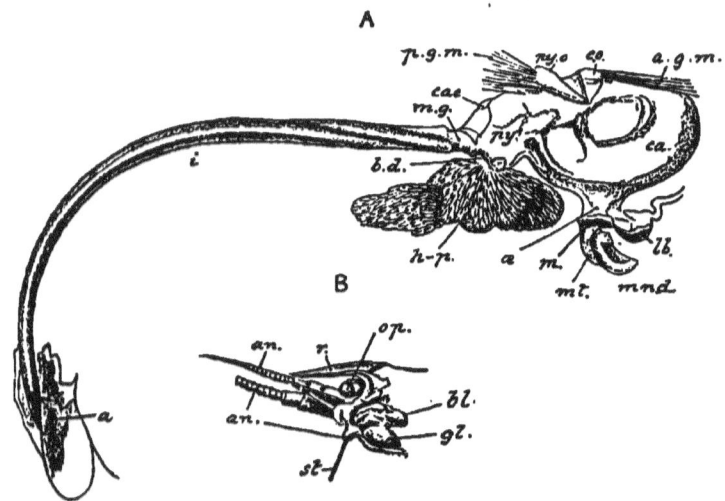

Fig. 28.—ALIMENTARY CANAL AND EXCRETORY ORGANS OF CRAYFISH.— A, Alimentary canal; m, mouth; lb, labrum; mt, metastoma; $æ$, gullet; ca and py, cardiac and pyloric ends of stomach; $c.o$, cardiac sclerite; the horizontal shading just below reference-line shows the commencement of the ptero-cardiac sclerite, which tapers to a point below; $py.o$, pyloric sclerite, running into the zygo-cardiac sclerite, which joins the ptero-cardiac; $a.g.m$ and $p.g.m$, anterior and posterior gastric muscles; $m.g$, mid-gut; $cæ$, cæcum; $h.p$, digestive gland; $b.d$, duct of the same; i, intestine; a, anus. B, Left renal organ; gl, glandular part; bl, bladder; st, style passed into renal opening; an and an', antennule and antenna; op, eye; r, rostrum.

The **hind-gut** or *intestine* (i), which succeeds the mid-gut, is a narrow thin-walled tube running back in the middle line, between the abdominal flexor and extensor muscles, to open by the *anus* (a). It is lined by cuticle, and six slightly-twisted longitudinal ridges, covered by minute elevations (papillæ) project into its cavity.

The alimentary canal is lined throughout by epithelium one layer of cells thick, which in the fore and hind guts secretes a

ARTHROPODA. 95

cuticle similar to that covering the outside of the body. Outside the epithelium, connective tissue and muscular layers are present. Each digestive gland is made up of a very large number of short cæca, lined by glandular epithelium. In each lobe these open into a central duct, and the three central ducts unite to form the main duct.

The crayfish feeds upon various substances, vegetable and animal, which are seized by the chelate appendages and torn into small fragments. These are passed on to the third foot-jaws, by which, and the other mouth-appendages, they are still further reduced. All the jaws work from side to side, and not up and down, as in backboned animals. Mastication is completed by the gastric mill. The pyloric strainer prevents any but small particles from passing on to the mid-gut, and parts incapable of sufficient reduction are ejected from the mouth. During the action of the gastric mill, probably, and certainly after reaching the mid-gut, the food is acted upon by the secretion of the digestive glands. This is an alkaline fluid, containing ferments which convert starch into sugar, and proteids into peptones, also emulsifying fats (*i.e.*, converting them to a state of fine division).

The ridged and papillated lining of the intestine affords a large absorptive surface. Owing to the action of the muscle in the wall of the alimentary canal, the food is gradually passed backwards, and the undigested parts are ejected from the anus.

The digested parts of the food, which are either dissolved or emulsified, diffuse out of the digestive organs into the blood-system, by which they are distributed over the body.

5. The **Circulatory Organs** consist of a blood-system only, which can be divided into heart, arteries, capillaries, and more or less definite channels and spaces (sinuses) in various parts of the body. All these are in continuity, and they contain a nearly colourless blood. In those invertebrates which, like crustacea, insects, &c., possess jointed lateral appendages and are known as Arthropods, the cœlomic system seems to have been almost aborted as a result of exuberant development of the blood-system.

The **blood** is of a faint bluish tint, owing to the presence of *hæmocyanin*, a complex compound of the nature of a proteid and playing the same physiological part that hæmoglobin does in red blood. It contains, however, copper instead of iron. The blood consists of plasma, in which are suspended numerous *colourless corpuscles*, nucleated, and capable of performing amœboid movements. On the dorsal side of the thorax is a good-sized *pericardial sinus*, in which the **heart** is suspended by fibrous cords. This organ is a flattish, thick-walled sac, into which open three chief pairs of valvular apertures (*ostia*), dorsally, laterally, and ventrally respectively. From the heart seven delicate **arteries**[*] proceed, five in front and two behind. From the front a small unpaired *ophthalmic artery* runs forwards and divides to supply the eyes, antennules, and cerebral ganglia, while on each side of this a much larger *antennary artery* arises which supplies the antenna, giving off branches to the stomach during its course. A *hepatic artery* takes origin a little behind each antennary artery, and supplies the mid-gut and digestive gland. The posterior end of the heart dilates into a kind of bulb which is continued backwards into a *dorsal abdominal artery* running in the middle plane above the intestine, to which it gives off numerous branches. Near the point of origin of this vessel a *sternal artery* leaves the heart, gives a branch to the reproductive organs, and, running downwards on one side or other of the intestine, pierces the nerve-cord, and divides into an anterior *ventral thoracic artery* and a posterior *ventral abdominal artery*, both of which run below the nerve-cord.

The arteries branch repeatedly, and some of the branches enter into capillary plexuses, as, for example, upon the cerebral ganglia. The finest ramifications end in minute spaces (*lacunæ*) found in all parts of the body and which communicate with larger spaces (**sinuses**) that are placed around the various organs. The most important of these is the *sternal sinus*, running along the ventral surface, and with which the others are directly or indirectly connected. There are external channels in the gills which communicate on the one hand with the sternal sinus, and on the other hand with internal gill-channels opening into *branchio-cardiac canals* which run up the sides of the thorax into the *pericardial sinus*.

The heart is chiefly made up of interlacing muscular cords, which in their turn are aggregates of elongated muscle-cells, the inner sides of which are metamorphosed into transversely striated contractile substance.

[*] An *artery* is a blood-vessel carrying blood away from the heart.

Circulation of the blood is effected by the rhythmical contraction of the heart, which acts as a central pump. The valves allow the blood to enter the heart, but prevent it from passing out again into the pericardial sinus during contraction.

Course of the Circulation.—The gills are placed in the course of the blood-current returning to the heart, this passing from the internal gill-channels to the branchio-cardiac canals, and thence into the pericardial sinus. The heart in its *diastole, i.e.*, when expanding, sucks in this purified blood through its ostia, and then undergoing *systole* (contracting) forces it into the arteries. From these it passes into venous spaces, and ultimately reaches the sternal sinus, loaded with carbon dioxide, and poor in oxygen. From the sternal sinus it enters the external gill-channels. A heart, which, like that of the crayfish, distributes oxygenated blood to the body at large is said to be *systemic*.

6. **Special respiratory organs** are present on the thorax in the form of *gills* (branchiæ), adapted for breathing the oxygen dissolved in water. These are vascular outgrowths covered by a very delicate cuticle. Those on each side are contained in a gill-chamber covered over by the branchiostegite, and communicating with the exterior by a narrow slit above the bases of the thoracic limbs. Parasites are particularly fond of fixing upon gills, since these are always delicate structures with an abundant supply of blood and fresh water, and, besides, are usually well sheltered. In this case such unwelcome guests, as well as particles of sand, &c., are largely prevented from entering by long *coxopoditic setæ* projecting from the basal joints of the third foot-jaws, forceps, and first three walking-legs. Each chamber contains 18 perfect, and 2 or 3 rudimentary gills. The free ends of all converge upwards. Six of the perfect gills, *podobranchiæ* (Fig. 27, G, H, I, *ep*), are constituted by the lamellar epipodites of the second and third foot-jaws, forceps, and first three walking-legs, which bear a large number of delicate gill-filaments on their anterior and external surfaces. Eleven more, *arthobranchiæ* (Fig. 27, N), are attached to the membranous junctions between the bases of the foregoing appendages and the body. Two are placed on each of these junctions except the first, which possesses only one. These gills are plume-like, with a central stem bearing numerous gill-filaments. The remaining perfect gill, *pleurobranchia*, similar in structure to one of the arthrobranchiæ, is attached to the epimeron

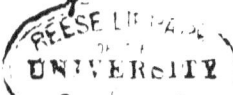

of the last thoracic segment. The two or three rudimentary gills are also pleurobranchiæ, and are attached to the epimera of the two or three preceding thoracic segments.

The number and position of the gills can be expressed by a branchial formula, as follows:—

	pd		a^a		a^p		pl		
VI	0 (ep) +		0	+	0	+	0	=	0 (ep)
VII	1	+	1	+	0	+	0	=	2
VIII	1	+	1	+	1	+	0	=	3
IX	1	+	1	+	1	+	0	=	3
X	1	+	1	+	1	+	0	=	3
XI	1	+	1	+	1	+	r	=	3 + r
XII	1	+	1	+	1	+	r	=	3 + r
XIII	0	+	0	+	0	+	1	=	1
	6 + (ep) +		6	+	5	+	1+2r	=	18+(ep) + 2r

[The numbers refer to the thoracic segments; pd = podobranchs; a^a = anterior arthrobranchs; a^p = posterior arthrobranchs; pl = pleurobranchs; ep = epipodite; r = rudiment.]

The gills are covered with a very thin cuticle, underneath which lies a delicate epithelium. The connective tissue forming the central parts of the filaments is traversed by a network of blood-passages connected with the external and internal channels of the gill-axis.

The essential part of respiration consists in diffusion of oxygen into the blood contained in the gills, while at the same time carbon dioxide diffuses out into the surrounding water. The hæmocyanin acts as an oxygen-carrier, since it is able to take a certain amount of that element into loose chemical combination, and parts with it as readily to the tissues. The water in the gill chamber is renewed chiefly by the action of the scaphognathite which bales it out in front, while fresh water enters below and behind. The movements of the thoracic limbs also assist in the production of currents.

In the young crayfish water is regularly taken into, and expelled from, the intestine by the anus. This *anal respiration* may possibly also occur in the adult.

7. **Excretory Organs** (Fig. 28, B).—A pair of *kidneys* ("green glands") are situated in the head, one on each side of the œsophageal connectives. Each is a small green body (gl) round and flattened, from the upper surface of which proceeds the *ureter*,

an oval thin-walled bladder (*bl*), which opens (*st*) externally on the basal joint of the antenna. The green part is made up of a single much-coiled tubule, lined by glandular epithelium. The kidney is richly supplied with blood, from which it separates nitrogenous waste in the form of guanin ($C_5H_5N_5O$), and uric acid ($C_5H_4N_4O_3$), by means of its glandular epithelium.

8. Reproductive Organs (Fig. 29).—The sexes are distinct, and the external differences between them chiefly consist in the greater breadth of the abdomen in the female, and the modification of the two first pairs of abdominal appendages in the male.

(1) In the **male** crayfish (A) a single yellowish-white *spermary* (testis) underlies the pericardial sinus. It possesses two short

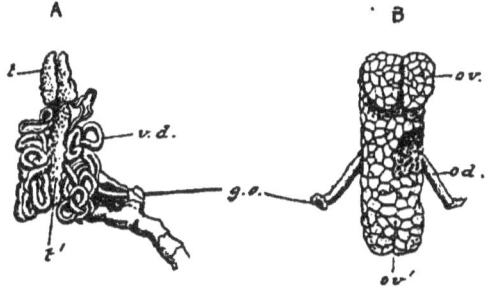

Fig. 29.—REPRODUCTIVE ORGANS OF CRAYFISH.—A, Male organs; *t* and *t'*, anterior and posterior lobes of spermary (testis); *v.d*, spermiduct (vas deferens); *g.o*, style inserted into genital aperture. B, Female organs; *ov* and *ov'*, anterior and posterior lobes of ovary; *od*, oviduct; *g.o*, genital opening. The ovary has been cut open on one side to expose its cavity.

anterior lobes (*t*), which broaden behind, and a longer and narrower posterior lobe (*t'*). A slender spermiduct (*vas deferens*) (*v.d*) arises on each side from the junction of the lobes, enlarges considerably, and, after many convolutions, runs downwards to one of the *male genital pores* (*g.o*), situated on the proximal joint of the last pair of walking-legs.

The spermiducts branch repeatedly within the spermary, their finest branches ending in minute lobules, each of which is formed by a small group of vesicles lined by mother-sperm-cells. Each of these divides, before the breeding-season, and gives rise to a number of spermatocytes, which develop into minute nucleated *sperms*, rounded and somewhat flattened, with a number of delicate, stiff processes, all curved in one direction, projecting from their edges. The epithelial lining of the spermiducts is glandular, and secretes a viscid fluid which hardens to form threads (spermatophores) in which the sperms are imbedded.

(2) The **female** (B) possesses an *ovary* (*ov, ov'*), similar in shape to the spermary, and occupying a similar position, but larger and broader. Its colour is reddish-brown, and its surface is raised up into rounded projections of various sizes, caused by the presence of ova. This is most obvious in the breeding-season, when the ovary becomes considerably larger. Within it is a cavity on either side into which ova project, and from which, near the junction of the lobes, a short, wide *oviduct* (*od*) leads to the basal joint of the second walking-leg (*g.o*).

The *ova* are reddish-brown spheres about $\frac{1}{12}$ of an inch in diameter, and enclosed in capsules, the *ovarian follicles*, which project into the cavities of the ovary. The wall of a follicle is constituted by a structureless membrane, underneath which is a layer of simple epithelium. The large *ovum* consists of a *vitelline membrane*, a *vitellus* containing a considerable quantity of food-yolk, and a *germinal vesicle* with numerous *germinal spots*. The ripe ova burst from their follicles, pass into the ovarian cavities, and thence by the oviducts to the exterior.

In the breeding-season (autumn) the male, by means of the two pairs of copulatory abdominal appendages, deposit spermatophores upon the posterior thoracic sterna of the female. Just before the eggs are laid the abdomen of the female is flexed, and a glairy secretion is poured out from the numerous cement-glands, probably dissolving the substance which unites the sperms together. The eggs are now passed out from the oviducts, fertilized, and each of them enclosed in a sort of capsule formed by this secretion, and fixed by a filament of the same substance to one of the swimmerets. Some two hundred eggs are thus attached, and during the winter gradually develop, the movements of the swimmerets keeping them surrounded by fresh water.

9. The **nervous system** consists of a nerve-ring passing into a ganglionated ventral cord, and of nerves connected with these. The ring is thickened dorsally into a large *cerebral ganglion* (brain) situated in the head between the origin of the eyestalks. It is divided into lobes, which indicate that it is composed of at least three pairs of ganglia fused together. A very long *œsophageal*

connective runs down on each side of the gullet to join an elongated *post-œsophageal ganglion*, notched on each side, and representing the six ganglion-pairs of segments 3-8 inclusive. The post-œsophageal ganglion is the first of the ventral chain, the thoracic part of which lies in the sternal canal. The rest of the cord consists of eleven well-marked pairs of ganglia united by double connectives. The first five of these are *thoracic ganglia* placed at unequal distances, the posterior ones being rather near together. The connectives between the third and fourth diverge to allow the sternal artery to pass between them. The last six are the rather smaller *abdominal ganglia* placed at regular intervals upon the ventral wall of the abdomen. The last is somewhat larger than the others, and probably represents two pairs fused together.

The ganglia supply nerves to their own segments. Two pairs run from the cerebral ganglion to the eyestalks, and one pair each to the antennules and antennæ. The anterior part of the post-œsophageal ganglion supplies the segments to which the mandibles, maxillæ, and first two pairs of foot-jaws belong. Its posterior part (the only partially fused ganglion of the third thoracic segment) supplies the third pair of foot-jaws, and behind it two pairs of interganglionic nerves run to the other parts of the segment. From the following ganglia two pairs of nerves arise in most cases, the anterior of which supply the corresponding pair of appendages. An interganglionic pair of nerves is found (in the majority of segments) behind the ganglion. The sixth abdominal ganglion supplies the last two segments.

The **visceral** (sympathetic) **nervous system** consists of an *anterior visceral nerve* (formed by the union of three trunks, one running back from the cerebral ganglion, and one arising from each œsophageal commissure), which branches out on the stomach, and a *posterior visceral nerve* which runs forwards on the ventral surface of the intestine from the last abdominal ganglion. Several small ganglia are present on the branches of the anterior visceral nerve.

The *ganglion-cells*, which are here confined to the ganglia, possess the usual parts, and are surrounded by special nucleated sheaths. Each *nerve-fibre* is invested by a firm *primitive-sheath*, in which are imbedded at irregular intervals, small *nerve-corpuscles*, with large nuclei and scanty protoplasm, and within the sheath is a clear *axis-cylinder*.

The nerve-fibres, considered from the physiological standpoint, are either *afferent* or *efferent*, transmitting molecular impulses to and from the ganglion-cells. These are confined to the ganglia, which are consequently the **central organs**. Most of the afferent nerves are *sensory*, connecting sense-organs with the ganglia, while a majority of the efferent ones are *motor*, running from the ganglia to muscles. Each ganglion has more especially to do with the segment it supplies, but can also work conjointly with others. An abdominal ganglion, for example, if the connectives which unite it with its neighbours are cut, still causes regular movements in the corresponding swimmerets. These movements are undoubtedly caused by efferent nerve impulses from the ganglion, as they cease on its extirpation.

If the nerve-cord is cut at the end of the thorax, the abdominal ganglia, working *as a whole*, cause all the swimmerets to work regularly *together*. And by allowing more and more ganglia to remain, connected movements of increasing complexity are rendered possible. The addition of the cerebral ganglion, however, makes far more difference than the addition of any other. Without this the movements are for the most part irregular and lacking in purpose. The cerebral ganglion, in fact, controls the body at large.

The actions may be reflex or spontaneous. A **reflex action** starts with an external stimulus, by which one or more afferent fibres are affected. These carry impulses *to* the central organ, *from* which other impulses pass along efferent fibres to the parts they supply, causing them, if muscular, to contract. Suppose, for instance, the abdomen, when nervously isolated from the rest of the body, be irritated, say, by tapping, it will "flap vigorously." Here the tapping causes afferent fibres to carry impulses to the abdominal ganglia (central organs). These "reflect" the impulses into motor fibres, down which they pass to the flexor and extensor muscles, and cause these to contract.

Spontaneous actions are not directly dependent on external stimuli. A crayfish from which the cerebral ganglion is removed exhibits constant movements, to all appearance "spontaneous," as above defined. They are, however, not *voluntary*, the effect of "will," spontaneous actions of this sort being probably the result of impulses passing from the cerebral ganglion. This, too, is the seat of consciousness and intelligence in so far as these exist.

The *visceral* nerves are concerned with the regulation of the movements of the alimentary canal. As small ganglia are developed upon their course in front, they acquire a certain independence.

10. The different **sense organs** have a certain community in structure. They all consist of elongated "end-organs," belonging to the ectoderm, and connected directly or indirectly with the cerebral ganglion by sensory nerve-fibres.

(1) **Tactile Organs.**—The numerous setæ scattered over the body, and the slender ectodermic processes with which they are connected, serve as end-organs of *touch*. The antennæ and, to a less extent, the antennules, are of great use to the animal in informing it as to the position of external objects, and well deserve the name of "feelers."

(2) **Olfactory Organs.**—Each joint of the exopodite of the antennule bears below two tufts of flattened, spatula-shaped setæ, which probably serve as organs of *smell*, and are denominated *olfactory setæ*.

(3) **Auditory Organs.**—The organs of hearing are two small *auditory sacs*, lodged one in the basal joint of each antennule, and opening on its upper surface by a three-cornered aperture fringed with setæ (Fig. 27, A, *au*). Each is somewhat pear-shaped, with a backwardly directed narrow end, into which an auditory nerve passes. The floor and posterior wall of the sac bear numerous *auditory setæ*, which project into a gelatinous mass, in which are imbedded numerous foreign particles (sand, &c.), the *otoliths*, introduced from the exterior. The auditory nerve branches in the wall of the sac, one of its fibres becoming continuous with the granular axis of each auditory seta.

(4) **Visual Organs** (Fig. 30).—The crayfish possesses a pair of *eyes*, each of which is placed on a two-jointed stalk traversed by an *optic nerve*, that dilates at its extremity into a rounded *optic ganglion*. On the end of the stalk is an oval area, the cornea, where the cuticle is transparent and uncalcified. Owing to underlying dark pigment it appears black. The corneal surface is divided into a large number of small square facets, each of which corresponds to one of the elements which build up the deeper part of this so-called "compound" eye. These visual elements (ommatidia, visual pyramids) are the slender, four-sided bodies whose bases are applied to the inner sides of the corneal facets,

while their slender apices abut against the optic ganglion. Each *ommatidium* is surrounded by a pigmented sheath, and consists of the following parts, commencing from the outside :—(a) A *corneal lens*, (b) modified epidermic *corneal cells* by which this is secreted, (c) a group of four *crystal cells*, which secrete refracting structures constituting a *crystal cone*, (d) a *retinula*, constituted by seven slender cells which surround a transversely ridged *rhabdom* (striated

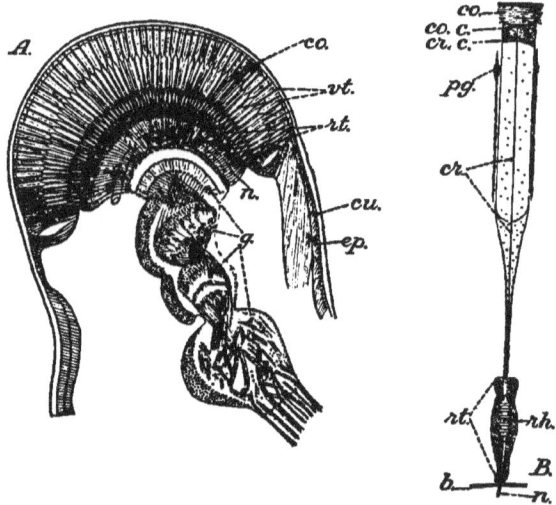

Fig. 30.—CRAYFISH EYE (after *Carrière*). Enlarged.—A, Longitudinal section; *cu*, cuticle, and *ep*, epidermis, of eyestalk; *co*, cornea; *vt*, vitreous body, the dark lines in its outer part indicate pigment; *rt*, retinulæ imbedded in pigment, and bounded internally by a basement membrane (indicated by a strong line); *n*, nerve-fibres running from retinulæ to the optic ganglion, *g*. B, An ommatidium; *co*, corneal lens; *co.c*, corneal cells; *cr.c*, crystal cells; *cr*, crystal cone; *rt*, retinula, enclosing rhabdom, *rh*; *b*, basement membrane; *n*, nerve-fibres.

spindle), secreted by them and continuous with the crystal cone. Fibres of the optic nerve are probably continuous with the cells of the retinula, which appear to be the end-organs for sight, while the corneal lens, crystal cone, and rhabdom are refracting elements. The crystal cones form collectively a "vitreous body."

The action of the eye is best explained by the theory of "mosaic vision," according to which each ommatidium is only affected by rays which correspond with it in direction, so that if the animal perceives the *form* of external objects, it must be by the *combined action* of the ommatidia, which almost certainly do not act like so many simple eyes.

DEVELOPMENT.

1. Early Stages (Fig. 31).—The ovum is centrolecithal, containing a central mass of food-yolk. Cleavage (segmentation) is *regular*, but *incomplete* and *peripheral*, chiefly concerning the external protoplasmic part of the oösperm. This is owing to the food-yolk; but this, though it impedes division, is for the most part affected by the cleavage, being divided into radiating

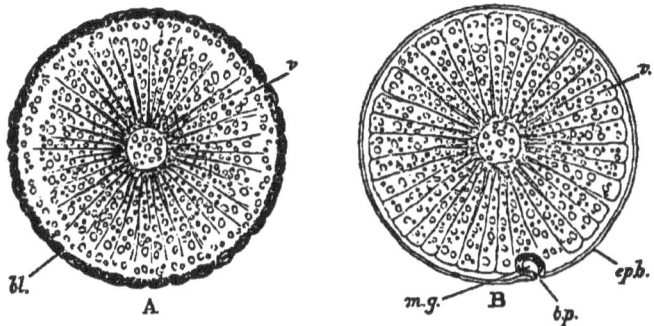

Fig. 31.—BLASTULA AND GASTRULA OF CRAYFISH (after *Reichenbach* and *Huxley*).—A, Blastula (segmented ovum); *bl*, blastoderm; *v*, yolk. B, Gastrula ; *ep.b*, ectoderm (epiblast) ; *m.g*, archenteron (which becomes the mid-gut), bounded by endoderm (hypoblast), shaded darkly ; *bp*, blastopore.

yolk-pyramids, continuous at first (A) with the segments of the peripheral protoplasm, but ·soon (B) becoming separated. The **blastula**, at the end of cleavage, consists of a central mass of yolk (*v*), enveloped by the *blastoderm*, a single layer of small cells (*bl*), which become thickened over a small oval area, the *germinal disc*, that marks the ventral surface of the embryo. The posterior part of this is invaginated (B) to form a small pouch (*m.g*) sunk in the food-yolk. The embryo has now reached the **gastrula** stage, and is practically a double-walled pouch, the inner wall of which is the **endoderm** (hypoblast) and the outer wall the **ectoderm** (epiblast), while its small cavity is the *archenteron*, and its wide mouth the *blastopore* (*bp*), which soon narrows, and eventually closes. Owing to the presence of food-yolk, which the ectoderm envelops, this layer is much more extensive than the endoderm. The **mesoderm** (mesoblast) is recognizable at an early stage as a solid mass of cells arising in front of the archenteron, and derived from its wall. It gradually extends, by division of its elements, until it forms a complete layer surrounding the body within the ectoderm.

2. General Growth.—The germinal disc is the first trace of the ventral surface of the embryo. From its hind end the abdomen grows out as a small knob, and on its front end and sides little elevations appear, the rudiments of the eyes, labrum, and appendages of the head and thorax. The swollen part containing food-yolk corresponds to the dorsal region of the cephalothorax, and it gradually becomes smaller as its contents are used up in the processes of growth.

The young crayfish are hatched at or near the commencement of summer. They resemble the adult in most respects, but the first and last abdominal appendages are wanting, and the ends of the forceps are sharply hooked, thus serving as a means of attachment to the cement with which the swimmerets are more or less covered.

Owing to the firm exoskeleton, growth cannot take place in the same way as in soft-bodied animals, and the adult size is only reached after a series of *moults* or *ecdyses*, by which the cuticle (including the endophragmal system and the linings of the fore- and hind-guts) is thrown off, a fresh one being subsequently secreted. Moulting occurs as often as eight times the first year, five the second, and after that less frequently.

3. Fate of the Germinal Layers.—(1) The **ectoderm** (epiblast) develops into the epidermis, with the exoskeleton and its various types of setæ. The epithelium and cuticle of the fore- and hind-guts are also derived from this layer. They both arise as deep pits which become connected with the archenteron to form a continuous alimentary canal. The nervous system is formed from ectodermal thickenings, and the auditory sacs from ectodermal pits. Each eye (except the pigment, which is mesodermal) is developed from (1) an ectodermal thickening, and (2) an optic pit or fold of ectoderm internal to this. All external to the retinulæ arises from (1), while (2) becomes isolated from the surface as a solid mass consisting of an outer layer, from which the retinulæ originate, and an inner layer which is converted into nerve-fibres connecting the eye with the optic ganglion.

(2) The **endoderm** (hypoblast) is converted into the epithelium of the mid-gut, and of the digestive gland which grows out from each side of this.

(3) The **mesoderm** (mesoblast) does not divide into regular mesoblastic somites (except in the tail at a comparatively late period). Irregular spaces are formed in it which become the venous sinuses. The circulatory and respiratory organs (except the epithelium and cuticle), the excretory (?), and reproductive organs, together with the muscular system and the connective tissue uniting the various parts, all arise from this layer.

Further Remarks on the Development of the Crayfish :—

It is a noteworthy fact that most Invertebrates which inhabit rivers have a *direct* development, that is the young, when hatched, closely resemble the adults in appearance. Marine forms, on the contrary, are commonly hatched as free-swimming *larvæ*, leading an independent existence for some time, and differing more or less from their parents, which they ultimately come to resemble by passing through a series of changes known as a metamorphosis. This is the case, for example, with crabs and lobsters, and a larval form is of obvious advantage to them since it affords a means of widening their area of distribution. But to most fluviatile animals free-swimming larvæ would be a positive disadvantage, since they would be very liable to be swept down to sea by the current and so perish. There is some danger of this even in the case of the young crayfish, but the risk is reduced to a minimum by their attachment immediately after hatching to the swimmerets of the mother, and it is stated that even after this she shelters them for a time under her tail, when danger threatens.

Other Crustacea.

The **Lobster** (*Homarus*) closely resembles the crayfish in structure, but differs in the following respects:—*Ext. Chars.*—Last thoracic segment completely fused with cephalothorax. Telson undivided. 1st abd. app. of ♂ two-jointed, with spoon-shaped end ; 2nd ditto with plate-like endopodite. Well-developed swimmerets on 1st abd. segment of ♀. Antenna with small squame. *Digestive Organs.*—Mid-gut with small bilobed cæcum. Intestine of a smooth anterior and ridged posterior part, with a dorsal cæcum at the junction of the two. Digestive gland extends far forwards. *Respy. Organs.*—Podobranchs arranged as in crayfish, but the gill-plume is not fused with the epipodite. Arthrobranchs one less in number, the one found on the second thoracic segment of the crayfish being absent here. Pleurobranchs four in number. *Reproductive Organs.*—Two tubular sper-

108 AN ELEMENTARY TEXT-BOOK OF BIOLOGY.

TABLE OF CRAYFISH SEGMENTS.

	No. of Segment.	Skeleton.	Appendages.	Apertures.	Gills. Pod.	Arth.	Pleu.	Ganglia.	Sense-Organs.
	Pro-sto-mium		...		...	...	...	Cerebral (Gang. 1)	Eyes Tactile Olfactory Auditory
Head.	I.		Antennules	Auditory	...	...	...		Tactile
	II.		Antennæ	Excretory	...	...	...		
	III.		Mandibles	Mouth	...	...	...		
	IV.		Mx. 1		...	...	...		The appendages of segments I. and II. are active tactile organs, but the entire surface of the body is also endowed with the sense of touch.
	V.		Mx. 2	...	Sca-phog	...	...	Post-œsopha-geal (Gang. 2)	
Thorax.	VI.	Cara-pace.	Mxpd. 1	...	Epi-pod	...	...		
	VII.		Mxpd. 2		1	1	...		
	VIII.		Mxpd. 3		1	2	...		
	IX.		Forceps		1	2	...	Gang. 3	
	X.		Leg 1 } Chel-ate	...	1	2	[r]	Gang. 4	
	XI.		Leg 2 }	♀ genital	1	2	r	Gang. 5	
	XII.		Leg 3	...	1	2	r	Gang. 6	
	XIII.		Leg 4	♂ genital	...	...	1	Gang. 7	
Abdomen.	XIV.		♀ [Sw. 1] ♂ Cop^r		...	...	...	Gang. 8	
	XV.		Sw. 2	App.	...	...	...	Gang. 9	
	XVI.		Swmt.		...	...	...	Gang. 10	
	XVII.	Free.	Swmt.		...	...	...	Gang. 11	
	XVIII.		Swmt.		...	...	...	Gang. 12	
	XIX.		Swmt. (very large)		...	...	...	Gang. 13	
	XX.		O	Anus	...	...	...		

r = rudiment; [ ] signify present or absent.

maries united by a transverse commissure; spermiducts comparatively short. *Nervous System.*—Post-œsophageal ganglion comparatively small. *Development.*—Hatched as a somewhat shrimp-like *Mysis* larva, the last five pairs of thoracic limbs of which are provided with slender exopodites, afterwards lost, and which has no abdominal appendages.

The Sea **Crayfish** or Norway Lobster (*Nephrops*) closely resembles the common lobster, but the antenna possesses a large squame, and the gill-plume of the first podobranch (that of mxpd. 2) is rudimentary or absent. The number of gills on each side in the three forms is, therefore, as follows, writing p, a, and pl for podo-, arthro-, and pleurobranchs, and omitting rudiments:—

$$\text{Astacus,} \quad 6p + 11a + 1pl = 18.$$
$$\text{Homarus,} \quad 6p + 10a + 4pl = 20.$$
$$\text{Nephrops,} \quad 5p + 10a + 4pl = 19.$$

CHAPTER VII.—MOLLUSCA.

§ 11. ANODONTA and UNIO (Fresh Water Mussels).

THESE are bivalve molluscs found in considerable numbers in the mud at the bottom of many ponds, streams, and canals, in this and other countries. Mature specimens may be as much as five or six inches long.

MORPHOLOGY AND PHYSIOLOGY.

1. External Characters.—The bilaterally symmetrical body is unsegmented, flattened from side to side, and entirely enclosed in a calcareous bivalve shell, cuticular in nature, and consisting of a right and a left piece (valve). These two valves are hinged together dorsally by the *ligament*, an uncalcified band of horny elastic fibres, which in the closed shell are on the stretch. In dead specimens there is nothing to keep the valves closed, and they consequently "gape," owing to the elasticity of the ligament, but in the living animal their edges can be closely approximated by means of two strong *adductor muscles*, running across from valve to valve. Each valve is oval with a rounded anterior and a more pointed posterior end, a strongly curved ventral margin, and a straight dorsal margin near the front end of which is the *umbo* or oldest part of the shell. Around the umbo *lines of growth* are concentrically disposed, which mark successive additions made during the life of the animal. The outside is olive green in colour, owing to the *periostracum*, a thin horny membrane, which also forms a flexible margin to the shell. Within the periostracum there is a *prismatic layer*, and the shell is lined by a *pearly layer*, marked by a number of *muscular impressions*. Near the anterior end there is an oval *anterior adductor impression*, and near the posterior end a similar but larger *posterior adductor impression*. The two are connected by a *pallial line* running parallel to the

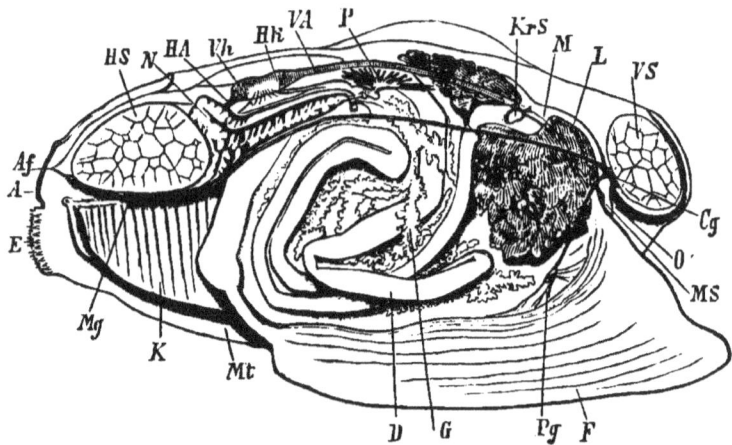

Fig. 32.—GENERAL DISSECTION OF UNIO (from *Claus*, after *Grobben*).— V.S and H.S, anterior and posterior adductors; M.S, labial palp; F, foot; Mt, mantle; K, gills; C.g, cerebro-pleural ganglion; P.g, pedal ganglion; M.g, visceral ganglion; O, mouth; M, stomach; L, digestive gland; Kr.S, crystalline style; D, intestine; Af, anus; G, genital gland; A, exhalent aperture; E, inhalent aperture; N, nephridium; Vh, auricle; Hk, ventricle; V.A, anterior aorta; H.A, posterior aorta; P, organ of Keber.

ventral edge of the shell. Immediately behind the anterior adductor impression is a much smaller *anterior retractor impression*, behind which again is a small *protractor impression*. Just in front and above the posterior adductor impression, there is a small *posterior retractor impression*. From the various places above mentioned lines of *shifting* converge towards the umbo, indicating the successive positions occupied by the impressions during earlier stages of growth.

In *Unio* the valves are connected at the hinge-line by small projections, *teeth*, which fit into corresponding *sockets*.

In removing the animal from its shell the various muscles attached to the impressions are cut through. They have corresponding names. The *adductor muscles* (Fig. 32, V.S and H.S) are broad bands of fibres which run transversely across from one valve to the other, and by their contraction keep the shell closed.

Each shell is lined by a flap, the right or left lobe of the **mantle**. These lobes have thickened edges which are attached to the shell along the pallial line by means of muscle-fibres (Fig. 32, Mt). They are continuous dorsally with the wall of the body, from which they grow out, are fused together above and behind the posterior adductor (H.S), and immediately beyond this are closely apposed to bound two oval openings. These are —a smaller *exhalent* or *cloacal aperture* (A) above, and a larger *inhalent aperture* (E) below, the edges of which are fringed with short tentacles. When at rest the mussel is completely imbedded in the mud, with its posterior end projecting, so that water can pass in at the inhalent and out at the exhalent aperture. In this way the animal is supplied with food and oxygen on the one hand, and gets rid of waste on the other. The mantle-lobes enclose between them a large space, the *mantle-cavity*, which is divided into two parts, a large *branchial chamber* below, and a small *suprabranchial chamber* above. It must not be supposed that the mantle lobes are fused posteriorly to bound the inhalent and exhalent apertures, or ventrally to bound the branchial chamber. If the animal is placed on its back the lobes can be separated without cutting anything, as in Fig. 33, where by this means the contents of the branchial chamber are exposed. In this figure the smooth sides of the exhalent aperture have been separated, and are seen just above A, on either side of which letter are the fringed right and left sides of the inhalent aperture. A large oval *visceral mass* (mesosoma), compressed from side to side, hangs down into the branchial cavity, but there is no distinct head, nor anything comparable to the lateral appendages of a crayfish. The lower edge of the visceral mass is produced into a yellow ploughshare-shaped muscular expansion, the *foot*, which projects forwards (Fig. 32, F, and Fig. 33, P), and can be protruded from the shell, serving as a locomotor organ. On either side of the

112 AN ELEMENTARY TEXT-BOOK OF BIOLOGY.

visceral mass is a *gill* (*ctenidium*), made up of two elongated plates (Figs. 32 and 33, K). These extend behind the visceral mass below the posterior adductor, and in this region the gills of opposite sides are united together. In Fig. 33 this union has been cut through in order to expose part of the suprabranchial chamber. At the front end of the visceral mass, and just beneath the anterior adductor, is a transversely elongated *mouth* (O), on each side of which are two triangular folds, the *labial palps* (Fig. 33, Se), which are united together at their bases to form ridge-like anterior and posterior lips. The exhalent aperture opens out of a *cloacal chamber*, which is the posterior part of the supra-branchial chamber, and is floored by the united gills. The last part of the intestine or *rectum*, which ends in the *anus* (Fig. 32, Af, and Fig. 33, A), runs dorsal to the posterior adductor, and projects into this chamber. Upon either side of the visceral mass, near the attachment of the gills, a small male or female *genital opening* is placed (Fig. 33, Oe′), and close to this is a rather larger *excretory* or *renal opening* (Oe″).

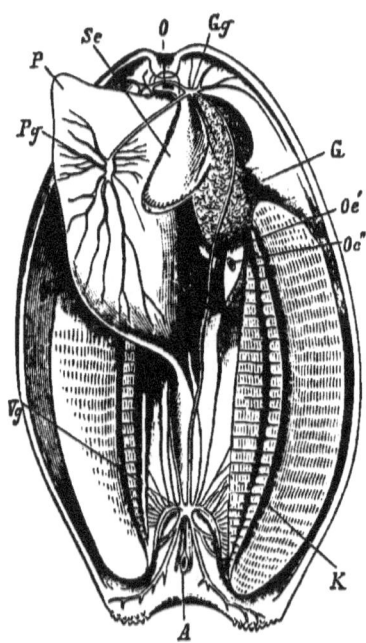

Fig. 33.—Nervous System of Anodonta (from *Claus* after *Keber*). —O, Mouth; A, anus; K, gills; P, foot; Se, labial palp; Gg, cerebropleural ganglion; Pg, pedal ganglion; Vg, visceral ganglion; G, part of genital gland; Oe′, opening of ditto; Oe″, opening of excretory organ.

2. **Skin.**—The most important part of this is the **mantle**, which forms a sort of flap drawn out on either side, and secretes most of the shell. The whole external surface of the mantle and dorsal region of the body can add to the pearly-layer, but the periostracum and prismatic layer can only be increased by the thickened edge of the mantle, and hence if worn away from

MOLLUSCA. 113

the older part of the shell, as commonly occurs in the region of the umbo, cannot be renewed.

Pearls are formed by the deposition of matter similar to that making up the pearly-layer around grains of sand, &c., that get into the shell. The pearl-fisheries of Britain were once famous, the pearls being obtained from a species of Unio. In China, small metal images of Buddha are placed within the shells of fresh-water mussels, and thus receive a pearly coating.

The *mantle* is covered externally and internally by a layer of simple columnar epithelium, which on the outer side contains many glandular cells, and on the inside is ciliated. The substance of the mantle between these layers of epithelium is made of connective tissue, traversed by muscle-fibres, blood-channels, and a complex network of nerves.

The skin covering the visceral mass and foot is very glandular and closely connected with underlying layers of muscle. The labial palps are, like the inside of the mantle, covered by ciliated epithelium.

The shell consists of an organic basis impregnated (except in the periostracum) with carbonate of lime. The prismatic layer is made up of polygonal prisms packed closely together, and arranged obliquely to the surface. The pearly-layer is composed of numerous thin laminæ, the edges of which constitute a series of delicate ridges which by diffraction produce an iridescent appearance.

3. The **digestive organs** consist of a convoluted tube running from mouth to anus and of a large digestive gland (Fig. 32). The *mouth* (O), on each side of which are the transversely striated labial palps (MS), leads into a very short *gullet* (œsophagus), which runs almost directly upwards, just behind the anterior adductor, and passes into a dilated *stomach* (M). This is succeeded by a narrow thin-walled *intestine* (D), which coils about (mostly in the median vertical plane) within the visceral mass, where it is closely surrounded by the genital gland (G), finally curving upwards and running back dorsally as the *rectum* through the ventricle of the heart, and over the posterior adductor to terminate in the anus. The cavity of the stomach is partially subdivided by irregular folds, and the ventral wall of the rectum is longitudinally infolded to form a typhlosole. A lobed *digestive gland* (L), dark brown in colour and made up of numerous branched

tubules, closely surrounds the stomach, into which it opens by several ducts.

The alimentary canal is lined throughout by simple columnar epithelium, which is partly ciliated, partly glandular. External to this are muscular layers. The tubules of which the digestive gland is made up are lined by glandular epithelium, the cells of which are cuboidal with brown granular contents. By the successive unions of these tubules the ducts are ultimately formed.

The ciliated epithelium, lining the mantle-lobes and covering the gills and labial palps, sets up currents in the mantle-cavity, and causes water to enter the branchial chamber by the inhalent orifice. Some of this is conducted along the groove between each pair of labial palps to the mouth, into which the small organisms it contains are carried. These constitute the *food*, and by the contraction of the muscular walls of the alimentary canal gradually pass backwards within it. The fluid secreted by the digestive gland contains ferments by the action of which fats are emulsified, while starch and proteid are respectively converted into grape-sugar and peptone. The length of the intestine increases the surface for absorption, and this is augmented by the typhlosole. The digested parts of the food diffuse into the blood-vessels of the intestinal walls, and the refuse is ejected at the anus, being afterwards carried out of the cloacal chamber by the currents which flow from the exhalent aperture as a result of ciliary action.

In specimens examined during autumn a transparent elastic rod, the *crystalline style*, will be found in the stomach. It is of albuminous nature and probably serves as a store of nutriment, for it is gradually used up during the winter. Some material of similar kind, only less compacted, is found in the intestine at the same period.

4. **Circulatory Organs** (Figs. 32 and 34).—A **blood system** alone is present, containing colourless blood in which are suspended numerous nucleated amoeboid corpuscles devoid of colour. This system is constituted by heart, arteries, veins, and irregular blood-spaces. The **heart**, which is contained within a thin-walled pericardial cavity situated on the dorsal side of the body, consists of a central oval muscular *ventricle*, into which a thin-walled *auricle* opens on each side. The auricles are funnel-shaped, with the narrow end attached to the ventricle, the broad end to the

ventro-lateral wall of the pericardial cavity. An *auriculo-ventricular valve* is placed at the point of union of each auricle with the ventricle. It is made up of two small flaps, which permit blood to pass into but not out of the ventricle.

The ventricle (through which the rectum runs) gives off two **arteries**, an *anterior aorta* in front running forwards above the rectum, and a *posterior aorta* behind running backwards below the rectum. The anterior aorta divides into branches, which supply the labial palps, the greater part of the alimentary canal, the foot, the anterior adductor, and adjacent parts; while the posterior aorta supplies the hinder part of the rectum, the posterior adductors, and the greater part of the mantle. The ultimate branches of these arteries end in minute irregular spaces (lacunæ) from which **veins** arise. These are vessels in which the blood is flowing towards the heart. The foot and organs of the visceral mass return their blood to the *vena cava*, a longitudinal vein lying in the middle line immediately below the floor of the pericardial chamber. The blood passes from the vena cava into the nephridium on either side, where it enters a close network of channels. From this network vessels run to the gills, opening on each side into an *afferent branchial vein*, which runs along the junction of the outer and inner gill-plate. Numerous branches pass into the gill-substance from this vein. The blood is returned to each auricle by an *efferent branchial vein*, which takes a course along the attachment of the outer lamella of the outer gill-plate on that side. Efferent veins from the mantle-lobes also enter the efferent branchial veins at each end.

Course of the Circulation.—The heart is systemic, receiving pure blood from the gills and mantle-lobes, which passes into the auricles, and thence into the ventricle. The former contract together, filling the ventricle, which then also contracts in a wave-like manner, and forces the blood through the arteries to the body at large. From the smallest arterial branches the blood passes into the lacunæ, and thence into venous channels, which are arranged as above described. The result is that the blood from the visceral mass and foot is purified in the nephridia, and then in the gills, before reaching the auricles, while the blood from the mantle-lobes is oxygenated there, and returned to the auricles without passing through either nephridia or gills.

4. **Respiratory Organs** (Figs. 32, 33, and 34).—These consist

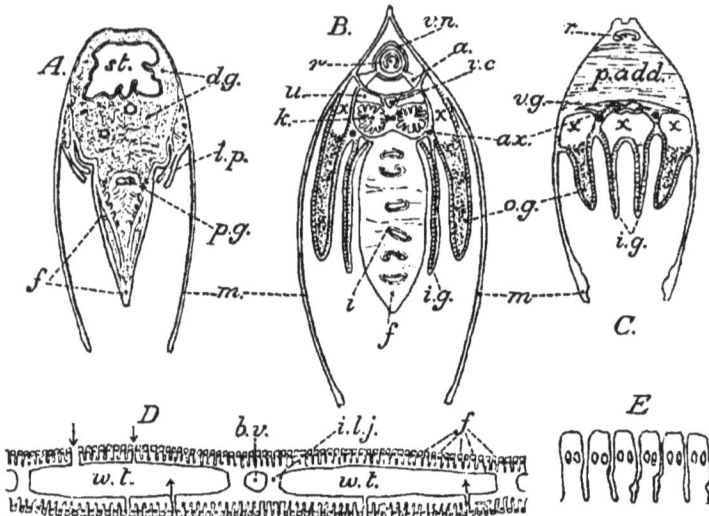

Fig. 34.—MUSSEL, ♀.—A, Section passing through pedal ganglia, *p.g*, showing stomach, *st*, surrounded by digestive gland, *d.g*; labial palps, *l.p*, and foot, *f*, projecting into branchial chamber, laterally bounded by mantle, *m*. B, Section passing through heart. Above is seen the ventricle, *vn*, traversed by rectum, *r*, with auricle, *au*, on each side. Below pericardium are seen ureters, *u*, with vena cava, *v.c*, between them; and kidneys, *k*, with cerebro-visceral connectives (black) between them. Projecting into branchial cavity is visceral mass showing intestine, *i*, cut through in six places, and foot, *f*; also inner and outer gill-plates, *i.g* and *o.g*, on each side; *ax*, gill axis on each side traversed by afferent branchial vessel (black); the outer gill-plates are full of larvæ, and χ is placed in each of the outer supra-branchial passages, the smaller inner passages are seen above the inner gill-plates. The branches of the ovary, which fill up the greater part of the visceral mass, are omitted. C, Section passing through visceral ganglia, *v.g*. The rectum, *r*, is seen above the posterior adductor, *p.add*. The inner gill-plates have united together behind the visceral mass; a χ is placed in each outer supra-branchial passage, and in the median passage formed by union of two inner ones. In a section still further back these three channels would be found to have coalesced to form the cloacal chamber. Gill axes as in B. D, Small part of longitudinal section through outer gill (enlarged), showing the two gill-plates formed by union of flattened gill-filaments, *f*, supported by chitinous rods (dark dots). Between the plates are seen two water-tubes, *w.t*, separated by an interlamellar junction, *i.l.j*, in which is a blood-vessel, *b.v*. The plates are perforated by numerous small apertures, ↓, ↑, leading into the water-tubes. E, Outer parts of six gill-filaments, further enlarged, showing supporting rods.

MOLLUSCA. 117

of the mantle and of a *gill* (*ctenidium*) on each side, composed of an inner and an outer gill-plate, each of these being again made up of an outer and inner *lamella* united together ventrally. The *outer* lamella of the outer gill-plate is attached by its dorsal edge to the inner side of the mantle, close to where it joins the body-wall. The *inner* lamella of the outer gill is united above with the outer lamella of the inner gill, and the common union of the two (gill-axis) is attached above to the wall of the body. The *inner* lamella of the inner gill-plate is attached anteriorly to the body-wall, is then free for a short distance, and behind the visceral mass unites with its fellow in the middle line to form the floor of the cloacal chamber. The gill-lamellæ are vertically striated, and between the striations numerous minute apertures are present, leading into an interlamellar space present in the interior of each plate, which is divided into a number of vertical *water-tubes* by the union of the two lamellæ along a corresponding number of narrow vertical strips by *interlamellar junctions*. The visceral mass divides the anterior part of the *supra-branchial chamber* into right and left halves. Each of these, owing to the dorsal attachment of the gill-axis, is again subdivided into two. There are, therefore, four channels above the gill-plates in this region, into which the four sets of water-tubes open. Behind the visceral mass the two inner channels coalesce with one another, and, finally, in the cloacal region, the supra-branchial chamber is undivided. These various relations will be understood by examining Fig. 34.

In Fig. 33 the cohering inner plates have been separated so as to expose part of the supra-branchial chamber.

The gills of Anodon are unusually complicated. They appear to have arisen in the following way:—From the gill-axis on either side two longitudinal rows of filaments grew out. The ends of the outer row turned sharply outwards and upwards; those of the inner row inwards and upwards. Thus in cross-section a W-shaped outline would be seen. These filaments then fused into two folded lamellæ, numerous apertures being left, however. The descending and ascending halves of each lamella then became connected to form a gill-plate. Union of the plates with one another and with surrounding parts increased the complexity, and brought about the distinction between branchial and supra-branchial chambers.

Each gill is covered by a single layer of ciliated epithelium. The external ridges (representing gill-filaments), which appear to the naked eye as vertical striæ, are supported by slender horny rods, two of which run vertically within each ridge. Beneath the epithelium the gills are composed of loose connective tissue, everywhere permeated by blood-containing lacunæ. The interlamellar junctions are traversed by more definite blood-channels.

Respiration is said to be effected more by the mantle than the gills, which are elaborate current-producing organs, and, as usual, essentially consists of gaseous interchange between the blood and the surrounding medium, oxygen being taken up and carbon dioxide eliminated. Water entering the branchial chamber by the inhalent aperture partly passes to the mouth and partly into the water-tubes of the gills through the numerous small apertures in the lamellæ. It then runs upwards into the supra-branchial passages, and backwards through the exhalent aperture to the exterior. All the currents are the result of ciliary action.

6. **Excretory Organs** (Figs. 32 and 34).—A pair of *kidneys* (organs of Bojanus) are present, each of which is essentially a tube folded upon itself, and communicating on the one hand with the pericardial cavity, which appears to represent a much-reduced cœlom, and on the other with the exterior by the renal opening noted above. The tube is divided into a thick-walled glandular part, the lining of which is raised into numerous ridges, and a thin-walled *ureter* (non-glandular part). The vena cava lies between the two ureters, which open in front by the renal openings, and communicate posteriorly with the glandular parts which underlie them. The glandular part is broadest behind, where it abuts upon the posterior adductor, and narrows in front where it opens into the floor of the pericardium. The kidneys should perhaps be regarded as nephridia, like the excretory organs of earthworm, since they are excretory tubes-opening out of a cœlomic body-cavity.

The glandular parts of the kidneys are lined by glandular epithelium, the cells of which are granular, and often contain concretions, in which guanin ($C_5H_5N_5O$), uric acid ($C_5H_4N_4O_3$), and urea (CH_4N_2O), have been stated to occur. The *organs of Keber*, consisting of glandular tubules, and situated one on each side of the pericardium, into which they have been said to open, may also have an excretory function.

7. **Reproductive Organs.**—The sexes are distinct, and the female is somewhat thicker from side to side. The genital glands (*spermaries* or *ovaries*, as the case may be) are much-

branched structures, surrounding the coils of the intestine in the visceral mass (Fig. 32, G). They are yellowish in the male, reddish in the female, and consist of an immense number of blind tubules, from which ducts pass on either side, uniting together into a short *spermiduct* or *oviduct* which opens to the exterior by the genital aperture (Fig. 33, Oe').

The tubules of the gonads are lined by germinal epithelium. These in the male produce large numbers of *sperms*, each of which possesses a cylindrical head and motile filiform tail. In the female *ova* are developed, each of which is covered by a vitelline membrane, which at one point, the *micropyle*, is incomplete. The *vitellus*, which contains a moderate amount of food-yolk, is separated when ripe by albuminous fluid from the vitelline membrane. It contains a large *germinal vesicle* with a *germinal spot*. Each ovum while still in the ovary is contained in a *follicle*, to the wall of which it is attached by a protoplasmic stalk passing through the micropyle.

The ova pass back from the genital openings to the cloacal chamber, and then forwards into the water-tubes of the outer gill-plates, where they are fertilized by sperms brought in by the inhalent current. The sperm enters the ovum by the micropyle. The breeding season is in the summer.

8. Muscular System.—The skin is intimately connected with underlying muscle, and there are also definite muscles, of which the adductors have already been described. The foot is almost entirely made up of muscular tissue, and slender *protractor, anterior retractor*, and *posterior retractor* muscles, which take origin from the corresponding impressions on the shell are inserted into it.

The muscles are composed of fusiform muscle-cells, exhibiting somewhat indefinite transverse striations. The foot serves as an organ of locomotion by which the animal can move forwards in the mud, leaving a furrow-like trail behind it. The retractor muscles by their contraction draw the foot back, and the protractors pull it forwards.

9. The nervous system (Figs. 32, 33, and 34) consists of a wide œsophageal nerve-ring, a long visceral loop connected with this on the dorsal side, and numerous nerves. Two pairs of ganglia (cerebro-pleural and pedal) are developed upon the ring, and one pair (visceral) upon the loop. It is convenient to use the term *commissure* for a nerve-cord connecting ganglia of the same name,

120   AN ELEMENTARY TEXT-BOOK OF BIOLOGY.

and the term *connective* for a nerve-cord connecting ganglia of a different name. Both kinds are here exemplified, but most of the nerve-cords are connectives, not commissures. On each side of the mouth, just beneath the skin, a *cerebro-pleural ganglion* is situated, and the dorsal part of the nerve-ring constitutes a short commissure above the mouth connecting these two ganglia. A pair of *pedal ganglia* lie close together at the junction of the foot and the visceral mass, resting upon the muscle of the former nearer its anterior than its posterior end. Each side of the nerve-ring may be regarded as a cerebro-pedal connective, since it unites a cerebro-pleural ganglion with the pedal ganglion of the same side. The visceral loop runs back along the dorsal side from the cerebro-pleural ganglia to the under side of the posterior adductor, where it thickens into a closely apposed pair of *visceral (parieto-splanchnic)* ganglia. Each side of the loop constitutes a very long cerebro-visceral connective running back from one cerebro-pleural ganglion to the visceral ganglion of the same side, to reach which it first traverses the digestive glands and then takes a course along the inner side of the kidney.

The cerebro-pleural ganglia supply the lips, labial palps, anterior adductor, and probably the otocysts. They also give off anterior pallial nerves which break up into a network in the margin of the front part of the mantle. The pedal ganglia supply the muscular tissue of the foot, and from the visceral ganglia three chief pairs of nerves run out to the gills and mantle. These are (1) in front, branchial nerves, then (2) lateral pallial nerves, and (3) posterior pallial nerves, which supply the tentacles of the inhalent aperture, and break up to form a complex nervous network in the margin of the hinder part of the mantle, continuous with the similar network arising from the anterior pallial nerves. The visceral ganglia also send nerves to the posterior adductor and the rectum.

A visceral or gastric nerve runs to the digestive gland and stomach from each cerebro-visceral connective, not far from its anterior end.

As usual, *ganglion-cells* and *nerve-fibres* are the essential parts of the nervous system, the former being entirely confined to the outer part of the ganglia and nerve-origins.

10. Sense Organs.—(1) **Tactile Organs.**—The labial palps act as feelers, and their epithelium comes into close relation with fibres from the cerebral ganglia. The tentacles surrounding the

MOLLUSCA. 121

inhalent aperture are well supplied with fibres from the visceral ganglia, and appear to be specially sensitive. The palps, tentacles, and mantle margin are provided with tactile cells, each of which is elongated and provided at its external end with a bunch of delicate setæ.

(2) **Gustatory Organs.**—The labial palps (?) possibly subserve the function of taste.

(3) **Olfactory Organs.**—Each of the branchial nerves is covered at its beginning by a patch of sensory epithelium known as an *osphradium*, and usually considered as a sort of olfactory organ which perhaps tests the quality of the water entering the pallial chamber.

(4) **Auditory Organs.**—Two minute vesicles, the *otocysts*, are found just behind the pedal ganglia, each of them being placed at the end of an auditory nerve which comes off from the lower end of the corresponding cerebro-pleural connective, and the fibres of which probably run up the connective to the cerebro-pleural ganglion of the same side.

The otocysts are lined by ciliated columnar epithelium of sensory nature, and each of them contains a rounded otolith made up of concentric layers of carbonate of lime.

The otocysts are difficult to study in Anodon, but are readily seen under the microscope in a living specimen of Cyclas, a small freshwater bivalve abundant in many streams and canals.

The mussel never at any time possesses eyes.

LIFE-HISTORY.

The ova are fertilized and the oösperms developed, up to a certain point, in the water-tubes of the outer gills of the female, the interlamellar junctions of which secrete a nutritive substance. It is an interesting fact that the shell commences as a single saddle-shaped plate on the dorsal surface. As the mantle-lobes are formed, this is divided into the two valves, which remain connected by the ligament, so that this must be regarded as an uncalcified part of the shell. Each valve is triangular, and its lower pointed end is bent inwards into a sharp spine. In the hinder part of the body a pit is formed, the *byssus gland*, which secretes a long sticky filament or *byssus*.

Many shell-fish, *e.g.*, the salt-water Mussel, are attached during life to surrounding objects by threads forming the *byssus*.

The embryo now passes out of the gill of the parent, being ejected from a small opening bounded by the coalesced mantle-lobes, and situated some distance above the exhalent aperture. The young mussel is very unlike the adult. It was formerly thought to be a distinct mollusc, and named *Glochidium*. This, however, was afterwards found to be only a *larva—i.e.*, a free-living embryo, very dissimilar to the mature form, and undergoing a series of changes known collectively as a **metamorphosis** before becoming adult.

The Glochidium has special larval sense organs, by which it is enabled to detect the presence of fish, and it swims freely by flapping the valves of the shell. The long byssus trails behind, and if it comes into contact with the skin of a fish adheres to it, and the sharp spines on the valves then effect a firmer hold. In this position the larva is covered over by outgrowths from the skin of the fish, and undergoes metamorphosis. The byssus and larval sense organs are lost, while the foot, gills, and internal organs are gradually developed. The host is then left.

This peculiar parasitic habit of the larva prevents it from being washed down to sea by the current (*see* p. 107).

§ 12. HELIX (The Snail).

The two species of Helix most available in Britain for dissection are *Helix aspersa*, the common snail, and *Helix pomatia*, the Roman snail. *H. pomatia* is characterized by its much larger size, and lighter colour. In the essential features of its organization, however, it differs but little from *H. aspersa*, which, as the commoner kind, will be described here.

MORPHOLOGY AND PHYSIOLOGY.

1. External Characters.—The most striking feature is the possession of a *shell*, into which the animal can entirely withdraw itself. It consists of one piece only; in other words, it is *univalve*. If a hollow cone with a very small angle is thought of as wound round and round into a right-handed spiral with closely adherent coils, a good notion will be formed of its structure. By the juxtaposition of the inner edges of the coils or *whorls*, a hollow pillar, the *columella*, is produced, which forms the axis of the spiral, and opens to the exterior below. The opening of the shell (*mouth*, or *peristome*) has a prominent smooth rim, and the blind end is termed the *apex*.

During winter the snail passes into a torpid condition (hybernates) when the aperture of the shell is closed by means of a thick calcareous plate, the *epiphragm* or *hybernaculum*. When the animal is fully expanded, the part of the body protruding from the shell is bilaterally symmetrical, terminated in front by a distinct head, and produced ventrally into the broad sole-like *foot*, an elongated muscular expansion, tapering behind to a point, and ending in front just below the mouth. The animal crawls by wave-like contractions of this organ. The dorsal part of the body projects as a spirally coiled *visceral hump*, always contained within the shell, and attached to the columella by a special muscle (spindle muscle) by the contraction of which the animal can be completely withdrawn into its shell. The anterior boundary of the visceral hump is marked by a thickened edge, the *collar*.

The head bears two pairs of *tentacles* or *feelers*, which are hollow, and can be retracted into the body much as the finger of a glove can be turned into the glove's interior. Each of the posterior longer pair, the *optic tentacles*, bears a small black dot, the *eye*, on the outer side of its tip. The anterior shorter pair are the *olfactory tentacles*. In the front of the head is the *mouth*, guarded by an inferior and two lateral *lips*. Below the mouth there is a small pore, the opening of the *supra-pedal gland*. Not far below the right optic tentacle the rounded *genital opening* is situated. In the thickened anterior edge of the mantle, on the right side, there is a deep depression, into the left side of which the large valvular *respiratory aperture* opens, and on the right side of this a smaller *anus*. The respiratory opening leads into a spacious mantle-cavity, which acts as a lung and is roofed in by the *mantle*, a vascular flap-like outgrowth of the body-wall.

2. **Skin.**—The skin is smooth on the under side of the foot,

and on the visceral hump, where it is so thin and transparent that many of the internal organs can be seen through it. Elsewhere the surface of the body is corrugated. The skin is very glandular and secretes the characteristic slime.

The skin consists of an *epidermis* made up of a single layer of cells, and a connective tissue *dermis*, closely united with underlying muscular tissue. The epidermal cells are flat on the visceral

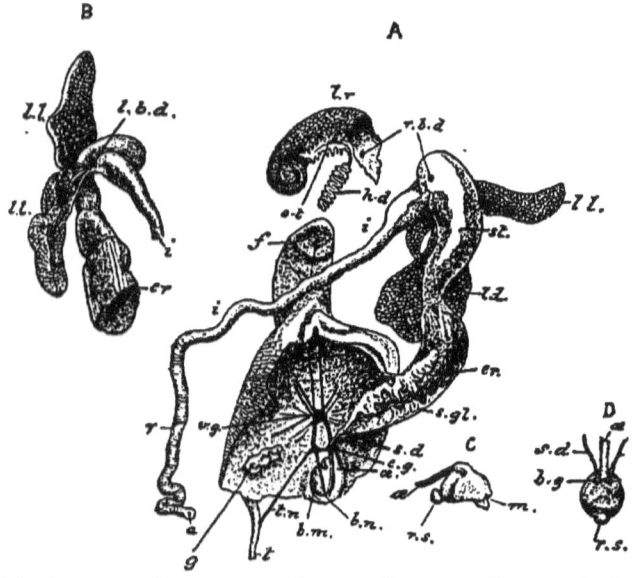

Fig. 35.—DIGESTIVE ORGANS AND NERVOUS SYSTEM OF SNAIL.—A, General view; B, left side of stomach, &c.; C and D, side and hind views of buccal mass; *f*, foot; *t*, tentacle; *m*, mouth; *b.m*, buccal mass; *r.s*, radular sac; *œ*, gullet; *cr*, crop; *st*, stomach; *i*, intestine; *r*, rectum; *a*, anus; *s.gl*, salivary glands; *s.d*, salivary ducts; *l.l*, *l.r*, left and right lobes of digestive gland; *l.b.d*, *r.b.d*, ducts of ditto; *c.g*, cerebral ganglion; *v.g*, ventral ganglionic mass; *b.g*, buccal ganglion; *t.n*, tentacular nerve; *b.n*, buccal nerve; *o.t*, hermaphrodite gland; *h.d*, hermaphrodite duct; *g*, place from which genitals have been cut away.

hump, cylindrical elsewhere. Many of them constitute unicellular mucous and pigment glands, while few-celled pigment and calcareous glands are also present. These various glands project into the dermis, which is also traversed by muscle-fibres, blood-spaces, and nerves.

The *shell* consists of three layers. There is a thin external chitinous *periostracum*, which is pigmented, and gives colour to the shell. Below this is a dense *prismatic layer*, and internally a thinner *pearly layer*, composed of numerous laminæ, and with a smooth and polished internal surface. The last layer can be secreted by all parts of the epidermis covering the visceral hump, but the two others can only be formed by the collar, which, as the animal grows, adds successive increments to the mouth of the shell, the boundaries between which are indicated by *lines of growth*.

3. The **digestive organs** (Figs. 35, 36) consist of a convoluted gut running from the anterior ventral mouth to the unsymmetrically placed anus, and receiving the secretions of salivary and digestive glands. The gut is divisible into buccal mass (pharynx), gullet, crop, stomach, intestine, and rectum.

The *mouth* leads into a large mouth-cavity, contained in an oval muscular *buccal mass*. Immediately within the lips, dorsally, is a crescentic, horny, toothed structure, the *jaw*. But the most important organ connected with the mouth is the *odontophore*. This consists of an elevation rising up like a tongue from the floor of the mouth, on which a horny ribbon (the *radula*), bearing innumerable minute pointed teeth, is spread out from back to front, passing behind into a pouch, the *radular sac*, which lies at the back of the buccal mass, and forms a small rounded protuberance of whitish colour. Into the outside of the buccal mass special muscles are inserted, *the buccal protractors* and *retractors*. From the upper side of the buccal mass a narrow, thin-walled gullet (œsophagus) passes back, and merges into a spindle-shaped crop, the thin walls of which are marked by longitudinal striations. The crop narrows behind, and is succeeded by a rounded *stomach* with moderately thick walls. This turns sharply upon itself, and is then followed by the narrow, thin-walled *intestine*, which first bends ventrally, and then, after coiling a little, passes into the *rectum*, which takes a straight course along the right side of the lung-chamber to open by the *anus*. Projecting into the intestinal cavity is a longitudinal fold.

The *salivary glands* are paired branching structures, placed one on each side of the crop, to which they are attached by connective tissue. A slender *salivary duct* runs from each of them down the side of the gullet, and, dilating slightly, opens into the side of the mouth-cavity.

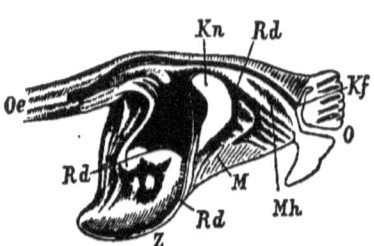

Fig. 36.—SECTION THROUGH BUCCAL MASS OF SNAIL (from *Claus*, after *Keferstein*).-O, Mouth; Mh, mouth-cavity; M, muscles; Rd, radula; Kn, supporting cartilage; Z, radular sac; Kf, jaw; Oe, œsophagus.

The *digestive gland* ("liver") is a large brown organ making up a considerable part of the visceral hump. It is divided into right and left lobes, of which the latter ($l.l$) is deeply three-cleft, and much larger than the other, which occupies the final coils of the visceral hump. Two ducts (bile-ducts) run from the corresponding lobes to the right and left sides of the stomach. The left duct is extremely short, and formed by the union of three branches from the subdivisions of its lobe.

The gut is lined by a single layer of epithelial cells, which are, for the most part, of the simple columnar type. The radula is a cuticular ribbon, developed by the epithelium lining the radular sac. Upon its upper surface are numerous longitudinal rows of minute pointed, backwardly-projecting "teeth." Teeth of the same age are at the same level, so that a clear arrangement into transverse rows is also seen. The central row contains symmetrical median teeth (*uncini*), while the lateral (*rachidian*) teeth of the other rows are asymmetrical. The arrangement may be expressed by the formula $\infty . 1 . \infty$, in which the $1 =$ the single longitudinal row of median teeth and $\infty =$ the numerous longitudinal rows of lateral teeth on each side of this. As the radula is worn away in front, it grows forwards (like a finger-nail on its bed) upon, and connected with, the *sub-radular membrane*, which covers the projection from the floor of the mouth. This membrane is formed by epithelium, together with underlying connective tissue, and is to some extent movable upon the central part of the odontophore, which is supported by two masses of gristle, the *odontophoral cartilages*. These serve as the origins of minute muscles, which are inserted into the sub-radular membrane in front and behind. The cartilages consist of a clear *matrix*, in which are imbedded numerous branched cartilage cells. The wall of the alimentary canal, outside the epithelium, contains muscular layers, an internal longitudinal and an external circular. The salivary glands are aggregates of unicellular glands, each of which is sharply marked off from its neighbours, and has its own minute duct. The digestive gland is composed of branched tubules, ending blindly, and lined by glandular epithelium, in which three chief types of cell can be distinguished, (1) *granular cells* (liver-cells), containing yellowish granules, (2) pear-shaped *ferment cells*, (3) large *calcareous cells*.

The snail chiefly feeds upon the fresh leaves, stems, &c., of plants, from which it rasps off small fragments by means of its radula, in the following way:—The buccal mass is pulled forwards by means of its protractor muscles, when the front of the radula on its cushion projects a little from the mouth. Appropriate contractions of the small muscles within the cushion move the sub-radular membrane, and with it the radula, backwards and forwards, the jaw meanwhile holding the food firmly, and serving as a relatively fixed part against which the radula works. The particles of food scraped off pass back into the mouth-cavity, partly by the agency of the flexible lips, and partly as a result of the backward movement of the odontophore, which acts to some extent like a suction-pump. The salivary secretion has been stated to contain a ferment which converts starch into sugar, and the acid secretion of the digestive gland is known to bring about fermentative changes of this kind and others involving the conversion of proteids into peptone. The food gets mixed up and gradually passed backwards by the contractions of the muscular walls of the gut, the length of which gives a considerable absorbing surface, augmented by the longitudinal fold in the intestine.

The digestive gland has other functions besides that of aiding digestion. The granules in its granular cells are most likely of excretory nature, and the material secreted by its calcareous cells is used in the construction of the epiphragm.

4. Circulatory Organs.—As in the mussel, a **blood system** alone is present, and the reduced cœlom is represented by the pericardial cavity. Heart, arteries, and venous system can be distinguished. The *blood* is of a bluish tinge owing to the presence of hæmocyanin, and consists of plasma and amœboid colourless corpuscles.

The muscular **heart** is situated in the posterior part of the lung-chamber, in close proximity to the kidney, and contained in a pericardial cavity bounded by a firm translucent *pericardium*. It is oval in form, and made up of an anterior thin-walled *auricle*, which communicates by a valve with a posterior *ventricle*.

Arteries.—The ventricle is continued into a large artery, the *aorta*, which almost immediately gives off an important visceral artery to the visceral hump, then runs forwards, supplying the body-wall, muscles, viscera, &c., finally perforating the ventral nerve-mass, and breaking up into branches for the head.

Venous System.—The smallest arteries form networks (? capillaries), from which the blood passes into minute spaces (lacunæ), which communicate on the other hand with large *venous sinuses*. Of these the most important are—the spacious *body-cavity* surrounding those viscera which are not contained in the visceral hump, two *lateral sinuses* in the foot, a *visceral sinus* along the inner edge of the coiled visceral hump, and a *pulmonary sinus* with which this communicates, running round the floor of the lung-chamber.

From the pulmonary sinus numerous *afferent pulmonary vessels* are given off, which branch in the roof of the lung, and from these branches *efferent pulmonary vessels* arise, which unite together to form the *pulmonary vein* opening into the auricle. A number of the afferent trunks enter the kidney, and form a network within it, from which one large and several smaller *renal veins* run to the pulmonary vein.

Course of the Circulation.—Blood, oxygenated in the lung, and (part of it) purified in the kidney, enters the auricle during its diastole by the pulmonary vein. It then passes into the ventricle during the auricular systole. The ventricle, owing to its muscular walls, contracts more vigorously, and the blood, prevented by the auriculo-ventricular valve from returning to the auricle, is forced into the arteries. From the fine ramifications of these it passes into the lacunæ, and thence into the venous sinuses, ultimately reaching the pulmonary sinus. From this the blood, now carbonated and loaded with nitrogenous waste, passes by the afferent pulmonary vessels into the roof of the lung. Here it is oxygenated, and passes into the efferent pulmonary vessels from which the pulmonary vein arises. Part of the blood traverses the kidney before entering the pulmonary vein, and thus gets rid of nitrogenous waste, some of which has been previously eliminated in the digestive gland. The hæmocyanin of the blood is a copper-containing proteid which readily takes up oxygen from the air in the lung into loose chemical combination, and parts with it as readily to the tissues. It acts, therefore, as an oxygen-carrier.

5. The **Respiratory Organs** are represented by a true air-breathing *lung* (pulmonary sac) placed on the upper side of the visceral hump, behind the collar, and communicating with the exterior by a rounded valvular opening on the right side of the body. The delicate roof and side-walls of the lung are formed by the mantle, which presents internally a network of ridges, in which the vessels already described ramify. The floor is thin, but muscular, and immediately overlies the crop and the bulk of the reproductive organs. When at rest it is strongly convex upwards, but it becomes flattened by contraction so that the lung-cavity is increased in size, and air consequently passes in. This is *inspiration*, the converse of which, *expiration*, is effected by the floor simply ceasing to contract. The pulmonary opening is valvular, and thus the supply of air is regulated and desiccation prevented. The essential part of respiration consists in carbon dioxide diffusing *out of* and oxygen diffusing *into* the vessels ramifying in the lung-roof. The part played by hæmocyanin is explained above.

6. **Excretory Organs.**—A large, cream-coloured *kidney*, somewhat triangular in form, with the apex forwardly directed, is closely united to the posterior part of the lung-roof. It contains a cavity (the surface of which is increased by the projection into it of numerous lamellæ) which communicates with the pericardium by a minute opening, and with the exterior by an *ureter*. This arises from the anterior end of the kidney, passes along its right side, and then runs along the side of the lung above the rectum to open close to the anus. The kidney is equivalent to one of the renal organs in the mussel, and is perhaps to be regarded as homologous with one of the nephridia in such a form as the earth-worm. It is plentifully supplied with blood, from which its glandular epithelium separates nitrogenous waste in the form of ammonium and calcium urates. The digestive gland has also been shown to take part in the work of nitrogenous excretion.

7. **Reproductive Organs** (Fig. 37).—The snail possesses a very complicated set of *hermaphrodite* reproductive organs, mostly of a whitish colour. (1) The *hermaphrodite gland* is a small kidney-

130   AN ELEMENTARY TEXT-BOOK OF BIOLOGY.

shaped organ imbedded in the inner side of the coiled right liver-lobe. From it a much convoluted *hermaphrodite duct* proceeds.

(2) **Special Male Organs.**—The hermaphrodite duct splits as it were into male and female ducts, or spermiduct and oviduct, which convey the sperms and ova respectively. The *spermiduct*

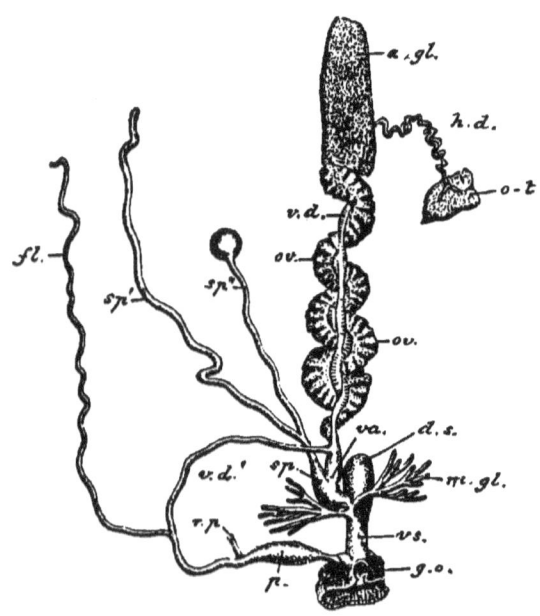

Fig. 37.—REPRODUCTIVE ORGANS OF SNAIL.—*o-t*, Hermaphrodite gland; *h.d*, hermaphrodite duct; *a.gl*, albumen gland; *v.d*, spermiduct (vas deferens); *ov*, oviduct; *va* and *vs*, upper and lower parts of vagina; *d.s*, dart-sac; *m.gl*, mucous glands; *g.o*, genital opening; *sp*, spermotheca; *sp'* and *sp"*, left and right branches of ditto; *p*, penis; *r.p*, retractor penis; *fl*, flagellum.

(vas deferens) is a narrow tube running at first along the side of the much larger oviduct, the cavities of the two being incompletely separated.* In this part of its course it is beset with numerous small eminences collectively constituting a *prostate gland*. Further forwards the spermiduct separates from the

* The two together often receive in this region the name of **common duct.**

oviduct, and, after pursuing an independent course for some time, opens into a hollow muscular tube, the *penis*, which receives at the same point the *flagellum*, a long and hollow filament. The penis opens into the *genital atrium*, a shallow depression common to both male and female ducts, and opening to the exterior below the optic tentacle by the genital opening. A narrow band-like muscle, the *retractor penis*, takes origin in the lung-floor, and is inserted into the penis.

(3) **Special Female Organs.**—The *oviduct* is a wide, somewhat twisted tube, with folded and pouched walls, along one side of which the spermiduct runs. The tongue-shaped *albumen-gland*, which varies very much in size, according to the time of year, opens into its commencement. Where the spermiduct assumes an independent course, the oviduct merges into a muscular, smooth-walled tube, the *vagina*, which opens into the genital atrium. A tubular organ, the *spermotheca* (receptaculum seminis), consisting of right and left branches, opens into the hinder end of the vagina. The right and shorter branch ends in a berry-like dilatation.

In *Helix pomatia* the left branch is only represented by a small projection.

Two tuft-like *mucous glands* open into the side of the vagina somewhat further forwards, and in front of this an extremely muscular pouch, the *dart-sac*, which can be everted from the genital opening, communicates with it. This sac contains an elongated calcareous body, the *spiculum amoris*, which is sharply pointed and possesses four slightly twisted lateral ridges.

It is doubtful whether the dart-sac really belongs to the female organs, but its position would seem to indicate this.

The hermaphrodite gland is made up of numerous branched tubules lined by germinal epithelium. Some of these germinal cells develop into rounded *ova*, which possess a well-marked *germinal vesicle* with *germinal spot*, but are devoid of a *vitelline membrane*. Other germinal cells pass into the cavities of the tubules, and as mother-sperm-cells divide repeatedly to produce bunches of *sperms* (spermatozoa), each of which has an irregularly oval head, and long vibratile tail.

The breeding season is early summer. As in most hermaphrodite animals cross-fertilization takes place, and self-fertilization is prevented, in this case by the sperms maturing before the ova.

The ripe sperms pass down the spermiduct and into the flagellum, by a viscid secretion of which they are bound into a thread-like packet (spermatophore). Two individuals mutually fertilize each other, the eversible penis being used as a copulatory organ, by which the spermatophore is conveyed into the spermotheca. A preliminary stimulus is given by the ejected darts, which may be found sticking in the skin. After some days the spermatophore disintegrates and the liberated sperms fertilize the ova as they pass down the oviduct. Each oösperm is surrounded by an albuminous investment secreted by the albumen gland, and, external to this, by a tough calcareous shell, with the formation of which the mucous glands appear to be concerned. The eggs, thus constituted, are about a quarter of an inch long in *H. pomatia*. They are laid in damp earth during June or July.

8. **Muscular System.**—The foot is almost entirely made up of muscle-bands, arranged in a complicated way, and bringing about creeping movements by their contraction. The *spindle-muscle* serving to pull the body into the shell is a firm band taking origin in the columella, and, dividing into right and left halves, again subdivided into numerous slips which are inserted into the foot. The retractor of the buccal mass, by which the snail's head is drawn in, is a branch of this muscle. The contrary movement is effected by protractors of the buccal mass which take origin in the foot, and there are also depressors, having a similar origin, which pull the buccal mass down. A branch of the spindle-muscle on each side constitutes a tentacular retractor which bifurcates into two slips traversing the corresponding tentacles and inserted into their tips. When these muscles contract they draw the tentacles back into the body-cavity, invaginating them. The retractor penis has been mentioned above (p. 131).

The muscle-fibres are composed of slender spindle-shaped cells, which, except in the odontophore, are unstriated.

9. **Nervous System** (Fig. 35).—This is remarkable for its great concentration, and is chiefly localized in the head, where a nerve-ring enclosed in a firm sheath is found surrounding the gullet, immediately behind the buccal mass in the extended state.

When the animal retracts itself, this is drawn backwards through the nerve-ring, which is, therefore, then found further forwards than usual.

The ring is thickened dorsally into two *cerebral ganglia*, connected together by a broad commissure, and ventrally into a ganglionic mass, formed by the coalescence of several pairs of ganglia. This ventral mass is divisible into postero-dorsal *pleuro-visceral ganglia*, and antero-ventral *pedal ganglia*, respectively connected with the cerebral ganglia by a posterior cerebro-pleural connective and an anterior cerebro-pedal connective, on each side.

The head is innervated by branches from the cerebral ganglia, which give off five pairs of nerves. (1) The *tentacular nerves*, which supply the optic tentacles. Each runs within the corresponding tentacle, gives off an *optic nerve* to the eye, and ends in a *tentacular ganglion*, from which branches run to the skin covering the end of the tentacle. (2) Two pairs of *labial nerves*, one of which gives off to the short tentacle a *tentacular nerve* ending in a *tentacular ganglion*. These nerves supply the lips and neighbouring parts. (3) An *auditory nerve* passes down on each side between the two connectives to the otocyst. (4) A *buccal nerve* comes off from the front of each cerebral ganglion, and runs forwards to a *buccal ganglion* placed in the angle where the gullet joins the buccal mass. The buccal ganglia innervate the pharynx, gullet, and salivary glands. They are connected together by a commissure ventral to the gullet. An unpaired nerve runs to the penis from the right cerebral ganglion. The body-walls and viscera are supplied by the *pleuro-visceral ganglia*, while the nerves of the foot come off from the *pedal ganglia*.

The nervous elements are as usual *ganglion-cells* and *nerve-fibres* —the former are confined to the outer parts of the ganglia, and usually possess only one process.

10. **Sense Organs**—(1) **Tactile Organs.**—The sense of touch is possessed by the surface of the body generally, but is specially localized in the head, tentacles, and sides of the foot. The tactile cells are narrow cylinders with tapering external ends formed by an aggregation of hair-like processes.

(2) **Olfactory Organs.**—The snail possesses a keen sense of smell which enables it to detect the presence of various kinds of food and to avoid certain strongly odorous liquids, such as turpentine. This discriminative power is lost if the tentacles are removed, and there is a patch of modified epithelium at the tip of each of them, apparently of olfactory nature and containing numerous flask-shaped cells closely related to the corresponding tentacular ganglion. An olfactory function has also been ascribed to the *supra-pedal gland*, a tubular organ, lined with columnar

epithelium, lodged in the foot, and opening to the exterior by a small pore beneath the mouth.

(3) **Auditory Organs.**—A minute spherical *otocyst* or *auditory sac* is placed on each side of the ventral nerve-mass, and is connected by an auditory nerve with the corresponding cerebral ganglion. The sac possesses an outer firm investment, and is lined by ciliated columnar epithelium, composed of auditory cells. It is filled with numerous calcareous particles or *otoliths*, suspended in fluid.

(4) **Visual Organs.**—Each optic tentacle bears on the outer side of its tip an *eye*, which appears as a black dot, and is innervated by the optic nerve. It is placed immediately below a small transparent area of the epidermis, and essentially consists of a spherical vesicle, enclosed in a firm sheath, and containing a large globular *lens*, devoid of structure. The vesicle is made up of a single layer of cells, which in front are short and transparent, constituting a *cornea*, while behind they are much more elongated and form a sensitive *retina*. The retinal cells are of two kinds, unpigmented and pigmented. Each of the former is produced into a flask-shaped visual rod next the cavity of the vesicle, and is surrounded by several of the latter, which are pigmented in their outer parts only, while internally they taper to transparent processes ensheathing the visual rod.

It has been shown that the eyes of the snail can only clearly distinguish the form of external objects when at a distance of from $\frac{1}{25}$ to $\frac{2}{25}$ of an inch. The pigmented skin appears to be sensitive to light, helping the animal to distinguish between light and darkness.

CHAPTER VIII.—VERTEBRATA ACRANIA.

§ 13. AMPHIOXUS (The Lancelet).

AMPHIOXUS is a semi-transparent somewhat fish-like animal, not exceeding two inches in length, and laterally flattened. There is a free-swimming larva, but the adult animal is a shallow water marine form generally found vertically buried in the sand, from which only its anterior end projects, but also capable of swimming by eel-like movements of its body. It is abundant round many coasts, as, for example, the Mediterranean, and has been found off our own shores.

All the types hitherto described belong to the **Invertebrata**, or animals devoid of backbone, while Amphioxus, Dog-fish, Frog, Pigeon, and Rabbit, the remaining animals to be dealt with, are examples of the **Vertebrata**, or backboned group, using the word in a very broad sense, and by "backbone" understanding a firm rod, not necessarily of bone, underlying and supporting the central nervous system. To avoid confusion it is perhaps advisable to drop the name Vertebrata as used in the broader sense, replacing it by the term **Chordata**, and styling the Invertebrates **Non-Chordata** for the sake of uniformity. All Chordate animals possess, temporarily or permanently—

(1) A tubular *central nervous system* running along the **dorsal** side of the body, and not perforated by the gut as in many higher Non-Chordates (*e.g.*, Earthworm, Leech, Crayfish).

(2) A firm elastic rod, the **notochord**, underlying the central nervous system for part or all of its length.

(3) [Except in one instance] A laterally perforated pharynx, serving in the lower forms as a respiratory organ.

The Chordata are classified as follows:—

CHORDATA (= VERTEBRATA in wider sense).

I. **Hemichorda.** Small notochord in anterior part of body.—A small group of worm-like forms.

II. **Urochorda.** Small notochord in tail of larva, generally absent in adult.—A group of degenerate forms, the *Ascidians* or Sea-Squirts, which are usually fixed when adult.
} = VERTEBRATA ACRANIA.
No brain-case, limbs, or jaws.

III. **Cephalochorda.** Notochord extending from one end of the body to the other.—*Amphioxus*.

IV. **Vertebrata** (in narrower sense). Notochord ends anteriorly below the middle of the brain. A more or less complete vertebral column ("backbone") in which segmentation is always indicated, and which usually encroaches more or less upon the notochord.-Dogfish, Frog, Lizard, Fowl, Rabbit.
} = VERTEBRATA CRANIOTA.
A brain-case. Limbs, when present not more than two pairs.

MORPHOLOGY AND PHYSIOLOGY.

1. External Characters (Fig. 38).—The scientific and popular names of Amphioxus are alike derived from the fact that the flattened body is pointed at both ends. There is no distinct division into head, trunk, and tail, as is the case in a fish, and, at the first glance, it does not seem easy to say which is the anterior and which is the posterior end. The former, however, is recognizable on a cursory examination by the presence of the ventral *mouth*, an oval opening bordered by numerous stiff ciliated processes, the *buccal cirri*. Further back along the ventral surface is seen another median aperture, the *atriopore*, situated on a prominent papilla. It is the outlet of a large *atrial cavity*, by which the perforated pharynx is surrounded. A thin, laterally flattened *fin* runs from the mouth round the front end of the body along the dorsal surface, round the posterior end, and forwards along the ventral surface as far as the atriopore. About the middle of this ventral section of the fin a small opening, the *anus*, is to be found on the left of the median line. This is one of several particulars in which Amphioxus deviates from strict bilateral symmetry. The fin is rather larger round the posterior end, constituting a *caudal fin*, while that part of it which runs along the upper side of the body in front of this is known as the *dorsal fin*, and the part between anus and atriopore as the *anal fin*. All these are perfectly continuous. The ventral surface between mouth and atriopore is broad, gently convex, and marked by a series of longitudinal ridges. It is bounded on each side by a longitudinal fold (metapleural fold), the *lateral fin*, which unites with its fellow just behind the atriopore, at the point where the anal fin begins.

Amphioxus is a segmented animal, and this is indicated externally by a number of > shaped lines on the sides of the body, corresponding to a division of the lateral muscles of the body into *muscle-segments* (myomeres, myotomes) which in *Amphioxus lanceolatus*, the species commonly used in laboratories, are 61 or 62 in number. The atriopore corresponds to the 36th myomere, while the anus is situated between the 51st and 52nd.

There is a small ciliated pit (? olfactory) on the left side of the head above the anterior end of the mouth.

VERTEBRATA ACRANIA. 137

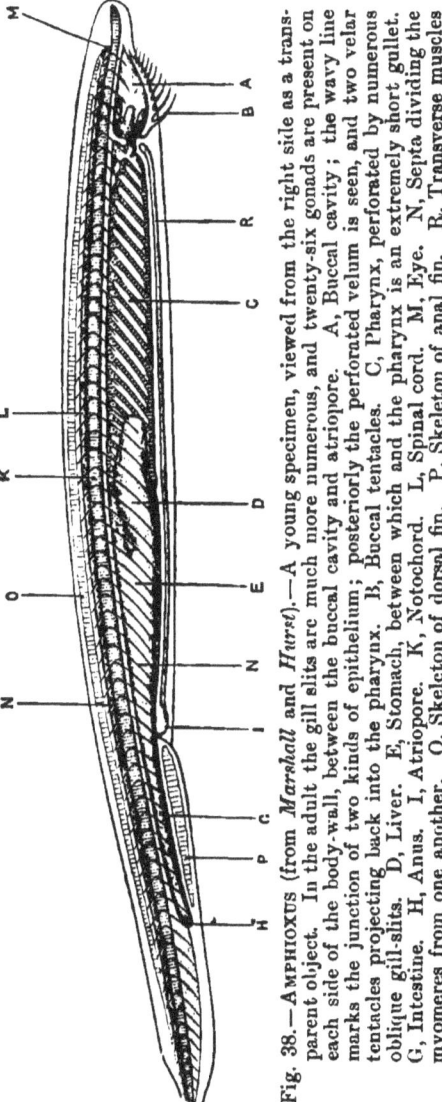

Fig. 38.—AMPHIOXUS (from *Marshall* and *Hurst*).—A young specimen, viewed from the right side as a transparent object. In the adult the gill slits are much more numerous, and twenty-six gonads are present on each side of the body-wall, between the buccal cavity and atriopore. A, Buccal cavity; the wavy line marks the junction of two kinds of epithelium; posteriorly the perforated velum is seen, and two velar tentacles projecting back into the pharynx. B, Buccal tentacles. C, Pharynx, perforated by numerous oblique gill-slits. D, Liver. E, Stomach, between which and the pharynx is an extremely short gullet. G, Intestine. H, Anus. I, Atriopore. K, Notochord. L, Spinal cord. M, Eye. N, Septa dividing the myomeres from one another. O, Skeleton of dorsal fin. P, Skeleton of anal fin. R, Transverse muscles in floor of atrial cavity.

2. **Skin.**—This consists of the epidermis and dermis. The *epidermis* is made up of a single layer of columnar cells (ciliated in the larva), among which are a number of scattered sensory cells, each of which is somewhat rod-like and terminates externally in a stiff tapering process, while internally it is continuous with a nerve-fibre. These cells are most numerous in the anterior part of the body.

The *dermis* presents a firm external layer, beneath which is a much thicker gelatinous stratum again succeeded by a very thin nucleated layer.

3. **Skeleton** (Figs. 38 and 39).—As in all Chordates the most important part of this is internal, constituting an *endoskeleton*, which contrasts strongly with the cuticular *exoskeletons* of such forms as Crayfish and Mussel. Amphioxus, however, possesses only a feeble endoskeleton, of which the most characteristic part is the **notochord**. This is an elastic rod which runs from one end of the body to the other, above the gut and below the central nervous system. It gives a certain amount of firmness to the body, and serves for the attachment of the lateral muscles. The notochord is of cellular nature, as is readily seen in young specimens, but most of the constituent cells become, later on, much vacuolated, so that their outlines are obscured. They are arranged so as to form a succession of thin vertical discs.

The notochord is surrounded by a firm connective-tissue sheath which is continued dorsally into a tube investing the spinal cord, and laterally into septa running between the myomeres and joining the dermis. The notochordal sheath resembles the dermis in structure, the dense layer being in this case internal, while the other layers are continued into the neural sheath and septa.

Amphioxus possesses other skeletal structures besides those already described—*i.e.*, a buccal skeleton, a branchial skeleton, and fin-rays.

The *buccal skeleton* consists of a series of short rods jointed together so as to form an incomplete ring stiffening the margin of the mouth, with processes extending into the buccal tentacles. These parts resemble the notochord in minute structure.

The *branchial skeleton* supports the pharynx, and will be described in the next section.

The *dorsal fin* is supported by a very large number of minute vertical fin-rays, consisting of little columns of gelatinous material attached below to a ridge running along the top of the neural sheath, and projecting above into box-like spaces full of lymph. The anal fin is supported by a double series of such fin-rays.

4. **Digestive and Respiratory Organs** (Figs. 38 and 39).—The gut is a straight tube running from mouth to anus, and consisting of buccal cavity, respiratory pharynx, gullet, stomach with liver, and intestine. It is ciliated throughout.

The *buccal cavity*, into which the wide jawless mouth opens, is somewhat funnel-shaped, and it is lined by two kinds of epithelium, the boundary between which is marked by a series of lobes. The kind which occurs in the posterior part of the cavity is distinguished by the presence of pigment and specially long cilia. At the back of the buccal cavity there is a muscular partition, the *velum*, which is perforated by an aperture leading into the pharynx and guarded by a circlet of twelve delicate backwardly projecting tentacles. Between buccal cavity and atriopore the gut is suspended from the sheath of the notochord in a spacious *atrial cavity*.

The *pharynx* is the largest and most characteristic part of the gut, extending for about half its length as a wide tube laterally perforated by numerous oblique gill-slits, opening into the surrounding atrial cavity. The first formed gill-slits in the young larva open at first directly to the exterior, and then into a longitudinal groove, the sides of which unite to form a tube open behind. This tube, the opening of which persists as the *atriopore*, gradually becomes more extensive, sinking into the body, so to speak, and surrounding the gut (except on the dorsal side) behind the buccal cavity. There is also a narrow prolongation of the atrial cavity extending back on the right side of the intestine between atriopore and anus. It is obvious from the above outline of its development, that the atrial cavity is, morphologically, a part of the exterior.

The cavity of the pharynx is much wider in its anterior than in its posterior portion, where it is somewhat flattened from side to side. A well-marked *epibranchial groove* runs along the middle of the roof of the pharynx, while in the median line of the floor there is a thickening known as the *endostyle*, flat or convex anteriorly, groove-like posteriorly. Both these median regions are characterized by the presence of elongated columnar epithelial cells provided with very long cilia.

140 AN ELEMENTARY TEXT-BOOK OF BIOLOGY.

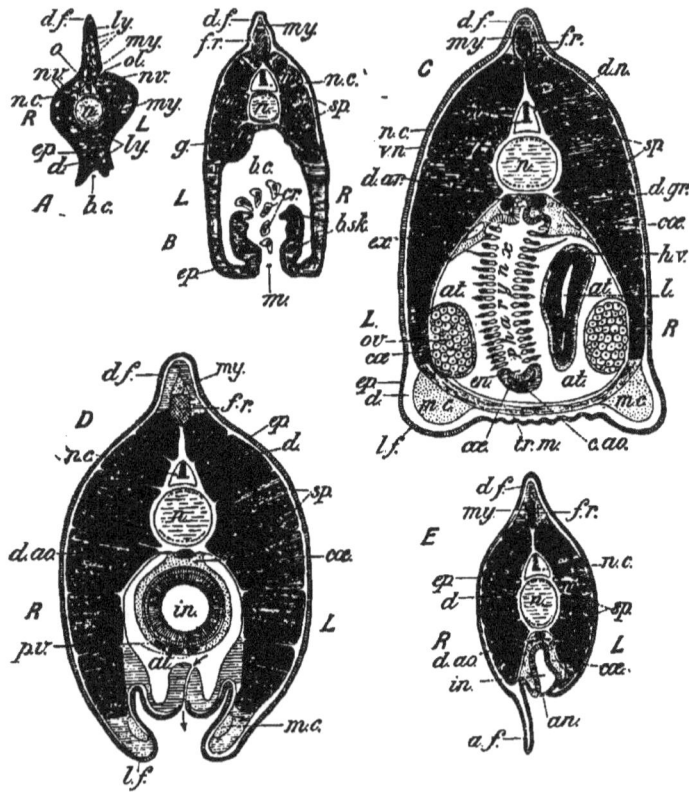

Fig. 39.—AMPHIOXUS, reduced. Transverse sections through regions of—
A, olfactory pit; B, buccal cavity; C, pharynx and liver; D, atriopore;
E, anus. All enlarged to same scale. Lymph-spaces dotted. R and
L, right and left sides. *a.f*, Anal fin; *an*, anus; *at*, atrial cavity
(atriopore shown diagrammatically in D); *b.c*, buccal cavity (just
beginning in A); *b.sk*, pieces of buccal skeleton (others seen within
sections through buccal cirri); *c.ao*, cardiac aorta; *cœ*, sections of
cœlom; *cr*, buccal cirri; *d*, dermis; *d.ao*, dorsal aorta; *d.ar*, dorsal
artery; *d.f*, dorsal fin; *d.gr*, dorsal groove of pharynx; *d.n*, dorsal
nerve; *en*, endostyle; *ep*, epidermis; *ex*, excretory tube (brown tube
of Lankester); *f.r*, fin ray; *g*, possible gustatory organ; *h.v*, hepatic
veins; *in*, intestine; *l*, liver; *l.f*, lateral fin; *ly*, lymph-spaces; *m*,
myomeres and (in B) mouth; *m.c*, metapleural lymph-canal; *my*,
myocœle, including fin-ray spaces and (in A) spaces connected with
myomeres; *n*, notochord, the dots represent nuclei—around it is
seen its sheath (left white) continued into septa and neural sheath;
n.c, nerve cord; *nv*, nerves (in A); *o*, eye; *ol*, olfactory pit; *ov*, ovary;
p.v, portal veins; *sp*, septa; *tr.m*, transverse muscles; *v.n*, ventral
nerve (represented as coming off too high up).

The *gill-slits* slope downwards and backwards, so that a number of them are cut through by a single transverse section. They are separated by lath-shaped *gill-arches*, the flat surfaces of which face one another and are covered by very long cilia. New gill-slits are continually being added, in growing animals, at the posterior end of the series as the pharynx increases in length. Each gill-slit is at first a simple oval aperture, but soon becomes horseshoe-shaped and finally divided into anterior and posterior parts by the downgrowth of a tongue-like secondary gill-arch from its dorsal margin. The 1st, 3rd, 5th, &c., arches are consequently secondary, and the alternate ones primary. It is also to be noted that the gill-slits are bridged over by numerous short horizontal bars, so that the lateral walls of the pharynx resemble open basketwork in structure.

There is a somewhat complex *branchial skeleton* supporting the parts described, and composed of firm material, conveniently termed chitinoid, since it resembles horn or chitin in physical respects, though its exact chemical nature is not known. Each gill-arch is traversed by an internally grooved rod of this kind, which in the primary arches is solid and forked at both ends, but in the secondary arches hollow and simple ended. These rods are connected by short horizontal pieces which run through the bars that bridge the gill-slits, and ventrally they come into relation with a double series of small chitinoid plates supporting the floor of the pharynx.

The pharynx is succeeded by an exceedingly short *gullet*, and this again by a fairly wide *stomach*, from which a simple blindly ending tube, the *liver*, runs forwards along the right side of the pharynx, extending further in adult than in young specimens, such as the one drawn in Fig. 38. The stomach passes gradually behind into a tubular *intestine* running straight to the anus.

Amphioxus feeds chiefly upon small organisms suspended in the surrounding water, and in the normal position of the animal —*i.e.*, vertically imbedded in the sand, with its anterior end projecting, these are swept into the mouth by the currents which the ciliated lining of the gut sets up. A certain amount of sand also appears to be swallowed for the sake of the contained organic matter. Since the anterior part of the Chordate gut has to do with respiration as well as nutrition, it is commonly specialized so that the food takes a different path from the respiratory current. In this case the epibranchial groove, judging from its contents, serves as a channel along which food passes back to the gullet.

The ciliated lining of the buccal cavity and pharynx sets up currents which not only bring in food but also the oxygen required in respiration. These currents continually stream into the pharynx, through the gill-slits into the atrial cavity, and out at the atriopore. The blood contained in the vessels of the gill-arches is thus oxygenated and at the same time gets rid of its carbon dioxide. The current emerging from the atriopore also serves to carry the sperms or ova, as the case may be, out of the body.

5. **Circulatory Organs.**—These are in a degenerate condition. There is no heart, but this is compensated for by the contractile nature of some of the blood-vessels. A distinction can be drawn between blood and lymph-systems, but these appear to communicate with one another and consequently contain the same circulatory fluid, which may be termed **blood** or **lymph** indifferently. It is colourless, and chiefly consists of coagulable plasma in which a few amœboid corpuscles are suspended.

(1) **Blood System.**—A *cardiac aorta* runs along the floor of the pharynx, giving off branches, the aortic arches, which traverse the primary gill-arches, uniting above to form a *dorsal artery* running along each side of the epibranchial groove. These two arteries are indirectly connected in front, while posteriorly they form by their union a *dorsal aorta*, which runs below the notochord. The aortic arch which is furthest forward on the right side is larger than the rest and supplies the front end of the body. A *lateral artery* runs along each side of the body, just within the gonads. It is connected by transverse vessels with the corresponding dorsal artery.

The remaining important blood-vessels are the *portal* and *hepatic veins*. The first of these run along the under side of the intestine, passing to the liver and breaking up into capillaries from which the hepatic veins arise. These run along the dorsal side of that organ and unite at its origin to form the cardiac aorta.

Course of the Circulation.—This is only imperfectly known, but is probably as follows:—Impure blood passes from the hepatic veins to the cardiac aorta, thence through the aortic arches (and connected vessels in the secondary gill-arches), where it is oxygenated, and on into the dorsal arteries and dorsal aorta for general distribution. There also appears to be an *hepatic portal system*—that is to say, the impure blood from the gut passes by portal veins to the liver and enters a capillary system drained by hepatic veins.

The motive power by which circulation is effected appears to be the contractility of the chief vascular trunks, especially of the aortic arches, each of which commences in a small contractile bulb which lies in the ventral fork of the corresponding primary gill-bar.

(2) **Lymph System.**—This consists of a body-cavity or cœlom, and of other lymph-spaces. These are continuous with one another and with the blood-system, but their exact relations are complex and but ill understood.

Cœlom (Fig. 39).—In a cross-section taken between anus and atriopore this is easily recognised as a fairly wide space surrounding the gut, but in front of this region the arrangement is complicated by the large development of the atrial cavity. Between atriopore and pharynx it is seen as a narrow space surrounding the gut and continued round the liver. In the pharyngeal region the most obvious parts of the cœlom are the *dorsal cœlomic canals* which run one on each side of the upper part of the pharynx. The floors of these canals are obliquely fluted to form suspensory folds connected with the primary gill-arches, and each containing a cœlomic pouch running down the outer side of its arch.

A ventral cœlomic canal runs below the endostyle and receives cœlomic tubes which traverse the chitinoid rods of the secondary gill-arches. Besides this there are special sections of the cœlom surrounding the gonads.

The most important lymph-spaces, in addition to the cœlom, are (1) a *metapleural canal* running along each lateral fin, (2) spaces round the fin-rays, (3) spaces in the myomeres of the head.

6. Excretory Organs.—It has been shown by experiment that the organs which, under ordinary circumstances, excrete nitrogenous waste, can also get rid of certain pigments artificially introduced into the system. In this way a means is afforded of recognizing such organs in doubtful cases. This method has been employed for Amphioxus. The living animals were kept in sea-water full of suspended carmine until they became pink in colour, after which they were transferred to clean sea-water till the colour faded somewhat owing to the action of the excretory tissue. Sections cut at this stage showed the presence of carmine in the cells of (1) the atrial epithelium, more particularly that part of it covering the outside of the secondary gill-arches and the atrial floor, (2) a series of small tubules, previously overlooked, opening (?) from the dorsal cœlomic canals into the atrial cavity at the tops of the primary gill-arches. These tubules, and part of the atrial epithelium have probably, therefore, an excretory function.

Besides this, a pair of *atrio-cœlomic funnels* (pigmented canals) occur in the 27th segment, which are possibly of similar nature. Each of them is a short funnel-shaped tube, situated in the dorsal cœlomic canal of its side (Fig. 39), opening by its wide posterior end into the atrial cavity and (?) by its narrow front end into the cœlom. Owing to observation being rendered difficult by the abundant pigment naturally found in the walls of these tubes, the carmine method just described gave no positive results in this case.

7. **Reproductive Organs** (Fig. 39).—Although the sexes are separate there are no distinctive external characters, and the reproductive organs are of the simplest possible kind, consisting of 26 pairs of gonads, **spermaries** (testes) or **ovaries**, as the case may be, imbedded in the outer wall of the atrial cavity. Those of one side are not exactly opposite those of the other side. The gonads are squarish sacs, each of which is surrounded by a special section of the cœlom. The **ova** are just visible to the naked eye, and when mature are dehisced into the atrial cavity, out of which they are carried by the respiratory current flowing through the atriopore. The very much smaller **sperms** (spermatozoa), which reach the outside of the body in a similar manner, are tadpole-shaped, each of them possessing a short conical head and a vibratile tail.

8. The **muscular system** (Figs. 38 and 39) exhibits well-marked segmentation, for the great lateral mass of muscle on each side is divided into a series of (in *A. lanceolatus*) 61 or 62 > shaped *muscle-segments* (myomeres, myotomes), separated by connective-tissue septa. Those on opposite sides do not correspond. The constituent muscle-fibres run longitudinally. Swimming movements, consisting in bending the body first to one side and then

to the other, can be effected by alternate contraction of the lateral muscles on each side.

Transverse muscles run across the floor of the atrial cavity, and these no doubt assist in the expulsion of water through the atriopore.

The muscle-fibres are transversely striated rhomboidal plates.

9. The **nervous system** consists of a dorsal nerve-cord and of nerves connected with this.

The **nerve-cord** (spinal cord) constituting the central part of the nervous system is a thick-walled tube exactly fitting the neural sheath above the notochord, and extending the greater part of the length of the body. In front, however, it does not reach so far forwards as the notochord, but terminates bluntly just above the front end of the mouth. The cavity of the nerve-cord forms an exceedingly narrow central canal, except anteriorly, where it dilates into a much larger *ventricle* which opens to the exterior on the left-hand side within the olfactory pit.

Behind the ventricle, a narrow slit, the *dorsal fissure*, divides the upper part of the cord, above the central canal, into right and left halves.

The nerve-cells are grouped round the central canal and dorsal fissure; in a stained section they appear as a deeply-coloured dorso-ventral streak, readily seen under the low power. As in other cases they are produced into processes by which union with one another and with nerve-fibres is effected. The smallest nerve-cells have but one process (*i.e.*, are *unipolar*), the larger ones generally have several processes (*i.e.*, are *multipolar*). A double longitudinal series of small irregular masses of black pigment-cells is imbedded in the floor of the central canal. The greater part of the nerve-cord is made up of slender *nerve-fibres*, most of which take a longitudinal direction, while others pass out into the nerves.

Numerous segmentally-arranged **nerves** take origin from the nerve-cord and constitute a *peripheral* nervous system. Most of them correspond to myomeres, being consequently arranged asymmetrically and not in regular pairs. The nerves are of two kinds—(1) single-rooted and (2) multiple-rooted. The *single-rooted nerves* (except the first two) arise from the dorsal surface of the cord as single bundles of nerve-fibres, and the first six of them constitute three regular pairs. Behind this, however, those

146   AN ELEMENTARY TEXT-BOOK OF BIOLOGY.

on one side alternate, like the myomeres, with those on the other side. The first pair have a ventral origin from the extreme front of the nerve-cord, and, together with the second pair, which arise dorsally behind the ventricle supply the parts in front of the mouth, while the third pair supply the mouth-region. These single-rooted nerves are partly *sensory*, innervating the skin, and partly *motor*, giving off twigs to the muscles. Their sensory branches divide repeatedly, and their finest ramifications ultimately unite together to form a delicate *nerve-plexus* below the epidermis. Such a nerve-plexus, investing the whole body, is characteristic of certain non-chordates, and must be regarded here as a primitive feature inherited from non-chordate ancestors.

The *multiple-rooted nerves* all arise from the ventral side of the nerve-cord for the greater part of its extent, alternating with the dorsal nerves. Each of them consists of a longitudinal series of slender bundles (roots) which remain independent of one another, and branch out in the adjacent muscles. These nerves, therefore, are of *motor* nature.

10. The **sense-organs** of Amphioxus are ill-developed and for the most part of problematical nature.

A considerable number of the epidermal cells, especially in the anterior part of the body, probably serve as *tactile organs*. These cells are comparatively slender, continuous internally with sensory nerve-fibres and produced externally into a stiff process.

A *gustatory* function has been ascribed to the following structures :—(1) Small aggregates of sense-cells, similar to those described in the last paragraph and situated upon the buccal cirri. The external processes of each group form a conical projection. (2) Circlets of sense-cells occurring in the velar tentacles. (3) A sac lined by sensory epithelium and opening on the left side of the roof of the buccal cavity.

It is exceeding doubtful whether the so-called *olfactory pit*, situated on the left side of the head, has anything to do with smell. It is a depression lined with ciliated epithelium and communicating with the ventricle.

A pigmented mass imbedded in the front wall of the ventricle has received the name of *eye*.

DEVELOPMENT.

The breeding-season commences about the end of March and lasts throughout the summer. The eggs are laid about an hour after sunset, and are fertilized at once by sperms shed over them by the male. Development begins an hour later and at first goes on very rapidly, so that by sunrise next morning free-swimming flagellated embryos escape from the eggs. Twenty-four hours later the flagellated embryo has become an elongated unsymmetrical larva, with mouth, anus, and one gill-slit. This completes what may be termed the *embryonic development.*

The *larval development* which now succeeds extends over a much longer period. The free-swimming larva is for some time extremely unsymmetrical, but gradually becomes less so. At the same time the adult structure is slowly assumed; ultimately the free-swimming life is given up, and this closes the larval period.

I. EMBRYONIC DEVELOPMENT. **Cleavage** (Segmentation).— The ovum, in spite of its small size ($\frac{1}{250}$ of an inch in diameter), contains a considerable number of yolk-granules. It is covered by a delicate vitelline membrane, which becomes separated from it as soon as fertilization has been effected. Only one polar body has been observed, and this rests on the upper pole of the oösperm. Cleavage is complete (holoblastic) and nearly regular. It occupies from three to four hours. The first division takes place in a vertical plane and results in two cells of equal size. These are then equally halved by a second vertical division at right angles to the first. The third division is horizontal (equatorial) and separates four rather smaller upper cells from four rather larger lower cells in which more numerous yolk-granules are present. These eight cells are now bisected by two vertical divisions making angles of 45° with the earlier vertical ones. The 16-celled stage so constituted becomes a 32-celled one by two new equatorial divisions, the planes of which are respectively above and below the first equatorial division. From this point cleavage proceeds more irregularly. It ultimately results in a spherical **blastula** (blastosphere), which may be compared to a hollow ball with a large central cavity (blastocœle, segmentation cavity) and a wall composed of a single layer of cells, which in the upper half are smaller than in the lower half (Fig. 40). The former constitute the **ectoderm** (epiblast), the latter the **endoderm** (hypoblast).

Gastrulation (Fig. 40), by which the single-walled blastula is converted into the double-walled **gastrula**, follows cleavage and occupies about six hours. The endoderm cells undergo a gradual inpushing or invagination (emboly, embolic invagination) until the blastocœle is obliterated, and the embryo has become cup-shaped, with a central digestive cavity (archenteron) opening by a wide blastopore. The endoderm cells now line the archenteron and are covered by the ectoderm, each cell of which develops a flagellum. The cup-shaped gastrula soon assumes an ovoid shape with a flattened dorsal and a convex ventral surface. Meanwhile

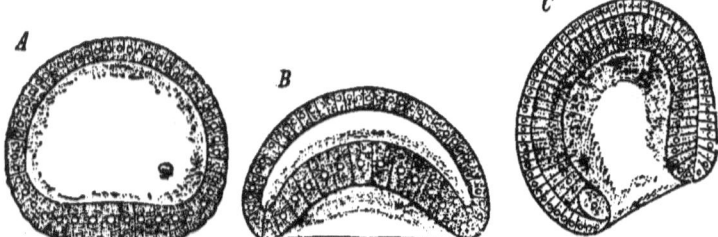

Fig. 40.—BLASTULA AND GASTRULA OF AMPHIOXUS (from *Claus*, after *Hatschek*).—A, In optical section. A, Blastula with flattened lower pole of larger cells. B, Commencing invagination. C, Gastrulation completed; the blastopore is still widely open, and one of the mesodermic teloblasts is seen at its ventral lip. The flagella of the ectoderm cells are not represented.

the blastopore has narrowed to a small rounded aperture, and is now situated at the posterior end towards the dorsal surface. Two endoderm cells on the ventral side of the blastopore are distinguished by their relatively large size. These are the mesodermic teloblasts, which subsequently originate a part of the mesoderm.

The completed gastrula escapes from the vitelline membrane and swims freely, front end first, by means of the ectodermal flagella.

Origin of the Mesoderm, Cœlom, Muscles, Notochord, and Nervous System (Figs. 41 and 42).—The foundations of all these are simultaneously laid during the twenty-four hours which succeed hatching.

The **mesoderm** in the anterior part of the body is constituted by the walls of *myocœlomic pouches* which successively grow out from the archenteron. The first of these pouches is unpaired and median, the remainder are paired and lateral. They develop in order from before backwards. All these outgrowths become separated from the gut (which may now be called the mesenteron), and are known as *mesodermic* (mesoblastic) somites, since they indicate the segmentation of the body. Other such somites are added later as endodermic outgrowths, but their cavities never communicate with that of the gut. The mesodermic teloblasts give rise to part of the mesoderm of the tail-region. The somites soon become divided into dorsal sections, the cavities of which collectively constitute the *myocœle*, and ventral sections, the cavities of which collectively constitute the *splanchnocœle*. The outer wall of each section consists of *somatic* mesoderm, and its inner wall of *splanchnic* mesoderm. The splanchnic walls of the dorsal sections are converted into the lateral muscles, and their somatic walls into the dermis. The sheath of the notochord, neural sheath, and septa are formed from outgrowths of these dorsal sections. The myocœle is mostly obliterated in the adult, except in the

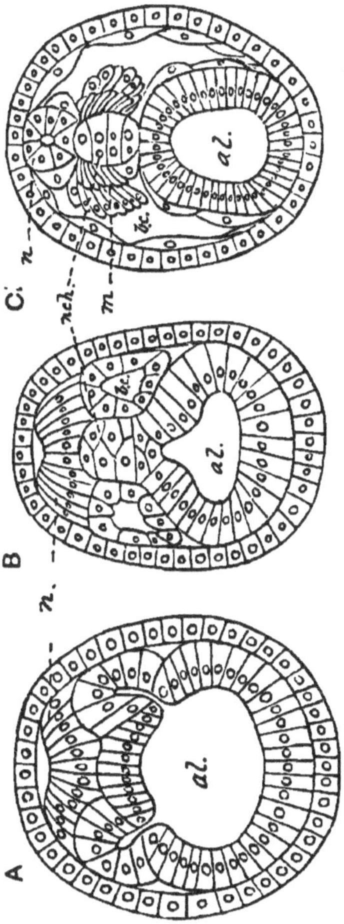

Fig. 41.—TRANSVERSE SECTIONS OF AMPHIOXUS EMBRYOS (from *Hadlon*, after *Hatschek*).—A, Section through the first mesodermic somite of an embryo in which the 5th pairs of m. somite are being formed. B, Section through the same region of an embryo with 8 pairs of m. somites. C, Section through the centre of the body of an embryo with 11 somites. *al*, Gut; *b.c*, body cavity, not yet divided into myocœle and splanchnocœle ; *n*, neural plate and canal; *nch*, notochord.

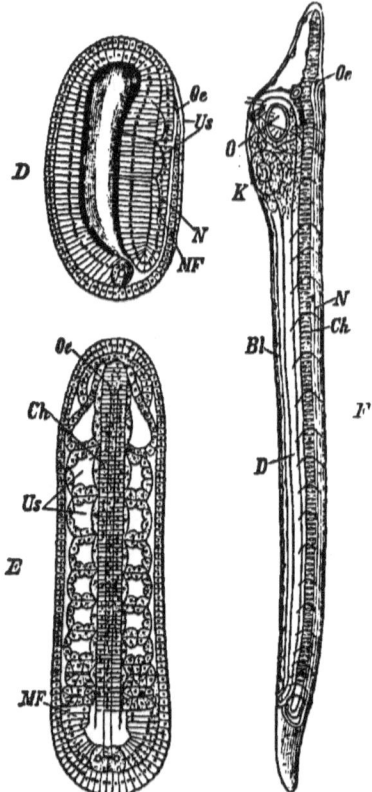

Fig. 42.—THREE LARVAL STAGES OF AMPHIOXUS (from *Claus*, after *Hatschek*).—D, Stage with 2 pairs of mesodermic somites, seen in optical longitudinal section, dorsal side to right. E, Stage with 9 pairs of mesodermic somites, seen from above. F, Living larva with mouth and first gill-slit, seen from the left side; the 2nd, 4th, and 6th bent lines represent respectively the posterior boundary of the 1st, 2nd, and 3rd somite of the opposite side. *Bl*, ventral blood-vessel; *Ch*, notochord; *D*, intestine; *K*, gill-slit; *MF*, unsegmented mesoderm fold, behind which one teloblast is seen in D, both in E; *N*, neural canal; *O*, mouth; *Oe*, anterior opening of neural canal; *Us*, mesodermic somites.

head (see Fig. 39), but portions of it are converted into the fin-ray lymph-spaces and the metapleural canals. The ventral sections of the somites grow downwards, ultimately fusing below the gut, the wall of which, outside the epithelium, is formed from their splanchnic layer. The splanchnocœle becomes the **cœlom**, which is later on split up into sections. A body-cavity, which, like this, is derived by the outgrowth of myocœlomic pouches, is said to be *enterocœlic*, and may be called an *enterocœle*. Such a simple mode of origin is probably a very primitive one.

The anterior unpaired archenteric pouch divides into right and left halves, of which the latter becomes the problematic sense organ that, in the adult, opens on the roof of the mouth.

The **notochord** is developed as a longitudinal fold of the archenteric wall in the mid-dorsal line.

The **central nervous system** arises in a way which, in its more general features, is characteristic of all Chordates. The ectoderm covering the flattened dorsal surface of the completed gastrula becomes marked off by a slight furrow from the lateral ectoderm on each side, and constitutes a *neural* (medullary) *plate*, which becomes depressed in the centre so as to originate a *neural* (medullary) *groove* bounded by *neural* (medullary) *folds*. These folds gradually approach each other, and ultimately unite, so that the trough-shaped neural plate, now V-shaped in transverse section, is completely covered. It ultimately folds up into a tube, which is converted into the nerve-cord by thickening of its walls. A small opening, however, is left in front, and this *neuropore* is apparently converted into the olfactory pit of the adult. Posteriorly the union of the neural folds involves roofing over the blastopore with consequent formation of a short *neurenteric canal* by which the cavity of the nerve-tube communicates with the digestive cavity, which now no longer opens directly to the exterior.

The embryonic development is brought to a close by the appearance of the larval mouth, anus, and first gill-slit.

The **larval mouth** (= velar opening of adult) appears as a small round ciliated opening on the left side of the head, in the region of the first myomere. The **anus** develops soon afterwards. The **first gill-slit** is formed on the right-hand side, close to the median line, as a ciliated rounded aperture leading out of the pharyngeal region.

II. LARVAL DEVELOPMENT.—The appearance of the larva at the beginning of this period may be gathered from Fig. 42, F. The points of most importance in the further history are connected with the development of the adult mouth and buccal cavity, gill-slits, and atrial cavity.

Adult Mouth and Buccal Cavity.—The larval mouth becomes wide and oval, and two folds grow out, one above it, the other below it, which respectively become the left and right boundary walls of the buccal cavity. Later on the mouth shifts round to a median ventral position, while the two folds become at the same time more prominent, fusing together so as to enclose a cavity, the buccal cavity, outside the larval mouth, and provided with an oval opening, the adult mouth, from the margins of which buccal cirri grow out.

Gill-Slits.—It is convenient to classify these, according to the order of their appearance, as primary, secondary, and tertiary.

(1) *Primary Slits*.—The first of these, already mentioned, is succeeded by 13 others, developed on the right side of the mid-ventral line, and passing well up the right side of the body.

Later on, the first and some of the posterior ones close up, while the others gradually shift round across the mid-ventral line to the left-hand side of the body and become the anterior slits of that side.

(2) *Secondary Slits*—7, 8, or 9 in number—appear on the right-hand side of the body *above* the primary ones, the first slit corresponding with the gap between the 2nd and 3rd primaries. So many of the primaries close up as is necessary to make their number equal to that of the secondaries. Thus if there are 8 of these last, 6 of the primaries close, as follows, using Roman numerals for the primaries, arabic numerals for the secondaries, and brackets to indicate closure:—

$$8 \quad 7 \quad 6 \quad 5 \quad 4 \quad 3 \quad 2 \quad 1$$
[XIV] [XIII] [XII] [XI] [X] IX VIII VII VI V IV III II [I]

The secondary slits become the anterior slits of the right side.

(3) *Tertiary Slits* are formed on each side behind those already developed. New ones are added as long as the animal continues to grow in length.

Primaries, secondaries, and tertiaries alike become divided into two by the downgrowth of secondary bars, as described on p. 141.

Atrial Cavity.—The primary gill-slits at first open directly to the exterior, but later on into a groove which is bounded by a ridge on each side. The groove is then converted into a narrow canal, open at both ends, by the fusion of two shelf-like subatrial folds which grow out from the ridges. This canal, which becomes the atrial cavity, sinks as it were into the body, gradually extending round the gut so that the new gill-slits open into it. The anterior opening soon closes, but the posterior one remains as the atriopore.

CHAPTER IX.—PISCES (Fishes).

THE lowest group of Vertebrates, using that term in the more restricted sense (*see* p. 135), is constituted by Fishes, which are again arranged in smaller subdivisions, as follows:—

1. *Cyclostomata*—lampreys and hags.
2. *Elasmobranchii*—dogfish, shark, skate.
3. *Ganoidei*—sturgeon.
4. *Teleostei*—herring, perch, cod, eel.
5. *Dipnoi*—mudfishes.

The last group is a small and highly specialized one, including a few fresh-water forms which breathe by gills and also by a lung-like swim-bladder. It includes *Ceratodus*, in Australia; *Protopterus*, in Africa; and *Lepidosiren*, in S. America.

The large majority of recent fishes belong to the *Teleostei*, some of the most obvious characteristics of which are—an externally symmetrical tail, terminal mouth, thin scales, comb-like gills protected by a firm gill-cover, and a well-ossified endoskeleton.

The *Ganoidei* are represented at the present time by a small number of genera widely distributed in the fresh waters and estuaries of the globe. The group was once large and important, but is now approaching extinction. The recent Ganoids, of which the best known is the one (Acipenser) including the sturgeon, form a very heterogeneous assemblage, not closely related, and difficult to include in a common definition.

Cyclostomata are limbless, jawless fishes with a suctorial mouth.

Elasmobranchs are a very ancient type, and though common at the present day are relatively far less abundant than they were in former geological epochs. The skates are a good deal specialized, but dogfishes and sharks present the features of the fish-type in a comparatively unmodified condition, and are, therefore, better general illustrations of the group of fishes than members of the teleostei. Dogfishes, which may be regarded as small sharks, are represented on the British coasts by several genera, of which **Scyllium**, the Spotted Dogfish, is perhaps the commonest. **S. canicula**, the one usually dissected in laboratories, has an average length of about two feet; **S. catulus** is much larger. The following account will apply to either.

§ 14. SCYLLIUM (Dogfish).

MORPHOLOGY AND PHYSIOLOGY.

1. External Characters.—The elongated spindle-shaped body, eminently adapted for rapid progression through water, exhibits complete bilateral symmetry and is divisible into head, trunk, and tail, between which there are no sharp lines of demarcation. The head is flattened from above downwards, and ends anteriorly in a rounded snout. The trunk and tail are laterally flattened, while the latter is extremely long and very narrow in its posterior part.

A number of thin flat **fins** are present, some unpaired and situated in the median plane, others paired and lateral. They are all supported by an internal skeleton. The **unpaired fins** are four in number—two dorsals, a caudal, and an anal. They are to be looked upon as surviving portions of a continuous expansion which in ancestral forms probably ran along the dorsal surface, round the tail, and forwards for some distance along the ventral surface (*cf.* Amphioxus, p. 136). The *first dorsal* is a small triangular flap commencing about half-way back along the upper surface; not far behind it is a similar but smaller *second dorsal*. The *caudal fin* fringes the tail and is markedly asymmetrical (heterocercal). It consists of a square-ended upper lobe into which the upwardly bent end of the body is continued, and a rather broader lower lobe. The *anal fin* projects from the ventral surface opposite the space between the first and second dorsals.

The **paired fins** are four in number, and are homologous to the fore and hind limbs of terrestrial Vertebrates. They are probably specialized portions of continuous lateral fins which existed in ancestral forms (*cf.* Amphioxus, p. 136). The anterior pair, or *pectoral fins*, project horizontally from the sides of the broadest region of the body, and mark the junction of head and trunk. Each is a broad flat plate with dorsal and ventral surfaces, and (when it is pulled slightly outwards) anterior (pre-axial), posterior (post-axial), and external margins.

The much smaller *pelvic fins* are attached to the ventral side of the body, half-way between the snout and beginning of the caudal fin. Their post-axial margins touch each other in the female and are fused together in the male. A ready means of distinguishing the sex is thus afforded, and further, in the male, a part of each pelvic fin is converted into a grooved rod or *clasper*, which functions as a copulatory organ.

Apertures.—The *mouth* is a large crescentic slit on the under side of the head; the *cloacal aperture* is an elongated opening between the pelvic fins. There is a small *abdominal pore* on each side of this aperture leading into the body-cavity. The remaining external openings are all in the anterior part of the body. Upon the under side of the snout the *external nares* or *nostrils* are seen as large rounded openings, from each of which a groove, covered by a fold of skin, leads back to the mouth.

The oblique eyes, provided with upper and lower eyelids, are placed on the sides of the head above the posterior corners of the mouth. Just behind each of them is a small round hole, the *spiracle*, opening out of the pharynx. It is of the same nature as five oblique *gill-slits* which are seen further back immediately in front of the pectoral fin.

Two minute apertures on the top of the head communicate with the auditory organs. A large number of regularly-arranged pores can be seen upon the head, especially in its anterior part. They are the openings of *sensory tubes* (jelly-tubes, mucous canals), which lie under the skin. Another sensory structure underlies the groove-like *lateral line* which runs along each side of the body.

The dogfish is of a whitish colour ventrally, grey with dark brown spots dorsally and laterally. The fins are spotted as well as the body. This colouration must make the animal extremely inconspicuous when seen from above in its natural surroundings.

The body is entirely covered by small sharp, placoid scales, imbedded in the skin, but with projecting, backwardly directed points. The scales near the mouth closely resemble the teeth.

2. The **skin** consists of an *epidermis* composed of stratified epithelium, and an underlying *dermis* made up of connective-tissue traversed by blood-vessels, lymphatics, and nerves.

The only important *glands* connected with the skin, if the jelly-tubes are excluded, are the clasper-glands of the male, each of which is a pouch underlying the skin between the pelvic fins and opening backwards into the groove of the corresponding clasper.

The small *placoid scales* are developed in the skin. Examined with a powerful lens, or under a low power of the microscope, each of them is seen to consist of a four-rayed basal plate, and of a much larger spine attached to it. The end of the spine is leaf-shaped and directed obliquely backwards with its flat side uppermost. The basal plate is bone-like, the spine composed of hard *dentine*, capped by exceedingly hard *enamel* secreted by the epidermis. The rest of the scale is developed by the dermis.

The very young scales are covered by the epidermis, through which the spines later on force their way.

3. Endoskeleton.—In an embryo dogfish a firm cellular rod, the notochord, underlies the central nervous system, much as in Amphioxus, but in this case only extending as far forwards as the middle of the brain. A cartilaginous sheath is soon formed round the notochord; in the head-region the floor of a firm cartilaginous brain-case is developed in connection with this; in the trunk and tail the sheath is transversely segmented into a series of joints, flexibility being thus much increased. Further modifications and additions result in a complicated endoskeleton, composed almost exclusively of cartilage.

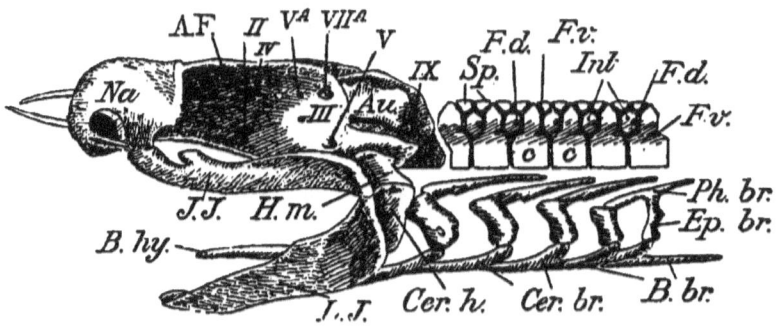

Fig. 43.—DOGFISH. Skull and part of vertebral column (reduced).—*Au*, Auditory capsule; *Na*, nasal capsule; *II.*, *III.*, *IV.*, &c., nerve-exits; *J.J*, upper jaw; *L.J*, lower jaw; *H.m*, hyo-mandibular; *Cer.h*, ceratohyal; *B.hy*, basi-hyal; *Ph.br*, pharyngo-branchials; *Ep.br*, epibranchials; *Cer.br*, cerato-branchials; *B.br*, basi-branchials; *c*, centra, continued up into neural plates; *Sp*, neural spines; *Int*, intercalary pieces; *F.d*, *F.v*, foramina for dorsal and ventral roots of spinal nerves, indicated for two nerves.

It is convenient to consider the skeleton under two headings: I. Skeleton of the body—axial skeleton; II. Skeleton of the fins —appendicular skeleton.

I. *Axial Skeleton* (Fig. 43).—This is divisible into (1) Skull, (2) Vertebral Column and Ribs.

(1) The **skull** is characteristic of Vertebrata proper (Vertebrata Craniota). It includes a cranium or brain-case—olfactory and auditory capsules; and the visceral skeleton, consisting of jaws and respiratory skeleton.

The **cranium** is a somewhat rectangular box of cartilage enclosing the brain, and incomplete dorsally, where there is a large gap or fontanelle closed by membrane. Posteriorly the cranium articulates immovably with the vertebral column by two rounded projections (condyles), between which is a large aperture (foramen magnum), where the spinal cord and brain are united. There are nerve-exits in front and at the sides, while the cranial floor is continued forwards as a median nasal septum bearing in front a slender pointed rod.

The **olfactory capsules** are large thin-walled structures separated from one another by a median septum, open below, and fused with the front of the cranium. The **auditory capsules** are much firmer. They enclose the organs of hearing, and are fused with the sides of the cranium in its hinder region, each of them appearing as a squarish projection.

Visceral Skeleton.—In the embryo dogfish seven thickenings, *visceral arches*, appear on each side of the neck, and between them six openings, *visceral clefts*, placing the cavity of the pharynx in communication with the exterior. The arrangement may be indicated as follows for the left side, using strokes for the arches and numbers for the clefts; the arrow points to the front.

⟵   \ 1 \ 2 \ 3 \ 4 \ 5 \ 6 \

Beginning in front, the arches are termed mandibular, hyoid, 1*st*, 2*nd*, 3*rd*, 4*th*, and 5*th* branchials. The first cleft is the hyomandibular, and becomes the spiracle; the rest are named like the arches which bound them behind, and become the gill-clefts.

Curved supporting rods of cartilage are developed in the visceral arches, and become converted into the visceral skeleton.

The *mandibular bars* are converted into two cartilages which support the lower jaw. From the upper end of each a forward outgrowth is developed, which becomes separated off and supports the corresponding half of the upper jaw.

The *hyoid bars* are segmented into upper pieces, the *hyomandibular cartilages*, which suspend the jaw-cartilages from the auditory region of the skull, and lower pieces (*cerato-hyals*) which unite with a median ventral *basi-hyal* cartilage. A skull like this, in which the jaws are suspended by means of hyomandibulars, is said to be *hyostylic*. The upper jaw is also connected with the skull by two strong fibrous bands (ligaments), one in front (ethmopalatine ligament) and one behind (pre-spiracular ligament) running in front of the spiracle and containing a small pre-spiracular cartilage.

Each of the five branchial rods on each side becomes jointed into a series of segments, named, from above downwards, *pharyngo-, epi-, cerato-,* and *hypo-branchials*. A median ventral *basi-branchial* cartilage lies between and connects together the arches of opposite sides.

Slender cartilaginous *gill-rays* for the support of the gill-folds radiate backwards from the posterior margins of the cerato-hyals and first four pairs of cerato-branchials.

The visceral skeleton includes a few other cartilages besides the above. They are (1) a pair of rod-like *labials* at each corner of the mouth, (2) three flattened rods, the *extra-branchials*, on each side, external to the three middle branchial arches.

(2) **Vertebral Column** and **Ribs**.—The vertebral column consists of a series of joints or vertebræ, which are united together to form a flexible rod. The vertebræ are of two kinds, trunk vertebræ and tail vertebræ; the latter are the more typical, and each of them consists of a deeply biconcave (amphicœlous) *body* or *centrum* lined by bone, a *neural arch* under which the spinal cord runs, and a *hæmal arch* protecting the blood-vessels of the tail, and produced downwards into a *hæmal spine*. The trunk vertebræ chiefly differ in the absence of complete hæmal arches; those in front possess horizontal projecting *transverse processes* which bear short **ribs**; those behind have downwardly projecting *hæmal processes* (= transverse processes of those in front, and sides of hæmal arches of tail vertebræ). The *neural arches* of both kinds of vertebra are similar. Each consists of four pieces, a lateral *neural plate* projecting from the centrum on each side, and two small rounded dorsal cartilages (*neural spines*) placed one behind the other. The spaces between the neural plates are filled up by hexagonal *intercalary cartilages*.

The *notochord* persists throughout the entire extent of the vertebral column, but by the development of the vertebral centra is deeply constricted at regular intervals, though not quite divided into a series of separate sections.

II. *Appendicular Skeleton*.—Each **unpaired fin** is in typical cases supported by a series of cartilaginous rod-like *fin-rays*, bearing small plates of cartilage at their distal ends. The fin-skeleton is completed by a large number of horny fibres developed in the skin and running in the same direction as the fin-rays.

The skeleton of either anterior or posterior **paired fins** is divisible into a proximal part, the *limb-girdle*, connected with the body, and a distal part which supports the free limb. (1) *Pectoral fins* (Fig. 44).—Each *shoulder girdle* is a curved flat cartilage

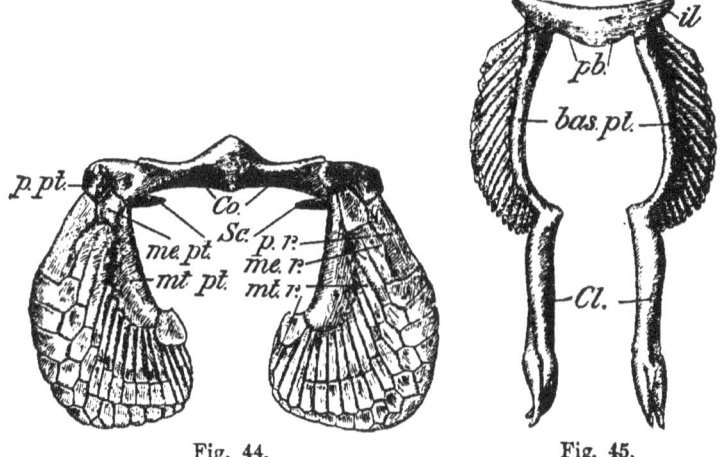

Fig. 44.   Fig. 45.

Fig. 44.—DOGFISH. Skeleton of pectoral fins seen from below (reduced).— *Co*, Coracoid region; *Sc*, scapular region; *p.pt*, propterygium; *me.pt*, mesopterygium; *mt.pt*, metapterygium; *p.r*, *me.r*, *mt.r*, corresponding fin-rays.

Fig. 45.—DOGFISH. Skeleton of pelvic fins seen from below (reduced).— *Pb*, Pubic region; *il*, iliac process; *bas.pt*, basipterygium, continued back into *Cl*, clasper skeleton. Fin-rays seen externally.

consisting of a dorsal half, the *scapular region*, and a ventral half, the *coracoid region*, the junction of the two being marked by the attachment of the free limb. The two girdles are fused together ventrally. The base of the free fin is supported by three cartilages, named from before backwards *propterygium, mesopterygium*, and *metapterygium*, the last being much the largest. To these elements a number of fin-rays succeed, one propterygial, one mesopterygial split into two or three distally, and about a dozen metapterygial. The fin-skeleton is completed by rows of polygonal plates and by horny fibres like those of the unpaired fins. (2) *Pelvic fins* (Fig. 45).—The two *pelvic girdles* are fused together into a transverse bar of cartilage lying a little way in front of the cloaca. Most of the bar (*pubic region*) lies between the attachment of the free fins, but there is a small projection (*iliac process*) external to this on each side.

The inner side of the free limb is supported by an elongated *basipterygium* (= metapterygium) which in the male is continued into the clasper. A series of fin-rays are attached to the outer side of this cartilage, and one ray directly to the girdle. The fin is completed by small plates of cartilage and by horny fibres.

4. The **digestive organs** (Fig. 46) consist of the gut or alimentary canal running from mouth to cloacal aperture, and of appended glands. The sections of the gut are mouth-cavity, pharynx, stomach, intestine, and cloaca. The glands are the liver and pancreas.

The margins of the mouth are beset with several rows of small, sharply-pointed *teeth*, which must be regarded as modified placoid scales. The **mouth-cavity** is spacious, and upon its floor there is an ill-developed *tongue*, supported by the basi-hyal cartilage, and with a forwardly-directed rounded end. The **pharynx** or respiratory section of the gut, which next succeeds, communicates with the exterior by means of the spiracles and gill-slits, and merges into a short, wide **gullet** (œsophagus) which enters the abdominal cavity, and is there continuous with a large U-shaped **stomach**. This is followed by the **intestine**, which is divided into (*a*) a short, moderately-large *bursa Entiana;* (*b*) a much larger and longer section, into which a shelf-like spiral valve projects; and (*c*) a short narrow *rectum* opening into a good-sized **cloaca** which also receives the excretory and genital ducts.

The **liver** is a large brown organ attached to the front end of the abdominal cavity and divided into two long backwardly-directed lobes. The secretion of the liver (bile) is carried away by a *bile-duct* which opens into the middle section of the intestine, on the right side, not far from the beginning of the spiral valve. A large *gall-bladder* connected with the duct is imbedded in the left lobe of the liver near its origin.

The **pancreas** is a small, pale, flattened gland situated in the angle between the stomach and the bursa Entiana. The *pan-*

creatic duct is a short tube carrying off the pancreatic secretion and opening into the left side of the intestine about the same level as the bile-duct.

A short tube with thickened walls, the **rectal gland**, opens into the dorsal side of the rectum.

The **abdominal cavity**, in which most of the digestive organs are contained, is lined by a thin membrane, the *peritoneum*, which leaves the body-wall in the median dorsal line to form a double sheet, the *mesentery*, the halves of which diverge and wrap round the gut, liver, &c., constituting suspensory folds, which, however, are for the most part very incomplete.

The dogfish is a very voracious animal, feeding upon other fishes, crustacea, and molluscs. In some districts, at any rate, it is especially abundant during the herring season. The rows of sharp backwardly-pointed teeth assist in securing the prey. By means of contractions of the muscular walls of the gut the food is gradually passed backwards, and the force expended during this process, combined with the softening and chemical action of the digestive juices, serves to disintegrate it. The chief digestive juices are the gastric juice, pancreatic juice, and bile, of which the first is secreted by small glands in the wall of the stomach and contains a ferment which converts proteids into soluble diffusible peptones. The pancreatic juice, also by ferment action, completes the digestion of proteids, converts starch into sugar, and emulsifies fats. Bile assists in the last kind of digestion.

The digested food diffuses into the blood-vessels and lymphatics which ramify in the wall of the gut. An increased absorptive surface is given by the spiral valve, which also prevents the contents of the intestine from passing backwards too rapidly. The comparative shortness of the gut is correlated with the easily digestible animal diet.

5. The **circulatory organs** of the dogfish comprise (I.) a blood system, and (II.) a lymphatic system.

(I.) The **blood system** (Fig. 46) is a closed set of tubes containing red **blood**, consisting of coagulable plasma, in which *colourless corpuscles* and *red corpuscles* are suspended. The former are amœboid and nucleated, the latter are oval discs, containing a well-marked nucleus, and coloured red by hæmoglobin.

A heart, arteries, veins, and capillaries can be distinguished.

The **heart** lies in a pericardial cavity which is separated from the abdominal or peritoneal cavity by a transverse septum, but the two cavities communicate with each other by a pair of pericardio-peritoneal canals. The pericardium is immediately above the ventral portion of the pectoral girdle, and its dorsal wall is supported by the basi-branchial cartilaginous plate.

The heart, which contains only impure blood (*i.e.*, blood poor in oxygen, and loaded with CO_2), consists of sinus venosus, a single auricle, a single ventricle, and a conus arteriosus. The *sinus venosus* is a transverse tube which receives blood at its two ends from the chief venous trunks. From this the blood passes through a valved aperture into the *auricle*, which is by far the largest division of the heart, and occupies the dorsal half of the pericardial cavity. The thin walls of the auricle are provided with a plentiful meshwork of muscles, which by their contraction drive the blood through a valved auriculo-ventricular aperture into the ventricle which lies on the ventral wall of the pericardium. The *ventricle* is a nearly globular sac with a very thick muscular wall, which by its contraction drives the blood forwards into the conus through a valved aperture. The *conus arteriosus* is a muscular tube running horizontally forward from the ventricle to the anterior wall of the pericardium, from which point it is continued forwards by the cardiac aorta. Within the conus are two series of valves, viz.:—the series already referred to guarding the aperture from the ventricle, and a

---

Fig. 46.—DOGFISH. General dissection of ♀, semi-diagrammatic (reduced).—*na*, Nostrils; *s.t*, openings of sensory tubes; *gl*, gill-slits, widened in both directions by means of scissors; *p.f*, right pectoral fin; *pl.f*, right pelvic fin—the left one is cut away; *cl*, cloacal aperture; *ab.p*, abdominal pore; the tail has been cut off. The floor of the mouth-cavity and pharynx has been cut through and its left half removed, and the walls of these cavities have been dissected to show blood-vessels; *g*, cut end of gullet; *st*, stomach; *int*, intestine cut open to display spiral valve; *rct*, rectum; *lr*, liver, the bile-duct, *b.d*, is seen crossing the bursa Entiana; *pa*, pancreas, the pancreatic duct is seen at ×; *rct.gl*, rectal gland. Heart, *ht*, with ventral aorta, *v.ao*, and afferent branchial arteries (shaded with transverse lines). The dorsal aorta is seen to be formed by the union of efferent branchial arteries (darkly shaded), arising from loops surrounding gill-clefts, × × × × ×; *sp*, internal opening of spiracle; *ca*, carotid; *a.ca*, anterior carotid; *scl*, subclavian; *cœ*, cœliac; *a.m*, anterior mesenteric. The ovary has been removed; *ovd*, right oviduct with oviducal gland, *od*; *ovd'*, common abdominal opening of oviducts.

PISCES.

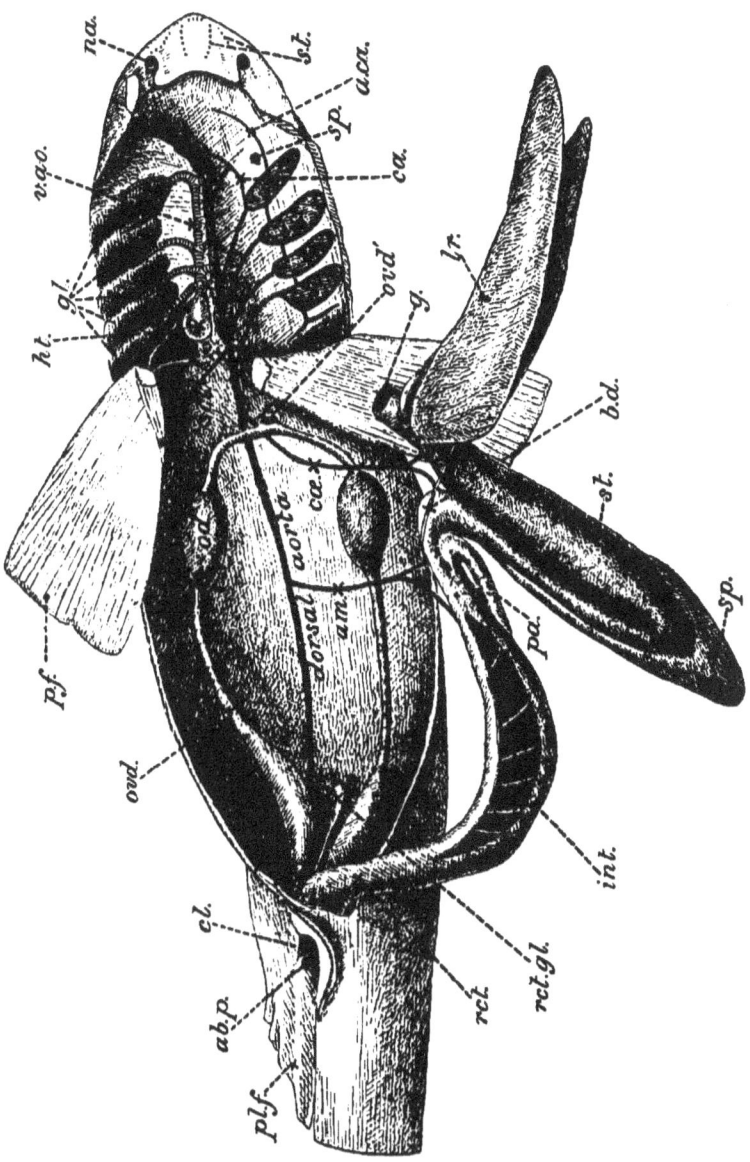

Fig. 46.
[From a Dissection by S. T. Parkinson.]

second set about the middle of its length. Each series consists of three pouch-like flaps (pocket-valves) attached to the wall by their posterior edges, their anterior edges being free. All the valves of the heart are so arranged as to allow blood to flow freely in the direction described, but not in the opposite direction. The *arteries* are well-defined tubes with largely muscular walls. The blood leaving the heart passes through a mid-ventral cardiac aorta to be distributed to the gills, where the blood receives a supply of oxygen from the water by which the gills are bathed, and gets rid of its CO_2. It is then collected by a series of efferent branchial vessels into a median dorsal aorta, from which arise arteries distributing blood to all parts of the body with the exception of the head.

The head is supplied with blood by arteries arising direct from the foremost of the efferent branchial vessels.

The blood-vessels of the gills are:—(1) the *afferent branchial arteries* bringing the blood from the cardiac aorta, of which one supplies the gills of each gill-arch, the three hinder pairs arising separately from the cardiac aorta, while those supplying the two foremost pair of arches arise by one pair of vessels, each of which then divides into two. (2) The *efferent branchial arteries*, carrying pure blood into the dorsal aorta, and arising from loops which surround each of the gill-clefts except the last. The blood from the gill of the last (or fifth) gill-cleft is received by an efferent vessel which opens into the efferent loop of the fourth cleft. Each efferent loop communicates also by means of a short vessel with the loops in front of and behind it.

The *dorsal aorta*, formed by the union of the efferent branchial arteries, runs back below the vertebral column to the posterior end of the body, becoming the *caudal artery* in the tail. It also has a forward continuation which divides into two branches that are connected with the carotid arteries.

Branches of the Dorsal Aorta.—(1) A pair of *subclavian arteries* to pectoral fins. (2) A *cœliac artery*, supplying liver, anterior end of stomach, beginning of intestine, liver, and pancreas. (3) An *anterior mesenteric artery* to intestine and reproductive organs. (4) A *lieno-gastric artery* to stomach, spleen, and pancreas. (5) A *posterior mesenteric artery* to rectal gland. (6) Numerous pairs of small *parietal arteries* to body walls. (7) Numerous pairs of small *renal arteries* to kidneys.

Each half of the head is supplied with pure blood by:—(1) a *carotid artery*, running forwards from the top of the first efferent loop and dividing into (a) *external carotid* to upper jaw and snout; (b) *internal carotid* to brain.

(2) A *hyoidean artery*, running from the middle of the same loop to supply pseudobranch and brain. (3) Small vessels to the floor of the mouth from the ventral ends of the efferent loops.

The **veins** are partly regular tubes, partly irregular sinuses, all of which have thin walls. They may be divided into :—(1) Systemic veins. (2) Portal systems.

(1) **Systemic Veins.**—The impure blood from the anterior part of the body is brought back on each side by an *anterior cardinal sinus* running above the gill-clefts to the level of the sinus venosus, where it unites with a much larger *posterior cardinal sinus* bringing forwards the impure blood from the body behind the pectoral fins. By the union of the anterior and posterior cardinal sinuses on each side a very short *Cuvierian sinus* is formed, which, after receiving a *jugular sinus* from the floor of the mouth and pharynx, is merged in the corresponding side of the sinus venosus.

Factors of Cardinals.—An orbital sinus surrounding the eye opens behind into a short *post-orbital sinus*, which again communicates with the anterior cardinal. A *hyoidean sinus* opens into the front end of the anterior cardinal. The posterior cardinals begin between the kidneys, and each of them, before it unites with the corresponding anterior cardinal, receives a *lateral vein* from the body-wall, a *genital sinus* from the gonad, and a *subclavian vein* from the pectoral fin.

(2) **Portal Systems.**—Hepatic portal and renal portal systems are present, which respectively supply the liver and kidneys with impure blood.

Hepatic Portal System.—The blood from the abdominal digestive organs and spleen, containing most of the products of digestion, enters the *hepatic portal vein*, which breaks up into branches in the substance of the liver. This organ returns its impure blood direct to the sinus venosus by means of two *hepatic sinuses*.

Renal Portal System.—The impure blood of the tail passes into a *caudal vein* which runs forwards to the posterior ends of the kidneys and divides into right and left *renal portal veins*, which divide up in the kidneys, from which the blood is returned to the posterior cardinals by numerous small *renal veins*.

The **capillaries** are minute tubes arranged in plexuses, in which the smallest arteries end and the smallest veins begin.

(II.) The **lymphatic system** consists of minute spaces, small lymphatic vessels, and large cavities, all containing **lymph**, which resembles blood in many respects but possesses no red corpuscles. There are also certain "ductless glands" connected with the lymphatic system, of which the largest is the **spleen**, a reddish body attached to the bend and distal limb of the stomach.

The largest lymph-space is the **body-cavity** or **cœlom**, which includes the abdominal and pericardial cavities.

6. The **respiratory organs** of the dogfish are *gills*, adapted for breathing the oxygen dissolved in the surrounding sea-water. They consist of vascular folds arranged upon the posterior side of the hyoid arch, and both sides of the first four branchial arches. Water is taken in at the mouth and expelled through the gill-slits to the exterior, so that a continual stream passes over the gills, in the small vessels of which the blood is purified. The hæmoglobin of the red blood-corpuscles acts as an oxygen-carrier, taking up a certain amount of free oxygen from the exterior into a state of loose chemical combination, and parting with it again to the tissues.

A rudimentary gill (pseudobranch) is found on the anterior wall of the spiracle.

7. The **excretory** and **reproductive organs** are so closely connected that it is best to consider them under the common heading of **urino-genital organs**.

The **male** dogfish possesses a pair of elongated narrow kidneys extending nearly the whole length of the abdominal cavity and situated close together below the vertebral column and above the peritoneum. Each kidney is divided into a number of segments, and its anterior half is distinguished as *mesonephros* (Wolffian body) from its posterior half or *metanephros*, the excretory products of these being carried off by distinct *mesonephric* (Wolffian) ducts, and *metanephric ducts* (ureters).

Each mesonephric duct, which, since it also acts as a spermiduct, may be termed *urino-genital duct*, is a convoluted tube running along the ventral side of the corresponding mesonephros, dilating into a *vesicula seminalis*, and finally opening into an *urino-genital sinus* that communicates with the cloaca by a small aperture placed on the end of a dorsally situated *urino-genital papilla*.

The *metanephric duct* on each side is formed by the union of several smaller ducts and opens into the dorsal side of the urino-genital sinus.

The *spermaries* (testes) are two soft flattened bodies, each connected by a number of small ducts (vasa efferentia) with the front end of the corresponding mesonephros.

The kidney is made up of numerous glandular tubules, and the sperms have to traverse some of these before they can reach the urino-genital duct.

The urino-genital sinus is forwardly produced into two blindly-ending *sperm sacs* situated on the ventral side of the kidneys.

The *claspers* serve as copulatory organs by which the sperms are introduced into the oviducts of the female.

Two short tubes, with a common opening into the abdominal cavity, can be seen on the ventral side of the gullet. These are rudimentary *Müllerian ducts*, equivalent to the oviducts of female specimens.

The kidneys in the **female** exhibit the same regions as in the male, but the mesonephros is not so well developed. The two mesonephric ducts are straight and unite to form a *urinary sinus*, which receives a number of distinct metanephric ducts and opens into the cloaca on a dorsal *urinary papilla*.

The reproductive organs of the female are not so intimately connected with the urinary organs as in the male. There is a large unpaired *ovary* from which large ova in various stages of development can be seen projecting. When ripe these may exceed half an inch in diameter, their large size being due to the presence of abundant food-yolk.

The *oviducts* (Müllerian ducts) have a common anterior opening into the abdominal cavity, situated on the ventral side of the gullet in front of the liver. Each of them curves back and soon dilates into an ovoid *oviducal gland*, after which it runs back as a good-sized tube towards the cloaca, just before reaching which it unites with its fellow to open by a median dorsal aperture.

The ova are fertilized in the oviduct, after which each of them is surrounded by an albuminous fluid and enclosed in a horny case secreted by the oviducal gland, and with four corners produced into tendril-like threads. In this condition the eggs are laid, the threads serving to attach them to seaweeds, &c. Development takes place at the expense of the food-yolk, which after a time is found stored in a vascular sac, the *yolk-sac*, attached to the ventral side of the embryo.

In most dogfishes the entire embryonic development takes place in the oviduct, and in one species (Mustelus lævis) the vascular yolk-sac is thrown into folds which interlock in corresponding folds of the oviduct, so that a kind of *placenta* is formed—*i.e.*, an embryonic structure by which the blood-systems of parent and embryo are brought into close relation. Scyllium, therefore, is *oviparous*, a term applied to cases where most or all of the development takes place external to the body—other dogfish are *viviparous*.

8. The most obvious part of the **muscular system** is a great lateral mass extending along each side of the body from the neighbourhood of the spiracle, and segmented into a succession of *myomeres*, with zigzag boundaries. The lateral muscles are also divisible into dorsal and ventral sections along the boundary between which runs the lateral line. These muscles effect swimming movements, during which the body is not merely bent from side to side but thrown into sinuous curves. There are also special muscles for moving the fins, lower jaw, &c.

9. The **nervous system** (Fig. 48) consists of (1) cerebro-spinal axis, (2) cranio-spinal nerves, and (3) sympathetic system.

(1) The **cerebro-spinal axis** is a thick-walled tube invested in a delicate vascular membrane (pia mater) and contained in the neural canal of the skull and vertebral column. The canal is lined by a firm membrane (dura mater) between which and the pia mater there is a large lymph-space. The front end of the cerebro-spinal axis is dilated into a brain, lying within the cranium and a spinal cord, running along the spinal canal above the vertebral centra. The cavity of the neural tube constitutes an extremely small central canal in the spinal canal and larger spaces known as ventricles in the brain.

The **brain** at an early stage exhibits three successive swellings, the anterior, middle, and posterior cerebral vesicles. These, together with outgrowths from them, become the fore-, mid-, and hind-brains of the adult.

The *fore-brain* consists of a central thalamencephalon, with a large antero-dorsal outgrowth, the prosencephalon, with which are connected two olfactory lobes.

The *thalamencephalon* is somewhat cylindrical and contains a large cavity (3rd ventricle) with thin roof but thickened floor and side walls. The floor is produced downwards into a projection (infundibulum) in which are two oval swellings (lobi inferiores). Two structures of non-nervous nature are connected with the thalamencephalon. They are known as the pineal and pituitary bodies. The *pineal body* is a stalk-like structure with a swollen end, running forwards and upwards from the roof of the 3rd ventricle. It corresponds to what appears to have been an unpaired dorsal eye (pineal eye) in ancestral forms (see p. 2). The *pituitary body* is a problematic organ attached to the infundibulum, and consisting of two thin-walled sacs (sacci vasculosi) and a central tube.

The *prosencephalon* (= cerebral hemispheres of higher vertebrates) is the largest part of the brain. It is a rounded mass presenting indications of divisions into right and left halves, and containing a ventricle connected with the 3rd ventricle. An expanded *olfactory lobe* abuts on the posterior wall of each olfactory capsule, and is connected behind with the prosencephalon by a short stalk. It contains an olfactory ventricle which communicates with the brain-cavity last mentioned.

The short, thick *mid-brain* which next succeeds contains a ventricle (iter or Sylvian aqueduct), the roof of which is raised up into two ovoid *optic lobes* containing optic ventricles.

The axis of the *hind-brain* is constituted by a long cylindrical *bulb* (medulla oblongata) containing a cavity (4th ventricle) with a thin roof. A large hollow body, the *cerebellum*, of elongated oval shape is attached by the middle of its length to the dorsal side of the bulb, which is also produced on each side into a conspicuous lobe (restiform body).

The **spinal cord** is cylindroidal in shape and somewhat flattened from above downwards. It is deeply furrowed by two median fissures—one dorsal, the other ventral.

(2) **Cranio-spinal Nerves.**—These may be divided into cranial nerves and spinal nerves, which respectively take origin from the brain and spinal cord.

There are ten pairs of **cranial nerves**, as follows:—

I. *Olfactory*, each of which consists of two bundles of fibres arising from the corresponding olfactory lobe and supplying the lining of the olfactory sac on the same side.

II. *Optic*, running to the eyeballs. They unite to form an X-shaped chiasma on the ventral side of the thalamencephalon. The posterior limbs of the X (optic tracts) end in the optic lobes.

The III., IV., and VI. nerves are known as the eye-muscle nerves.

III. *Oculomotor.*—These arise from the ventral side of the mid-brain and supply most of the eye-muscles.

IV. The *pathetic* nerves are the only cranial ones which have a dorsal origin. They arise nearly in the middle line just behind the optic lobes and supply the superior oblique muscles. All the remaining nerves take origin from the bulb.

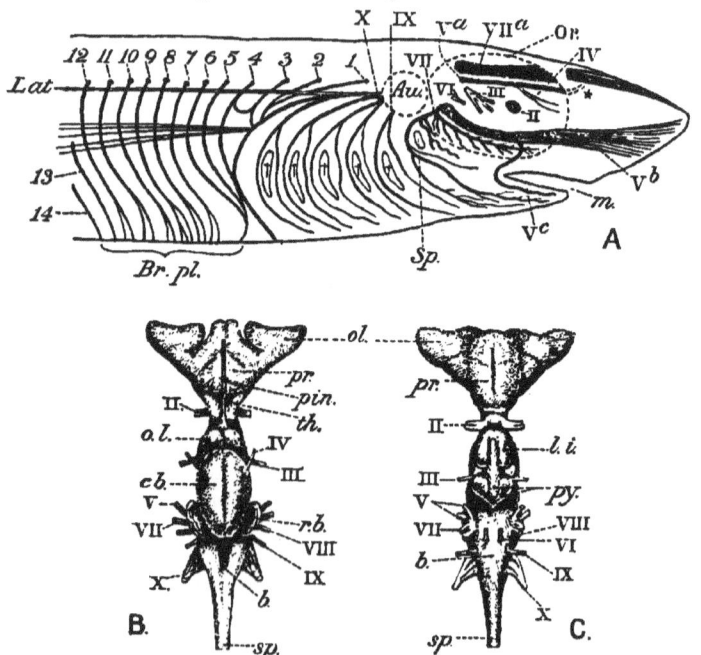

Fig. 47.—DOGFISH.—A, Diagram of cranio-spinal nerves (slightly altered after *Wiedersheim*).—*M*, Mouth; *Or*, margin of orbit (indicated by dotted line); *Au*, auditory capsule; *Sp*, spiracle; †††††, gill-slits; II., optic; III., oculomotor; IV., pathetic; V^a, ophthalmic branch of trigeminal; *, its course in front of orbit to join VII^a; V^b and V^c, maxillary and mandibular branches of trigeminal; VI., abducent; VII^a, ophthalmic branch of facial; VII., main trunk of facial, where palatine branch runs forward and hyomandibular backward behind spiracle, giving off prespiracular branches; IX., glossopharyngeal, forking over first gill-slit; X., vagus, giving off branchial branches which fork over gill-slits, a lateral line branch (*Lat*), and then supplying viscera; 1-14, first fourteen spinal nerves, forming brachial plexus (*Br.pl*), from which pectoral fin is supplied. B, Dorsal view of brain. C, Ventral view of brain.—*ol*, Olfactory lobes; *pr*, prosencephalon; *th*, thalamencephalon; *pin*, pineal body; *py*, pituitary body; *l.i*, lobi inferiores; *o.l*, optic lobes; *cb*, cerebellum; *r.b*, restiform body; *b*, bulb; *sp*, spinal cord; II.-X., cranial nerves.

VI. The *abducent* nerves, which innervate the external rectus muscles, arise almost mid-ventrally from the anterior part of the bulb. [*N.B.*—These nerves are mentioned out of their order for the sake of convenience.]

V. The *trigeminal*, which, like all its successors, arises from the side of the bulb, is a large nerve which has 3 chief branches:—

α. *Ophthalmic*, to sensory tubes on upper side of snout;

β. *Maxillary*, to sensory tubes on under side of snout;

γ. *Mandibular*, to muscles of the lower jaw.

VII. The *facial* nerves arise close behind the trigeminals, and like them have three chief branches:—

α. *Ophthalmic*, which first runs parallel to the similarly named division of the fifth, then fuses with it and has the same distribution;

β. *Palatine*, to roof of mouth;

γ. *Hyoidean* (post-spiracular), which runs down behind the spiracle, supplying muscles, and gives off small pre-spiracular branches.

VIII. The *auditory* nerves run into the auditory capsules to supply the membranous labyrinths.

IX. The *glossopharyngeal* nerve on each side arises just behind the eighth, traverses the floor of the auditory capsule, and forks over the first branchial cleft into an anterior (hyoidean) branch and a posterior (branchial) branch.

X. The *vagus* (pneumogastric) is a large nerve arising by several roots and taking a backward course. It gives off a *lateral line nerve*, which supplies the similarly named sense-organs, four *branchial nerves*, which fork over the last four branchial clefts, and finally divides into branches for the heart and viscera.

The **spinal nerves** arise segmentally from the spinal cord, each by two roots, a dorsal (upon which is a small ganglion) and a ventral, which pierce the wall of the spinal canal and then unite together. Each segment is supplied by the corresponding pair of nerves, the fibres of the dorsal root going to the skin, those of the ventral root to the muscles. The regularity is somewhat disturbed by the presence of the paired fins, which are supplied from plexuses formed by certain of the anterior and posterior nerves.

(3) The **sympathetic system** consists of two longitudinal cords situated ventral to the vertebral column and dilated into segmentally-arranged ganglia which are connected with the adjoining spinal nerves. Each cord is connected in front with

the corresponding vagus. The sympathetic ganglia supply the viscera and vascular system.

Histologically, the nervous system is essentially made up of the two usual elements, *nerve-cells* (ganglion cells) and *nerve fibres*. In the brain and spinal cord these elements are respectively aggregated into what are known as grey and white matter. The former exists in the spinal cord and bulb as an axial core, in the prosencephalon and cerebellum as an external crust. Masses of it are also found in the optic lobes, side-walls of the thalamencephalon, basal part of the prosencephalon, and in the olfactory lobes.

10. The chief **sense organs** of the dogfish are those connected with the skin, the olfactory sacs, the ears, and the eyes. All of these essentially consist of end-organs connected with nerve-fibres.

The **skin** is abundantly provided with groups (end-buds) of projecting sense-cells, probably of tactile nature, and it also possesses lateral line organs and sensory tubes. The indistinct groove termed lateral line in describing the external characters corresponds pretty much in position with a tube which underlies the skin and opens to the exterior at intervals. It is lined by epithelium, many of the cells of which are mucus-secreting goblet cells, while others are sense-cells (hair-cells), each provided with a slender hair-like process.

The *sensory tubes* (jelly tubes) which open on the snout underlie the skin of the head and each of them ends in a rounded sac (ampulla) divided into compartments, and partly lined by hair-cells. These tubes contain a gelatinous substance, secreted by the lining epithelium. The ampullæ are supplied by the ophthalmic branches of the fifth and seventh cranial nerves, the fibres of which run to the hair-cells.

The large **olfactory sacs**, contained in the olfactory capsules opening ventrally by the nostrils, have their lining raised into numerous transverse folds, the epithelium of which contains elongated olfactory cells, supplied by the olfactory nerve.

The **ear** (Fig. 48) on each side is a membranous sac (labyrinth) of complicated shape contained within the auditory capsule, and surrounded by a large lymph-space. It is lined by epithelium, part of which is sensory, and is filled with fluid in which are suspended numerous calcareous particles. The membranous labyrinth originates as an ectodermic pit, and in the adult still retains a connection with the exterior by means of a narrow tube (aqueductus vestibuli) opening on the top of the head.

The labyrinth is divided into a central region (*vestibule*) and three curved tubes, the *semicircular canals*. The vestibule is again divided into an anterior *utriculus*, and a posterior sacculus with which the aqueductus vestibuli communicates and which is produced into a process, the *cochlea*. The semicircular canals

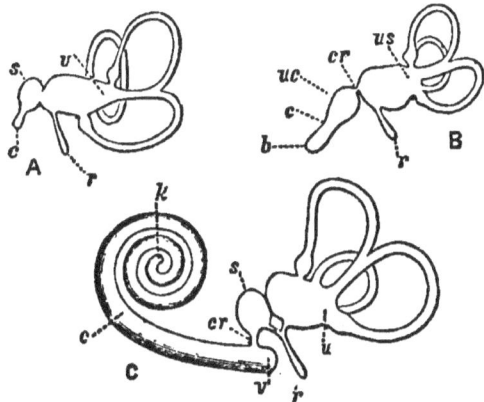

Fig. 48.—DIAGRAMS OF THE MEMBRANOUS LABYRINTH (from *Bell*, after *Waldeyer*).—A, Fish; B, Bird; C, Mammal. Internal side of left labyrinth.—*us*, Utriculus and sacculus; *u*, utriculus; *s*, sacculus; *c*, cochlea.

are known from their position as anterior vertical, posterior vertical, and external horizontal. They open at both ends into the utriculus, and each has a small swelling (ampulla) at one end. This is anterior in the first and last named, posterior in the horizontal canal. The ampullæ and parts of the vestibule possess patches of auditory hair-cells, with which fibres of the auditory nerve are connected.

Each of the **eyes** is flattened externally and rounded internally. Its wall consists of three coats, the most external of which is the firm *sclerotic*, supported by cartilage and exhibiting externally a transparent oval area, the *cornea*. The middle coat (*choroid*) is a pigmented vascular membrane lining the rounded part of the eye, and continued into a partition, the *iris*, by which the cavity of the eye is divided into a small outer and a large inner chamber. The iris is perforated by an oval slit (the *pupil*) for the transmission of light. The innermost coat of the eye is a delicate membrane, the *retina*, which contains the end-organs for sight (rods and cones). The optic nerve perforates the sclerotic and choroid on the inner side of the eye and branches out in the retina, its fibres becoming indirectly connected with the rods and cones. The interior of the eye is occupied by refracting structures, by which the light is focussed on the sensitive retina. These are :—a watery fluid (*aqueous humour*) external to the iris, a jelly-like substance (*vitreous humour*) occupying the internal chamber, and a spherical transparent *lens* suspended on the inner side of the iris.

The eyelids have already been mentioned (p. 155). Other important accessory parts are the six small band-like muscles by which the eyeball is moved. Four of these take origin from the hinder part of the orbit, and are inserted into the upper, lower, anterior, and posterior sides of the eyeball, being respectively known as superior, inferior, internal, and external *rectus muscles*. The other two are known as superior and inferior *oblique muscles*, taking origin from the front end of the orbit, and inserted respectively into the upper and lower sides of the eyeball.

CHAPTER X.—AMPHIBIA.

§ 15. RANA (The Frog).

THE two commonest kinds of Frog are *Rana temporaria*, the Common Frog, and *Rana esculenta*, the Edible Frog. Both are found on the Continent, but only the former in this country. The following description applies to both, any important differences being noted :—

MORPHOLOGY AND PHYSIOLOGY.

1. External Characters.—The bilaterally symmetrical body exhibits no external trace of segmentation, and is divided into *head* and *trunk*, between which no neck intervenes. There is also no tail. Fore and hind limbs are present, and these, unlike the paired fins of dogfish, to which they are homologous, are transversely jointed, and split at their distal ends into digits. The surface of the body is soft and moist, and there is no general investing exoskeleton. Owing to the presence of pigment in the skin the body is of a yellowish-brown, and is mottled dorsally. The ventral surface is much smoother and paler than the dorsal. The colour varies with the surroundings.

The flattened **head** is bluntly triangular, with a forwardly-directed apex. The mouth is extremely wide, and extends backwards to the posterior angles of the head. On the dorsal surface, near the front and widely separated, are two small valvular apertures, the nostrils or *external nares*. Behind these are the large projecting eyes, with small, immobile upper eyelids, and delicate, semi-transparent lower eyelids, capable of considerable movement.

The space between the eyes is broader in *R. temporaria* than in *R. esculenta*, and while flat or convex in the former, is concave in the latter.

Behind each eye is a circular space, the *tympanic area* (larger in *R. esculenta*), which in *R. temporaria* is inside a dark patch of pigment, that tapers to a point behind. In the male *R. esculenta* a pair of *vocal sacs* are found, which, when inflated, appear as rounded projections near the angles of the mouth.

The **trunk** is somewhat oval, tapering to a blunt point behind, where a small rounded *cloacal aperture* is found. Hard parts can be felt through the skin along the entire dorsal surface, but this is only the case with the *anterior* part of the ventral surface. In this way a *thoracic region* in front can be distinguished from an *abdominal region* behind.

The **fore-limb** commences immediately behind the head, and is divided into (1) *Brachium* (arm), (2) *Antebrachium* (fore-arm), and (3) *Manus* (hand).

If the fore- or hind-limb of an animal is spread out in the *primitive position*—*i.e.*, at right angles to the body, with the "palm" or "sole" side downwards, *dorsal* and *ventral surfaces* can be distinguished. There are also anterior and posterior edges, which, being in front and behind an axial line running down the centre of the limb, are termed *preaxial* and *postaxial edges*. (*Cf.* dogfish, p. 154.)

The manus possesses four well-developed digits, and, in addition, a rudimentary one concealed under the skin, and corresponding to the *pollex* (thumb) or 1*st digit* of other forms. The remaining fingers are 2*nd*, 3*rd*, 4*th*, and 5*th* respectively, reckoning from the preaxial side (thumb side). None of the digits possess nails or claws. In the male Frog a thickened pad, especially prominent during the breeding-season, is found on the preaxial side of the palm of the hand. It is black in *R. temporaria*.

The hind-limb is also divisible into three parts—(1) *Femur* (thigh), (2) *Crus* (leg), (3) *Pes* (foot), consisting of a short cylindrical *tarsus* (ankle), and five slender *digits*, united together by a delicate web. The *hallux* (great toe), or 1*st digit*, is preaxial, and on its outer side there is a small, horny elevation, the "*calcar*," which is a rudimentary "sixth toe" (præhallux).

All the preceding regions, especially the pes, are much longer than the corresponding parts of the fore-limb, but there is obvious serial homology between them. The typical number of digits for manus and pes in terrestrial *Vertebrates* is five, so that the pentadactyle pes of the frog must be regarded as more typical than the tetradactyle manus. Using * for rudiment, the state of things in the frog may be represented by 4 */5 *.

Position of Body.—When at rest the frog assumes a squatting position, with the ventral surface near the ground posteriorly. The back is humped, marking the union of the sacrum and ilia. The elbow is directed backwards and outwards, and the hand rests with its preaxial side turned sharply inwards. In the hind-limb the knee is directed outwards and slightly forwards, while the crus is bent back parallel to the femur, by which the ankle-joint is thrown behind the body. The pes runs outwards and forwards with its preaxial side internal.

2. **Skin** (Fig. 49).—The skin is thin, and raised on the dorsal surface into an immense number of minute wart-like protuberances. It is only loosely united to the body, by bands of connective tissue along certain definite lines. The intervening subcutaneous spaces, over which the skin is baggy, are filled with lymph.

A superficial epidermis can be distinguished from an underlying dermis, which is considerably thicker.

The *epidermis* is made up of several layers of cells, and hence is classified as *stratified** epithelium. The most external cells are very flat, without evident nuclei, and make up a *horny layer*, below which are rounded cells with well-marked nuclei and granular protoplasm. Those abutting upon the dermis, forming the *Malpighian layer*, are columnar. A small number of irregular *pigment cells*, containing dark colouring-matter, are scattered through the epidermis. Numerous glands open upon the surface

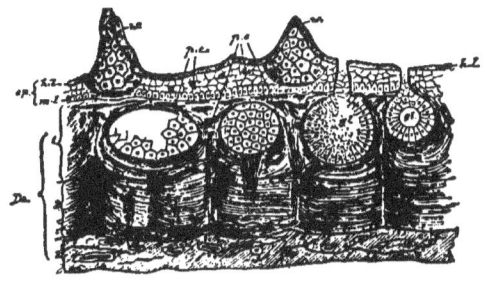

Fig. 49.—VERTICAL SECTION OF SKIN OF FROG (after *Wiedersheim*), much enlarged.—*ep*, Epidermis; *h.l*, horny layer; *m.l*, Malpighian layer; *w*, warts; *gl*, cutaneous glands; *p.c*, pigment cells; *De*, 1, 2, 3, layers of dermis.

of the epidermis. They are of two kinds, *serous* and *mucous*. The former chiefly occur in the skin of the back, and appear to secrete

* The chief kinds of **Epithelium** may be thus classified:—

I. **Squamous** (Pavement).— Composed entirely, or mostly, of flattened, cubical, or polyhedral cells.

1. *Simple.* One cell thick. *Endothelium* is a variety of this, composed of flattened cells, and lining heart, vessels, body-cavity, &c.
2. *Stratified.*— More than one cell thick. Epidermis, &c.

II. **Columnar.**—Composed of cells more or less elongated at right angles to surface. May be ciliated.

1. *Simple.*—Stomach and intestine.
2. *Stratified.*—Mouth-cavity of frog.

III. **Glandular.**— Composed of spherical or cubical cells, which elaborate a secretion or excretion.

For the most part *Simple.*—Peptic glands, kidney tubules, &c., &c.

2   12

a substance of irritant nature. The latter, which are more numerous and widely distributed, elaborate a slimy secretion which makes the skin moist and slippery. These glands are rounded vesicles, lying in the outer part of the dermis, but lined by glandular epidermal cells. They are large and numerous.

The *dermis* is mainly composed of fibrous* connective tissue, the fibres of which are mostly parallel to the surface, but also form vertical bands. The glands are imbedded in the external part (1) of the dermis, which also contains a large number of pigment-cells (chromatophores) that give to the skin its characteristic colours. These vary owing to the fact that the chromatophores contract under the influence of the nervous system. In the fully contracted condition the pigment is limited to a relatively small area, and the skin appears light. The reverse is true when the chromatophores are non-contracted. Since these colour-changes protect a frog by causing it to harmonize with its surroundings, thus making it inconspicuous to its enemies, we have here a case of *protective general resemblance*. But since also the arrangement conceals the frog from its prey to a greater or less extent, the general resemblance is *aggressive* as well as protective.

The glands are surrounded by unstriated muscle-fibres, which also form a layer in the deeper part of the dermis, where also occur networks of blood-vessels, lymphatics, and nerves. Many of the nerve-fibres end in *touch-corpuscles;* small oval flattened bodies mostly forming groups underneath the epidermal warts.

3. The **endoskeleton** in the frog is mainly made up of gristle, or *cartilage*, and *bone*, both of which are modifications of connective tissue. The bones are largely connected at the joints by fibrous bands (*ligaments*), which resemble tendons in structure (see p. 208). Those parts of the endoskeleton belonging to the head and trunk

---

* *Connective Tissue*—This permeates the whole body, binding together the other tissues. In higher animals generally it consists of three elements, associated together in different proportions. These are—(1) *Connective-tissue corpuscles*, nucleated cells, often branched (of which pigment-cells are modifications), most abundant in young tissues. (2) *White fibres*—delicate, and wavy, yielding gelatin on boiling. (3) *Yellow elastic fibres*, much branched, or forming networks. Unaffected by boiling.

The fibres are developed from the cells. All are imbedded in a structureless *ground substance* or *matrix*, which is semifluid and albuminous.

All these elements are found in the skin and in the bands uniting it with the body-wall.

may be conveniently termed *axial*, while those supporting the limbs are *appendicular*.

(1) The *axial endoskeleton* consists of the skull, backbone or vertebral column, and breastbone or sternum.

(*a*) The groundwork of the **skull** is made up of cartilage, constituting the *chondrocranium* or *primordial cranium*. With this are connected bones of two kinds, named, from their mode of development, *cartilage-bones* and *membrane-bones*. The former replace pre-existing cartilage, the latter pre-existing connective tissue. It is convenient to consider the skull under the separate headings of cranium, sense-capsules, jaws, and hyoid apparatus. The membrane bones are flattened structures investing the other parts, and will be taken last in each section.

The **cranium** or *brain-case* is a narrow cartilaginous tube, the upper side of which is broader than the lower, and possesses three spaces, *fontanelles*, a larger anterior and two smaller posterior, where cartilage is wanting. These are filled in by connective tissue. The cavity of the brain-case opens behind by the large *foramen magnum*. The region around this is known as *occipital*, either side of which is occupied by an *ex-occipital* bone, upon which is a projection, or *occipital condyle*, with an oval smooth surface. The two ex-occipitals do not completely bound the foramen magnum, a small cartilaginous strip being left above and below. Each of these bones is perforated by a *vagus foramen* through which the vagus and glossopharyngeal nerves leave the skull.

The cartilaginous side-wall of the brain-case is perforated by an *optic foramen*, and smaller foramina are also present. The cartilage forming the front end of the brain-case is replaced by a bony ring, the *sphenethmoid* or girdle bone, which also supports the hinder part of the *olfactory capsules* that lie in front of the skull. The sphenethmoid somewhat resembles a dice-box in shape, and its cavity is divided into two halves by a transverse partition marking the anterior boundary of the brain-case, and is perforated by an *olfactory foramen* on each side for the passage of the corresponding nerve. The front half of the bone is divided by a vertical longitudinal partition, into right and left halves.

Membrane Bones.—The roof of the brain-case is invested by two long, flat *parieto-frontals*, which are bent over behind so as to protect the upper part of its side walls. The floor of the brain-case is covered by the *parasphenoid*, a dagger-shaped bone, with its "blade" running forwards.

Sense Capsules.—Two pairs of these are fused with the skull, (a) auditory capsules behind and (β) olfactory capsules in front.

(a) The *auditory capsules* are closely fused with the hind end of the brain-case, one on each side. They are largely cartilaginous, but each is invaded behind by the ex-occipital of its side, while in front the cartilage is replaced by a large *pro-otic* bone, which also assists to form the side-wall of the brain-case, and is notched below by the *trigeminal foramen*, through which the trigeminal and facial nerves pass. The auditory capsule contains a complex cavity, in which the essential organs of hearing are contained. On its outer side is a depression, in which is a small opening, the *fenestra ovalis*, filled up with membrane in the recent state. Another and smaller opening on the hinder border, the *fenestra rotunda*, is filled up in the same way.

Fig. 50.—ENDOSKELETON OF FROG (A, A', and C, after *Ecker*). Cartilage dotted.—A and A', Skull from above and below, membrane bones being stripped off on one side; f, f', fontanelles; $f.m$, foramen magnum; $ex\text{-}oc$, ex-occipital; c, condyles; $v.f$, vagus foramen; $pr\text{-}o$, pro-otic; $t.f$, trigeminal foramen; $op.f$, optic foramen; $g\text{-}b$, girdle bone; orb, orbit; ol, nasal capsules; $e\text{-}p.b$, ethmopalatine bar; $pt.b$, pterygoid bar; sp, suspensorium; $q\text{-}j$, quadrato-jugal; pt, pterygoid; $pa\text{-}f$. parieto-frontals; $p\text{-}s$, parasphenoid, half cut away; sq, squamosal; na, nasals; v, vomer; $pr\text{-}mx$, pre-maxilla; mx, maxilla; pl, palatine; $e.n$, external naris; $i.n$, position of internal naris. B, *Columella*, enlarged.—p and d, Ends fitting into fen. ovalis and fixed to tymp. membrane, respectively. C, *Left half of Mandible*, from outside.—$m\text{-}mk$, Mento-meckelian; c, condyle; $a\text{-}sp$, angulo-splenial; d, dentary. D, *Hyoid apparatus*.—$b\text{-}hy$, Body of hyoid; $a.c$, anterior cornua; br, remains of branchial arches. E, *Spinal Column and Pelvis* (top view). Numbers refer to vertebræ, and are placed near transverse processes.—ust, Urostyle; il, ilium; is, ischium; ac, acetabulum. F, *Various Vertebræ*.—1, Atlas, front view; 2, side view of two adjacent vertebræ; 3, back view of vertebra; c, centrum; $n.s$, neural spine; $t.p$, transverse process; $a\text{-}z$, pre-zygapophysis; $p\text{-}z$, post-zygapophysis; $n.c$, neural canal; $i.f$, inter-vertebral foramen. G, *Side View of Pelvis*.—il, Ilium; pb, pubis; is, ischium; ac, acetabulum. H, *Sternum, Shoulder Girdle, and Fore-limb* (*Ventral View*).—$o\text{-}s$, Omo-sternum; st, sternum proper; $x\text{-}st$, xiphi-sternum; gl, glenoid cavity; $s\text{-}sc$, supra-scapula; $ep\text{-}co$, epi-coracoid; $pr\text{-}co$, pre-coracoid sheathed by clavicle; h, humerus; hd, head of ditto; $d.r$, deltoid ridge; $r\text{-}u$, radio-ulna; ra, radiale; ul, ulnare; cn, centrale; c 1-5, carpalia; I-V, metacarpals; 1-5, phalanges. I, *Hind Limb, &c.* (ust, urostyle, &c., as before).—f, Femur; hd, head of ditto; $t\text{-}f$, tibio-fibula; as, astragalus; ca, calcaneum; t, 1-3, tarsalia; I-V, metatarsals; 1-5, phalanges; *, placed by calcar.

AMPHIBIA.

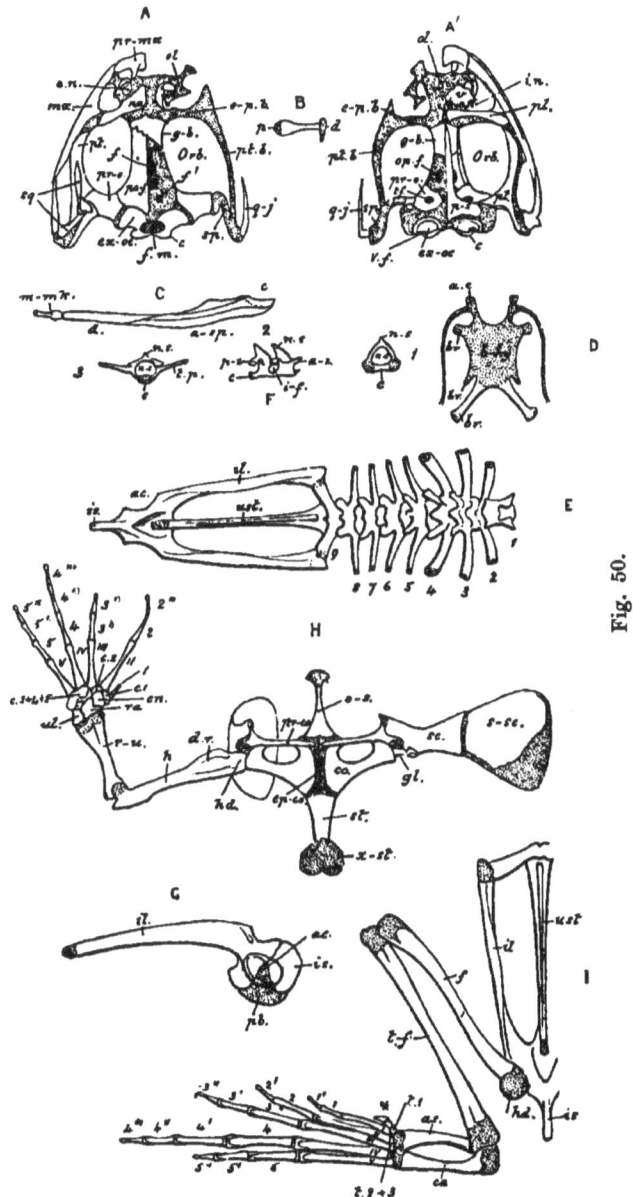

Fig. 50.

(β) The *olfactory capsules*, situated in front of the cranium, are separated from one another by a vertical plate of cartilage continuous with the longitudinal partition of the sphenethmoid. The walls of the capsules are mostly composed of cartilage.

Membrane Bones.—On the upper side of the olfactory capsules are two triangular *nasal bones*, which form the posterior boundary of the external nares. The cartilaginous floor of each olfactory capsule is partly covered by a somewhat triangular *vomer*, which is widely separated from its fellow. Each vomer bears a patch of sharp *vomerine teeth*, and partly bounds the internal naris of its side.

Jaws.—Both upper and lower jaws consist of a cartilaginous basis and of several bones, which in the case of the former are all membrane bones.

(a) *Upper Jaw.*—Owing to the projection of the sense-capsules a sort of bay, the *orbit*, in which the eye lies, is left on each side of the brain-case. It is bounded in front by a flat piece of cartilage, the *ethmo-palatine bar*, which runs out transversely from the floor of the olfactory capsule, sends a process forwards, and becomes continuous with a slender *pterygoid bar*, which passes back on the outer side of the orbit, and fuses behind with the *quadrate cartilage* (suspensorium). This is a cartilaginous rod directed outwards, downwards, and backwards. Its proximal end is forked, the short limbs of the fork being attached, above and below, to the outside of the auditory capsule. The lower jaw is "suspended" to its distal end, which presents an articular hollow. These parts constitute the cartilaginous basis of the upper jaw and its supports. Its actual margin is formed by a series of membrane bones, of which the most anterior are the two *premaxillæ*, small tooth-bearing elements meeting in the middle line in front of the olfactory capsules. On each side an elongated curved *maxilla*, which bears most of the teeth, runs back from the pre-maxilla and unites behind with a narrow *quadrato-jugal* extending posteriorly to the quadrate cartilage. Three bones on each side connect the upper jaw with the cranium and keep it firmly in position. These are the palatine, pterygoid, and squamosal.

The *palatine* is a slender transverse bone moulded on the hinder edge of the ethmo-palatine bar, while the pterygoid bar is largely ensheathed by the anterior ray of the three-rayed *pterygoid* bone, which touches the external end of the palatine in front. The other two rays are shorter. One of them runs inwards, the other along the under side of the quadrate cartilage. A T-shaped bone, the *squamosal*, the stem of which covers the quadrate cartilage, partly covers the auditory region. This bone supports the annular *tympanic cartilage*, over which the tympanic membrane is stretched.

(β) The *lower jaw* (mandible) consists of two strongly-curved halves, each of which is traversed by an axial cartilage, *Meckel's cartilage*, ossified in front into a small *mento-meckelian* bone, uniting with its fellow in a median *symphysis*. Behind it presents an oval projection, the *condyle*, which articulates with the quadrate cartilage. Meckel's cartilage is strengthened below and on its inner side by a long *angulo-splenial* bone, from the posterior part of which a small elevation, the *coronoid process*, projects upwards.

Each half of the mandible possesses a membrane-bone, the *dentary*, which is a thin splint covering the outside of Meckel's cartilage for rather more than its anterior third.

A series of structures forming the **hyoid apparatus** are connected with the skull, and usually described with it. They are partly related to the auditory apparatus, and partly to the floor of the mouth. The *columella* is a small rod, bony in the centre, which is fixed into the fenestra ovalis by one end, while the other is attached to the inside of the tympanic membrane. The remainder of the hyoid consists of a quadrangular plate of cartilage, the *body of the hyoid*, supporting the floor of the mouth. Its angles are produced into anterior and posterior processes. In front of the former two slender, curved rods of cartilage, the *anterior cornua*, arise, each of which runs backwards round the angle of the mouth to be attached to the auditory capsule just beneath the fenestra ovalis. From the posterior end of the hyoid body two short bony *posterior cornua* or *thyrohyals* run back, which diverge and enclose between them the laryngo-tracheal chamber.

(*b*) The **vertebral column** is a hollow rod running back from the skull along the dorsal side of the body. The *spinal cord* lies in its cavity. The anterior part of the column is segmented, being made up of nine rings, the *vertebræ*, while the posterior part, termed the *urostyle*, is unsegmented. All are cartilage bones.

The vertebræ are very similar, except the first and last. Each is a ring, the thickened ventral part of which is the *body* or *centrum*, while the rest forms the *neural arch*. The successive centra are united firmly together, and are *procœlous*—*i.e.*, concave in front and convex behind. Both surfaces are covered by a thin layer of cartilage. (The 8*th* vertebra is *amphicœlous*, concave on both faces.) Dorsal and lateral spaces are left between adjacent arches. The former are filled up by connective tissue, while through the latter, or *intervertebral foramina*, spinal nerves take exit. From the upper side of each arch a small *neural spine* projects upwards and backwards in the middle line, while on each side of the arch a stout, cartilage-tipped bar, the *transverse process*, runs outwards. Four small projections with smooth articular surfaces project from the front and back of the arch. By these *articular processes* or *zygapophyses* the adjoining arches are linked together. Two, the *præ-zygapophyses*, are anterior, and their articular surfaces face upwards and inwards, while the other two, *post-zygapophyses*, are posterior, and their articular surfaces face downwards and outwards, overlapping the præ-zygapophyses of the following vertebra.

The 1*st* vertebra, or *atlas*, is devoid of tranverse processes and præ-zygapophyses, and the neural spine is rudimentary. The centrum is somewhat thin, and projects forwards between the occipital condyles, which articulate with two large concave facets on the front of the vertebra.

The 9*th* vertebra, the *sacrum*, has very large and strong transverse processes directed outwards and backwards. The centrum is convex in front, and presents a pair of articular tubercles behind, which articulate with corresponding concavities on the front of the *urostyle*. This is a bony rod, somewhat trihedral, and tapering behind to a point tipped with cartilage. It may be regarded as representing a number of vertebræ fused together, of which a ridge running along its dorsal surface corresponds to the united neural spines. Two small foramina open, one on either side, into the canal of the urostyle, a short distance behind its anterior end. Through these the coccygeal nerves take exit, and they, therefore, correspond to intervertebral foramina. The part of the urostyle in front of them is the first of the fused vertebræ, and may possess more or less distinct transverse processes.

(*c*) The **sternum** consists of several cartilage bones and cartilages placed in the middle line on the ventral side of the thoracic region. In the extreme front a flat, narrow bone, the *omo-sternum*, is found, to the anterior end of which a rounded flap of cartilage (the episternum) is attached, while posteriorly it broadens out and abuts upon a pair of narrow cartilaginous *epi-coracoids*. These are succeeded by the *sternum proper*, a flattened rod of bone, with a core of cartilage. To its hinder end a large, thin, deeply-notched piece of cartilage, the *xiphi-sternum*, is attached.

(2) *Appendicular Skeleton.*—The endoskeleton of either the fore- or hind-limb is divisible into—(1) *Limb-girdle*, which is firmly attached to the trunk, and (2) *Free limb*, articulated to the girdle.

(a) **Fore-Limb.**—The two *shoulder-girdles* form an incomplete ring almost encircling the body just behind the atlas vertebra. They are firmly fused to the epi-coracoids, ventrally, while dorsally they are attached by muscles to the skull and vertebral column. Each girdle is made up of a dorsal and a ventral moiety. At their point of union is a shallow, articular, *glenoid cavity*, affording attachment to the free limb. The dorsal part is made up of— (a) The *supra-scapula*, a quadrangular plate of cartilage, more or less calcified and ossified, broad above, and narrowing downwards to join (b) the *scapula*, an hour-glass-shaped bone, the lower end of which partly forms the glenoid cavity. The ventral part is composed of—(a) The *coracoid* bone, similar in shape to the scapula, and completing the glenoid cavity. In front of the coracoid there is an oval space filled with connective-tissue, the *coracoid fontanelle*. This is bounded in front by (b) the *præ-coracoid* (*pr-co*), a transverse bar of cartilage largely ensheathed by (c) the *clavicle*, which is *the only membrane bone in the appendicular skeleton*.

The *free limb* is made up of bones supporting the upper arm, fore-arm, and hand. The *humerus* is a long bone belonging to the first of these. Like the long bones of the limbs generally, it consists of a hollow marrow-containing *shaft*, and an enlarged *epiphysis* at each end, covered by a thin layer of cartilage. The epiphyses remain for a long time distinct from the shaft, with which, however, they ultimately fuse. The proximal end of the humerus forms a rounded *head*, articulating with the glenoid cavity. A well-marked *deltoid ridge*, more prominent in the male, runs from this half way along the anterior (ventral) surface of the bone. The distal end presents a spheroidal surface with which the next bone articulates, and above and below this (pre- and post-axially) are *condylar ridges*. The ante-brachium is supported by a short, stout *radio-ulna*. This is a compound bone, made up of a pre-axial *radius*, fused with a post-axial *ulna*. The boundary between these is marked by a groove at the distal end. The proximal end is excavated to receive the corresponding projection on the humerus, with which it forms the elbow-joint, and produced back behind the last into the *olecranon process*, belonging to the ulnar half of the bone. There are two articular projections, one radial, the other ulnar, on the distal end, which help to make up the wrist-joint. The endoskeleton of the manus is made up of the wrist or *carpus*, and the bones of the *digits*. The typical or theoretical carpus, deduced from comparison of numerous cases, consists of 9 elements,—3 proximal, 1 central, and 5 distal, the relative position of which is as follows:—[R = radial or pre-axial side ; U = ulnar or post-axial side.]

R		U
radiale	intermedium	ulnare
	centrale	
carpale 1 carpale 2	carpale 3	carpale 4 carpale 5

In the frog there are six small bones in the carpus, three proximal and three distal. Two of the former, corresponding to *radiale* and *ulnare*, articulate with the radial and ulnar facets respectively. The third proximal bone, the displaced *centrale*, is on the inner (pre-axial) side of the radiale. The three distal bones correspond to the *carpalia*. The first supports the rudimentary 1st digit and = *carpale* 1, the second supports the 2nd digit and = *carpale* 2, while the much larger third bone represents *carpalia* 3, 4, 5, fused together, and supports the remaining digits. Following the wrist-bones are five slender *metacarpals*, one to each digit. The 1st metacarpal is very small, but is all that represents the 1st digit, while digits 2, 3, 4, 5 are terminated by 2, 2, 3, 3 slender *phalanges* respectively.

(*b*) **Hind-Limb.**—The *hip-girdles* are closely united to form the *pelvis*, which resembles in shape a two-pronged fork with an extremely short handle. Its posterior part is a rounded plate, laterally compressed, and presenting on either side a deep oval cup, the *acetabulum*, with a prominent margin. This cup is for the articulation of the free limb. Nearly half of the plate and acetabulum are formed antero-dorsally by the broad hinder ends of the two "prongs" or *ilia* fused in the middle line. Each ilium is continued forwards to the sacrum, to the corresponding transverse process of which its cartilage-tipped end is united. This part of the ilium is laterally flattened, with somewhat concave ventral and convex dorsal edges. Two other elements on either side, *pubis* and *ischium*, of which the first is cartilaginous, unite with their fellows to form the middle and posterior parts of the plate and acetabulum. The pubes are triangular, and their apices extend to the upper margin of the acetabulum, above which the ilia and ischia unite.

The *free limb* is composed of bones belonging to the thigh, leg, and foot. The first is supported by a long bone, the *femur*, the slender shaft of which possesses a slight double or sigmoid curve. Its proximal end is enlarged into a rounded *head*, which articulates with the acetabulum to form the hip-joint, and its distal end also presents an articular expansion. The bone of the leg, *tibio-fibula*, like that of the fore-arm, is compound. It is made up by the fusion of a pre-axial *tibia* with a post-axial *fibula*, the boundaries of which are indicated by grooves and a double marrow-cavity. The shaft is curved, and the ends are expanded into transversely elongated pulleys, which assist in the formation of the knee- and ankle-joints respectively. The bones of the foot are partly those of the ankle, *tarsus*, and partly those of the *digits*. The typical or theoretical tarsus contains the same number of elements as the theoretical carpus, and these are similarly arranged, as follows :—
[T = tibial or pre-axial side ; F = fibular or post-axial side.]

T		F
tibiale	intermedium	fibulare
	centrale	
tarsale 1 tarsale 2	tarsale 3	tarsale 4 tarsale 5

The tarsus of the frog is very much elongated, and composed of four bones, two of which are proximal and two distal. The proximal ones, astragalus and calcaneum (= *tibiale* + *intermedium*, and *fibulare*), articulate with the tibial and fibular sides of the articular surface on the distal end of the tibio-fibula. They are united to each other at either end by their epiphyses. The distal bones, equivalent to the *tarsalia*, are extremely small. One, representing *tarsale* 1, supports the calcar and 1st digit, the other represents *tarsale* 2 + *tarsale* 3, and supports the 2nd and 3rd digits. Tarsalia 4 and 5 are not distinctly represented, and there is no centrale. The tarsus is succeeded by the *metatarsus*. The base of the calcar is formed by a minute extra metatarsal, while the slender *metatarsals* 1-5 belong to the corresponding digits. A small flattened phalanx completes the calcar, and 2, 2, 3, 4, 3 slender *phalanges* terminate digits 1, 2, 3, 4, 5, respectively.

Cartilage and bone may both be considered as modifications of connective tissue in which the matrix is very plentiful.

The most typical kind of *cartilage* is made up of a clear, homogeneous *matrix*, in which are numerous small cavities, lacunæ, connected with one another by fine channels, and containing nucleated *cartilage-cells*. By the division of these, and the deposition of fresh matrix, cartilage grows, especially near its surface, which is covered by a connective-tissue membrane, the *perichondrium*, in which nerves, lymphatics, and blood-vessels run.

The matrix of bone is densely fibrous connective tissue, impregnated with lime salts, principally carbonate and phosphate. In this are imbedded numerous much-branched *bone-cells*, which lie in *lacunæ* connected by fine tubules, *canaliculi*, into which the cell-processes are continued. The matrix is traversed by blood-vessels, contained in *Haversian canals*. Bone may be either *spongy*, as in the epiphyses of long bones and the interior of flat and short bones, or *compact*, as seen in the shafts of long bones and the exterior of flat and short bones. A membrane, the *periosteum*, similar to perichondrium, covers the outer surface of bones. Its inner part is made up of rounded cells, *osteoblasts*, which during growth add fresh layers of bone to the outside. Ossification always starts from definite centres. *Membrane* bones commence as calcified networks of connective-tissue fibres covered by osteoblasts, while *cartilage* bones are pre-formed in solid cartilage, covered by periosteum. Into this solid cartilage processes of the osteoblastic part of the periosteum penetrate to form the *marrow*, which first absorbs the cartilage, and then replaces it by spongy bone. In the shafts of long bones this is absorbed in its turn to form a continuous marrow-cavity.

Cartilage bones can continue to *lengthen* by growth at their ends. The new bone is formed in the way just described, from the cartilage found there.

Bones *thicken* by the addition of layers to the outside, formed by the periosteum.

The vertebral centra contain peculiar cellular cores, the last remnant of an important embryonic structure, the *notochord*.

4. The **digestive organs** (Fig. 51) consist of a convoluted tube, the gut or alimentary canal, and of glands connected with this. The gut presents the following regions:—mouth-cavity, gullet, stomach, small intestine, and large intestine opening into a cloaca which also receives the excretory and reproductive ducts. The most important annexed glands are the liver and pancreas.

The wide *mouth*, which reaches back as far as the tympanic area, and possesses a narrow *upper lip*, leads into a spacious **mouth-cavity**, the back part of which is termed the *pharynx*. More than 100 minute double-pointed *teeth* are affixed to the inner side of the upper jaw. They are attached to the premaxillæ and maxillæ, and are placed in a furrow bounded by the upper lip externally, and a fold of mucous membrane internally.

The *mucous membrane* is a pale, soft, extremely glandular layer which lines the alimentary canal.

On the roof of the mouth are two small patches of *vomerine teeth*, forwardly placed, and borne by the vomers. They are similar to the others. No teeth are present in the lower jaw. On the roof of the mouth, in the extreme front, a number of minute pores are present, the openings of the *intermaxillary glands*. The vomerine teeth are a little way behind this, and external to each patch is a small, transversely oval opening, the *internal naris*. Still further back two large rounded prominences, caused by the eyes, project into the mouth. Near the angle of the jaw, on each side, is a good-sized opening, that of the *Eustachian tube*, which leads to the tympanic cavity. On the floor of the mouth is a narrow, elongated muscular *tongue* attached in front. In the quiescent state its forked end is directed backwards. Numerous small elevations cover its surface, some of which, *filiform papillæ*, are narrow-ended; others, *fungiform papillæ*, broad-ended.

Near the angle of the mouth in the male *R. esculenta* an oval opening is present, which leads into a rounded, dilatable *vocal sac*.

Posterior to the end of the tongue there is a longitudinal chink, the *glottis*, with firm swollen edges. It leads into the respiratory organs.

The mouth-cavity passes behind into a short, wide **gullet** or **œsophagus**, and this merges without sensible demarcation into a narrower tube, the **stomach**. This passes backwards, gradually narrowing as it does so. The lining of both œsophagus and stomach is raised into longitudinal ridges. At the posterior or *pyloric* end of the latter there is a well-marked constriction

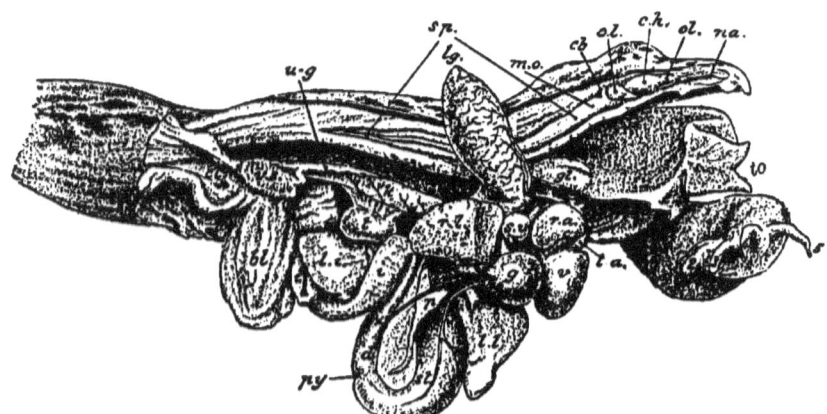

Fig. 51.—SIDE-DISSECTION OF MALE FROG.—2, 3, 4, 5, Digits; *to*, tongue; *st*, stomach, upon which are blood-vessels; *py*, pylorus; *d*, duodenum; *i*, rest of small intestine; *l.i*, large intestine; *cl*, cloaca, opened; *r.l*, and *l.l*, right and left liver-lobes; *g*, gall-bladder; *x*, opening of bile-duct, which is represented by the black line; *p*, pancreas. *s.v*, Sinus venosus, near which is cut end of right precaval; *r.a*, right auricle, above which are the cut ends of carotid arch, systemic arch, and cutaneous artery, while the pulmonary artery is seen running along lung; *v*, ventricle; *t.a*, truncus arteriosus. *gl*, Glottis; *lg*, right lung. *k*, Right kidney, the dark space above which represents the subvertebral lymph-sinus; *u-g*, right urinogenital duct, which opens on a papilla in the cloaca (just above the *l* in *cl*); *bl*, bladder (its opening is seen just below *l* in *cl*). *t*, Spermary (testis) above which the vasa efferentia are seen running in the mesorchium; *v.s*, vesicula seminalis. *ol*, Right olfactory lobe, from which right olfactory nerve is seen running forwards; *ch*, right cerebral hemisphere; *o.l*, right optic lobe, passing below and in front into optic tract and nerve; *cb*, cerebellum; *m.o*, medulla oblongata; *sp*, spinal cord. *na*, Right nasal sac.

marking the position of the *pylorus* or point at which the cavity of the stomach communicates with that of the **small intestine**. This is a narrow thin-walled tube, the first part of which, the *duodenum*, forms, together with the stomach, a ∪-shaped loop, while the rest is thrown into several coils. Its lining is raised into transverse folds connected by fainter longitudinal ridges. The end of the small intestine suddenly dilates into a short thin-walled **large intestine**, which narrows behind and becomes continuous with a chamber, the **cloaca**, opening externally by the small circular *cloacal aperture*. The cavity of the large intestine is separated by an annular valve from that of the small. Its lining is at first raised into delicate intersecting ridges, and then into longitudinal folds which pass back to the end of the cloaca, the posterior part of which is lined by ordinary skin.

Two important glands are connected with the alimentary canal—the liver, and pancreas.

The **liver** is a very large reddish-brown organ, occupying a considerable space near the front of the body-cavity. It arises as an outgrowth from the alimentary canal, which quickly becomes bilobed. In accordance with this the adult liver is divided into right and left halves, connected by a narrow strip of liver-substance. The left half is again subdivided into two, so that altogether three *lobes* are present. These are convex ventrally, somewhat concave dorsally, blunt-ended in front, and elsewhere thinning off into edges. The liver secretes the *bile*, a bright yellowish-green fluid. Closely attached to the dorsal surface of the right lobe is a rounded thin-walled sac, the *gall-bladder*, full of bile, in the fresh state. From it a short tube, the *cystic duct*, proceeds, which is connected with a fine network of *hepatic ducts*, arising from the liver. From this a *bile-duct* is given off, which is reinforced by other hepatic ducts, and runs in the U formed by stomach and duodenum, finally opening into the latter about half an inch beyond the pylorus on the inside of the U.

The **pancreas** is an elongated yellowish mass lying in the U-shaped loop, and produced into several tapering processes. With it the bile duct is closely connected, receiving from it a number of small *pancreatic ducts*. This gland secretes the *pancreatic juice*.

The digestive and other organs of the frog are contained in the large **pleuro-peritoneal-** or **body-cavity** (cœlom). This narrows in front, where it is bounded by the heart in its pericardium, and a kind of muscular partition, the *diaphragm*. It also narrows behind. Its dorsal wall is formed by the backbone and associated muscles, its ventral and lateral walls, in the thoracic region, by hard parts and muscles, by muscles alone in the abdominal region. The body-cavity is lined ventrally and laterally by a thin pigmented membrane, the *pleuro-peritoneum*, which dorso-laterally becomes free, leaving a space, the *subvertebral lymph-sinus*, beneath the vertebral column. The two lateral halves of the pleuro-peritoneum run to the middle line, forming a floor to this sinus, and then unite to make a double sheet, the *mesentery*, the halves of which diverge, surround the alimentary canal, and then become continuous. The alimentary canal is thus suspended from the dorsal wall of the body-cavity, and, strictly speaking, is *outside it*. The same may be said of the other contained organs.

The walls of the œsophagus, stomach, and intestines are composed of four coats, which are, commencing from the outside:—

(1) serous coat; (2) muscular coat; (3) sub-mucous coat; (4) mucous membrane.

(1) The *serous coat* is in reality the pleuro-peritoneal investment of the gut. It consists of a layer of simple squamous epithelium, with a thin underlying stratum of connective tissue in which run the blood-vessels, lymphatics, and nerves that supply the other coats.

(2) The *muscular coat*, best developed in the stomach, is divided into an outer *longitudinal* and an inner circular layer. It is made up of unstriated muscle-fibres. These are spindle-shaped cells (Fig. 52), each containing an elongated nucleus, which dovetail together by their tapering ends.

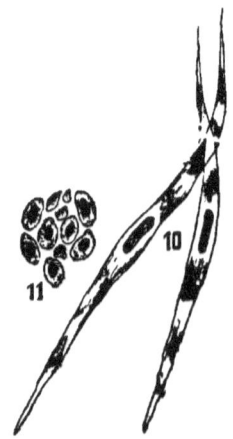

Fig. 52.—UNSTRIATED MUSCLE-FIBRES (from *Landois* and *Stirling*), much enlarged. — 10, Isolated; 11, in cross-section.

(3) The *sub-mucous coat* is also best developed in the stomach, and consists of loose connective tissue traversed by numerous blood-vessels and lymphatics.

(4) The *mucous membrane* is the part which is raised into folds internally. It is exceedingly glandular, and is made up of (*a*) a thin external layer of unstriated muscle, the *muscularis*

mucosæ; (*b*) a layer of connective tissue with glands, bloodvessels, and lymphatics; (*c*) a layer of simple columnar *epithelium*, adjoining the cavity of the gut.

The epithelium is of most interest. It is ciliated in the œsophagus and beginning of the stomach, and many of its cells, known from their shape as *goblet-cells*, are unicellular mucus-secreting glands. Besides this there are tubular multicellular glands lined by the epithelium and lying in the connective-tissue layer of the mucous membrane. They are of three kinds:—a, *œsophageal glands* in the gullet; β, *peptic* or *gastric glands* in the stomach; γ, *glands of Lieberkühn* in the small intestine.

The deeper parts of the peptic glands are lined by cuboidal cells, which secrete the gastric juice.

There are several points of interest in the histology of the mucous membrane lining the mouth. Its epithelium is stratified near the margins, elsewhere simple, columnar, and ciliated. Numerous goblet-cells are present. The tubules of the intermaxillary glands are lined by mucus-secreting epithelium.

The teeth are developed in the mucous membrane of the mouth. The projecting part or *crown*, which is forked, is separated by a constriction from the rest of the tooth, or *socket*, which contains a *pulp-cavity*, in which is the *pulp*, a small mass of vascular connective tissue. The tooth is mostly made up of *dentine*, the constituents of which are wavy tubules filled by prolongations of the pulp. The crown is covered by a cap of extremely hard *enamel*, exhibiting a layered structure, and containing prolongations of the dentine tubules. It is secreted by the epithelium of the mouth. The socket is covered by a layer of *cement*, resembling bone in structure. The teeth are all very similar, and during life are constantly replaced by new ones, which grow up from their bases.

The liver (*cf.* Fig. 51) is mostly made up of polyhedral glandular *hepatic cells*, granular, nucleated, and containing fat drops. Between these cells minute tubes, the *bile-capillaries*, run, forming a complex network from which the ultimate branches of the hepatic ducts arise. The vessels supplying the liver with blood break up into capillaries within it, and in these the branches of the hepatic veins, which carry blood *from* the liver, take their origin.

The ducts of the **pancreas** subdivide considerably, and finally terminate in blind tubules lined by glandular epithelium.

Nutrition.—The food of the adult frog consists mainly of insects. These are secured by the tongue, which can be rapidly protruded and retracted. Its peculiar mode of attachment enables it to be put out to some distance, and being covered with a viscid secretion, insects are easily secured. The secretion of the inter-maxillary gland is especially sticky, and the tip of the tongue brushes past its openings when protruded. The teeth are not used for chewing, but to prevent the escape of prey. The cilia of the mouth-cavity work so as to carry the food to the gullet, and from this point the contractions of the muscular wall of the alimentary canal propel it backwards. The chemical agents acting upon the foods are principally—(1) Gastric juice; (2) Pancreatic juice; and (3) Bile.

The **gastric juice** is secreted by the peptic glands, and in virtue of a ferment, *pepsin*, which it contains, converts some of the proteids into peptones. The ferment can only act in an *acid* solution, and the secretion contains a small amount of free hydrochloric acid. The cuboidal cells of the peptic glands secrete pepsin, while the acid is formed by the ovoidal cells. The **pancreatic juice** contains ferments which (*a*) convert starch into grape-sugar; (*b*) proteids into peptones; and (*c*) split up fats. Alkalinity is necessary for this action, and the **bile** neutralizes any acid passing over from the stomach. It also emulsifies fats, and facilitates their absorption.

The ridges and folds into which the lining of the alimentary canal is raised largely increase its absorptive surface. Some of the products of digestion pass at once into the blood-system, others reach it indirectly by way of the lacteals, subvertebral sinus, and lymph-hearts. The refuse is ejected at the cloacal aperture.

Most of the products of digestion, by way of the portal vein, pass through the liver, and this organ absorbs the carbo-hydrates, stores them up (as *glycogen*), and returns them to the system as required.

5. The **circulatory organs** consist of a closed *blood-system* with which a *lymph-system* communicates.

(1) **Blood System.**—The *blood* is a bright-red, coagulable fluid, composed of a clear *plasma* in which float *colourless corpuscles* of

the usual type, and larger and more numerous *red corpuscles*, which owe their colour to hæmoglobin. They are oval and flat, with a large central nucleus of similar shape, projecting somewhat on both surfaces of the corpuscle. The blood circulates in a closed system of tubes, consisting of heart, arteries, veins, and capillaries.

The **heart** (Figs. 51 and 53) is situated in the anterior part of the thoracic region, on its ventral side. It is enclosed in a membranous bag, the *pericardium*, which is composed of an inner layer closely adherent to the heart, and a loose outer layer. A space, the *pericardial cavity*, which is a separated part of the cœlom, is found between the two. The heart is roughly conical

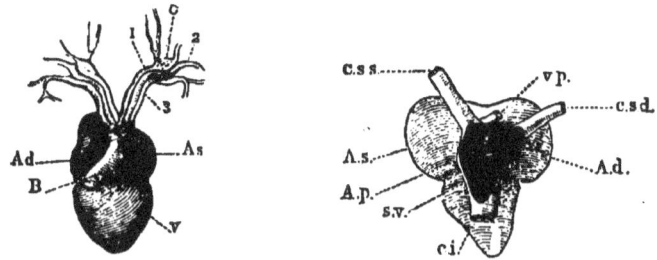

Fig. 53.—HEART OF FROG (after *Ecker*), slightly enlarged.—*a*, Ventral view; *b*, dorsal view, sinus venosus opened; *s.v*, sinus venosus; *Ap*, its opening into *Ad*, right auricle: *As*, left auricle; *V*, ventricle; *B*, truncus arteriosus; 1, 2, 3, carotid, systemic, and pulmocutaneous arches; *c*, carotid gland; *c.s.d* and *c.s.s*, right and left precavals; *c.i*, postcaval; *v.p*, pulmonary vein.

in shape, and its backwardly-directed apex fits into a notch in the liver. A dark coloured triangular sac, the *sinus venosus*, with thin walls, lies on the dorsal side of the heart, and its base is formed by two thin-walled, dark-red *auricles*, right and left, while a muscular, paler *ventricle* makes up its apical region. A muscular tube, the *truncus arteriosus*, leading from the right side of the ventricle, is closely applied to the ventral surface of the right auricle.

The cavity of the right auricle is separated by a thin muscular partition, the *auricular septum*, from that of the left auricle. Into the dorsal wall of the former cavity, near the septum, the sinus venosus opens by an oval valvular aperture. There is a somewhat similarly situated rounded aperture in the left auricle, that of the pulmonary veins. The auricles have a common opening into the ventricle, guarded by two flaps, a dorsal and a ventral, which project into the ventricular cavity, and are attached to its walls by numerous minute fibrous strings. This auriculo-ventricular valve is imperfectly divided into two by the free posterior edge of the auricular septum, which is attached above and below to the flaps.

The relatively small cavity of the ventricle is transversely elongated, and the ventricular wall is spongy. The auricles open into the left side, and the truncus arteriosus out of the right side of its front end. The latter opening possesses three *semilunar valves*, small membranous pouches with their concavities towards the truncus. This is divided into two parts—(1) the *pylangium* (into which the aperture just mentioned opens), separated by another set of semilunar valves from (2) the excessively short *synangium*. The cavity of the pylangium is imperfectly divided into right and left halves by a sinuous longitudinal flap attached to its dorsal wall, and to one of the anterior semilunar valves, but possessing a free ventral

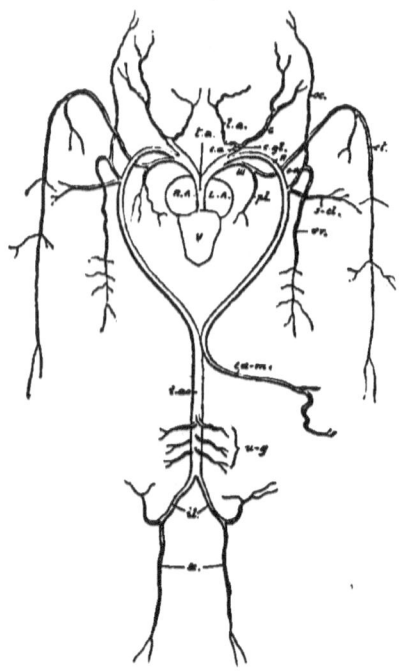

Fig. 54—ARTERIES OF FROG (after *Ecker*). —R.A, Right auricle; L.A, left auricle; V, ventricle; *t.a*, truncus arteriosus; *c.a*, carotid arch; *c.gl*, carotid gland; *c*, carotid artery; *l.a*, lingual artery; II., systemic arch; *o-v*, occipito-vertebral artery; *o*, occipital artery; *vr*, vertebral artery; *s.cl*, subclavian artery; III., pulmo-cutaneous arch; *pl*, pulmonary artery; *ct*, cutaneous artery; *d.ao*, dorsal aorta; *cœ-m*, cœliaco-mesenteric artery; *u·g*, urinogenital arteries; *il*, iliac arteries; *sc*, sciatic arteries.

edge. The passage on the left of this flap leads to a small pulmo-cutaneous aperture just behind the anterior valves, the passage on the right into the synangium. This is continuous on either side with a systemic arch and in front two small carotid apertures, placed on a small elevation, open out of it.

Arteries (Fig. 54).—From the truncus arteriosus two apparently single trunks arise, one on either side, and form with it a Y-shaped figure. Each of these trunks is in reality triple, its cavity being divided by two longitudinal partitions into three cavities—anterior, middle, and posterior. These belong to three corresponding *aortic arches*—carotid, systemic, and pulmo-cutaneous, into which the trunk soon splits. Into the first and last of these the apertures of the same name lead, and the synangial cavity is directly continuous with the cavities of the systemic arches.

(1) The **carotid arches** supply the brain, orbit, and walls of the mouth-cavity with pure blood.

Each gives off a small *lingual artery* to the tongue, dilates into a small, rounded body, the *carotid "gland,"* and then becomes the *carotid artery*, which divides into α, *external carotid* supplying the orbit, and roof and side-walls of mouth-cavity; β, *internal carotid artery* supplying the brain.

(2) The **systemic arches** supply the rest of the body with partially purified blood. The arch of each side runs upwards and backwards, giving off branches to the larynx, œsophagus, fore-limb, and other parts, and uniting with its fellow, just ventral to the backbone, near the front ends of the kidneys, to form the *dorsal aorta*, a median trunk running back in the subvertebral lymph sinus. The dorsal aorta gives off branches to the viscera and body-walls, ultimately dividing, at the posterior end of the body, into two *iliac arteries*, continuations of which run into the hind-limbs.

Chief Branches.—(1) From each arch in front of dorsal aorta—(a) *laryngeal* to larynx; (b) one or two *œsophageal* to œsophagus; (c) *occipito-vertebral*, dividing into (α) *occipital* for side of head and jaws, (β) *vertebral*, running along one side of the vertebral column, supplying adjacent muscles and sending branches to the spinal cord through the intervertebral foramina; (d) *subclavian* to shoulder and fore-limb. (2) From dorsal aorta—(a) *cœliaco-mesenteric*, coming from near junction of two arches, or from left arch before the union. It divides into α, *cœliac artery*, (which has two branches, 1, *gastric artery* to stomach, 2, *hepatic artery* to liver and gall-bladder), and β, *mesenteric artery*, which breaks up into

1, *anterior mesenteric artery* to front part of intestine, 2, *posterior mesenteric artery* to back part of intestine, 3, *splenic artery* to spleen ; (*b*) 4 to 6 small *urinogenital arteries* which at once bifurcate, and run to urinogenital organs and fat-bodies ; (*c*) small paired *lumbar arteries* to adjoining muscles ; (*d*) a small *hæmorrhoidal* artery to the large intestine. (3) Each iliac artery gives off (*a*) a *hypogastric artery* to the bladder, and (*b*) *epigastric arteries* to the ventral body-wall. It then becomes the *sciatic artery* of the thigh, which divides into *peroneal* and *tibial arteries* for the leg and foot.

(3) The **pulmo-cutaneous arches** carry impure blood to the lungs and skin. Each of them quickly divides into a *pulmonary artery* running down the outer side of the lung and a *cutaneous artery* which ramifies in the skin.

The **veins** (Fig. 55) bring back blood from the various parts of the body to the heart. They may best be considered under the following headings:—(1) Caval system ; (2) Portal systems ; (3) Pulmonary veins.

The **caval system** consists of three caval veins with their branches—*i.e.*, two precavals (anterior or superior venæ cavæ) in front, and a postcaval (posterior or inferior vena cava) behind. These pour their blood into the sinus venosus, which may be regarded as formed by the fusion of their ends (see Fig. 55). The *precaval* on each side is formed by the union of three veins —(1) The *external jugular* bringing back blood from the lower jaw, floor of the mouth, and tongue; (2) The *innominate* formed by the union of the *internal jugular* and *subscapular veins* which respectively return blood from the brain, and shoulder with back of arm ; and (3) The *subclavian*. This is a large vein made up of the *brachial vein* from the fore-limb, and the *musculo-cutaneous vein*, which brings back blood from the muscles and skin of the sides and upper surface of the body and head.

The *postcaval* lies ventral to the dorsal aorta. It commences between the kidneys, and is made up by the union of several pairs of *renal veins* from those organs, and veins from the genital glands and fat bodies. Just before entering the sinus venosus it receives two *hepatic veins* from the liver.

Portal Systems.—A *portal vein* is one which, instead of pouring its blood into a larger trunk, breaks up into capillaries within the substance of some organ, supplying it with impure blood. Both kidneys and liver in the frog possess such a supply, and in accordance with this, (1) *renal portal* and (2) *hepatic portal* systems can be distinguished.

AMPHIBIA.

(1) *Renal Portal.*—The principal veins of the hind-limb are the *femoral* and *sciatic*, running along the front and back of the thigh. The femoral vein is connected by a cross-branch (*) with the sciatic just before reaching the trunk, and then divides into a *pelvic vein* which runs to the middle ventral line to unite with its fellow, and a *renal portal vein* which runs to the outer side of the kidney, and divides into small veins breaking up in the kidney-substance. The renal portal vein receives the sciatic vein, *dorso-lumbar* veins from the dorsal abdominal walls, and, in the female, small veins from the oviduct.

(2) *Hepatic Portal.*—The liver is supplied with impure blood by two veins, (*a*) the anterior abdominal and (*b*) the portal. (*a*) The *anterior abdominal* is formed by the union of the two pelvic veins, receives veins from the bladder, and runs forwards in the middle line of the abdominal wall as far as the liver. Here it bifurcates, the two divisions going to the right and left halves of the liver. (*b*) The *portal vein* is formed by the

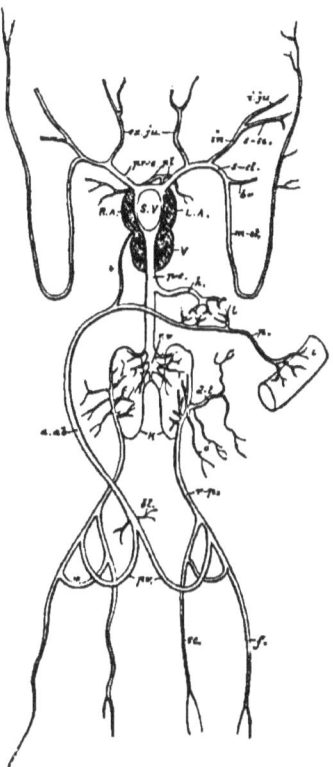

Fig. 55.—VEINS OF FROG (after *Ecker*), semi-diagrammatic.—*S. V*, Sinus venosus; *R.A*, right auricle; *L.A*, left auricle; *V*, ventricle; *pr-c*, precaval; *ex.ju*, external jugular veins; *i.ju*, internal jugular vein; *s-sc*, sub-scapular vein; *in*, innominate vein; *s-cl*, subclavian vein; *br*, brachial vein; *m-ct*, musculo-cutaneous vein; *p-c*, postcaval; *sc*, sciatic vein; *f*, femoral vein; *, cross-branch connecting sciatic and femoral; *p.v*, pelvic veins; *r.p*, renal portal vein; *d-l*, dorso-lumbar veins; *o*, veins from oviduct; *r.v*, renal veins; *a.ab*, anterior abdominal vein; *bl*, veins from bladder; +, small vein from truncus arteriosus; *p*, portal vein; *h*, hepatic veins; *k*, kidneys; *i*, represents alimentary canal with its capillaries; *l*, represents capillaries of liver; *pl*, pulmonary veins.

200 AN ELEMENTARY TEXT-BOOK OF BIOLOGY.

union of the *gastric vein*, bringing blood from the stomach, with the *lieno-mesenteric vein* returning it from the spleen and intestines. It runs to the left half of the liver, and is connected by a cross-branch with the point of bifurcation of the anterior abdominal.

A **pulmonary vein** runs up the inner side of each lung, and unites with its fellow to form a short common trunk, which opens into the left auricle.

The **capillaries** are excessively fine tubes forming networks in nearly all parts of the body, by means of which the ultimate branches of veins and arteries are united. A vein *commences*, an artery *ends*, and a portal vein both commences and ends, in capillaries.

(2) **Lymph System.**—The **lymph** is a colourless fluid which resembles blood, differing, however, in the absence of red cor-

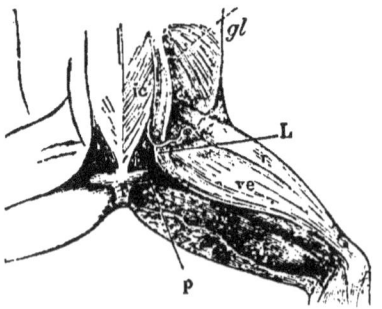

Fig. 56.—POSTERIOR LYMPH-HEARTS OF FROG (after *Ecker*).
L, Lymph-heart.

puscles. The lymph-system consists of **lymphatic vessels** of larger and smaller size, which ramify in the skin, intestinal wall, and other parts of the body, and **lymph-spaces** with which these vessels are connected. The lymph-spaces include the irregular lacunæ between the tissues, and also larger cavities, such as the pleuro-peritoneal, pericardial, subvertebral, and subcutaneous spaces. The lymphatics of the intestine, which receive the special name of **lacteals**, traverse the mesentery and open into the subvertebral sinus. The lymph is propelled by two pairs of **lymph-hearts**, small oval sacs, with rhythmically-contractile walls. The anterior lymph-hearts are situated one on each side between the transverse processes of vertebræ 3 and 4, and are connected with the subscapular veins. The posterior lymph-hearts (Fig. 56) are placed one on each side of the urostyle, not far from its posterior end, in a triangular space between the muscles. Each communicates with a small vein which opens into the cross-branch between the sciatic and femoral veins of its side.

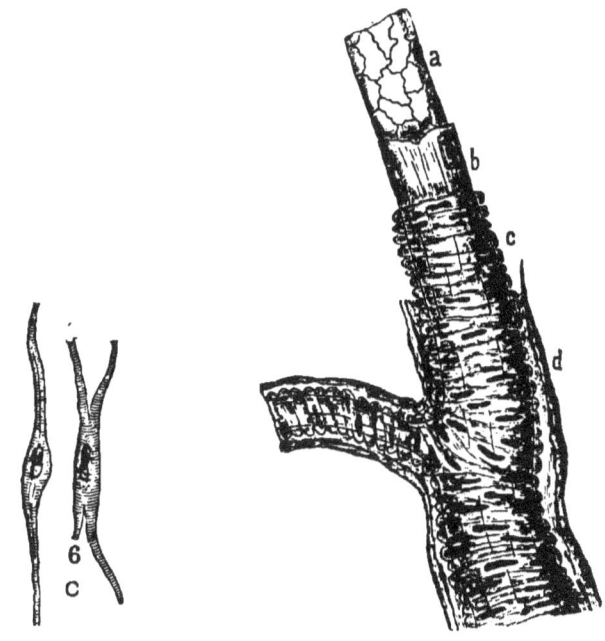

Fig. 57. Fig. 58.

Fig. 57.—MUSCLE-FIBRES FROM HEART OF FROG (from *Landois* and *Stirling*), much enlarged.

Fig. 58.—SMALL ARTERY SHOWING THE COATS (from *Landois* and *Stirling*).— *a*, Endothelium; *b*, elastic membrane; *c*, muscular coat; *d*, connective-tissue coat.

Closely connected with the circulatory system are several small organs known as ductless glands. These are the *spleen* and the *thymus* and *thyroid* "*glands.*" The **spleen** is a small rounded body, of dark red colour, lying in the mesentery, near the commencement of the large intestine. The **thymus glands** are two small bodies situated one on either side, near the end of the quadrate cartilage, and the **thyroid glands** are similar structures, two or more in number, placed near the posterior hyoid cornua.

The cavities of the heart and blood-vessels are lined by *endothelium*, which alone forms the capillary walls (Fig. 59). The large lymph-cavities are similarly lined, and patches of granular *germinal* epithelial cells occur in them, from which colourless corpuscles are budded off. The heart is mainly composed of unicellular, spindle-shaped muscular fibres (sometimes branched), which are transversely striated, but possess no sarcolemma

Fig. 59.—CAPILLARIES (from *Landois* and *Stirling*), much enlarged.— The outlines and nuclei of the endothelial cells making up their walls are clearly shown.

(Fig. 57). The veins and arteries possess three coats outside their endothelial lining (Fig. 58). These are, beginning from the inside—(1) a membrane formed by elastic connective tissue; (2) a muscular coat made up for the most part of unstriated muscle-fibres arranged circularly; and (3) a connective-tissue sheath. These coats are thicker in arteries than in veins, and in large vessels than small. They gradually thin out as capillaries are approached, till at last only endothelium is left (Fig. 59).

Circulation.—The heart receives oxygenated blood from the lungs by the pulmonary veins, which open into the left auricle. The caval veins pour impure blood from the general body into the right auricle.

The blood in the precavals is partly oxygenated, as they receive the cutaneous veins from the skin, while that in the postcaval is largely free from nitrogenous waste.

The blood is pumped out of the heart by the contraction of its walls. The sinus venosus first undergoes systole, then the auricles together, then the ventricle, and lastly the truncus arteriosus. Systole is followed by diastole, and the various valves prevent regurgitation.

Although the ventricle is single, and receives both kinds of blood, yet these do not completely mix, and the carotid, systemic, and pulmo-cutaneous arches are supplied by oxygenated, mixed, and impure blood respectively. This is explained as follows:— The cavity of the ventricle is transversely elongated. Into its left side the auricles open (their common aperture being rendered *practically* double by the free edge of the auricular septum), and out of its right side the truncus arteriosus. Moreover, its spongy wall absorbs much of the blood received by the auricular systole, and renders mixing more difficult. Just previous to the ventricular systole, the three arches offer different degrees of resistance to the passage of blood—the pulmo-cutaneous least, owing to the small extent of their branches, the carotid most, on account of the spongy carotid glands. When the ventricle contracts, the blood nearest the truncus is impure, as the right auricle fills the right side of the ventricle. This blood takes the direction of least resistance—*i.e.*, runs on the left of the longitudinal septum of the truncus into the pulmo-cutaneous arches. As these are filled, their resistance increases, and blood (now mixed, as sufficient time has elapsed for mixture to commence in the ventricle) flows on the right side of the septum, flapping it over the pulmo-cutaneous aperture, and enters the systemic trunks. Their resistance also increases, and the last portion of blood runs into the carotid arches. This is oxygenated blood from the left auricle, which has been stored up in the spongy wall of the left side of the ventricle.

The large amount of elastic and muscular tissue in the arterial walls serves two useful purposes. (1) Part of the force of the systole of the ventricle and truncus is expended in dilating the arteries, which subsequently contract, and continue the pumping work during diastole, so that jerking is avoided. (2) The blood-supply to any part can be regulated. Owing to the influence of the sympathetic system the muscle is usually kept in a semi-contracted (tonic) state. The same system can cause them to contract more or to contract less, thus diminishing or increasing their calibre.

The thin walls of the capillaries readily allow the blood-plasma to soak out of them, and waste products to enter them in the tissues, and leave them in the excretory organs.

The *lymph-hearts* pump lymph into the blood-system. Blood-plasma exuding in excess through the capillaries is thus carried back again, and some of the products of digestion also enter the blood in this way.

6. Respiratory Organs (Fig. 51).—The opening of the *glottis* leads into a small *larynx*, which is firmly fixed between the posterior hyoid cornua. Its walls are strengthened by two triangular *arytenoid cartilages* bounding the glottis, and below these by a ring-like *cricoid cartilage* with several processes. Projecting from the sides of the cavity into its interior are two elastic folds, the *vocal cords*, one on each side. The larynx opens behind into the two *lungs*, which lie freely in the dorsal part of the thoracic region. Each is an elongated thin-walled bag, which first dilates and then tapers to a smooth tip. Its inner surface is raised into a prominent series of ridges, which form a close honeycombing.

The *skin* must also be considered as a respiratory organ.

The larynx is lined by a continuation of the ciliated epithelium of the mouth-cavity, and the lungs, for the most part, by simple squamous epithelium. The basis of these organs is fibrous connective tissue, with many elastic fibres and a good deal of unstriated muscle. A network of capillaries immediately under-lies the epithelium.

Respiration is said to be " buccal," because *inspiration* (*i.e.*, intaking of air) is effected by the agency of the mouth-cavity, as follows :—The mouth being shut, and the walls of the gullet contracted, appropriate muscles lower the floor of the mouth, and thus cause air to rush in by the nostrils. These are then closed, the floor of the mouth is elevated, and air is thus forced into the lungs. *Expiration* is mainly brought about by the elasticity of these organs, and the contraction of the muscle-fibres they contain. The contraction of the abdominal walls may also assist.

The thin-walled lung-capillaries are only separated from the air contained in the lung-cavities by a layer of epithelium, and, therefore, carbon dioxide can readily diffuse out of, and oxygen into, the blood which they contain. A similar interchange takes place between the blood in the capillaries of the skin and the oxygen in the surrounding air or water. This cannot take place in air unless the skin is damp.

The *hæmoglobin* of the red corpuscles plays the usual important part in respiration.

Voice is produced by the *vocal chords*, the edges of which can be brought parallel to one another by certain muscles, and then thrown into vibration by the expired air. The vocal sacs in the male, *R. esculenta*, serve as resonators.

7. Urinogenital Organs, including excretory and reproductive organs.

The **excretory organs** (Figs. 51 and 60) are two elongated, flattened, reddish-brown *kidneys*, symmetrically disposed in the posterior part of the subvertebral lymph-sinus, and covered with pleuro-peritoneum on their ventral surfaces only. A slender tube, *urinary duct* in the female, *urinogenital duct* in the male, runs from the outer side of each kidney to open into the dorsal side of the cloaca, close to its fellow, by a minute slit. Directly opposite this there is a rounded aperture in the ventral cloacal wall which leads into the *urinary bladder*, a large bilobed sac with delicate membranous walls.

<small>Closely connected with the excretory organs are two anomalous structures, the *adrenals* and *fat-bodies*. The former are two narrow bodies, of yellowish colour, one of which is closely imbedded in the ventral surface of each kidney. The *fat-body* (corpus adiposum) is a bright, orange-coloured tuft of finger-like processes attached to the front end of the kidney of either side. It consists of a network of connective tissue, in the meshes of which numerous *fat-cells* are imbedded. These are spherical bodies full of fat, and covered by a thin layer of protoplasm with a nucleus on one side. They result from the metamorphosis of ordinary connective-tissue cells.</small>

The kidney is mostly made up of a large number of *uriniferous tubules*. Each of these commences in a thin-walled *Bowman's capsule*, lined by delicate squamous epithelium, into which has been pushed from the outside, so to speak, a tuft of capillaries, the *glomerulus*, formed by the breaking up of a branch of one of the renal arteries. The capsule and glomerulus together are known as a *Malphigian body*. A convoluted tubule in which four regions can be distinguished succeeds Bowman's capsule. These regions are (1) a short *neck*, lined with ciliated epithelium; (2) a much convoluted *glandular* part, lined by columnar epithelium which is pigmented and granular; (3) a second ciliated section; and (4) a collecting portion.

The *urinary duct* is formed by the union of the collecting portions, and is at first imbedded in the kidney on its outer side. The renal-portal vein breaks up into capillaries which closely surround the glandular parts of the uriniferous tubules, and are joined by the small veins coming off from the glomeruli.

The nitrogenous waste of the body (as *urea*, CH_4N_2O), together with much water and various salts, is excreted by the kidneys. The water and salts are filtered from the glomeruli into Bowman's capsules, while the urea is excreted by the glandular parts of the kidney-tubules. The urinary bladder serves as a temporary receptacle for the urine.

Male Reproductive Organs (Fig. 51).—An oval, yellowish, *spermary* (testis), enveloped in pleuro-peritoneum, is closely connected by a double fold of the same, the *mesorchium*, with the inner side of each kidney. In this fold a number of delicate tubules, the *vasa efferentia*, run from one organ to the other. The two spermaries are commonly unequal in size. In *R. esculenta* the *urinogenital duct* is considerably dilated just as it leaves the kidney, and in *R. temporaria* an oval glandular mass, the *vesicula seminalis*, is placed on its outer side immediately before it enters the cloaca.

The male Frog is distinguished externally by the characters mentioned on pp. 175 and 176.

The spermary is essentially composed of seminiferous tubules, lined by epithelium, and opening into a central cavity. From this the vasa efferentia arise, and, after traversing the mesorchium, enter a longitudinal canal imbedded in the inner edge of the kidney, and giving off numerous transverse tubules, which join the collecting parts of the uriniferous tubules just before they enter the urinary duct. The *sperms* (spermatozoa) are thread-like bodies, with narrow cylindrical heads and vibratile tails. They are developed from spindle-shaped sperm-mother-cells, which largely make up the epithelial lining of the seminiferous tubules. Each sperm-mother-cell originates a tuft of sperms, an unused stump (sperm-blastophor) being left over.

Female Reproductive Organs (Fig. 60).—The *ovaries* are two flattened sacs, varying much in size according to the season, and situated similarly to the spermaries. The pleuro-peritoneum invests each of them and forms a double suspensory fold, the *mesovarium*. Ova, of various sizes, enclosed in follicles, project on the outer surface of the ovary. The *oviducts* are two much-coiled tubes invested in pleuro-peritoneum, and each with a narrow anterior end opening near the root of the lungs by a ciliated funnel. The two funnels are close together (*the organs in the Fig. have been taken out of the body*), and open into a common ciliated depression. The oviducts widen behind, and each of them finally dilates into a thin-walled *uterus*, opening on a prominent papilla on the dorsal wall of the cloaca. The two papillæ are situated close together, just in front of the urinary apertures.

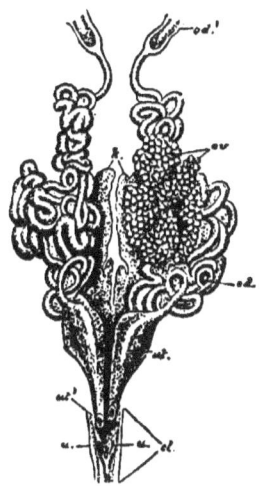

Fig. 60.—FEMALE REPRODUCTIVE ORGANS OF FROG (after *Ecker* and *Wiedersheim*). Right ovary removed.—*ov*, Left ovary; *od*, oviduct; *od'*, funnel of ditto; *ut*, uterus; *ut'*, opening of ditto in cloaca; *k*, kidneys; *u,u*, openings of ureters; *cl*, cloaca.

The *ovary* contains numerous ova enclosed in *follicles*, by the rupture of which they escape into the body-cavity. The mature *ovum* is about $\frac{1}{12}$ of an inch in diameter. It is covered by a delicate *vitelline membrane*, and possesses granular pigmented *vitellus*, and a large *germinal vesicle* with numerous *germinal spots*. Two *polar cells* are formed before fertilization.

The wall of the *oviduct* contains numerous branched tubular glands which secrete a glairy substance, capable of swelling up immensely by imbibition of water. The wall of the uterus contains numerous unstriated muscle-fibres.

At the commencement of the breeding-season (early spring for *R. temporaria*), the males affix themselves to the backs of the

208   AN ELEMENTARY TEXT-BOOK OF BIOLOGY.

females, in which process the roughened pads on the hands afford firm attachment. This occurs in the water. The ova burst out from the ovaries, are taken up by the ciliated funnels of the oviducts, and pass down those structures to the uteri, receiving on their way gelatinous investments. Here they are aggregated into "clumps," and expelled by the contraction of the muscle in the uterine and cloacal walls. As they are extruded from the body, sperms are shed over them by the male. Fertilization occurs in the usual way by fusion of a sperm with each ovum.

8. Muscular System.—In a tadpole there are lateral muscles divided into myomeres, as in a fish, but only traces of this segmentation are to be found in the adult. The *muscles* are very numerous, and have a very complicated arrangement. The following table will give a rough notion of their classification (after Huxley):—

A. **Muscles of Head and Trunk.** { *Episkeletal*, superficial to the endoskeleton.
{ *Hyposkeletal*, within the endoskeleton.

B. **Muscles of Limbs.** { *Intrinsic*, taking origin in the limbs themselves (including girdles).
{ *Extrinsic*, taking origin outside the limbs.

The firm endoskeleton affords points of attachment, and this is often effected by the intermediation of cord-like *tendons*, almost entirely made up of white connective-tissue fibres. The firm sheaths, *aponeuroses*, by which muscles are covered, may also serve for the attachment of other muscles. This is best seen in *flat* muscles, such as some of those forming the abdominal wall.

The muscles are made up of bundles of transversely *striated fibres*, each of which is invested by a delicate *sarcolemma*, and contains numerous *muscle-corpuscles*—*i.e.*, nuclei surrounded by small quantities of protoplasm. These fibres also exhibit longitudinal striations.

From a physiological point of view the muscular tissue of Vertebrates is divided into—(1) **Involuntary**, not under the control of the will, including the unstriated fibres of the viscera and blood-vessels, as well as the striated fibres of the heart; and (2) **Voluntary**, under the control of the will, including the striated fibres of the ordinary muscle.

The minute structure of muscle and its physiological import have long been much vexed questions. The views here adopted are those of Melland and C. F. Marshall, as set forth in the following summary given by the latter :—

" 1. In all muscles which have to perform rapid and frequent movements, a certain portion of the muscle is differentiated to perform the function of contraction, and this portion takes on the form of a very regular and highly modified intracellular network.

" 2. This network, by its regular arrangement, gives rise to certain optical effects, which cause the peculiar appearances of striped muscle.

" 3. The contraction of the striped muscle-fibre is probably caused by the active contraction of the longitudinal fibrils of the intracellular network; the transverse networks appear to be passively elastic, and by their elastic rebound cause the muscle to rapidly resume its relaxed condition when the longitudinal fibrils have ceased to contract; they are possibly also paths for the nervous impulse.

" 4. In some cases where muscle has hitherto been described as striped, but gives no appearance of the network on treatment with the gold and other methods, the apparent striation is due to optical effects, caused by a corrugated outline in the fibre.

" 5. In muscles which do not perform rapid movements, but whose contraction is comparatively slow and peristaltic in nature, this peculiar network is not developed. In most if not all of the invertebrate unstriped muscles there does not appear to be an intracellular network present in any form, but in the vertebrate unstriped muscle a network is present in the form of longitudinal fibrils only; this possibly represents a form of network intermediate between the typical irregular intracellular network of other cells and the highly modified network of striped muscle.

" 6. The cardiac muscle-cells contain a network similar to that of ordinary striped muscle."

The *motor nerve-fibres* may terminate in *end-plates* (compare Fig. 90), within the sarcolemma of striated muscle-fibres, but more frequently the axis-cylinder breaks into a brush of fibrils which run longitudinally in the muscle-substance.

10. The **nervous system** (Fig. 51) may conveniently be subdivided into—(1) cerebro-spinal axis; (2) cranio-spinal nerves; and (3) sympathetic system.

(1) The **cerebro-spinal axis** is made up of the brain and spinal

cord, contained in the neural canal. They are invested by a delicate pigmented vascular membrane, the *pia mater*, and the neural canal is lined by a firm fibrous membrane, the *dura mater*. Between the two is a lymphatic *arachnoid space*, the walls of which are formed by a delicate *arachnoid membrane*.

The **brain** is the anterior enlargement of the cerebro-spinal axis, contained in the cavity of the brain-case. At an early stage of development it consists of three vesicles, anterior, middle, and posterior. The adult parts derived from these constitute the fore-, mid-, and hind-brains. That part of the **fore-brain** which corresponds to the original anterior vesicle is known as the *thalamencephalon*, and contains a vertical slit-like cavity, the *third ventricle*, the anterior boundry of which is a thin lamella of brain-substance, the *lamina terminalis*. The thin roof of the third ventricle, is covered by a vascular membrane, the *choroid plexus*, and in the young tadpole is connected with a small rounded body, the *pineal "gland,"* by a hollow *pineal stalk*, directed upwards and backwards. The formation of the roof of the skull pinches off the pineal gland, which in the adult underlies the skin covering the top of the head. The floor of the third ventricle is produced downwards in its hinder part into a funnel-like projection, the *infundibulum*, with which a flattened sac, the *pituitary body*, is connected. The much-thickened side-walls (*optic thalami*) are connected together behind by a small band, the *posterior commissure*. A similar *anterior commissure* runs transversely across through the substance of the lamina terminalis, and connects the corpora striata (*see below*).

The rest of the fore-brain is made up of the structures developed from a pair of lateral outgrowths of the anterior cerebral vesicle. These are—(1) the *cerebral hemispheres* continuous with the antero-lateral parts of the thalamencephalon, which structure they partly overlap, and in front of which they extend; and (2) the *olfactory lobes* into which the hemispheres pass anteriorly.

The cerebral hemispheres are smooth, ovoid bodies, broadest behind, and closely approximated in the middle line. Each contains a *lateral ventricle*, the inner wall of which presents an elevation, the *corpus striatum*, and which communicates with the third ventricle behind by a small aperture, the *foramen of Monro*. The olfactory lobes are two small bodies in front of and continuous with the cerebral hemispheres, from which they are not very clearly marked off. They are closely united in the middle line, and each contains an *olfactory ventricle* continuous with the lateral ventricle of its side.

The **mid-brain** is a small axial part lying immediately behind the thalamencephalon, with which it is directly continuous. It contains a narrow canal, the *Sylvian aqueduct*, opening in front into the third ventricle, and with roof and side-walls swollen into two large ovoid *optic lobes*, each of which contains an *optic ventricle* opening into the aqueduct, while its thickened floor is formed by the *crura cerebri*, two masses of longitudinal fibres.

The **hind-brain** is mainly constituted by the *bulb* (*medulla oblongata*) which is continuous in front with the mid-brain, and behind with the spinal cord, which it much resembles in structure. The bulb contains a relatively large *fourth ventricle*, connected by the aqueduct with the third ventricle. Its roof is shaped like a triangle with forwardly-directed base, and is constituted by a vascular membrane. On the under side of the thickened floor is a median groove, the *ventral fissure*. The *cerebellum* is a small, solid, transverse ridge, just behind the optic lobes, which arises as a dorsal outgrowth from the posterior vesicle, and together with the bulb makes up the hind-brain.

The **spinal cord** is a thick-walled tube, somewhat flattened from above downwards, which is contained in the spinal canal. It merges into the bulb in front, and from this point gradually tapers backwards (dilating, however, where the limb-nerves come off), and ends in a filament, the *filum terminale*, which occupies the canal of the urostyle. The small *central canal* of the spinal cord opens into the fourth ventricle. Dorsally, the cord exhibits a deep cleft, the *dorsal fissure*, and there is a similar *ventral fissure* which passes on to the base of the hind-brain.

(2) There are twenty pairs of **cranio-spinal** nerves—ten cranial belonging to the brain, and ten spinal belonging to the spinal cord.

Cranial Nerves.—The two first pairs belong to the fore-brain. I. The *olfactory* arise from the outer and front sides of the olfactory lobes, pass through the foramina in the transverse septum of the sphenethmoid, and supply the olfactory epithelium of the nasal sacs. II. The *optic* nerves. On the ventral side of the thalamencephalon there is an X-shaped structure, the *optic chiasma*. The posterior thickened limbs of this, the *optic tracts*, curve backwards and upwards to fuse with the optic lobes. The anterior thinner limbs are the optic nerves. They pass through the optic foramina to reach the eyeballs.

The connection with the optic lobes is a *secondary* one—that is, the optic structures do not originally grow out of them, but fusion occurs later.

The very small third and fourth pairs belong to the *mid-brain*. III. The *oculo-motor* nerves arise, near the middle line, from the front of its floor, and pierce the side-walls of the skull by special foramina to supply most of the eye-muscles. IV. The *pathetic* (trochlear) nerves are given off from the dorsal surface of the brain just behind the optic lobes. They pass through foramina in the cranial wall, and supply the superior oblique muscles of the eye.

The remaining six pairs belong to the *hind-brain*, and, with the exception of the sixth, arise from the sides of the bulb. V. The *trigeminal* are the largest cranial nerves. Each dilates into a *Gasserian ganglion*, leaves the skull by the *trigeminal foramen*, and immediately divides into two—(1) The *ophthalmic* branch, which runs along the inner side of the orbit, and then divides to supply the olfactory capsule and skin of the snout; and (2) The *maxillo-mandibular* branch. The latter at once bifurcates into— (a) The *maxillary nerve*, supplying the margin of the upper jaw; and (b) The *mandibular nerve*, running round the angle of the mouth and along the outside of the mandible, supplying the skin of that region. Branches from this nerve are also distributed to the muscles of the mouth-floor and to certain muscles which elevate the lower jaw. VI. The very delicate *abducent* nerves arise close together from the ventral surface of the bulb near its front end. Each comes into close connection with the Gasserian ganglion, separates from it, and leaves the skull by a small aperture in front of the trigeminal foramen to supply the external rectus and retractor bulbi muscles of the eye, and also the iris. VII. The *facial* nerve on each side arises just behind the trigeminal, and passes out of the skull with it, having previously united closely with the Gasserian ganglion. It divides at once into—(1) The *palatine nerve*, which runs along the floor of the orbit, on its inner side, and supplies the mucous membrane of the mouth-roof; and (2) The *hyomandibular* nerve. This runs back round the auditory capsule, crosses the columella, and then descends in the posterior wall of the tympanic cavity to the angle of the mouth, supplying the regions of the tympanic membrane and jaw-articulation. It then divides into—(a) The *mandibular* nerve, running along the inner side of the mandible, giving branches to its superficial muscles; and (b) The *hyoid nerve*, which passes along the anterior hyoid cornu and innervates its muscles. VIII. The *auditory* nerves arise immediately posterior to the facials, and enter the auditory capsules to supply the essential organs of hearing. IX. The *glossopharyngeal* nerve on each side takes origin a little way behind the root of VIII., and in close connection with that of X. After leaving the skull by the vagus-foramen, it is connected by a commissure with the facial, and then runs down to the floor of the mouth, along which it runs with a tortuous course, supplying the mucous membrane of the pharynx and tongue. X. The *vagus* (pneumogastric) nerve of each side arises in close connection with IX., passes through the vagus-foramen and dilates outside the skull into the vagus-ganglion. It takes a downward and backward course, and gives off branches to the larynx, heart, lungs, and stomach.

The ten pairs of **spinal nerves** have this in common, that each arises from the spinal cord by two roots—a dorsal, upon which is a ganglion, and a ventral. These unite together to form the nerve-trunk in the corresponding intervertebral foramen, which serves as a point of exit. The roots of the anterior nerves pass directly outwards, but the posterior ones slope more or less backwards within the spinal canal before reaching their foramina. The last few roots thus form with the filum terminale, a brush-like *cauda equina*. The ganglia on the roots are covered ventrally by calcareous sacs. Each nerve sends off, near its commencement, a small *dorsal ramus* to part of the muscles and skin of the back, while the remaining part, the *ventral ramus*, is connected by a delicate commissure, *ramus communicans*, with a sympathetic ganglion.

The first or *hypoglossal* nerves run forwards in the throat-muscles and innervate the muscles of the tongue. The second and third unite to form a *brachial nerve*, which supplies the fore-limb. Nerves 4, 5, and 6 supply the body-wall in their region, while nerves 7, 8, and 9 unite to form a network, the *sciatic plexus*, which gives off branches to the posterior viscera, and a very large *sciatic nerve* to the hind-limb. The small tenth or

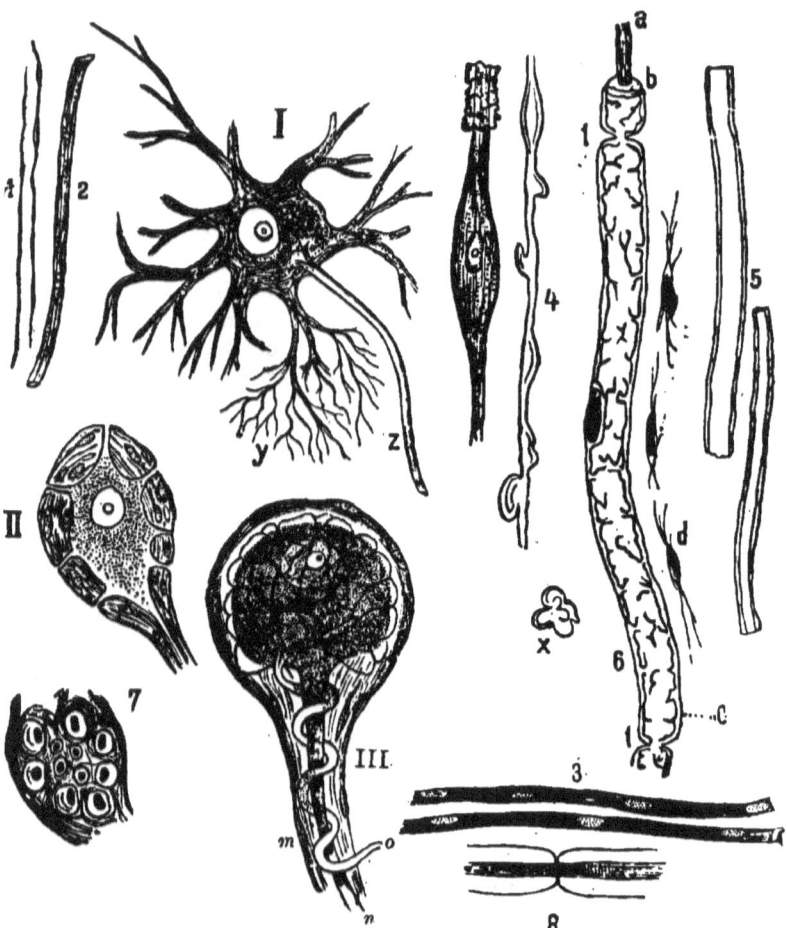

Fig. 61.—HISTOLOGY OF NERVE (from *Landois* and *Stirling*).—1, Primitive fibrillæ. 2, Axis-cylinder. 3, Non-medullated fibres. In their course are seen numerous nerve-corpuscles. 5, 6, Medullated fibres; *c*, primitive sheath, internal to which the medulla, *b*, is seen; *t.t*, nodes of Ranvier—between the two a single nerve-corpuscle; *a*, axis-cylinder. 7, Eleven medullated fibres in cross-section, surrounded by connective tissue; axis-cylinders dark, surrounded by white areas representing medulla. 8, Medullated fibre at node of Ranvier; medulla not shown. I., Multipolar ganglion-cell from spinal cord; *z*, process passing into axis-cylinder; *y*, branched process — to right of this ganglion-cell is a bipolar one. III., Bipolar ganglion-cell from sympathetic of Frog, surrounded by sheath, *m* : *n*, *o*, the two processes.

coccygeal nerve proceeds from the foramen in the urostyle, and supplies the posterior viscera. It is also connected with the sciatic plexus.

(3) The **sympathetic system** is made up of a slender cord on each side, lying just beneath the backbone. It is dilated at intervals into ten sympathetic ganglia, each connected by one or more *rami communicantes* with the corresponding spinal nerve. In front it passes through the vagus-foramen, and is united with the Gasserian ganglion, while in the region of the dorsal aorta the two cords are closely bound to the dorsal side of that vessel. The internal organs, including the vascular system, are supplied by nerves coming off from the sympathetic ganglia, and breaking up into plexuses.

Histology (Fig. 61).—The *ganglion cells* of the cerebro-spinal axis are mostly multipolar. They make up the greater part of the *grey matter*, which forms a central core to the spinal cord, medulla oblongata, optic lobes, and optic thalami, and an external crust to the cerebral hemispheres. The corpora striata are also composed of grey matter. Bipolar ganglion-cells of peculiar type, in which one process is twisted spirally round the other, have been found in the sympathetic ganglia, and unipolar cells in the spinal ganglia. The processes of the ganglion cells either become continuous with the axis cylinders of nerve-fibres or branch, and form networks.

The nerve-fibres are either *non-medullated* or *medullated*. The non-medullated (3) are made up of *axis-cylinder* and *primitive sheath* with *nerve-corpuscles*. They are the only kind found in the sympathetic system and olfactory nerves, besides which they are common in the other parts of the nervous system. The medullated fibres possess a layer of fatty matter, the *medullary sheath*, between the axis-cylinder and primitive sheath. It is broken up into short segments at the *nodes of Schmidt*, and at less frequent intervals *nodes of Ranvier* occur where the primitive sheath is constricted and touches the axis-cylinder. Each *internodal segment*, bounded at each end by one of these last, generally possesses a single nerve-corpuscle. Medullated fibres make up most of the *white matter* of the cerebro-spinal axis, which forms the external part of the spinal cord, medulla oblongata, optic lobes, and optic thalami, and the internal part of the cerebral hemispheres. The crura cerebri are also of white matter.

Both sorts of nerve-fibre are connected *centrally*, directly or indirectly, with ganglion-cells. Peripherally their axis-cylinders may break up into fine plexuses, or those of medullated fibres may terminate in various *end-bodies*, such as *sense-cells*, and, in the case of striated muscle, *end-brushes* and *end-plates* (Fig. 90).

Non-nervous structures are intimately connected with the nervous system, especially connective tissue, which forms the investing membranes, and binds the nerve-fibres into smaller and larger bundles, while the fibres and cells of the central organs are imbedded in an excessively delicate connective-tissue network, the *neuroglia*. The pineal gland and pituitary body are both non-nervous. The former is a rudimentary eye.

The nerve-fibres are physiologically divisible into *afferent* and *efferent*, along which impulses respectively pass *to* and *from* the central organs. The majority of the former are *sensory*, since they are connected with end-organs, the stimulation of which lead, in many cases, to a *sensation*, judging from analogy. A large number of efferent fibres, since they supply muscles, are known as *motor*. The brain and spinal cord are, in the first place, centres for **reflex actions**—*i.e.*, those which are independent of volition, and dependent on external stimuli. The apparatus involved in such an action normally consists of (*a*) end-organs, (*b*) an afferent nerve, (*c*) a nerve-centre, (*d*) an efferent nerve terminating commonly in (*e*) muscular or glandular tissue. The spinal cord alone, after the removal of the brain, enables fairly complicated and purposeful movements of the body to be effected, in answer to external stimuli. Thus, for example, pinching a foot causes the corresponding leg to be drawn up, and the placing on the skin of the back a bit of blotting-paper dipped in acid leads to leg-movements directed to its removal. The cord also serves as a channel by which afferent impulses can travel to the brain, and efferent impulses from it. If the cerebral hemispheres only are removed, still more complicated reflex actions can be evoked by external stimuli, such, for example, as swimming, croaking, and leaping. The medulla oblongata, next to the hemispheres, is the most important part of the brain. It regulates respiration, and has much to do (through the sympathetic) with the alimentary canal and circulatory organs. It is also concerned with the co-ordination of many muscular movements.

Spontaneity entirely resides in the cerebral hemispheres. When these are removed, movements only occur after the application of stimuli. They are also the seat of consciousness and intelligence.

		Afferent.	Efferent.
I.	Olfactory.	Sensory fibres for smell.	... ...
II.	Optic.	Sensory fibres for sight.	... ...
III.	Oculomotor.	... ...	Motor fibres for eyeball muscles.
IV.	Pathetic.		
VI.	Abducent.		
V.	Trigeminal.	Sensory fibres to skin of head and nasal sacs.	Motor fibres to floor of mouth, and certain muscles raising mandible.
VII.	Facial.	Sensory fibres to roof of mouth.	Motor fibres to superficial muscles of mandible, and some hyoid muscles.
VIII.	Auditory.	Sensory fibres for hearing.	... ...
IX.	Glossopharyngeal.	Sensory fibres for taste.	... ...
X.	Vagus.	Afferent fibres (some sensory).	Efferent fibres (some motor).
		To parts supplied.	

The *cranio-spinal nerves* are made up of *afferent* and *efferent* fibres, of which either or both may occur in the same nerve. The preceding table exhibits the leading features of the cranial nerves.

All the *spinal nerves* are of mixed character, the fibres derived from the *dorsal* roots being *sensory*, those from the *ventral* roots *motor*. The sensory-fibres supply the skin, the motor the voluntary muscles of the body; in the case of the hypoglossal some of those belonging to the tongue. The preceding facts have been ascertained by cutting the roots and stimulating the ends, it being remembered that fibres only carry impulses one way, and hence no demonstrable results follow on stimulating the peripheral ends of sensory, or the central ends of motor, fibres.

Therefore, the fibres of the dorsal root are sensory, those of the ventral, motor. (See next page.)

The **sympathetic system** supplies and largely regulates the internal organs. By its means *involuntary* muscular contrac-

218 AN ELEMENTARY TEXT-BOOK OF BIOLOGY.

Severed.	Result.	Central end stimulated.	Peripheral end stimulated.
Dorsal Root.	Loss of sensation in area supplied, but muscles can be moved at will, or reflexly.	Pain (comes to same thing as over-stimulation of parts supplied).	No effect.
Ventral Root.	Sensation remains in parts supplied, but muscles cannot be moved at will, or reflexly.	No effect.	Muscles supplied contract.

tions, such as those of the alimentary canal, circulatory organs, &c., are mainly effected. It is, however, quite subordinate to the brain and spinal cord.

Sense Organs.—(1) **Tactile Organs.**—The sense of touch is possessed by the skin generally, and the mucous membrane of the olfactory sacs and mouth-cavities. The nerve-fibres connected with this sense are derived from the trigeminal in the head and spinal nerves in the rest of the body. They break up into plexuses, and in the dermis numerous groups of oval flattened *touch-corpuscles* are present. These are most numerous beneath the wart-like elevations, and nerve-fibres are continuous with them.

(2) **Gustatory Organs.**—Scattered throughout the mouth-cavity, and especially numerous around the vomerine teeth, and on the fungiform papillæ of the tongue, are small groups of elongated forked *gustatory cells.* They belong to the epithelium, and fibres of the glossopharyngeal nerves terminate in them.

(3) **Olfactory Organs.**—Two *olfactory sacs* (Fig. 51), separated by a median partition, the *nasal septum,* are lodged in the front of the head. Each contains a complicated internal cavity, communicating with the exterior and mouth-cavity by the *external* and *internal naris* respectively, and with glandular walls largely lined by *olfactory epithelium,* containing numerous spindle-shaped olfactory cells (Fig. 62), in which fibres of the olfactory nerve end. From these cells a stiff process or bundle of olfactory hairs projects into the nasal cavity.

(4) **Auditory Organs.**—The *ear* on each side is made up of two parts—(1) Middle ear, and (2) Internal ear.

The *middle ear* consists of a small *tympanic cavity*, communicating by the short *Eustachian tube* with the back of the mouth-cavity, and lined by a continuation of its epithelium. It is closed externally by the rounded *tympanic membrane*, which is stretched over a cartilaginous ring connected with the squamosal and covered externally by skin. The rod-like *columella* (Fig. 50, B) is attached at one end to the tympanic membrane, and stretches across the cavity to the *fenestra ovalis*, into which its other end is inserted.

The *internal ear* is made up of the *membranous labyrinth* (Fig. 62), contained in the cavity of the auditory capsule which, however, it does not fill. Numerous strands of connective tissue unite it with the walls of this cavity, which is filled by a clear fluid, the *perilymph*. The labyrinth is of complicated shape, and consists of vestibule and semicircular canals:—(1) The *vestibule* is divided into (α) the *utriculus*, an irregular bag, the cavity of which is imperfectly bisected by a fold and communicates by a small aperture with that of (β) the *sacculus*, an underlying oval bladder. Three small swellings on the posterior side of the sacculus receive collectively the name of *cochlea*. (2) The *semicircular canals* are three curved tubes—anterior vertical, posterior vertical, and external horizontal, connected with the utriculus. They lie in planes mutually at right angles. The first and last dilate into swellings (*ampullæ*) in front, where they join the utriculus. The posterior vertical unites in front with the anterior vertical to form a common tube opening into the utriculus, and dilates behind into an ampulla.

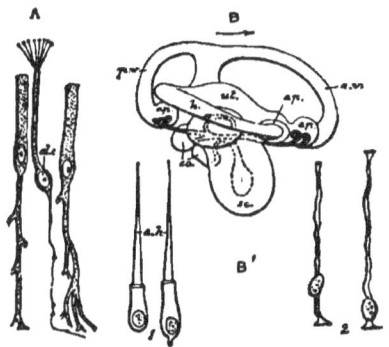

Fig. 62.—SENSE ORGANS OF FROG (after *Ecker* and *Wiedersheim*), various scales. A, Cells from olfactory epithelium; *ol.c* olfactory-cell. B, Right Membranous Labyrinth, viewed from outside. The arrow points to the front. The branches of the auditory nerve are shown (partly dotted). *ut*, Utriculus; *sc*, sacculus; *a.v*, *p.v*, and *h*, anterior vertical, posterior vertical, and horizontal semicircular canals; *ap*, ampullæ; *co*, cochlea. B', 1, auditory cells, with auditory hairs, *a.h*; 2, supporting-cells.

220   AN ELEMENTARY TEXT-BOOK OF BIOLOGY.

The labyrinth is filled with clear *endolymph*, in which are suspended, especially near the sensory patches, a large number of

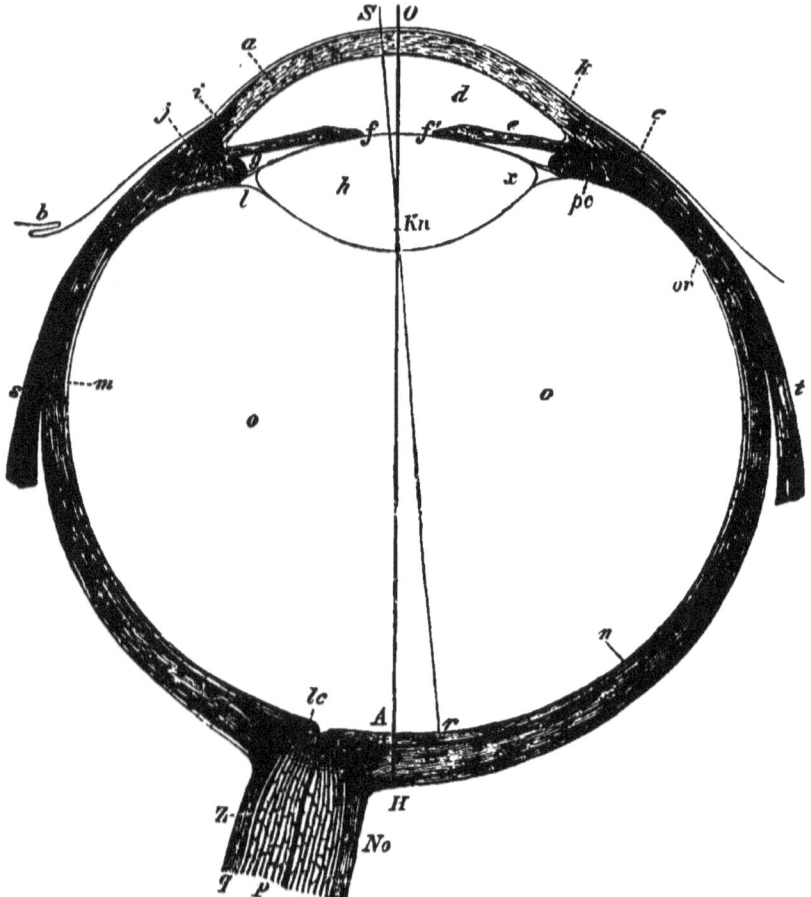

Fig. 63.—Diagrammatic Horizontal Section of Mammalian Eye (from *Landois* and *Stirling*).—*a*, Cornea; *b*, conjunctiva; *c*, sclerotic; *d*, external chamber, with aqueous humour; *e*, iris; *f, f'*, pupil; *h*, lens; *k*, junction of cornea and sclerotic; *m*, choroid; *n*, retina; *o*, vitreous humour; *No*, optic nerve; *q*, nerve sheath; *p*, nerve fibres, piercing outer coats and becoming continuous with retina.

calcareous *otoliths*. It is lined throughout by simple epithelium, mostly squamous, but in eight *sensory patches*, largely made up of

AMPHIBIA. 221

elongated *auditory-cells*, closely connected with fibres of the auditory nerve. From each a tapering *auditory* "*hair*" projects into the labyrinth-cavity. The sensory patches are disposed as follows:—One in each ampulla on a projecting fold, *crista acustica,* and the remaining five,*maculæ*, in utriculus, sacculus (2), and three cochlear projections.

(5) **Visual Organs** (Fig. 63).—The *eye* is relatively large, and projects not only externally but also internally into the mouth-cavity. It is lodged in the *orbit*, on the side of the skull. The eyeball is rounded internally but flattened externally, and its wall is made up of three concentric coatings. The outermost of these, the firm, white *sclerotic*, strengthened by cartilage, exhibits externally a circular transparent area, the *cornea*. The second much thinner coating, the *choroid*, is pigmented externally and very vascular internally. This layer does not line the cornea, but, at its margin, becomes continuous with a vertical partition, the *iris*, which presents a central perforation, the *pupil*. By the iris the eye is divided into a small external and a large internal chamber. The latter is lined by the *retina*, which is the third and most delicate coating. In it eight distinct layers (Fig. 64) can be distinguished. The optic nerve pierces the sclerotic and choroid coats on the inner side of the eyeball to reach the retina, in which it breaks up into a feltwork of fibres. These form a

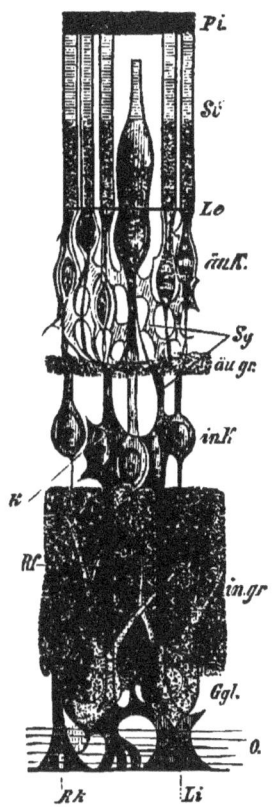

Fig. 64. — DIAGRAM OF LAYERS OF RETINA (from *Landois* and *Stirling*). — *Pi*, Pigment-cells; *St*, rods and cones continuous by various intermediate structures with *Ggl*, ganglion-cells connected with *o*, fibres of optic nerve; *Li*, internal limiting membrane which comes just outside vitreous humour (not shown).

layer on the internal side of the retina (*i.e.*, the side next the eye-chamber), and are connected through various intermediate layers with elongated sense-cells, the *rods* and *cones*, which occupy the outside of the retina. Refracting structures are contained within the chambers of the eye. A clear, watery *aqueous humour* fills the external chamber, while a gelatinous *vitreous humour* is contained in the internal chamber. Immediately behind the iris and connected with its outer margin is a firm spheroidal body, the *lens*.

A number of *accessory* parts are connnected with the eye, of which the two *eyelids* have already been mentioned. Continuous with the lining of these is a delicate transparent membrane, the *conjunctiva*, which closely covers the cornea. A small *Harderian gland*, which secretes a fatty substance, is situated in front of and below the eye.

The eye is moved by a number of small muscles, the most important of which take origin in the walls of the orbit, and are inserted into the eyeball. They are—(1) The *retractor bulbi*, a muscular sheath surrounding the optic nerve; (2) the four *recti* muscles, *superior*, *inferior*, *internal*, and *external rectus*, inserted into the upper, lower, anterior, and posterior sides of the eyeball. All the preceding take origin from the inner wall of the orbit, near its posterior end; (3) the two *oblique* muscles, *superior* and *inferior*, are inserted into the upper and lower sides of the eyeball, taking origin close together in the anterior part of the inner wall of the orbit.

DEVELOPMENT.

The frog is a good example of an animal in which there is a free-living embryo or larva, very unlike the adult stage, into which it is converted by a gradual *metamorphosis*. Frog larvæ are familiar to every one as fish-like tadpoles, which hatch out during March and April from the frog "spawn," which consists of fertilized ova surrounded by gelatinous matter.

1. Early Stages.—Cleavage (segmentation) (Fig. G5) is *complete* but *unequal*. The ovum is pigmented on one side which is termed the *upper pole*, the unpigmented end being the *lower pole*. Division may therefore be spoken of as "meridional" or "equatorial." The two first divisions are meridional, dividing the oösperm into two and four. Then division into eight is effected

AMPHIBIA.

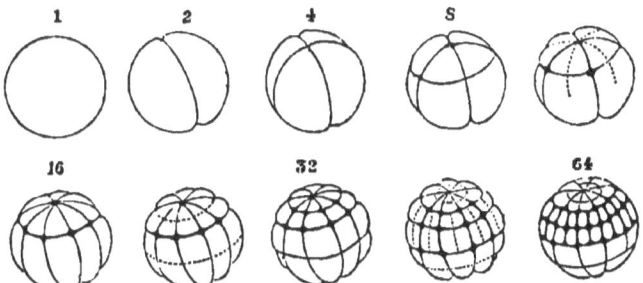

Fig. 65.—CLEAVAGE OF OÖSPERM OF FROG (from *Haddon*, after *Ecker*).—The numbers indicate the number of segments. Dotted lines show position of succeeding planes of segmentation.

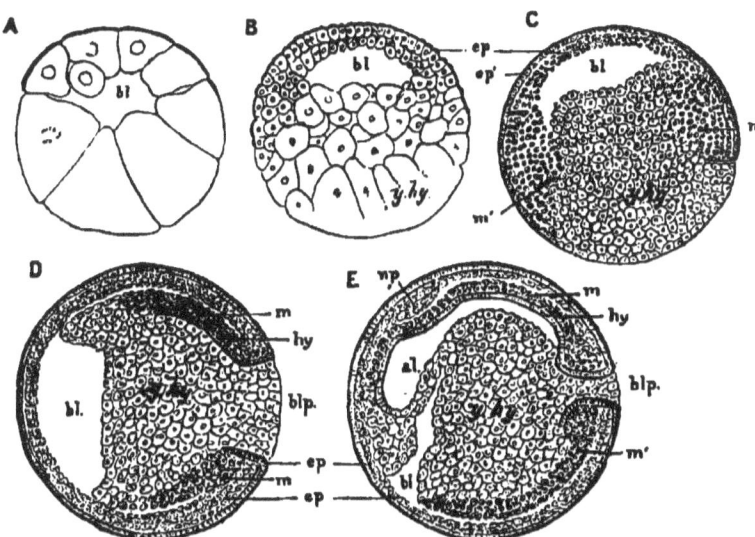

Fig. 66.—BLASTULA AND GASTRULA OF FROG (after *Götte*). All are sections.—A, Partly segmented oösperm. B, Completely segmented oösperm. C, D, E, Stages in formation of gastrula; *al*, archenteron (mesenteron); *bl*, segmentation cavity; *blp*, blastopore; *ep* and *ep′*, nervous and superficial layers of ectoderm; *hy*, endoderm; *m* and *m′*, mesoderm; *y.hy*, lower layer cells.

by an equatorial furrow nearer the upper than the lower pole. Divisions of both kinds now proceed, the final result being a *blastula* (blastosphere) (Fig. 66, B), with a small *blastocœle* (segmentation cavity) near the upper pole. This is roofed by *small cells*, while its floor and most of the blastula are made up of large *yolk-cells*, containing food material.

Invagination (C, D, E) is partly effected by an inpushing of the *small cells* at one point, and partly by their growing over the yolk-cells elsewhere, leaving, however, for a long time an uncovered circular area, the *blastopore*. Meanwhile the blastocœle is gradually pushed aside by the *archenteron*, which occupies the upper pole of the ovum, and is formed partly by the inpushing of the small cells, as above stated, and partly by an absorption of the yolk-cells to form a cavity, and their differentiation to give it definite walls. The small cells, now almost covering the embryo, may at this juncture be termed **ectoderm** (epiblast), the contained yolk-cells **endoderm** (hypoblast). The **mesoderm** (mesoblast) is formed as a layer several cells thick, commencing at the edges of the blastopore and gradually extending over the rest of the ovum below the ectoderm. It is apparently derived from the endoderm. The notochord is formed at the same time, as a median dorsal thickening of the roof of the archenteron.

2. General Growth.—The ectoderm on the upper side of the embryo thickens into a *neural* (medullary) *plate*, broad in front, and narrowing behind to the blastopore. The axis of this plate corresponds to the longitudinal axis of the future tadpole, the blastopore being posterior. A delicate *neural* (medullary) *groove* appears on the medullary plate. Its side-walls, the *neural* (medullary) *folds*, grow up and unite together to form a *neural tube*, the rudiment of the brain and spinal cord. The dorsal side of the embryo is thus indicated, and its head and tail soon become evident, the former being marked by the dilatation of the neural tube into the three *brain vesicles* in front. The ventral surface bulges out, owing to the food material contained in the endoderm, but it gradually becomes less and less conspicuous. The blastopore closes up, and the cavities of neural tube and gut are then connected by a short *neurenteric canal*. As growth proceeds, the embryo acquires a strong curve, the dorsal side being concave. The sense organs become established, as also do an oval suctorial mouth with horny beaks, and a cloacal aperture. On each side of the throat six thickened bars, the *visceral arches*, passing from above downward, are developed. These are termed *mandibular*, *hyoid*, 1st, 2nd, 3rd, and 4th *branchial arches* respectively. From branchial arches 1 and 2 a branched tuft-like *external gill* grows out on either side, and after hatching, which occurs about this period, another proceeds from the third branchial arch. The *tail* now assumes great size and importance as a locomotor organ. In front of each branchial arch a *visceral cleft*, leading into the throat, makes its appearance, and a fold of skin, the *operculum*, grows back from the hyoid arches over the external gills. By

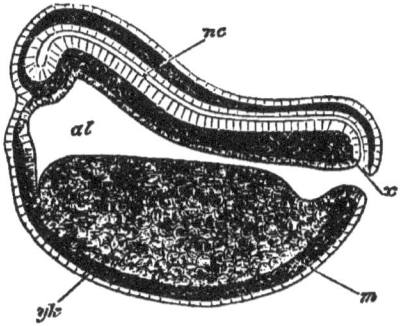

Fig. 67.—DIAGRAMMATIC LONGITUDINAL SECTION THROUGH A FROG EMBRYO (after *Götte*).—Ectoderm unshaded, endoderm shaded, mesoderm (*m*) darkly shaded. *nc*, Neural canal contained within neural tube, which on the left is dilated into the rudimentary brain already exhibiting flexure; *al*, alimentary canal, with mass of yolk (*yk*) beneath; *x*, placed at blastopore, the reference line runs to posterior end of neural canal, which is here continuous with alimentary.

the union of this fold with the body-wall a *branchial chamber* is formed on each side. The right one soon closes, but remains connected by a cross-passage under the throat with the left, which opens to the exterior by a small rounded opening. The external gills now shrink and disappear, being succeeded by *internal gills*, which are vascular folds on the walls of the clefts. As the lungs develop, these too disappear, and the visceral clefts close up.

The limbs have been arising meanwhile as bud-like outgrowths, the anterior pair being hidden at first by the operculum. They appear at the surface after a shedding of skin which follows the atrophy of the gills. At the same time the eyes become evident, and the horny jaws are cast off. The adult form is now rapidly assumed, the tail gradually shrinking, while vegetable food is given up, and exclusively aquatic life abandoned.

3. Fate of the Germinal Layers.—(1) The **ectoderm** is early differentiated into superficial and nervous layers (Fig. 66). The former gives rise to the horny layer of the epidermis, the deeper part of which is formed from the nervous layer. The epithelium of the mouth-cavity (stomodæum) and posterior part of the cloaca (proctodæum) are also ectodermic, these structures arising as pits of the external surface which ultimately communicate with the mesenteron to form a continuous alimentary canal.

The neural plate from which the cerebro-spinal axis originates arises as a thickening of the nervous part of the ectoderm. The essential parts of the sense organs are ectodermic. The nasal sacs commence as pits in this layer, while the membranous labyrinths are originally depressions of the nervous part of it, which are separated off as vesicles and grow into their complicated adult form. The lens of the eye (*cf.* Fig. 80, *l*) is formed as a similar vesicle, which becomes solid. An outgrowth from the anterior cerebral vesicle, the *optic vesicle*, grows towards the embryonic lens, which meets and pushes in its end, forming a double-walled *optic cup*. The part connecting this with the brain narrows to constitute the optic nerve, while the inner wall of the cup becomes the essential part of the retina, the outer part its pigment layer. The rest of the eye is mesodermic.

(2) The **endoderm** gives rise to the epithelium of the mesenteron and its outgrowths (Fig. 67, *al*)—*i.e.*, lungs, liver, and pancreas. The visceral clefts are also formed as pouches of the mesenteron. Four of them open as above described to the exterior. In front of these is a pouch which never opens on the surface and becomes the tympanic cavity. From the endoderm the first rudiment of the endoskeleton is formed. This is a cellular rod, the *notochord*, which underlies the nervous system as far forwards as the pituitary body. Remains of it are found in the centra of the adult vertebræ.

(3) The **mesoderm** at first forms a sheet on each side of the notochord below the ectoderm, and also extends into the front of the head. Each sheet splits into an outer *somatic layer*, and an inner *splanchnic layer*, the two remaining united dorsally. The

AMPHIBIA. 227

layers of the two sides become continuous ventrally, the split between them forming the body-cavity. The mesoderm on each side becomes divided, posterior to the head, into two longitudinal parts. The upper of these, the *vertebral plate*, lies alongside the neural tube and notochord, and becomes divided transversely into *mesodermic somites* (protovertebræ). The lower part is known as the *lateral plate*, the *somatic* mesoderm of which unites with the ectoderm to form the *somatopleure* (wall of body), while its splanchnic layer unites with the endoderm to form the *splanchnopleure* (wall of alimentary canal). Both mesodermic somites and lateral plate are made up of somatic mesoderm externally, and splanchnic mesoderm internally.

The dermis is developed from somatic mesoderm, mostly from that of the lateral plates. The axial endoskeleton is formed from splanchnic mesoderm, that of the mesodermic somites giving rise to the vertebral column, which, with the hinder part of the base of the brain-case, is moulded round the notochord.

The visceral arches become supported by rods of cartilage, similarly named. Their fate is as follows:—

I. Mandibular—Suspensoria and cartilaginous basis of the jaws.
II. Hyoid, ⎫   ⎧Columellæ and anterior
                     ⎪   cornua.
III. Branchial 1, ⎮ Fuse ventrally to ⎨ ⎰ Anterior   angles   of
IV. Branchial 2, ⎬ form hyoid ap-   ⎪ ⎱   hyoid plate.
V. Branchial 3, ⎮ paratus.   ⎪ Posterior angles.
VI. Branchial 4, ⎭   ⎩Posterior cornua.

The rest of the skeleton is also mesodermic.

The alimentary canal and its outgrowths are formed (except the epithelium) from the splanchnic layer of the lateral plate.

Important changes take place in the *circulatory organs* in connection with the successive appearance of gills and lungs. The heart at first possesses an undivided auricle, and pumps impure blood to the gills by *afferent branchial arteries* corresponding to the last four visceral arches (branchial arches), and breaking up into capillaries. After the blood has been oxygenated in these gill-capillaries it passes into *efferent branchial arteries*, which open above into a loop from which the *dorsal aorta* arises. These vessels may be numbered 3, 4, 5, 6, like the visceral arches with which they are connected. As the lungs develop they are supplied by branches of the 6th afferent arteries, and, gradually becoming functional, pour their blood into the auricle, which becomes divided. At the same time the afferent and efferent vessels become directly united to form continuous arches. The internal gills meanwhile atrophy, and the 4th arches alone remain connected to form the dorsal aorta. These changes are best exhibited tabularly.

	1st Stage.	2nd Stage.	3rd (Adult) Stage.
3rd arches, 4th arches, 5th arches, 6th arches,	Supply external gills.	Supply internal gills.	Become carotid arches. Become systemic arches. Disappear. Become pulmo-cut. arches.

The adult *kidney* is preceded by a rudimentary excretory organ, the **pronephros** (head-kidney), consisting of a glandular convoluted tube, with three openings into the body-cavity. A longitudinal *pronephric duct* (= archinephric or segmental duct) runs back from this and opens into the cloaca. The adult kidney, more properly termed the **Wolffian body** or **mesonephros**, is developed as a number of tubules, at first comparatively simple, which become connected with the body-cavity on the one hand, and the pronephric duct on the other. This duct becomes the urinogenital duct of the male and the ureter of the female.

The *ovary* and *spermary* are developed as thickenings of the peritoneum lining the body-cavity. The *oviduct* (Müllerian duct) commences on the dorsal wall of the body-cavity as a groove which is subsequently converted into a tube.

The *voluntary muscles* are formed from the somatic layers of the mesodermic somites.

CHAPTER XI.—AVES.

§ 16. COLUMBA LIVIA (The Rock-Pigeon).

The common Rock-Pigeon is here selected for description, but it must be remembered that the very numerous fancy breeds have all been derived from this form. In some few places the structure of the Fowl (Gallus bankiva) is alluded to, where it differs markedly from the Pigeon.

MORPHOLOGY AND PHYSIOLOGY.

1. **External Characters.**—The bilaterally symmetrical body, in addition to head and trunk, possesses a long flexible neck, and a short stumpy posterior process, the tail (*uropygium*). Fore- and hind-limbs are present, the former modified into wings, the latter set on to the body very far forwards, thus securing equilibrium, and facilitating bipedal progression. The body is largely, but not entirely, covered with feathers, arranged in definite tracts (*pterylæ*), there being intervening bare spaces (*apteria*). There are several kinds of feathers, the largest, *quill-feathers*, present on the wings and tail, are those principally used in flight. The wing-quills (*remiges*) are 19 in number on each side. They are overlapped, above and below, by smaller *wing-coverts*. The tail-quills (*rectrices*), 12 in number, are attached to the uropygium, and are overlapped on both sides by *tail-coverts*. The small soft feathers, which invest all parts of the body, and confer the characteristic outlines, are *contour-feathers*. In addition to these a very great number of minute rudimentary feathers are present, the *filoplumes*.

The head is well rounded behind, while in front it tapers into the beak, the upper and lower parts of which are encased in horn, and bound the large mouth. A bare elevated area, the *cere*, is present at the base of the upper division of the beak, and in front of this two slit-like nostrils (*external nares*). The large eyes possess upper and lower featherless eyelids, and each is provided as well with a thin translucent third eyelid (*nictitating membrane*) which can be quickly drawn over the eye, or folded up in its anterior angle. A little way behind and below the eye is a rounded auditory opening concealed by feathers. It opens into a short tube, the *external auditory meatus*.

The **trunk** is somewhat boat-shaped, presenting a prominent ridge along the middle of its ventral side, except for a small posterior area where soft abdominal walls can be left. A large transverse *cloacal aperture* with swollen edges, opens ventrally at the base of the tail. Upon the dorsal surface of the latter structure there is a small papilla, upon which the *oil-gland* opens (Fig. 70, *u.gl*).

The **fore-limb** (*cf.* Fig. 69) is divided into upper arm, forearm, and hand (*brachium, antebrachium*, and *manus*), which are of about equal length, and folded closely one upon another when not extended for flight, but, when so extended, approximating to the primitive position. The manus is *tridactyle*, but the small thumb (*pollex*) is the only digit well-marked externally. It bears a small tuft of feathers, the *bastard wing*. Eleven of the remiges are attached to the hand. These are the *primary quills*. The remaining eight remiges, *secondary quills*, are attached to the forearm.

The **hind-limb** (*cf.* Fig. 69) is divided into thigh, leg, and foot (*femur, crus*, and *pes*). The last is made up of a cylindrical *tarsometatarsus*, about as long as the thigh (the leg is much longer), and four *digits*, of which the first, or great toe (*hallux*), is directed backwards, the others forwards. The foot is covered by overlapping scales, and the digits are terminated by claws.

Position of Body.—The directions assumed by the divisions of the *fore-limb* have already been mentioned. In the standing posture the thigh, leg, and tibio-tarsus are arranged in a longitudinal vertical plane with the knee directed forwards and the mesotarsal ankle-joint backwards. The great toe is also turned backwards, and owing to its free metatarsal is more mobile than the other forwardly directed digits.

2. The extremely thin **skin** is divided into *epidermis* and *dermis*, the former possessing horny and deeper layers. The scales and claws are made up of coalesced epidermic cells, and the feathers are also modifications of this layer. A **feather** consists of a central stem, the proximal part of which is a hollow *quill*, and the distal part a solid *shaft*, forming the central part of the expanded *vane*, the rest of which is made up of a row of

narrow *barbs* on each side, flattened at right angles to the axis of the feather. Each barb possesses a proximal and a distal row of minute *barbules*. These overlap and hold the barbs together, and, for this purpose, the distal barbules are provided with diminutive hooklets. A minute aperture, the *inferior umbilicus*, leads into the proximal end of the quill, and a similar *superior umbilicus* into its distal end on the ventral side of the feather.

Ventral side=*below* in quill-feathers, *next body* in the others. The shaft is longitudinally grooved on this side.

In the fowl, but not in the pigeon, a small *after-shaft* which resembles the vane in structure is attached near the superior umbilicus.

The *filoplumes* possess a minute thread-like stem, but the barbs and barbules are only represented by a tuft of disconnected processes.

The *oil-gland* belongs to the skin. It is made up of numerous branched tubules, epidermic in origin, and lined by glandular-cells.

The most important histological point is the way of development of the feathers. A small papilla is formed on the surface, the base of which sinks down into the dermis and thus becomes enclosed in a pit, the *feather-follicle*, from which its pointed end grows out. The papilla is known as the *feather-germ*. It contains a vascular core of dermis, and the epidermis covering its surface is gradually moulded into the feather. The quill is formed around the base of the germ, the vane around the free part, from which it splits off and expands. The inferior umbilicus is the point where the vascular core entered the feather, the superior umbilicus shows where it left the unsplit quill-part to form the centre round which the vane was once folded.

This description refers to the first-formed feathers. The papillæ of new feathers are formed in connection with the old follicles, and never project on the free surface.

The feathers are renewed periodically during life, the old ones being cast off at moulting.

The *dermis* contains a network of muscle-fibres attached to the feather-follicles. By the contraction of this the feathers can be erected. Blood-vessels, lymphatics, and nerves are present, and a large number of ovoid *touch-corpuscles*.

3. The **endoskeleton** (Figs. 68 and 69).—This, as in the frog,

is mainly made up of cartilage, cartilage-bones,* and membrane-bones.†

There are also a few *sesamoid* bones, e.g., patella, developed in the course of tendons.

The bones are distinguished by their spongy character, and the shafts of the long bones contain, in the adult, air instead of marrow. The skeletons of the trunk and limbs may be distinguished as axial and appendicular.

(1) The **axial endoskeleton** consists of skull, vertebral column, ribs, and sternum.—(a) The mature **skull** contains comparatively little cartilage, and a chondro-cranium is only to be made out in the embryo. The bones in the adult are fused together so as to make the determination of boundaries a difficult matter. A rounded *cranial portion* behind may be distinguished from a tapering *facial* portion in front, the two being connected together by an imperfect joint which permits a small amount of up and down movement. A large rounded *orbit* is present on each side near the junction of the two.

The **cranial** part is mostly made up of the rounded **brain-case** or **cranium**. This exhibits a large *foramen magnum* behind, bounded by the occipital region, to form which four bones are fused together,—two *ex-occipitals* at the sides, a *supra-occipital* above, and a *basi-occipital* below. The last bears a median rounded *occipital condyle*, for articulation with the vertebral column. The roof of the brain-case is completed by two pairs of bones, the *parietals* behind and the *frontals* anteriorly. The side-wall of the brain-case is partly formed by the *squamosal*, below which is a depression, the *tympanic cavity*, bounded internally by the **auditory capsule**.

This is formed by the union of three elements. *pro-otic, epi-otic*, and *opisth-otic*, of which the first is most important. They are placed in front, above, and behind respectively.

About the centre of the capsular wall are two small openings, one above the other, separated by a very narrow interspace.

* *Cartilage-bones :—Basi-, ex-,* and *supra-occipitals, pro-, epi-,* and *opisth-otics, basi-, pre-, ali-,* and *orbito-sphenoids, mesethmoid, quadrate, articular, columella, basi-hyal, basi-branchial, posterior cornua* of hyoid. *Vertebræ. Sternum. Ribs. Appendicular skeleton* (except furcula).

† Membrane-bones :—Parietal, frontal, squamosal, parasphenoid, lachrymal, basi-temporal, vomer, nasal, premaxilla, maxilla, jugal, quadrato-jugal, pterygoid, palatine, angular, supra-angular, dentary, splenial. Furcula.

The upper is the *fenestra ovalis*, the lower the *fenestra rotunda*. Into the front of the tympanic cavity the Eustachian tube opens. Further forwards the brain-case is bounded laterally by an *orbital plate*, which also forms the hinder and upper parts of the wall of the orbit, and is made up below by the *ali-sphenoid*, and above by the *orbital process* of the frontal. The floor of the brain-case in front of the basi-occipital is formed by the *basi-sphenoid*. From this a pointed rod, the *para-sphenoid (basi-sphenoidal rostrum)*, projects forwards. It is fused with the lower edge of a thin bony plate, the *interorbital septum*, continuous behind with the ali-sphenoids. The posterior part of the septum represents three distinct bones, a median *pre-sphenoid*, and two lateral *orbito-sphenoids*. The upper and anterior parts of the septum are, in the young bird a distinct bone, the *mesethmoid*. The anterior margin of the orbit is formed at this point by the *lachrymal*. The basi-sphenoid is overlapped by a thin plate of bone, the *basi-temporal*, which forms the lower boundary of the tympanic cavity on either side. It tapers somewhat in front, and between it and the basi-sphenoidal rostrum is a median Eustachian opening, from which an Eustachian tube leads back on either side to the corresponding tympanic cavity.

The basi-temporal is probably equivalent to the posterior part of the frog's para-sphenoid.

The cavity of the brain-case closely corresponds to the shape of the brain. Its floor is very steep, rapidly ascending towards the front. The foramen magnum looks downwards in accordance with the upright position of the neck.

Nerve Exits.—The *olfactory foramen* (I.) is in the front of the brain-case. A vacuity in the dry skull at the upper margin of the inter-orbital septum marks the course of the olfactory nerves to the nasal capsules. Below the olfactory foramen is an *optic foramen* (II.), and at this point, in the dried skull, the two orbits are placed in communication by a hole. The oculomotor and pathetic nerves have exit by special small foramina near the optic, the fifth and sixth by a larger aperture behind the ali-sphenoid. The eighth nerve pierces the inner side of the auditory capsule, and, just in front of it, there is a small foramen for the seventh. Behind the tympanic cavity there is a small opening on each side at the posterior angle of the basi-temporal for the ninth, tenth, and eleventh (IX. and X.), while the twelfth passes out by a small *condylar foramen* situated in the ex-occipital near the condyle (XII.)

The **facial** part of the skull consists of the nasal capsules and jaws. The **nasal capsules** contain a good deal of cartilage, and are separated by a cartilaginous *nasal septum*, which is a continuation of the mesethmoid. A scroll-like fold, the *turbinal*, projects from it on either side. Below this is a slender *vomer*, just in front of the basi-sphenoidal rostrum. The capsules are partly roofed in by the **nasals**, which unite with the frontals behind, and are deeply notched in front. The notches partly bound the external nares.

The margin of the **upper jaw** is formed in front by the *premaxillæ*, each of which unites behind with a delicate rod-like *maxilla*, from the anterior end of which a flat *maxillo-palatine process* projects towards the middle line. The maxilla fuses behind with a styliform *jugal*. These two last bones form the greater part of a sub-orbital bar, which is completed posteriorly by a *quadrato-jugal*, the rounded posterior end of which is united by a fibrous band with the *quadrate*, a short thickened bone,

---

Fig. 68.—ENDOSKELETON OF PIGEON (Cartilage dotted).—A and A', Skull, side and under views; *s-oc*, supra-occipital; *ex-oc*, ex-occipital; *b-oc*, basi-occipital; *c*, condyle; *pa*, parietal; *sq*, squamosal; *a-s*, ali-sphenoid; *fr*, frontal; *fr'*, orbital process of ditto; *i-o.s*, inter-orbital septum; *la*, lachrymal; *b-t*, basi-temporal; *p-s*, para-sphenoid; *v*, vomer; *na*, nasal; *p-mx*, pre-maxilla; *mx*, maxilla; *mx-p*, maxillo-palatine process; *j*, jugal; *q-j*, quadrato-jugal; *pl*, palatine; *pt*, pterygoid; *qd*, quadrate; *d*, dentary; *ar*, articular; *an*, angular; *s-an*, supra-angular; *ty*, tympanic cavity (reference line runs to fenestra ovalis); *Eu*, Eustachian opening; *f-m*, foramen magnum. I., II., &c., Nerve exits. (V., just *above* foramen for fifth.) B, *Left columella*, enlarged; *pl*, plate fitting into fenestra ovalis. C, *Hyoid apparatus*, from above; *g-hy*, glosso-hyal; *b-hy*, basi hyal; *b-br*, basi-branchial; *a.c*, anterior cornua; *p.c*, posterior cornua. D and D', "*Sacrum*" and *pelvis*, under and side views, numbers refer to vertebræ; *t*, thoracic; *l*, lumbar; *S*, sacral; *c*, caudal; *c'*, free caudal; *p-sh*, plough-share bone; *r*, rib; *il*, ilium; *pb*, pubis; *is*, ischium; *ac*, acetabulum; *ant*, anti-trochanter; *ob*, obturator fissure; *i-s.f*, ilio-sciatic foramen. E, 1-6, Vertebræ. 1, Atlas: 2, axis: 3, two free cervical, from above: 4, ditto, from side: 5, ditto, from front: 6, first thoracic; *c*, centrum; *n.sp*, neural spine; *t.p*, transverse process; *hy*, hypapophysis; *a-z*, pre-zygapophyses; *p-z*, post-zygapophyses; *o.p*, odontoid peg; *i.l*, cervical rib; *n.c*, neural canal; *v.c*, vertebrarterial canal; *r*, vertebral rib; *c.p*, head; *tb*, tubercle; *un*, uncinate process: 7, sclerotic ring.

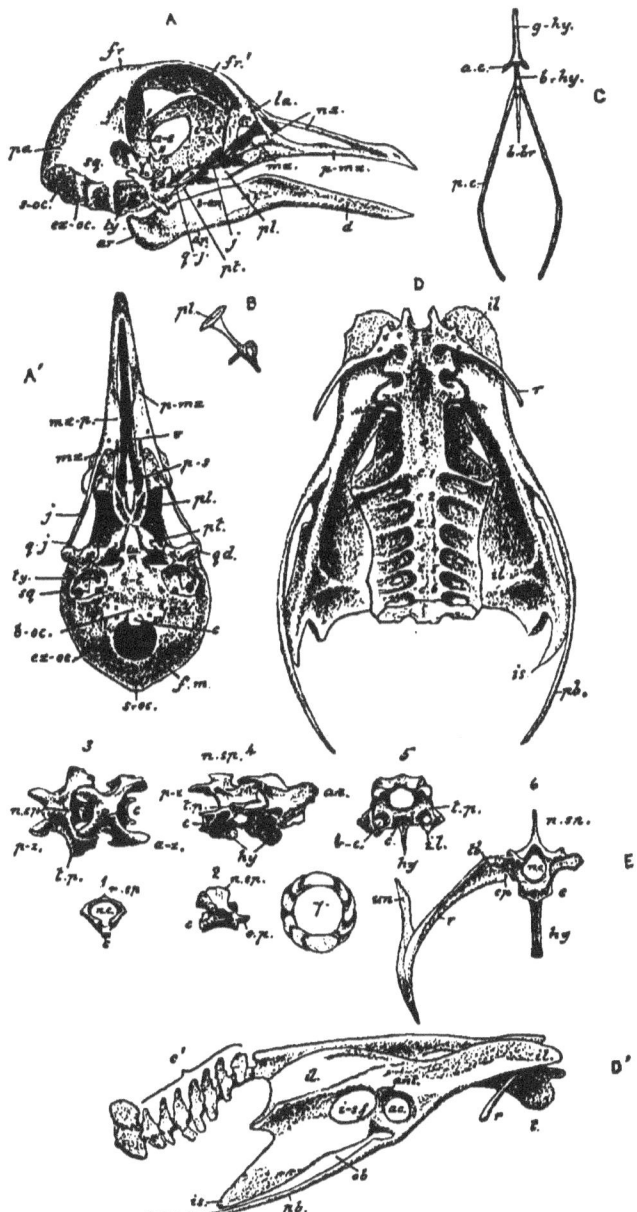

movably articulated at one end by two convex surfaces with the squamosal and pro-otic, while the other presents a transverse articular condyle for the mandible. Upon the ventral surface of the skull an **X**-shaped structure is formed by four bones, pterygoids behind and palatines in front. The *pterygoid* articulates behind with a facet on the quadrate, in front with a facet on the side of the parasphenoid, at its base. When the facial part of the skull is moved up and down, the pterygoids readily slide forwards and backwards on these smooth facets. The *palatine* is a flattened rod, broadest behind, where it meets its fellow and the pterygoid, and tapering to a point in front, where it fuses with the maxillo-palatine process.

The **lower jaw** or **mandible** consists of two almost straight halves (rami), each of which articulates behind with the corresponding quadrate, and unites with its fellow in front to form a median symphysis. The axis of the ramus is formed by Meckel's cartilage, which persists, more or less, throughout life, and is ossified proximally into the *articular*, which is jointed on to the quadrate. *Angular* and *supra-angular* elements ensheath the ramus below and above in its proximal half, the latter being produced above into a small *coronoid process*, while the distal half is covered ventrally and externally by the *dentary*, internally by the *splenial*.

Hyoid and First Branchial Arches.—The proximal end of the hyoid arch is formed by the *columella*, which helps to transmit sound-waves to the internal organs of hearing. Its inner end is bony and expanded into an oval plate fitting into the fenestra ovalis, while its outer cartilaginous end is firmly fixed to the inner side of the tympanic membrane. The " hyoid apparatus" consists of the *body* of the hyoid, supporting the base of the tongue, and *anterior* and *posterior cornua*. The body is composed of three median pieces:—(*a*) A cartilaginous *glosso-hyal*, shaped like an arrowhead, from which the short *anterior cornua* also of cartilage, diverge backwards; (*b*) a short bony *basi-hyal*; and (*c*) a styliform bony *basi-branchial*. Between (*b*) and (*c*) the slender three-jointed *posterior cornua* diverge backwards and upwards.

The basi-branchial and posterior cornua represent the *first branchial arches.*

(b) **Vertebral Column and Ribs.**—The vertebral column is divided into *regions*, arranged as follows :—

	No. of Vertebræ.	
1. *Cervical*,	13 or 14	Free.
1. *Thoracic*—		
(a) Anterior,	4	Fused.
(b) Posterior,	1	
3. *Lumbar*,	2 or 3	Fused to form
4. *Sacral*,	4 or 3	"*Sacrum.*"
5. *Caudal*—		
(a) Anterior,	6	
(b) Middle,	6	Free.
(c) Posterior,	Several	Fused to form *ploughshare* bone.

The *cervical* vertebræ are those supporting the neck. Each is made up of two parts :—(a) An elongated *centrum*, articulating with neighbouring centra by saddle-shaped surfaces conferring great flexibility to this region, and with a median ventral process (*hypapophysis*). (b) A *neural arch*, produced dorsally into a plate-like *neural spine*, and connected with adjacent arches by *pre-* and *post-zygapophyses* in the same way as in the frog. *Intervertebral foramina* are left between the arches for the transmission of the spinal nerves.

The *last* cervical vertebra possesses a strong *transverse process* running out from each side of the neural arch, and bears a pair of **cervical ribs**, each of which is a small curved bone, with a forked proximal end attached to the vertebra, and a free distal end. From its hinder edge a small flat *uncinate process* projects. The two limbs of the forked end terminate in articular projections termed *head* (capitulum) and *tubercle*. The former articulates with a small *capitular facet* on each side of the centrum near its front end, the latter with a similar *tubercular facet* on the under side of the transverse process. The last vertebra but one also bears free ribs, these, however, are devoid of tubercles and uncinate processes.

In the remaining vertebræ which possess them, the *cervical ribs* are very short, and completely fused with the transverse process and centrum, an aperture being left, however, which, with those of the other vertebræ, forms the *vertebrarterial canal* in which certain soft structures run.

The *first* and *second* vertebræ, *atlas* and *axis*, are very small and possess neither ribs nor transverse processes. The atlas (1) is a ring, without anterior zygapophyses. Over its centrum's upper surface a small peg, the *odontoid process*, projects from the front end of the centrum of the axis (2). The occipital condyle fits into an articular pit formed above by the odontoid process, and below by the atlas-centrum.

The odontoid process is at first a distinct bone. It is part of the atlas-centrum which fuses with that of the axis.

The *thoracic vertebræ* bear free **thoracic ribs**, and the commencement of the series is marked by the first vertebra bearing ribs which unite ventrally with the sternum. The four anterior vertebræ are closly united with one another by their centra and the processes of their arches. The thoracic ribs are made up of two segments, one, the *vertebral rib*, similar in character and mode of attachment to the last cervical rib, the other, *sternal rib*, articulated distally, with a facet on the sternum. The "*sacrum*" of the adult bird is made up of vertebræ belonging to four regions closely fused together. The single *thoracic* vertebra is easily recognised by its rib. It possesses a large bilaminar hypapophysis, and its stout transverse processes abut against the ilia. The *lumbar* vertebræ have short and strong, the *sacral*, lamellar transverse processes, while the first *caudal* possesses a ventral caudal rib as well as a dorsal transverse process.

The *free caudal* vertebræ are small, and their transverse processes flattened.

The *ploughshare bone* is a laterally flattened plate, bent up sharply on the rest of the vertebral column.

(c) The **sternum** (Fig. 69) is a very broad, plate-like bone, concave above, supporting the ventral wall of the thorax and part of the abdomen. It is produced in front into an obtuse process, the *manubrium*, and behind into a narrow plate, the *middle xiphoid process*. At the sides are paired *internal* and *external xiphoid processes*. On each side of the front of the sternum there is an excavation for the end of the coracoid, and behind this comes the *costal process*, bearing four facets for the articulations of the sternal ribs. The ventral side of the sternum is produced downwards into a prominent vertical plate, the *keel*, which is deepest in front.

(2) *Appendicular Endoskeleton.*—The long bones possess *epiphyses*, as in the frog. Girdle and free limb can be distinguished in both fore- and hind-limbs.

(*a*) **Fore-Limb**—*Shoulder-girdle* (Fig. 69).—The *scapula* is a backwardly-directed, blade-like bone, united closely in front with a rod-like *coracoid*, the other end of which articulates with the sternum. These two bones make with one another an acute angle, open behind, and the *glenoid cavity* is situated at their point of union. In front of this the slender *clavicle* is attached, which curves backwards and downwards, and unites with its fellow in the middle line to form the *furcula* ("merry-thought").

The union is effected (?) through the intermediation of a small *interclavicle*, which forms in the Fowl a small, laterally flattened disc.

Free Limb (Fig. 69).—The *humerus* possesses a proximal *head* for articulation with the glenoid, and a distal, pulley-like *trochlea*, which presents preaxial and postaxial articular surfaces for the radius and ulna. Two distinct bones, radius and ulna, support the antebrachium. The preaxial *radius* is straight and slender, the postaxial *ulna* relatively stout. The proximal end is produced into an *olecranon process*.

The *carpus* is made up of two small bones, the *radiale* and *ulnare*, which articulate with the radius and ulna respectively. First, second, and third *metacarpals* are present, closely united proximally. The first is small, and projects but little beyond the fused part. The second is long and stout, and the third long and slender. The three digits are completed by *phalanges*, the first and third possessing a single small one, while two large phalanges terminate the second digit.

(*b*) **Hind-Limb**—*Hip Girdle* (Fig. 68).—Each half, or *os innominatum*, is separate from the other, and made up of three elements, ilium, ischium, and pubis, all three of which contribute to the formation of the *acetabulum*, a rounded cup, not floored by bone, on the outer side of the girdle. The large *ilium* extends forwards and backwards above the acetabulum. It meets and fuses with the anterior ends of the ischium and pubis to bound the acetabular cavity, above which it presents a smooth surface, the *anti-trochanter*. The "sacrum" is closely connected with the

240   AN ELEMENTARY TEXT-BOOK OF BIOLOGY.

inner concave surface of the ilium. The *ischium* is a flat rod running back from the acetabulum parallel to the lower edge of

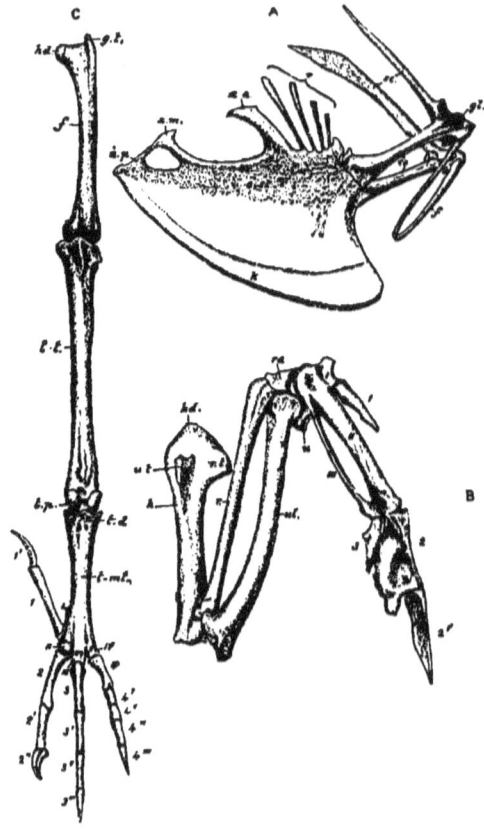

Fig. 69.—ENDOSKELETON OF PIGEON.—A, Sternum and shoulder-girdles; *c.st*, corpus sterni; *k*, keel; *x.a, x.m, x.p*, external, internal, and middle xiphoid processes; *r*, sternal ribs; *co*, coracoids; *sc*, scapulæ; *f*, furcula; *gl*, glenoid cavity. B, Left Wing—*h*, humerus; *hd*, head of ditto; *r.t* and *u.t*, radial and ulnar tuberosities; *r*, radius; *ul*, ulna, showing scars left by quills; *ra*, radiale; *u*, ulnare; I.-III., metacarpals; 1-3, phalanges. C, Left Leg (patella gone, fibula not seen)— *f*, femur; *hd*, head; *g.t*, great trochanter; *t-t*, tibio-tarsus; *t.p*, proximal part of tarsus; *t.d*, distal part of tarsus; *t-mt*, tarso-metatarsus; I.-IV., metatarsals; 1-4, phalanges.

the ilium, beyond which it extends. The *pubis* is a slender rod which, in the acetabulum, fuses with the ilium and ischium. It runs back below the latter, the *obturator fissure* separating the two, and extends for some distance behind it. A wide space separates the pubes and ischia of opposite sides.

Free Limb (Fig. 69).—The *femur* articulates with the acetabulum by a preaxial *head* at right angles to the shaft, while its proximal end is terminated by a facet which works against the anti-trochanter, and on the postaxial side of which is a considerable elevation, the *great trochanter*. The distal end of the femur is pulley-shaped. The main bone of the leg is the preaxial *tibiotarsus* with which the small rod-like postaxial *fibula* is partly fused. The tibio-tarsus is a stout bone, much longer than the femur, and with a pulley-shaped distal end. In front of the knee-joint and fibula there is a small irregular bone, the *patella*.

In the adult there is no separate *tarsus*, but in the embryo it is represented by two cartilages, the proximal of which unites with the preaxial tibia and forms the distal end of the tibiotarsus, while the distal fuses with the metatarsus. The anklejoint is thus between the two tarsal elements, and is said to be *mesotarsal*—*i.e.*, in the middle of the tarsus.

Four *digits* are present—first, second, third, and fourth. The *metatarsals* of the three last fuse with the distal part of the tarsus to form the *tarso-metatarsus*. This is a rod of bone, the distal end of which possesses three very distinct pulley-like surfaces for the three corresponding metatarsals. The small first metatarsal is attached to the preaxial side of this end. The four *digits*, of which the first is directed backwards, are completed by 2, 3, 4, and 5 *phalanges* respectively. The terminal phalanges support claws.

4. The **digestive organs** (Fig. 70) consist of gut and appended glands. The former is divided into mouth-cavity, gullet with crop, proventriculus, gizzard, small intestine, and large intestine opening into a cloaca. The glands are liver and pancreas.

The *mouth* is bounded by the horny margins of the beak, and, owing to the double joint by which the mandible is connected with the skull, can be opened very widely. It leads into a large **mouth-cavity**, upon the roof of which are two narrow slits, the *internal nares*, largely overlapped by folds of mucous membrane. Behind them is a small median *Eustachian aperture*, and posterior to this the end of the roof is marked by a backwardly directed fringe. The large *tongue* attached to the floor of the mouth, has a sharply-pointed tip directed forwards, and its hinder part is produced backwards into fringed processes, behind which is the *glottis*, a narrow slit, with edges also somewhat fringed. The back part of the mouth-cavity, or *pharynx*, is continued behind into a very long **gullet** (*œsophagus*), for the most part with thin walls, which runs back on the ventral side of the neck to reach the thorax. In the posterior part of the neck it dilates into the *crop*, a very large bilobed sac, with extensible walls. Behind this the gullet narrows, and its walls are much thicker. It passes in the thorax into a small oval chamber, the **proventriculus**, of somewhat larger calibre, and with thick soft walls, upon the inner surface of which are a number of small rounded apertures, those of the *peptic glands*. The proventriculus opens, on the left side, into the dorsal border of the **gizzard**, a large rounded, somewhat flattened structure with extremely thick hard walls, the lining of which is horny. It contains numerous small pebbles and other foreign bodies which have been swallowed. Proventriculus and gizzard together are equivalent to a stomach. A narrow tube, the **small intestine**, comes off from the right side of its dorsal border, forms a U-shaped loop, the *duodenum*, and, after making several loops and coils, merges into the short, straight, **large intestine**, the junction between the two being marked by a pair of small projections, the *intestinal cæca*. These are very large in the fowl. The large intestine is continued posteriorly into a small **cloaca** which opens externally by the *cloacal aperture*. The cloaca is divided by inwardly projecting folds into three compartments, internal, middle, and external (Fig. 71). The first receives the large intestine, the second the urinary and genital ducts, while into the dorsal wall of the third a small thick-walled pouch, the *bursa Fabricii*, opens in young birds. The lining of the small intestine is raised into innumerable delicate thread-like processes, the *villi*, which gradually die away posteriorly and are replaced by longitudinal ridges. The lining of the large intestine and cloaca is smooth.

The **liver** is a large, brown organ, lying on the ventral side of the body, and presenting depressions above and in front into which the heart, duodenum, and gizzard fit. It is divided into *right* and *left lobes*, of which the first is the larger. There are

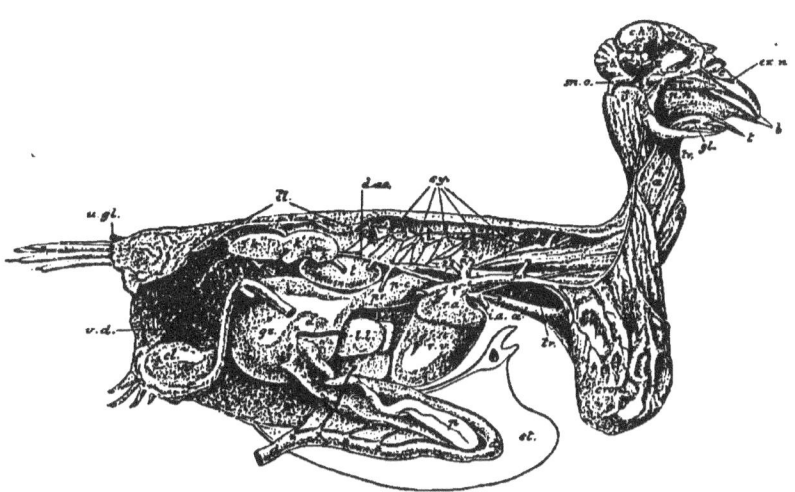

Fig. 70.—GENERAL DISSECTION OF MALE PIGEON.—c, Cere, close to which is external naris, ex.n.; u.gl, opening of uropygial gland; il, cut edge of ilium; st, keel of sternum. A large part of this bone has been removed together with ribs. b, Beak; t, tongue; p.n, internal nares; gl, glottis; œ, gullet, which swells into the crop in the middle of its course; pr, proventriculus; gz, gizzard; d, duodenum. Most of ileum has been cut away, at i its junction with large intestine is seen. This point is marked externally by the intestinal cæca, one of which is shown; cl, cloaca; l.l, placed on left liver-lobe at point where right lobe was removed; ×, ×, placed on points where bile-ducts enter duodenum; p, pancreas; ***, placed on points where the pancreatic ducts enter duodenum; r.a, right auricle of heart, into which the cavals run; pr-c, right precaval, formed by the union of (ju) jugular and subclavian, the short vessel just over pr-c, which the brachial (1) and pectoral (2, 2') veins combine to form; p-c, postcaval, the other cut end of which is close to t; r.v, right ventricle; i.a, right innominate artery bifurcating into common carotid above, and subclavian, the latter at once dividing into brachial (above) and pectoral (below). The dorsal aorta (d.ao) is seen running back from heart. The cut end of the cœliac artery is just above pr, and that of the anterior mesenteric to the right of t; tr, trachea; k, k', k", anterior, middle, and posterior lobes of right kidney cut through; v.d, right spermiduct (vas deferens), along inside of which right ureter runs; t, left spermary (testis) the right is removed; ol, placed over olfactory lobe, and olfactory nerve which runs forwards from it; c.h, cerebral hemisphere; o.l, optic lobe; cb, cerebellum; m.o, medulla oblongata. II., Optic nerve, just above which is optic tract; sy, right sympathetic ganglia of thorax, between which are cut ends of ribs. The ganglia are connected by double commissures, which, with connected nerve branches, are represented by black lines.

two *bile-ducts*, the left short and wide, running into the duodenum near its commencement, while the right, long and narrow, opens into its distal limb near the beginning of its distal third. There is no *gall-bladder*.

A gall-bladder is present in the Fowl. The bile and pancreatic ducts are also arranged differently.

The **pancreas** is a compact, elongated body, lying in the loop of the duodenum, and sending three *pancreatic ducts* into its distal limb.

As in the frog, the alimentary canal is suspended from the dorsal wall of the body-cavity by mesenteric folds. The membranous lining of the abdominal region (which, however, is quite continuous with that of the thorax) is the *peritoneum*. This closely lines the dorsal side of the cavity before leaving its walls to form mesenteric folds.

The wall of the alimentary canal is made up of the usual four coats—mucous, submucous, muscular, and serous. The *mucous membrane* possesses an epithelial lining varying in the different parts of the canal. That found in the mouth-cavity and parts of the œsophagus is *stratified squamous epithelium*, and that lining the crop *glandular*. The lower parts of the gullet, the proventriculus, and the intestines are lined by *simple columnar epithelium*, continued into the simple tubular *peptic glands* of the second, and there becoming glandular. *Glands of Lieberkühn* are present in the small intestine. The epithelium of the gizzard secretes a thick horny cuticle.

The *muscular coat* varies very much in thickness, and typically consists of two layers, an internal circular and an external longitudinal, the relative position of these being reversed, however, in the gullet. This coat is immensely thickened in the gizzard, and has there a complicated arrangement.

The *serous coat* is an epithelial and connective-tissue layer present from the proventriculus backwards, and formed by the investing mesentery.

The *liver* and *pancreas* present no important deviations from the structure of the same organs in the frog (pp. 193, 194).

The food, chiefly consisting of grain, accumulates in the crop, from which it passes on through the proventriculus to the gizzard, which grinds it up, the process being aided by the foreign bodies present. The three chief digestive juices are, gastric juice, bile, and pancreatic juice, secreted respectively by the peptic glands, liver, and pancreas, and acting as in the frog. An important addition to the absorptive surface is afforded by the villi.

5. **Circulatory Organs.**—*Blood* and *lymph systems* are present, and well developed.

I. **Blood System.**—The bright red *blood*, which coagulates very quickly, is made up of *plasma*, with amœboid *colourless corpuscles* and much more numerous *red corpuscles*. Both kinds are nucleated, and the latter are small, flattened, and oval, with ends somewhat pointed. The blood circulates in a closed system of tubes made up of heart, arteries, veins, and capillaries.

The large conical **heart** (Fig. 70), contained in its *pericardium*, lies in the anterior part of the thorax, and is made up of four chambers, two auricles and two ventricles. Its forwardly directed base is chiefly made up of the dark thin-walled *right* and *left auricles*, its apex by the thick-walled *right* and *left ventricles*, the latter being the firmer, and alone reaching the extreme apex. The cavities of the auricles are separated by a thin *auricular septum*, upon which is a depression, the *fossa ovalis*, where the septum is thinnest. Within the right auricle are the three openings of the caval veins, that of the postcaval being guarded by a muscular flap, the *Eustachian valve*, and, posteriorly, the crescentic *right auriculo-ventricular opening* leading into the right ventricle. From the right side of this opening a muscular flap, the *right auriculo-ventricular valve*, projects into the ventricular cavity, to the walls of which it is united by fibrous cords (*chordæ tendineæ*). The right ventricle is separated from the left by the *ventricular septum*, which bulges into its cavity, the walls of which are raised into muscular ridges (*columnæ carneæ*). The *pulmonary artery*, which arises from this division of the heart, has its origin guarded by three *pulmonary semilunar valves* (pocket-valves). Within the left auricle, dorsally, is the small opening of the pulmonary veins, and, posteriorly, the rounded *left auriculo-ventricular opening*. This is guarded by a *mitral* or *bicuspid valve*, made up of two membranous flaps, situated in the cavity of the left ventricle, and connected by chordæ tendineæ to *papillary muscles*, conical elevations, which may be considered as modified columnæ carneæ. The aorta arises from the anterior end of the left ventricle, and its origin is guarded by three *aortic semilunar valves* (pocket-valves).

Arteries.—A single aortic arch is present, the **aorta**, the first part of which (arch of aorta) curves round to the right, giving off branches which supply the head, neck, wings, and pectoral muscles, and then, reaching the middle dorsal line, becomes the *dorsal aorta*. This runs straight to the tail, where it becomes the *caudal artery*, giving off branches on its way to the viscera, hindlimbs, and body-wall.

From the *arch of the aorta*, large *right* and *left innominate arteries* are given off close to the heart, after which the calibre of the aorta diminishes considerably. Each innominate divides very quickly into *common carotid* and *subclavian arteries*. The common carotid runs along the neck, soon giving off the *vertebral artery* (which supplies the brain and spinal cord, and occupies the vertebrarterial canal), and, at the angle of the jaw, divides into *internal* and *external carotids*. The former supplies the brain, the latter the head generally. The *subclavian* divides almost at once into the small *brachial* and large *pectoral* arteries, which respectively supply the wing and muscles of the chest. The dorsal aorta gives off an unpaired *cœliac artery*, the branches of which run to the proventriculus, gizzard, spleen, duodenum, pancreas, and last loop of the small intestine; while, slightly behind this, an unpaired *anterior mesenteric artery* comes off, which supplies the rest of the small intestine. Still further back the small *anterior renal arteries*, for the anterior kidney-lobes; *the femoral arteries*, for the extensor muscles of the thigh; and the *sciatic arteries*, for the flexor muscles of the thigh and for the rest of the hind-limb, are given off in succession. Branches of the sciatic, the *middle* and *posterior renal arteries*, supply the corresponding kidney-lobes. Just behind the kidneys the dorsal aorta gives off an unpaired *posterior mesenteric artery* to the large intestine and cloaca, and paired *internal iliac arteries* to the hinder part of the pelvis, and then runs, as the *caudal artery*, into the tail.

The **pulmonary artery** arises from the right ventricle, and soon divides into right and left branches for the right and left lungs.

Veins.—These may be dealt with under the headings of caval system, hepatic portal system, and pulmonary veins.

(1) **Caval System.**—There are two precavals (superior venæ cavæ) and a postcaval (posterior vena cava), all opening into the right auricle. Each **precaval** is a large, short vein, formed by the union of three trunks—jugular, brachial, and pectoral, returning the blood from one side of head and neck, one wing, and chest-muscles of one side, respectively.

The *jugular* runs back from the base of the head (where it is united by a cross-trunk with its fellow), along one side of the neck, and not far from its union with the other two trunks, receives the *vertebral vein*.

The **postcaval** is a large vein returning the blood from the posterior part of the body, formed just in front of the kidneys, by the union of two *iliac veins*, and running through the liver, from which it receives two *hepatic veins*, to the heart.

The *iliac* on each side commences between the anterior and middle kidney-lobes where the *femoral vein*, bringing back blood from the greater part of the hind-limb, unites with the *hypogastric vein* to form it. The latter traverses the substance of the kidney, behind which it is connected with its fellow by a cross-branch, that receives the *caudal vein* from the tail, and *internal iliac veins* from the pelvis, and is also united with the hinder end of the *posterior mesenteric* vein. The *sciatic vein* carries blood from the hinder part of the leg into the hypogastric, at the junction of the middle and posterior kidney-lobes. These lobes return their blood by a large superficial *renal vein* to the iliac vein, which also receives a small vein from the anterior lobe.

(2) **Hepatic Portal System.**—This, the only portal system present in the pigeon, is entirely made up of vessels running to the liver from the alimentary canal and spleen.

Into the left liver-lobe two small *left gastric veins* take blood from the left side of the gizzard, while the large **portal vein** divides into two branches, one for each liver-lobe. This vein is formed by the union of three others—*gastro-duodenal*, returning blood from the right side of the gizzard, duodenum, pancreas, and last loop of the small intestine; *anterior mesenteric*, from the rest of the small intestine; and *posterior mesenteric* from the large intestine and cloaca. (The hypogastric veins receive part of their blood from this vessel.) The spleen pours its blood directly into the portal trunk.

(3) One or two short **pulmonary veins** from each lung unite together and open by a single aperture into the left auricle on its dorsal side.

Circulation.—The two sides of the heart do not directly communicate, and therefore no mixing of blood occurs, as in the frog. The impure blood from the body is poured into the right auricle by the caval veins, and passes into the right ventricle, by which it is forced into the lungs. Thence the oxygenated blood is returned to the left auricle, and, passing into the left ventricle, is pumped through the aorta to the body at large.

The movements of the heart are very vigorous. The auricles contract together, and their systole is followed by a ventricular systole, both ventricles contracting together. The auriculo-ventricular valves prevent the blood from passing back into the auricles, and the chordæ tendineæ prevent the valvular flaps from going too far. The chordæ would be slackened during the ventricular systole by the approach of the ventricular walls, were this not compensated for by the papillary muscles, which contract at the same time and pull the chordæ taut. The semilunar valves only allow blood to pass *out of* the ventricles.

II. The **lymph system** mainly differs from that of the frog (p. 200) in the absence of lymph-hearts, and the greater definiteness of the lymphatic vessels (*lymphatics* of general body, *lacteals* of gut). These resemble small veins in structure and ultimately open into two delicate tubes, the *thoracic ducts*, lying just beneath the vertebral column in the thoracic region, and communicating with the veins at the junction of the jugular and subclavian on each side. Minute lymph-spaces are found in all parts of the body, and there are also large lymph-spaces of which the most important is the **cœlom**, which includes the general body-cavity and the pericardial cavity.

Two *cervical lymphatic glands* are found at the base of the neck, and a small red ovoid *spleen* is attached to the right side of the proventriculus.

6. The **respiratory organs** (Fig. 70) consist of lungs, to which the air gains access through a trachea, and of accessory structures. An organ of voice results from modification of part of the air-passages.

The *glottis* leads into a small *larynx*, the walls of which are supported by several cartilages, and which is the commencement of the **windpipe** or **trachea**, a long tube which runs along the neck to the thorax. Numerous firm rings surround and support the trachea, which bifurcates in the thorax into a right and a left *bronchus*, one going to each lung. At the point of bifurcation is situated the **syrinx**, which is here the organ of voice.

The end of the trachea and commencements of the bronchi enclose the *tympanum* or syringeal cavity. Into this an elastic fold, the *membrana semilunaris*, projects from the point of bifurcation, and is supported by a slender bar of cartilage, the *pessulus*, running dorso-ventrally. The inner side of each bronchus, at its commencement, is membranous, and forms the *membrana tympaniformis interna*. Special muscles are connected with the syrinx.

The **lungs** are spongy bodies, of dark-red colour, which lie in the anterior part of the thorax. They are closely attached to the ventral side of the vertebral column and ribs, and a thin membrane, the *pleura*, continuous with the peritoneum, covers their ventral surface. The bronchus which runs to each lung enters it, becomes membranous, and, after dilating into a vestibule, runs towards the external side of its posterior end, bifurcating into two *secondary* bronchi. Other secondary bronchi come off from the vestibule. All these tubes are placed near the ventral surface of the lung, the dorsal part of which organ is mainly made up of their branches, delicate tubes with blind endings.

Closely connected with the lungs are a number of thin-walled **air-sacs** into which the secondary bronchi open.

On each side of the body, in the abdominal region, between the kidneys and the intestines, there is a large *posterior air-sac*, communicating with the hinder end of the lung. In front of this are two pairs of intermediate air-sacs, situated on the ventral side of the body-cavity. The *posterior intermediate air-sacs* overlap the hinder part of the lungs, and communicate with them by openings close to those of the posterior sacs, while the *anterior intermediate air-sacs* are situated ventral to the anterior two-thirds of the lungs, and open into them near the middle of their length. In the region of the clavicle a large, unpaired *interclavicular air-sac* is present (formed by the fusion of two *sub-bronchial air-sacs*), communicating with either lung by an aperture near the entrance of the bronchus. This air-sac is bilobed, and each lobe passes out of the body-cavity near the origin of the wing to communicate with the hollow shaft of the humerus, while dorsal to it there is a small *prebronchial air-sac*, connected with the anterior end of the lung.

The most important histological point to be noticed is the finer structure of the delicate tubules making up the greater part of the lungs. The walls of these are raised up internally into intersecting ridges, which form a complicated honeycombing, traversed by networks of capillaries, and covered by simple squamous epithelium.

The epithelium lining the bronchi and trachea is largely ciliated.

The lungs are very immobile and take but little part in the respiratory movements. *Expiration* is effected by a contraction of the body-walls, by which the highly elastic air-sacs are compressed. *Inspiration* is passively effected by the elasticity of the body-walls, which expand and cause air to rush into the air-sacs. Owing to the presence of these, a large amount of air passes through the main passages of the lung. The *essential* part of respiration takes place in the small tubes making up the bulk of the lung, the walls of which are raised into folds, and the air is kept pure in them by diffusion.

Respiration is very vigorous, in accordance with the rapid oxidation (metabolism) of the tissues. A great deal of heat results from this latter process, and the body is maintained at a very high temperature (103°-104° F.)

The organ of **voice** is here the syrinx.

7. The **urinogenital organs** (Fig. 71) include the excretory and reproductive organs.

Excretory Organs.—The reddish-brown **kidneys**, covered ventrally with peritoneum, are situated just behind the lungs, and closely applied to the under side of the ilia and "sacrum." Each

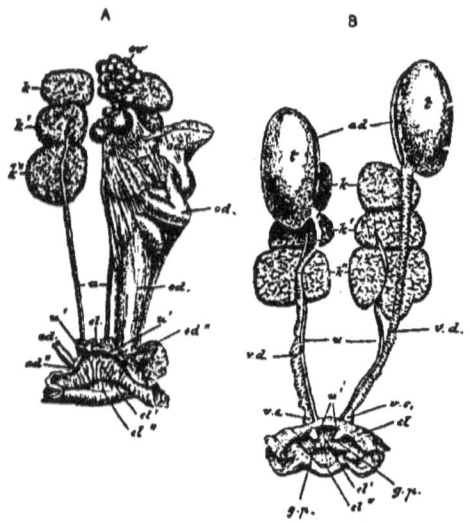

Fig. 71.—URINOGENITAL ORGANS OF PIGEON.—A, female; B, male; k, k', k'', anterior, middle, and posterior lobes of kidney; u, ureters; u', openings of ditto; cl, cl', cl'', internal, middle, and external divisions of the cloaca, which has been cut through on its ventral side and opened out; t, spermaries (testes); $v.d$, spermiducts (vasa deferentia); $v.s$, vesiculæ seminales; $g.p$, genital papillæ, on the end of which are the male genital openings; ov, ovary; od, oviducts (right a mere rudiment); od', funnel of left oviduct; od'', openings of oviducts into cloaca; ad, adrenals.

possesses three lobes, anterior, middle, and posterior. Between the two first, on the ventral side, a small thin-walled *ureter* arises, which runs to the middle division of the cloaca, into the dorsal side of which it opens.

The *adrenals* (Fig. 71, B, ad) are small elongated bodies, of yellowish colour, closely connected with the origin of the iliac vein on either side.

The *uriniferous tubules* are not ciliated. Each of them begins in a glomerulus-containing Bowman's capsule, which passes into a dilated glandular portion, that, after a somewhat convoluted course, is succeeded by a narrow looped part. The terminal portions of the tubules form the ureter by successive unions. The water and salts are strained off in Bowman's capsules, while the nitrogenous waste is excreted, mainly as *uric acid* ($C_5H_4N_4O_3$) by the glandular parts of the uriniferous tubules. The urine is semi-solid.

Male Reproductive Organs (Figs. 70 and 71).—The *spermaries* (testes) are two oval, whitish bodies, situated ventral to the kidneys, and near their anterior ends. They are invested and held in place by folds of peritoneum (*mesorchia*). From the inner side of each an opaque white tube, the *spermiduct* (vas deferens), repeatedly and sharply bent from side to side, runs on the outer side of the corresponding ureter to the cloaca, dilating at its end into a *vesicula seminalis*, and opening on a small *genital papilla* placed just external to the opening of the ureter.

The spermaries are made up of a large number of much-convoluted seminiferous tubules formed by the continued branching of the spermiduct. The tubules are lined by epithelium, many of the cells of which are sperm-mother-cells, producing sperms (spermatozoa), with cylindrical heads and motile tails.

Female Reproductive Organs (Fig. 71).—These are only developed on the left side. The *ovary* is a very irregular body (similarly situated to the spermaries), from the outer surface of which globular ova of various sizes project, enclosed in follicles. It is covered by peritoneum, which forms a suspensory fold (mesovarium). The *oviduct* is a thick-walled convoluted tube, communicating with the body-cavity at one end by a delicate membranous funnel situated near the ovary, and at the other entering the cloaca by a large aperture external to the opening of the left ureter. The *right* oviduct may be represented by a short blind tube having a similar situation.

The cloaca is larger in the female.

The *ovary* consists of a connective-tissue framework, the *stroma*, richly supplied with blood-vessels. The ova are contained in *ovarian follicles*, lined by *follicular epithelium*, which is again invested by the stroma.

These *ovarian ova* are, when ripe, yellow spheres rather less than an inch in diameter. Such an ovum (*cf.* Fig. 73) is covered by a delicate *vitelline membrane*, within which is the vitellus, containing a germinal vesicle placed close to the surface. That part of the vitellus in which the germinal vesicle is imbedded, the *germinal disc*, is a small lens-shaped mass of clear protoplasm, while the rest is mainly made up of food-yolk, through which protoplasm is sparingly diffused. The yolk is of two kinds— *yellow*, composed of granular spheres, and *white*, formed of smaller spheres enclosing highly refracting spherules. The yellow yolk is more abundant, while the white yolk is arranged in a flask-shaped mass running from the germinal disc to the centre of the ovum, and several thin lamellæ concentric to this.

The *oviducts* possess muscular walls, and are lined by glandular and ciliated epithelium.

The cloacal chambers of the two copulating individuals are partially everted and the sperm passed into the oviduct. The ova burst out of their capsules into the body-cavity, and are taken up by the funnel of the oviduct. Before passing down very far, they are fertilized by the fusion of a single sperm with each.

8. **Muscular System.**—The muscles may be classified as in the frog (p. 208), but their arrangement is more complicated. The presence of air-sacs largely increases the surface for attachment of the muscles of flight, the most important of which take origin in the sternum and its keel, and are inserted into the humerus. The specific gravity of the body is diminished by the presence of air-sacs, as also by the nature of the bones, and flight is thus indirectly aided.

Both unstriated and striated muscles are present, as in the frog, and their structure and distribution are substantially the same.

9. The **nervous system** (Fig. 70) consists of cerebro-spinal axis, cranio-spinal nerves, and sympathetic system.

(1) The **cerebro-spinal axis** consists of brain and spinal cord, enclosed in the neural canal and covered by membranes, but closely fitting the cavities in which they are placed.

The **brain** is large and rounded. As before (p. 210), it may be divided into fore-, mid-, and hind-brains:—(*a*) *Fore-brain.*— The *thalamencephalon* contains a large slit-like *third ventricle*, bounded by the *lamina terminalis* in front, and the *optic thalami* at the sides (united behind by a *posterior commissure*), while a stalked *pineal gland* is connected with its thin roof, and a *pituitary body* with the *infundibulum* formed by its floor. This part of the fore-brain is overlapped by the large smooth *cerebral hemispheres*. These are ovoid bodies closely applied to each other in the middle line, broadest behind and bluntly pointed in front. Each contains a large *lateral ventricle*, communicating with the third ventricle by a *foramen of Monro*, and having its floor raised into a considerable elevation, the *corpus striatum*. The two corpora striata are connected by the *anterior commissure* running in the lamina terminalis. A small pointed *olfactory lobe* is connected with the anterior end of each hemisphere, close to the middle line on the ventral surface, and it contains a small *olfactory ventricle*, continuous with the lateral ventricle.

(*b*) *Mid-brain.*—The *optic lobes* are very large. They are widely separated from one another, and situated laterally, but remain united by a dorsal *optic commissure*. Each contains a good-sized *optic ventricle*, opening into the *Sylvian aqueduct*, which connects the third and fourth ventricles, and possesses a thick floor formed by the *crura cerebri*.

(*c*) *Hind-brain.*—The thickened cylindrical *bulb* (medulla oblongata) possesses shallow *dorsal* and *ventral fissures*, and the roof of the *fourth ventricle* is extremely thin. The *cerebellum* is a large rounded projection, flattened laterally, which overlaps the midbrain and medulla. It is marked by deep transverse furrows, and on each side presents a small rounded elevation, the *flocculus*. The cerebellum is united with the bulb by a large cylindrical *peduncle* on each side.

The **spinal cord** is continuous in front with the bulb, a sharp ventral flexure marking their union, and tapers gradually back to the caudal region, exhibiting, however, considerable *brachial* and *lumbar enlargements* where the limb-nerves come off. *Dorsal* and *ventral fissures* are present, and a *central canal*, which in the lumbar enlargement expands into a *sinus rhomboidalis*, the lozenge-shaped roof of which is covered by membrane only.

(2) **Cranio-Spinal Nerves.**—There are twelve pairs of **cranial nerves**, the first ten of which correspond in origin and distribution to those of the frog (p. 211). The *optic chiasma* is very large, and the *optic tracts* very wide. The (V.) *trigeminal nerve* arises by two roots, upon the larger of which is the *Gasserian ganglion*. Owing to the elongation of the neck, a very long course is taken

by the (X.) *vagus* to reach the heart, lungs, &c. The (XI.) *spinal accessory*, a nerve not present in the frog, is made up of fibres arising from the side of the spinal cord and bulb, it leaves the skull with the vagus, and supplies some of the neck-muscles. The (XII.) *hypoglossal* nerve corresponds to the first spinal of the frog, and has a similar distribution, but it arises from the ventral surface of the bulb.

For the nerve exits see p. 233.

The **spinal nerves** arise by dorsal and ventral roots (the former ganglionated) from the spinal cord, and pass transversely outwards by the intervertebral foramina. They are named cervical, thoracic, &c., in the corresponding vertebral regions. The *brachial plexus*, from which the wing is supplied, is produced by the union of the last three cervical with the first thoracic. A *lumbar plexus* is formed by the last lumbar and first sacral nerves, and a *sciatic plexus* by the five succeeding nerves. These two plexuses supply the hind-limbs and their girdles.

(3) **Sympathetic System** (Fig. 70, *Sy*).—The main part of this is a delicate ganglionated cord running close to the vertebral column on each side. The ganglia are connected by *rami communicantes* with the commencements of the spinal nerves. The two cords unite behind in an unpaired ganglion (*ganglion impar*), while the anterior part of each of them is double, half of it running in the vertebrarterial canal.

The same histological elements are present as in the frog, and their structure and arrangement are similiar (*cf.* Fig. 61). There are, however, no bipolar ganglia in the sympathetic of the peculiar kind described on p. 215. The external part of the cerebellum is composed of grey matter, its internal part and peduncles of white matter.

The most important advance upon the frog is found in the large size of the *cerebral hemispheres*, and this is associated with increased intelligence. The large *cerebellum* has apparently the correlation of muscular movements for its main function.

10. **Sense Organs.**—(1) **Tactile Organs.**—Many of the nerves ramifying in the skin terminate in ovoid *touch-corpuscles*.

(2) **Gustatory Organs.**—*Gustatory cells*, supplied by the fibres of the glossopharyngeal nerve, are present in the hinder part of the tongue and roof of the mouth.

(3) **Olfactory Organs.**—The *olfactory sac* on each side, which communicates by an *external naris* with the exterior, and by an *internal naris* with the mouth-cavity, is partly lined by olfactory epithelium. This covers the projecting turbinal, the rolled shape of which increases the surface. The olfactory nerve breaks up below the olfactory epithelium to supply the *olfactory cells*, of which this is largely made up.

(4) **Auditory Organs.**—External, middle, and internal ears are present. The *external ear* consists of a short tube, the *external auditory meatus*, opening below and behind the eye. It is separated by the *tympanic membrane* from the *middle ear* or *tym-*

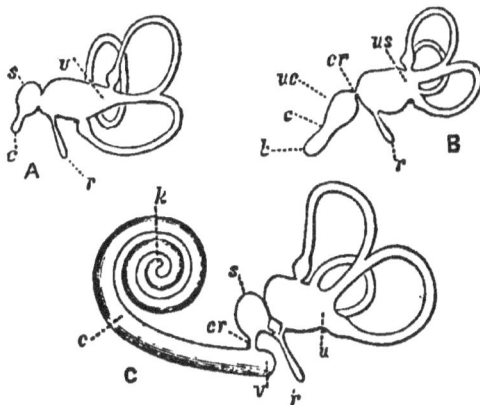

Fig. 72.—DIAGRAMS OF THE MEMBRANOUS LABYRINTH (from *Haddon*). Internal side of left labyrinth.—*A*, Fish. *B*, Bird. *C*, Mammal: *us*, utriculus and sacculus; *u*, utriculus; *s*, sacculus; *c*, cochlea.

panic cavity. This communicates with the mouth-cavity by an *Eustachian tube*, while the *columella* stretches across it, having one end attached to the tympanic membrane and the other inserted into the *fenestra ovalis*, below which is the *fenestra rotunda*. The *internal ear* is essentially composed of the *membranous labyrinth* (Fig. 72), enclosed in a somewhat larger cavity (filled with *perilymph*) of closely corresponding shape. This cavity is contained in the auditory capsule, and is surrounded by a thin, dense layer of bone, forming the *bony labyrinth*. The *utriculus* and *sacculus* are ill-marked off from one another. With the former three *semicircular canals* are connected, arranged as in the frog, with the exception that the posterior vertical takes an outward course, and crosses the horizontal before it dilates into an ampulla. A slightly-curved tube, the *cochlea*, is connected with the anterior part of the sacculus. The membranous labyrinth contains endolymph with otoliths, and is lined by epithelium, patches of which are largely made up of *auditory cells*, connected with auditory nerve-fibres.

It has been determined by experiment that the *semicircular canals* of the ear have to do with the perception of position in space. In this connection the fact that their planes are mutually at right angles deserves notice.

(5) **Visual Organs.**—The *eye* agrees in all essential particulars with the description on p. 221, but differs in certain points. The inner half of the eyeball is hemispherical, while its outer part is somewhat conical and terminated by the very convex *cornea* around which the *sclerotic* is strengthened by a circlet of small, flat *sclerotic plates* (Fig. 68). The *lens* is flattened, and a vascular, pigmented, longitudinally plaited fold, the *pecten*, projects into the vitreous humour below the entry of the optic nerve.

The *accessory* parts connected with the eye are the three *eyelids*, *conjunctiva*, two *glands*, and the *eye-muscles*. The glands are (1) the *Harderian gland* in front, and (2) the *lachrymal gland* behind and above. The eye-muscles are arranged much as in the frog.

DEVELOPMENT.

The common fowl has been most studied, and, as birds differ but little in their development, will be here described.

1. Early Stages.—The fertilized ovum, in passing down the oviduct, is covered by several structures secreted by its walls (Fig. 73). These are:—(1) The *white of the egg*, mainly composed of semifluid proteid material, and containing a somewhat convoluted cord, the *chalaza*, at each end, (2) the double-layered *shell-membrane*, covered by (3) the *shell*, formed of an organic matrix hardened by salts of lime. The entire egg is elongated, and broader at one end than the other.

Cleavage (segmentation) (Fig. 74) is *unequal* and also *partial* (meroblastic), being confined to the germinal disc. It commences in the lower part of the oviduct, and by a succession of furrows, some at right angles and others parallel to the surface, the germinal disc is converted into a many-celled *blastoderm*. This is placed on one side (Fig. 73, *Bl*), and always remains uppermost, the most favourable position for development, which requires a good deal of warmth, supplied in nature by the body of the hen. Before incubation (Fig. 75) the blastoderm consists of a superficial layer, one-cell thick, of columnar ecto-

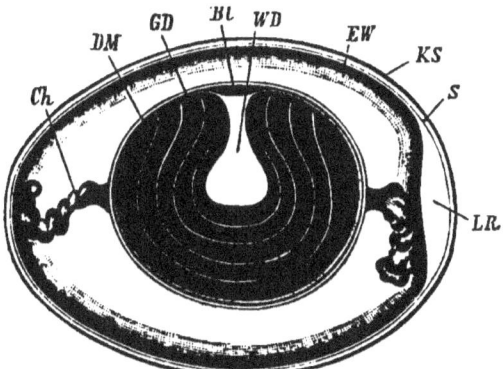

Fig. 73. - DIAGRAMMATIC LONGITUDINAL SECTION THROUGH UNINCUBATED HEN'S EGG (from *Claus*, after *Balfour* and *Allen Thomson*).—*Bl*, Blastoderm; *GD*, yellow yolk; *WD*, white yolk; *DM*, vitelline membrane; *EW*, "white;" *Ch*, chalaza; *S*, shell membrane; *KS*, shell; *LR*, air-chamber.

derm (epiblast), and a more irregular thickened mass of *lower layer cells*, which are rounded, granular, and of larger size. Below them is a space, the earliest rudiment of the alimentary canal.

Previous to laying, a **blastocœle** (segmentation cavity) can be seen between ectoderm and lower layer cells.

The blastoderm is at first circular, and the more transparent central part, known as the *area pellucida*, is surrounded by a darker rim, the *area opaca* (*cf.* Fig. 77). If the egg is placed with its broad end to the right, the diameter of the blastoderm, which is to become the long axis of the embryo, will run *across* the egg, the *posterior* end being *towards* the observer.

It now remains to describe the formation of the endoderm and mesoderm. These are first clearly differentiated in the *posterior* part of the blastoderm. Many of the lower layer cells here become much flattened and unite to form a thin membrane,

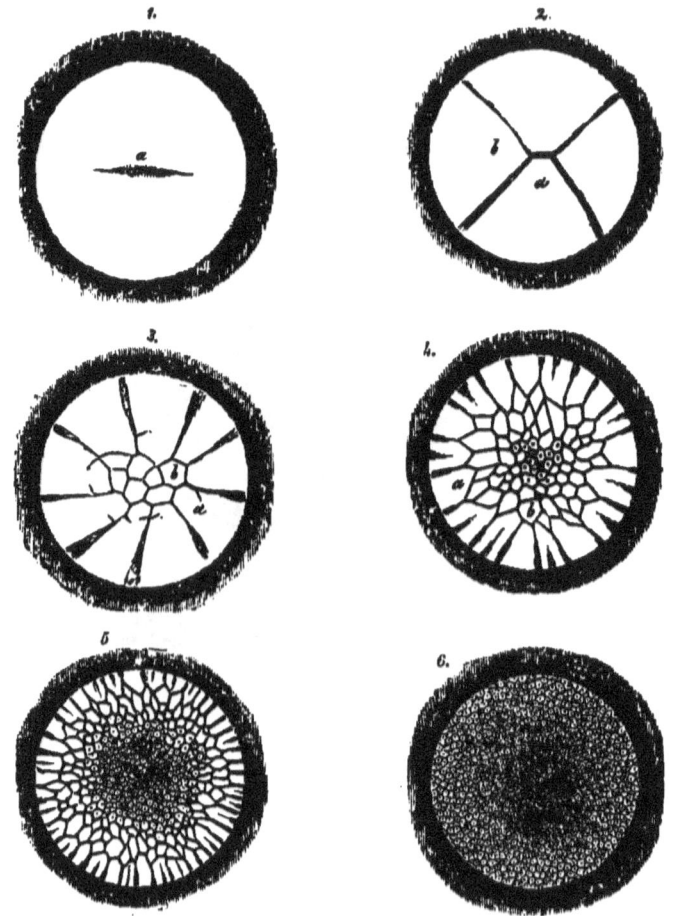

Fig. 74.—Surface Views to Show Cleavage in the Fowl's Oösperm (from *Kölliker*, after *Coste*).—Cleavage commencing in 1, completed in 6.

the **endoderm** (hypoblast) (Fig. 76, *hy*). In the hinder part of the area pellucida an opaque longitudinal strip, the *primitive* *streak* (Fig. 77, *Pr*), appears. This is a local thickening (Fig. 76), where ectoderm and endoderm are continuous, and from which a sheet of **mesoderm** (mesoblast) grows out on each side,

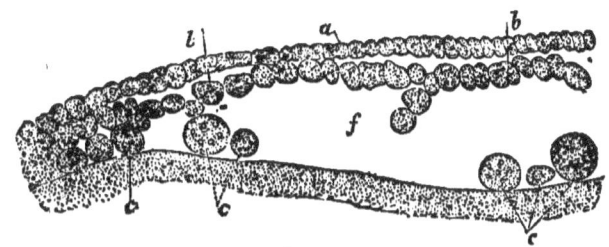

Fig. 75.—SECTION THROUGH PART OF UNINCUBATED FOWL'S BLASTODERM (after *Klein*).—*a*, Ectoderm; *b*, lower layer cells; *f*, archenteron; below this, yolk.

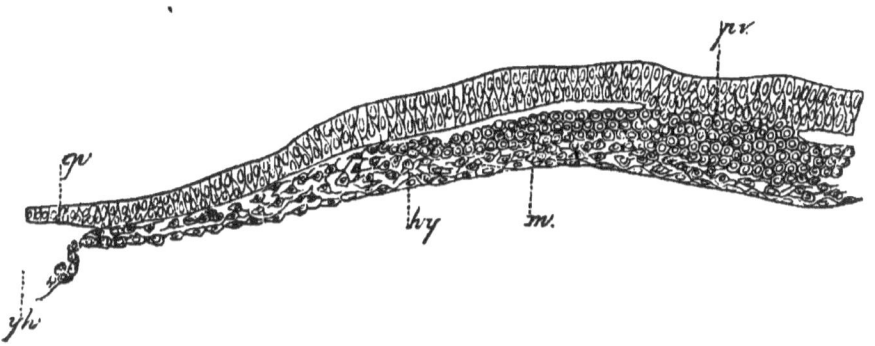

Fig. 76.—TRANSVERSE SECTION THROUGH FRONT END OF PRIMITIVE STREAK IN FIRST DAY CHICK (see Fig. 77)—(after *Balfour*).—Columnar ectoderm above, two cells thick. *Pr*, Region of primitive streak, where rounded mesoderm cells are originating from the ectoderm; *hy*, endoderm, above it, at the side, are stellate mesoderm cells of endodermal origin. N.B.—*Pr* is placed over the middle line.

being reinforced by those lower layer cells which have not united to form endoderm.

In the *anterior* part of the blastoderm, *i.e.*, in front of the primitive streak, the lower layer cell form—(1) a thin layer of **endoderm** (hypoblast), (2) two lateral sheets of **mesoderm** (mesoblast), (3) an axial rod of cells, the *notochord*, at first closely united with the hypoblast.

The blastoderm thus comes to be three-layered, both in front and behind.

The closure of the blastopore in a frog-embryo produces a small primitive streak, in which the germinal layers are continuous. There can be no doubt that the primitive streak in the chick-embryo is equivalent to a closed blastopore of slit-like form. If the slit were to re-open it would lead into the space figured at f in Fig. 75, and which = an archenteron. The gastrula of the fowl, therefore, is very much modified, owing to (1) closure of blastopore, (2) enormous accumulation of yolk on ventral side.

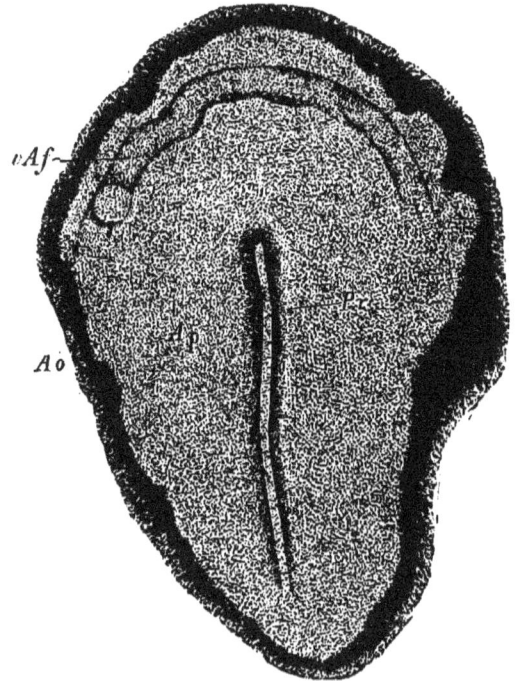

Fig. 77.—SURFACE VIEW OF FIRST DAY (20 HRS.) CHICK (from *Kölliker*).— *A.o*, Area opaca, bounding *A.p*, area pellucida; *Pr*, primitive streak; *Af*, head-fold.

2. **General Growth.**—The blastoderm is at first a watch-glass shaped plate resting upon the yolk. It generally grows in all directions, and ultimately completely encloses the yolk. This enclosure is mainly effected by the *area opaca*. The body of the embryo is formed in the *area pellucida*. This elongates (Fig. 77),

becoming somewhat pear-shaped, the broad (head) end being anterior. In front of the primitive streak a **neural** (medullary) **plate** is formed (*cf.* p. 224), upon which is a longitudinal *neural* (medullary) groove. This is bounded at the sides by *neural* (medullary) *folds* (Fig. 78, *Mn*), which unite above to form the **neural tube**. The folds first unite in the region of the mid-brain. The **head** now projects from the surface of the yolk, and makes with it, as seen from the left side, an **S**-shaped *head-fold*. A similar *tail-fold* is formed later, and then *lateral folds*. The embryo, at first flattened out on the curved yolk - surface, thus gradually comes to assume the form of the chick, and the ever diminishing yolk remains attached to its ventral surface (Fig. 82). In a chick at the end of the second day of incubation (Fig. 79), the *cerebral vesicles* (*Vh*, *Mh*, and *Hh*) are present, and the **eyes** and **ears** are commencing. The front part of the brain begins to bend round the end of the notochord. At the end of the third day, the front part of the embryo is twisted round so as to lie with its left streak; *vAf*, head-fold. Three either side of *Rf*.

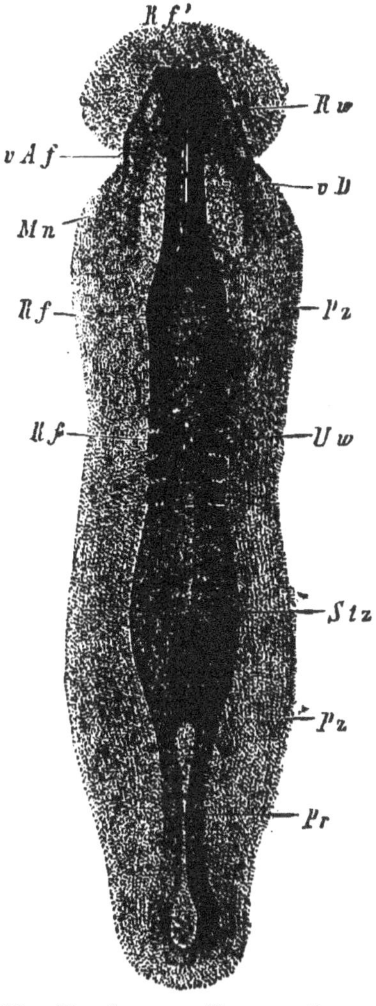

Fig. 78.—SURFACE VIEW OF CHICK, RATHER LATER THAN FIG. 77— Magnified 39 diameters (from *Kölliker*).—*Mn*, Medullary folds uniting in the head region; *Pr*, primitive mesodermic somites are visible on

side on the yolk, and the *cranial flexure* round the notochord is well-marked. Five **visceral arches** and four **visceral clefts** have made their appearance (Fig. 80), one less than in the frog. The most posterior of these are very small. The prominent **heart** is situated just under the throat, close to the

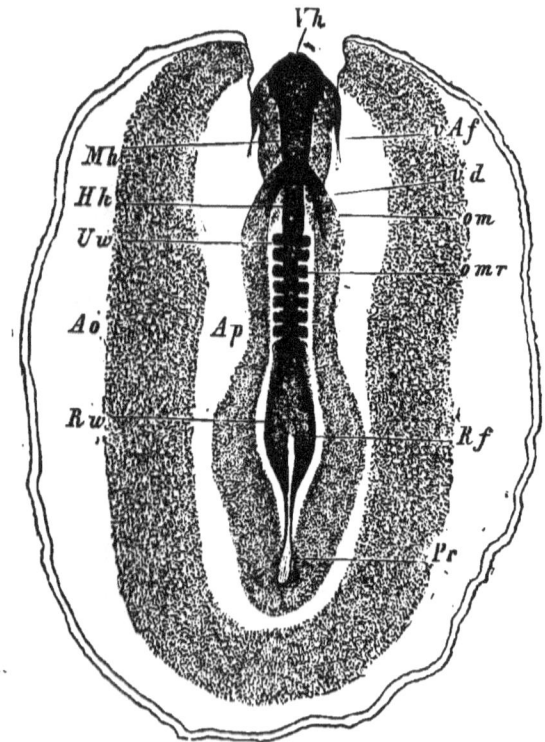

Fig. 79.—Surface View of Second Day Chick—Magnified rather more than 50 diameters (from *Kölliker*).—*Ap*, Area pellucida ; *Vh*, *Mh*, *Hh*, fore-, mid-, and hind-brains. The neural groove has closed as far back as *omr*. *Pr*, Primitive streak. Eight mesodermic somites are shown.

ventral ends of the visceral arches. The eye and ear are much more advanced, and the **olfactory sacs** are commencing as pits. On the fourth day the **limbs** appear as buds. The chick is hatched on about the twentieth day of incubation, breaking the shell with its beak, which possesses a small knob for the purpose.

3. **Fate of the Germinal Layers.**—These give rise to the same organs as in the frog (p. 226), but there are differences which will be here briefly noted. (1) The **ectoderm** (epiblast) does not become divided into superficial and nervous layers till comparatively late, and the membranous labyrinth commences as a pit open externally. (2) The **endoderm** (hypoblast) presents no important differences, but the notochord disappears more com-

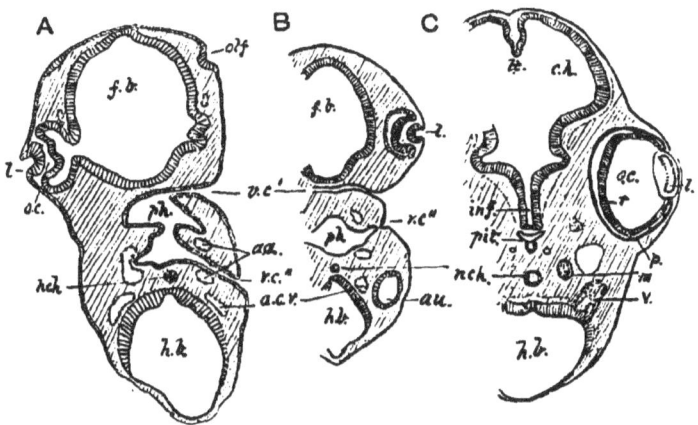

Fig. 80.—DEVELOPMENT OF THE EYE (from *Haddon*, A after *Marshall*).—
A and B sections through head of third day chick. In A the optic vesicle (*o.c*) from the fore-brain (*f.b*) is becoming cup-like, *l* is the developing lens. In B the cup and lens are more distinct. Stalk of the former not cut through. C. Later stage—Inner wall of optic cup (*o.c*) forming most of retina, *r*—outer wall, *p*, forming pigment layer of ditto; *l*, lens (quite separated off). These sections incidentally illustrate other points. In all, owing to the cranial flexure, both fore- and hind-brains have been cut through (*f.b*, *h.b*). In C, *c.h*, *l.t*, and *inf*, indicate cerebral hemispheres, lamina terminalis, and infundibulum. Near the last is the pituitary body, *pit*. In A and B the pharynx (*ph*) is cut through, and the first two visceral clefts (*v.c'* and *v.c''*) are shown. *a.a*, aortic arches; *a.c.v*, anterior cardinal vein; *nch*, notochord; *olf*, olfactory pit.

pletely in the adult. (3) The **mesoderm** (mesoblast) becomes divided (*cf.* p. 227) into *vertebral* and *lateral plates* in the region of the trunk, the former being again divided into *mesodermic* (mesoblastic) *somites* (Figs. 78 and 79). The cœlom (Fig. 81) is as before the split between the somatic and splanchnic layers of the lateral plates, the former uniting with the ectoderm to form the body-wall (*somatopleure*), the latter with the endoderm to form the wall of the gut (*splanchnopleure*). When the embryo is spread out flat on the yolk, the somatopleure and splanchnopleure belonging to opposite sides diverge widely, but, as the embryo is folded off, they gradually approach, unite, and form a continuous boundary to the body and gut respectively. The latter meets in front and behind with inpushings of ectoderm that form the mouth-cavity and posterior part of the cloaca (stomodæum and proctodæum).

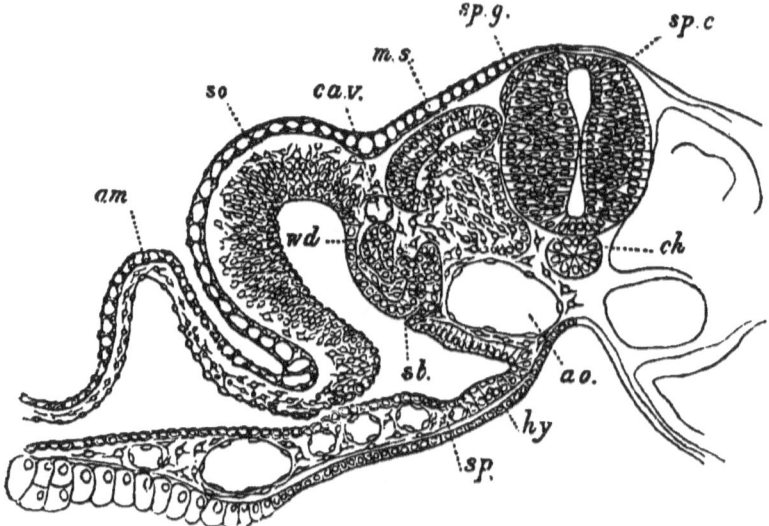

Fig. 81.—TRANSVERSE SECTION THROUGH EMBRYO DUCK AT THIRD DAY (after *Balfour*).—The external layer is ectoderm; *hy*, endoderm; mesoderm, a thick layer between the two; *am*, lateral amniotic fold; *so*, somatopleure; *sp*, splanchnopleure (the reference letters *wd* and *sb* are placed in cœlom); *sp.c*, spinal cord, at the side of which is section through mesodermic somite; *ch*, notochord (the reference letters *ao*, *hy*, *sp* are placed in the alimentary canal, the walls of which, at this stage, are widely divergent).

The first three visceral arches are supported by skeletal structures that become in the adult—

I. *Mandibular.*—Quadrates and Meckel's Cartilages, Palatines and Pterygoids.

II. *Hyoid*, } fuse ventrally { basi-hyal, { anterior cornua;
III. *Branchial* 1, } to form, { basi-branchial, } columellæ.
                                                 { posterior cornua.

The last two visceral arches, corresponding to branchials 2 and 3 of the frog, are small and have no skeleton.

The **circulatory organs** are not complicated by the presence of gills. The *heart* at first consists of two longitudinal vessels that fuse to form a simple tube, which first becomes twisted and then chambered. From its front end *aortic arches* run up the visceral arches (Fig. 80), and unite above to form the dorsal aorta.

1st and 2nd *Arches* become (in part)		.	carotids.
3rd *Arches*	,,	,, .	subclavians.
4th *Arches*	,,	{ *right*, .	aorta.
		{ *left*, .	aborts.
5th *Arches*	,,	. .	pulmonary arteries.

The *veins* of the embryo at first (3rd day) consist of paired *anterior cardinals* (= adult jugulars) and *posterior cardinals*, which bring back blood from the anterior and posterior parts respectively (*cf*. Dogfish, p. 165). The anterior and posterior cardinal of each side unite to form a *Cuvierian vein* (= adult precaval), which enters the heart. The posterior cardinals are supplanted later on by the development of a postcaval.

The course of the embryonic circulation will be discussed below.

Excretory Organs.—A very rudimentary **pronephros** is first developed, with a *pronephric duct* (= archinephric or segmental duct) which runs back and opens into the cloaca. It is succeeded by a **mesonephros** (Wolffian body), the duct of the pronephros becoming the *mesonephric* (Wolffian) *duct*. The mesonephros does not, however, become the excretory organ of the adult, but is succeeded by a **metanephros**, or true **kidney**, for which a special *metanephric duct* (ureter) is developed. The only conspicuous remnant of the mesonephros that persists after embryonic life is its duct, which becomes the spermiduct of the male.

4. **Embryonic Appendages** (Fig. 82).—These are important structures connected with, but not forming part of, the embryo. To understand them, it is not only necessary to remember that the blastoderm gradually grows round the yolk, but also that the mesoderm splits here, as elsewhere, so that continuations of somatopleure, splanchnopleure, and cœlom are present. As the embryo is folded off from the yolk, this, with the covering of splanchnopleure, comes to form a sort of bag, the **yolk-sac**, attached to the ventral side of the embryo by a stalk continuous with the wall of the gut. The yolk is gradually used up during development, so

266 AN ELEMENTARY TEXT-BOOK OF BIOLOGY.

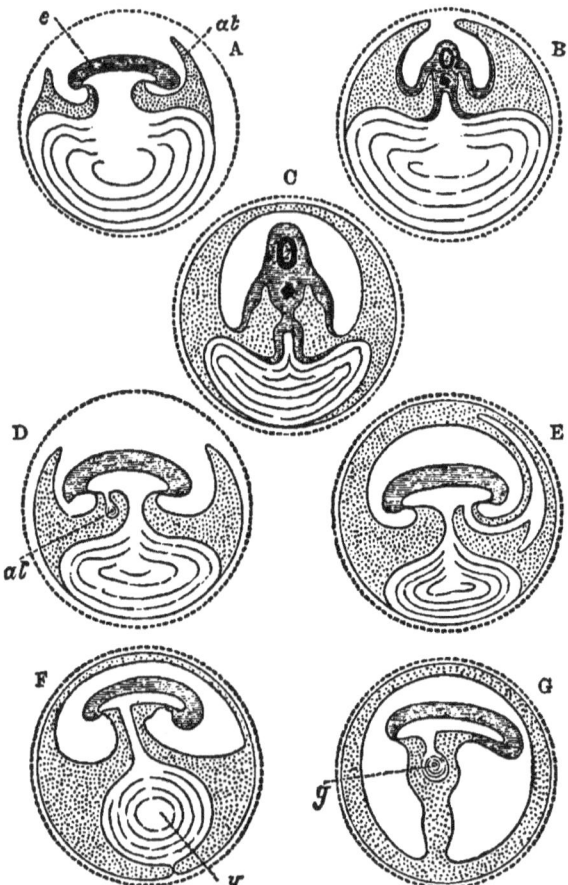

Fig. 82.—DIAGRAM TO ILLUSTRATE THE DEVELOPMENT OF THE AMNION
AND ALLANTOIS (after *Foster* and *Balfour*).—In A the embryo (*e*) is
being constricted off from the yolk-sac, and the folds of the amnion are
seen to be rising up at each end of the embryo, the anterior fold (*at*)
being the larger; in B (a transverse section) the lateral amniotic folds
nearly meet, and in C (also a transverse section) they have entirely
coalesced. In D (a rather later stage than A), the allantois (*al*) is
budding out from the alimentary canal; in E (side-view corresponding
to C), the allantois is seen extending round the embryo. In F, the
yolk-sac (*y*) is reduced in size, and in G it is being withdrawn into the
embryo's body. Allantois omitted in F and G. These diagrams only
roughly indicate the relation of parts. In all, the embryo is represented by horizontal shading, the cœlom and its extension are dotted,
and the yolk-sac marked with concentric lines. The dotted line represents the vitelline membrane.

that the yolk-sac becomes smaller, and at last passes into the body. The continuation of the *somatopleure* outside the boundary of the body rises into folds, which grow up, and, finally meeting above the body, fuse together. Their *inner* layers form the **amnion**, a membrane enveloping the embryo, while their outer layers (and the somatopleure prolongation in the yolk-sac region) unite with the vitelline membrane. As the somato- and splanchnopleures are prolonged outside the region of the embryo, the space there present between them must be a continuation of the body-cavity. A flattened sac, the **allantois**, grows from the posterior part of the alimentary canal into this space, and extends over the embryo.

The yolk-sac possesses at one time a system of capillaries, and functions as a respiratory organ. As the allantois develops it also becomes very vascular, first assists the yolk-sac in respiration, and then carries it on entirely. It also grows round, and absorbs the albumen ('white').

The urinary bladder of the frog is a rudimentary allantois.

Course of the Embryonic Circulation.—The dorsal aorta sends *vitelline* and *allantoic arteries* from its posterior part to the yolk-sac and allantois respectively. These structures return purified blood to the body by *vitelline* and *allantoic* * veins. These unite to form, with a vein from the gut, a trunk, the *ductus venosus*, which traverses the liver (giving off twigs in its course), and enters the now-developed postcaval. This communicates with the right auricle, and its (mostly) purified blood is directed by the Eustachian valve through the *foramen ovale*,† an aperture existing at this time in the auricular septum, into the left auricle. Thence it passes into the left ventricle, and so to the dorsal aorta. The right auricle also receives impure blood by the Cuvierian ducts. This passes into the right ventricle, whence, by the pulmonary artery, it reaches the lungs and also, to some extent, the dorsal aorta, by a cross branch, the *ductus arteriosus*. The lungs, at this time functionless, return impure blood to the left auricle. It will be seen that, in the embryo, the comparatively pure blood of the left ventricle is derived from the *right side* of the heart. As soon as *lung-respiration* commences, the foramen ovale closes,

* = *Anterior abdominal vein* of Frog.
† Its position is marked by the *fossa ovalis* of the adult.

the system of the ductus venosus becomes the *hepatic portal system* (that vessel being abolished as a continuous trunk), and the ductus arteriosus is reduced to a fibrous cord (ductus Botalli).

CHAPTER XII.—MAMMALIA.

§ 17. LEPUS CUNICULUS (The Rabbit).

THE wild Rabbit, which is here described, forms the parent-stock from which the different kinds of tame Rabbit have sprung, and which differ from it in no essential points of structure. It is, as everyone knows, a gregarious, burrowing animal, of a brownish colour, which harmonizes with the surroundings when it is out feeding, and so serves as a means of protection (protective general resemblance). The short tail is white on its under side, and is very conspicuous when the animal is moving, probably serving as a "danger signal."

MORPHOLOGY AND PHYSIOLOGY.

1. External Characters.—The bilaterally symmetrical body is divided into the same regions as in the pigeon—*i.e.*, head, neck, trunk, and tail. The fore- and hind-limbs are adapted for quadrupedal progression. Almost all the external surface is covered by hair, but a bare perineal space is present on each side near the root of the tail.

The elongated **head** tapers in front into the snout, at the end of which is the *mouth*, a transverse slit bounded by soft mobile *upper* and *lower lips*, the former of which is cleft. The *external nares* are two oblique slits near the tip of the snout, converging in front to the cleft of the upper lip. Long stiff hairs, the whiskers or *vibrissæ*, are present on the sides of the snout as well as in the neighbourhood of the eyes, which are large, and protected by *upper, lower,* and *third eyelids.* Each of the two first is fringed by a row of stiff *eyelashes*, while the last is a bare opaque white membrane, usually folded up in the anterior angle of the eye. The *auditory aperture* on each side is situated some distance behind the eye, and is guarded by a long backwardly and upwardly directed flap, the *pinna*, the base of which is supported by cartilage.

The pinna varies very much in size and position in tame rabbits, especially in lop-eared varieties.

The short **neck** forms a connecting region between head and trunk.

The **trunk** is somewhat flattened from side to side. It is divided into a relatively small *thorax* in front, bounded by hard parts, and a large soft-walled *abdomen* behind. There are, in the *female*, five to six pairs of *teats* on the ventral surface, belonging to the *milk glands* (mammary glands). Upon each teat are the small openings of the corresponding gland.

The intestine and urinogenital organs here open separately to the exterior, the former by a rounded *anus* situated immediately beneath the tail, and the latter by an *urinogenital aperture*, placed a little further forwards, and varying in character according to sex. The space between the anus and urinogenital aperture is known as the *perineum*, and on each side of this there is a bare, pouch-like depression, the *perineal space*, upon which is a small papilla bearing the aperture of a *perineal gland*, to the secretion of which the unpleasant odour of rabbits is due.

The **tail** is short, but distinct.

The **fore-limb** is divided into upper arm, fore-arm, and hand (*brachium, antebrachium*, and *manus*). The much longer **hind-limb** is divided into thigh, leg, and foot (*femur, crus*, and *pes*). The digits are clawed, five in number in the hand, four in the foot.

Position of Body. — In the standing posture the *elbow* is directed backwards, the *knee* forwards, and the *ankle-joint* backwards. Both limbs are disposed in longitudinal vertical planes, and the body is lifted well off the ground.

2. The thick **skin** is made up of *epidermis* and *dermis*, the hairs being developments of the former. The *mammary* and *perineal glands* belong to the skin, and they are imbedded in the underlying, *subcutaneous*, connective tissue, which is everywhere abundantly present. Beneath and closely connected with this is a

thin sheet of striated muscle, the *panniculus carnosus*, on the ventral side of the neck and trunk.

Each **hair** commences (in the embryo) as a thickening of the Malpighian layer of the epidermis projecting towards the dermis (Fig. 83). This *hair-germ* gradually becomes converted into a *hair-follicle*, at the bottom of which is a *hair-papilla*. A special

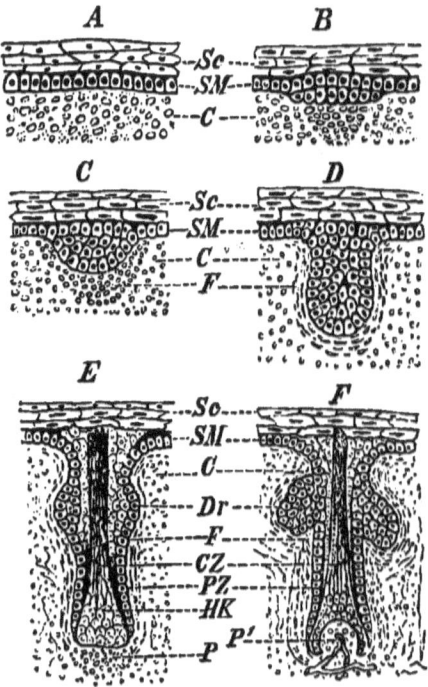

Fig. 83.—Six Stages in the Development of Hair (from *Haddon*, after *Wiedersheim*).—*Sc*, Horny layer; *SM*, Malpighian layer; *C*, dermis; *F*, hair-follicle; *Dr*, sebaceous glands; *CZ*, shaft of hair; *PZ*, hair sheath; *HK*, hair germ; *P'*, hair-papilla; *P*, commencement of ditto.

investing sheath is formed by the dermis, which also penetrates into the hair-papilla and vascularises it. The *hair*, developed from the epidermis of the hair-papilla, exhibits an external firmer *cortex*, covered by a scaly *cuticle*, and a softer internal air-containing *medulla*, which, in coloured hairs, is pigmented. Small branched *sebaceous glands*, secreting an oily substance, open into the follicle near its mouth, and with it are also connected bands of unstriated muscle, by contraction of which the hair can be erected. The hairs are continually being shed, and are replaced by new ones developed on papillæ connected with the old hair-follicles.

The **mammary** and **perineal glands** are made up of tubules, dilated at their ends, and lined by granular epithelium continuous with the epidermis.

The *dermis* is rich in blood-vessels, lymphatics, and nerves, and many nerve-fibres end in elongated ovoid *touch-corpuscles*. One of these is closely connected with the base of each vibrissa.

3. The **endoskeleton** (Figs. 84 and 85) is made up of the usual histological elements. The bones are much more compact than those of the pigeon, and the shafts of the long bones contain marrow.* As in other cases a distinction can be drawn between axial and appendicular skeletons, belonging respectively to trunk and limbs.

(1) The *axial endoskeleton* consists of skull, vertebral column, ribs, and sternum.—

(*a*) **Skull**.—As in the pigeon, the mature skull is mostly composed of bones, but these are comparatively little fused together, many of the boundaries remaining apparent, even in old animals. They are united, in many cases, by finely jagged edges, which interlock. Such unions are called *sutures*.

A small posterior *cranial* portion may be distinguished from a much larger *facial* portion, the boundary between the two being indicated by the large *orbits*.

It is convenient to consider the **cranium** or **brain-case** which constitutes most of the cranial region of the skull, as being made up of three rings (*not* segments), named from behind forwards—occipital, parietal, and frontal. These rings fit closely together, ventrally and dorsally, but certain gaps are left laterally. Each possesses an unpaired ventral element, and the floor of the brain-case is largely formed by these ventral elements, the line along which they are arranged being known as the *basi-cranial axis*.

* 1. *Cartilage bones:—Ex-, supra-,* and *basi-occipitals; basi-, pre-,* and *orbito-sphenoids; mesethmoid, periotic, malleus, incus, os orbiculare, stapes, turbinals, palatine, pterygoid, hyoids, Vertebræ, Ribs, Sternum, Appendicular endoskeleton* (except clavicle). 2. Membrane bones:—Interparietal, parietal frontal, squamosal, tympanic, vomer, lachrymal, nasal, premaxilla, maxilla, jugal, mandible, Clavicle.

I. *Occipital ring*, of four bones, surrounding foramen magnum at back of cranium—*basi-occipital* below, an *ex-occipital* each side, *supra-occipital* above.

II. *Parietal ring*, of six bones—*basi-sphenoid* below, a small *ali-sphenoid* each side, two large *parietals* above.

III. *Frontal ring*, of six bones—*pre-sphenoid* below, a small *orbito-sphenoid* each side, two large *frontals* above.

I.	II.	III.
s. occ.	pa. pa.	fr. fr.
ex. occ. ex. occ.	al. s. al. s.	or. s. or. s.
b. occ.	b. s.	pr. s.

The cranium narrows anteriorly, and its front end is closed by a median *mesethmoid*, which continues the direction of the basicranial axis into the facial region. Cranial axis + facial axis constitute the *cranio-facial axis*. The mesethmoid also assists in the formation of the olfactory capsules.

The **auditory capsule** is wedged in between the occipital and parietal rings, the lateral gap between which is partly filled by a *squamosal* bone.

<small>There are two smooth oval *occipital condyles*, one on each side of the foramen magnum, mainly formed by the corresponding ex-occipitals, but partly by the basi-occipital, which is a small, flat bone. Each ex-occipital is produced downwards into a long *par-occipital process*, closely applied to the hinder part of the tympanic bulla. The large, irregular supra-occipital is raised into a prominent, shield-shaped elevation. The roof of the braincase is completed by inter-parietal, parietal, and frontal bones. The *interparietal* is a small, unpaired bone, placed transversely, and partly shutting out the *parietals* from union with the supra-occipital. These are large, flattish bones, united together by a longitudinal *sagittal suture*, and with the supra-occipital and inter-parietal by the transverse *lamboidal suture*. A slender process runs downwards from the outer side of each. The *frontals*, which cover the front of the rapidly-narrowing brain-case, are united with each other by the median *frontal suture*, and with the parietals by the transverse *coronal suture*. Each is very irregular in shape, and possesses parts situated in very different planes. That part roofing the brain-case is broadest behind, and tapers anteriorly, to form with its fellow a sharp point. From the side of this part a plate-like *orbital process* turns sharply downwards and inwards to form the upper part of the inner orbital wall. Where it turns down, a large overhanging *supra-orbital process* sticks out, which partly roofs the orbit. A process from the frontal runs into the nasal region. The floor of the brain-case rapidly steepens in front of the basi-occipital, and is formed in the middle line by the basi- and pre-sphenoids, on each side by the ali- and orbito-sphenoids. The *basi-sphenoid* is a small bone with a broad posterior end separated from the</small>

basi-occipital by a thin plate of cartilage, and a narrow anterior end united by cartilage to the pre-sphenoid. Upon its upper surface is a pit, the *sella turcica* (for the pituitary body), bounded behind by the *posterior clinoid processes*. Firmly fused to each side of the basi-sphenoid is a thin wing-like expansion, the *ali-sphenoid*, from the under side of which a transverse lamella, the *external pterygoid process*, passes vertically downwards. The *pre-sphenoid* is a vertical plate, the upper surface of which is produced backwards into the *anterior clinoid processes* which bound the sella turcica in front. After limiting the *optic foramen* below, the pre-sphenoid ends in two diverging lamellæ which articulate above with the orbital processes of the frontal, and help to form the inner wall of the orbit. A wing-like *orbito-sphenoid* is firmly fused with the pre-sphenoid on either side, forming the hinder and upper boundaries of the optic foramen, and uniting with the orbital process of the frontal above, and the ali-sphenoid behind. The narrow front end of the brain-case is filled in by the *cribriform plate*, a lamella perforated by numerous holes and forming the party-wall between the cranial and nasal cavities. It is the hinder part of the *mesethmoid*. The side-wall of the skull is largely formed in its posterior region by the *squamosal*, which fills up the gap between parietal and ali-sphenoid, unites in front with the frontal and orbito-sphenoid, and behind overlaps the descending process of the parietal. A delicate backward process from the squamosal helps to keep the large, irregular *periotic* in place, and a large *zygomatic process* runs outwards and forwards from it just behind the orbit.

The **auditory capsule** is formed by the *periotic* bone which results from the early fusion of *pro-*, *epi-*, and *opisthotic* elements.

The outer and hinder *mastoid* portion of this bone is porous, and a tapering *mastoid process* runs down from it in front of the par-occipital process. The inner side of the mastoid part presents a deep rounded pit, the *floccular fossa*, for the flocculus of the cerebellum. The *petrous portion* which forms the rest of the periotic is very dense, and contains the membranous labyrinth. Upon its outer surface there is a smooth projection, the *promontory*, above which is the elliptical *fenestra ovalis*, and behind it the more irregular *fenestra rotunda*.

Closely applied to the outer surface of the periotic is the *tympanic*, a bone with a dilated lower part (*bulla*) and a short part directed upwards and backwards. The inner side of both is incomplete, for which the periotic makes up, forming the inner wall of the *tympanic cavity*, which is contained in the bulla. The tubular part supports the external auditory meatus, and, at its junction with the bulla, the *tympanic membrane* is stretched over a flattened rim which projects into the latter. Extending across the tympanic cavity are four minute **auditory ossicles**.

(1) The *malleus*, which possesses two slender processes (one, the *manubrium*, attached to the tympanic membrane), which are given off from a rounded head. This articulates by a saddle-shaped surface with the head of (2) the *incus* which gives off a small process backwards and a

slender one downwards. The latter is bent up at its tip, and connected by a minute disc (3), the *os orbiculare*, with (4) the stirrup-shaped *stapes*, the oval base of which fits into the fenestra ovalis. The Eustachian tube enters the tympanic cavity below and in front by a very short canal, formed by the apposition of a groove in the tympanic to the surface of the periotic. The par-occipital and mastoid processes overlap the tympanic behind, and hold it in place.

The cavity of the brain-case closely corresponds to the shape of the brain, and is divided into a small *olfactory fossa* in front, ill-marked off from the large *cerebral fossa*, which extends back to the anterior limit of the periotics and supra-occipital, where a sharp ridge separates it from the *cerebellar fossa*.

NERVE-EXITS.

I. *Olfactory.*	Numerous olfactory foramina.	Cribriform plate.
II. *Optic.*	Optic foramen.	A large unpaired aperture, bounded by pre- and orbito-sphenoids, which places hinder end of orbits in communication with each other and the cranial cavity.
III. *Oculomotor.* IV. *Pathetic.* VI. *Abducent.* V. *Trigeminal—* *a.* Ophthalmic. *b.* Maxillary.	Sphenoidal fissure (foramen lacerum anterius).	Between basi- and ali-sphenoid on either side.
c. Mandibular.	Foramen lacerum medium	Between ali-sphenoid and periotic.
VII. *Facial.* Enters periotic by	Aqueductus Fallopii.	Just below the floccular fossa.
Leaves skull by	Stylomastoid foramen.	Between mastoid process and tympanic.
VIII. *Auditory—* Enters periotic by	Meatus auditorius internus.	Just behind aqueductus Fallopii, and sunk in a depression with it.
IX. *Glossopharyngeal* X. *Vagus.* XI. *Spinal Accessory.*	Foramen lacerum posterius.	Between periotic and ex-occipital.
XII. *Hypoglossal.*	Condylar foramina (2).	In ex-occipital, near the condyle.

The **facial** part of the skull consists of the olfactory capsules and jaws with associated parts. The **olfactory capsules** are separated from the cranial cavity by the cribriform plate, from which a vertical partition, bony behind (*lamina perpendicularis*), cartilaginous in front (*septum nasi*), extends forwards and separates the two capsules. This partition, together with the cribriform plate, constitutes the *mesethmoid*, and its lower edge rests in a deep furrow on the upper surface of the elongated *vomer*.

The posterior wall of each capsule is partly formed by the small *lachrymal*, which also makes up part of the anterior orbital wall. Its outer side is deeply notched to transmit the *lachrymal duct*. The capsule is partly roofed in behind by the frontal, but mainly by an elongated flat *nasal*, which, with the premaxilla, largely bounds the external naris. The capsular side-wall and floor are principally formed by the premaxilla and maxilla. The nasal cavity contains the *ethmo-turbinal* behind and above, a bone made up of thin lamellæ disposed in a complex manner, which is fused with the cribriform plate, and a similarly constructed but more delicate *maxillo-turbinal* in front. From the nasal bone a delicate curved fold, the *naso-turbinal* projects into the cavity.

The margin of the **upper jaw** is formed by *premaxillæ* and *maxillæ*. With the latter a *zygomatic arch*, running along the lower side of the orbit, is connected externally, while internally they come into relation with *palatine* and *pterygoid* bones.

The *premaxillæ* contain sockets for the upper incisor teeth. Each sends an elongated *nasal process* upwards and backwards along the outer edge of the nasal, a delicate *palatine process* backwards along the roof of the mouth, and a stout *maxillary* process backwards and outwards to unite with the *maxilla*. This is a large, extremely irregular bone that forms the jaw-margin with the premaxilla, a little way behind which bone it dilates to contain the sockets of the grinding teeth, and projects into the front part of the orbit. The maxilla also forms a large part of the side-wall of the nasal capsule, and in this region is very loose in texture. From its external side, above the grinding teeth, a stout *zygomatic process* projects, which is connected by the laterally flattened *jugal* with the zygomatic process of the *squamosal* to make up the *zygomatic arch*, a bony bar forming the lower boundary of the orbit. A small, horizontal *palatine process* runs from the inner side of the maxilla, and unites with its fellow in the middle line to form the front part of the *bony palate*, a transverse bridge of bone between the first four grinding teeth of opposite sides. The inner surface of the maxilla behind this is covered by the *palatine*, a thin vertical plate uniting above with the orbital process of the pre-sphenoid. The bony palate is completed behind by the horizontal *palatine process* of this bone which unites with its fellow. The *posterior nasal chamber* is bounded laterally by the palatines, which are united behind with the downward processes of the ali-sphenoids, and with the *pterygoids*. Each pterygoid is a small plate, behind, and in the same plane as the palatine. It is produced below into a curved *hamular process*, and connected above with the junction between basi- and ali-sphenoids.

The **lower jaw** or **mandible** consists of two halves or rami, each of which unites in front with its fellow in the mandibular symphysis, and articulates behind by a longitudinally elongated *condyle* with the *glenoid fossa*, a smooth hollow on the under side of the zygomatic process of the squamosal. It is laterally flattened and composed of an anterior horizontal part bearing sockets for the lower teeth, and a posterior ascending part ending in the condyle above—in front of which is the *coronoid process*—and uniting with the horizontal portion in the rounded *angle*.

Hyoid and First Branchial Arches.—The proximal end of the hyoid arch is formed by the *stapes*. The "hyoid apparatus" consists of a small *basi-hyal* imbedded in the root of the tongue, and with which are connected short *anterior* and long *posterior cornua*. The anterior cornua represent the *hyoid arches* of the embryo, and are connected by ligaments with the skull, close to the mastoid processes. In this region a small part of the hyoid arch (*tympanohyal*) is fused to the skull. The posterior cornua represent the *first branchial arches*.

Fig. 64.—ENDOSKELETON OF RABBIT.—A, Skull (*jugal* bone removed); *s-oc*, supra-occipital; *ex-oc*, ex-occipital; *c*, condyle; *par*, par-occipital process; *s-m.f*, stylo-mastoid foramen; *ty*, tympanic; *per*, periotic; *i-p*, inter-parietal; *pa*, parietal; *a-s*, ali-sphenoid; *sq*, squamosal; *z.s*, zygomatic process of squamosal; *pt*, pterygoid; *fr*, frontal; *or.s*, orbito-sphenoid; *pr-s*, pre-sphenoid; *pl*, palatine; *mx*, maxilla; *z*, zygomatic process of ditto (with cut end); *p-mx*, pre-maxilla; *la*, lachrymal; *na*, nasal; *op-f*, optic foramen; × placed at junction of basi- and pre-sphenoid. A', Outside of periotic, after removal of tympanic; *ma*, mastoid portion; *m.pr*, mastoid process; *pr*, promontory; *f.ov*, fenestra ovalis; *f.ro*, fenestra rotunda. A", Auditory ossicles from inside; *m*, malleus; *i*, incus; *st*, stapes. B, *Mandible*; *c*, condyle; *a*, angle; *c.pr*, coronoid process. C, Hyoid apparatus; *b*, body; *a.c*, anterior cornu; *p.c*, posterior cornu. D, Various vertebræ; 1, atlas (front view); 2, axis; 3, typical cervical (front view); 4, typical lumbar; *c*, centrum; *n.s*, neural spine; *t*, transverse process; *a.f*, (in atlas), articular facet for condyle; *a.z*, prezygapophysis; *p.z*, post-zygapophysis; *n.c*, neural canal; *v.c*, vertebrarterial canal; *o.p*, odontoid peg; *i.l*, cervical rib; *ep*, epiphysis; *mt*, metapophysis; *an*, anapophysis. E, *Hip-girdles and Sacrum*. 1, 2, 3, Sacral vertebræ; *s.r*, sacral rib; *il*, ilium; *pb*, pubis; *is*, ischium (reference letter placed in obturator foramen); *ac*, acetabulum. F, Right tarsus; *as*, astragalus; *ca*, calcaneum; *cn*, centrale; *t*, 1-5, *tarsalia*; 2-5, metatarsals; *, placed by process of met. 2, representing hallux. G, *Left Scapula—s-sc.b*, *co.b*, *gl.b*, supra-scapular, coracoid, and glenoid borders; *sp*, spine; *ac*, acromion; *mtc*, metacromion; *co*, coracoid process; *gl*, glenoid cavity. H, *Left Carpus*—*r*, Radiale; *i*, intermedium; *u*, ulnare; *cn*, centrale; *c*, 1-5, carpalia; 1-5, metacarpals.

MAMMALIA.

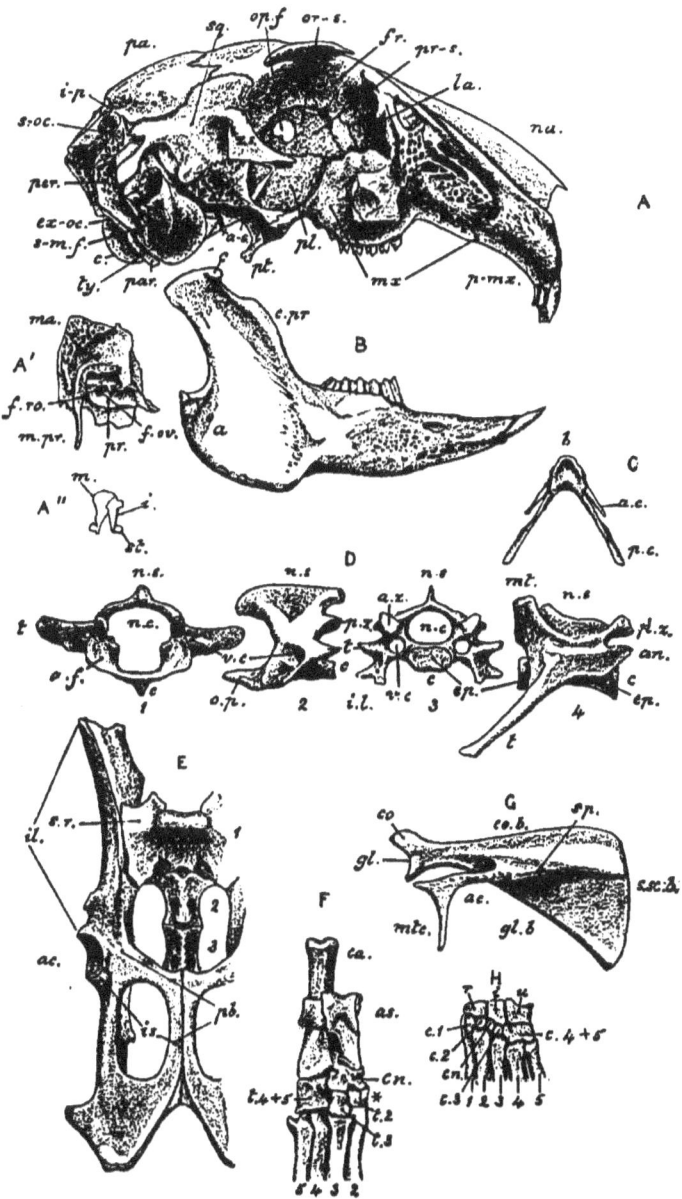

Skull of the Dog.—This agrees in all essential matters with the rabbit's skull, but there are differences in general shape and also in certain details. Most of these variations are correlated with the larger development of the jaw-muscles, as may be seen, for instance, in the presence of strong ridges on the brain-case and the width of the zygomatic arches. The skull is much stronger, and the face is in line with the cranium, whereas in the rabbit it is bent slightly downwards. The spongy areas present in the skull of the rabbit (especially on the outer side of the maxilla) are absent in the dog. Other differences are the following: — (1) *Cranium.*—Plane of occipital region vertical, instead of slanting somewhat downwards; interparietal represented by a process of supra-occipital. (2) *Olfactory capsules.*—Nasoturbinal represented by uppermost lamella of ethmo-turbinal. (3) *Bony palate* very large and long. (4) *Mandible*, with short ascending part, well-marked angle, large coronoid, and transverse condyle, articulating with glenoid fossa to form a well-marked hinge-joint. All these last features are connected with the great size of the muscles used in biting. The hinge-joint only allows of snapping movements, giving no lateral play. (5) *Nerve-exits.*—The 2nd and 3rd branches of the trigeminal exit by special openings, *foramen rotundum* and *foramen ovale*, situated one behind the other in the ali-sphenoid, between the sphenoidal fissure and the foramen lacerum medium. (6) *Teeth*, see p. 285.

(*b*) **Vertebral Column and Ribs** (Fig. 84). — The vertebral column is divisible into cervical, thoracic, lumbar, sacral, and caudal regions, corresponding to neck, thorax, loins, pelvis, and tail. Except in the sacral region the vertebræ remain distinct from one another, and the more or less flat ends of the centra are connected together by cartilaginous discs, intervertebræ, which come between them. The front and back of each centrum ossify independently as thin bony plates, *epiphyses*, which fuse later on with the middle part. The spinal nerves exit as in the other types by *intervertebral foramina* between adjoining arches, and *neural spines* and *zygapophyses* of the usual kind are present.

The cervical vertebræ are seven in number. Their centra are short, and their neural spines small. The cervical ribs are not free but fused, as in most of the pigeon's cervical vertebræ (*cf.* p. 237). The apparent transverse processes are thus perforated at their bases to form the *vertebrarterial canal*, and are divided into two parts, the upper of which is true *transverse process*, the lower (*inferior lamina*) the projecting part of the fused rib.

The *atlas* or first cervical vertebra is ring-like, possesses large transverse processes (but no inferior laminæ), and a thin narrow centrum. It presents in front two large oval articular surfaces for the occipital condyles, and behind two smaller concavities for the axis, but there are no true zygapophyses.

MAMMALIA.   279

The *axis* or second cervical vertebra has an *odontoid peg* (which ossifies separately), projecting forwards above the centrum of the atlas.

> A convex articular surface is present on the under side of the peg, which plays upon a corresponding surface on the upper side of the atlas-centrum, and there are also two large convexities on its base which fit into the two shallow cups on the hind end of the atlas. The neural spine of the axis is a prominent ridge which bifurcates behind.

The *last* (seventh) cervical vertebra has a half-facet on the side of the centrum, at its hinder end.

The **thoracic vertebræ** are 12 (or 13) in number, and are characterized by the possession of free thoracic ribs.

> They increase in size from before backwards, their centra, which are thick from above downwards, elongating. The neural spines of the first 9 are slender and backwardly directed, that of the tenth is vertical, and those of the remainder slope forwards like the lumbar spines. From the ninth backwards *metapophyses* appear, stout processes directed upwards and forwards in the region of the pre-zygapophyses. The transverse processes of the first 9 possess *tubercular facets* on the under side of their tips, and *capitular half-facets* on the sides of their centra, one in front, the other behind. The last 3 (or 4) have entire *capitular facets* on the sides of their centra, situated near the front.

The ribs are curved flattened rods, increasing in length up to the sixth, and then shortening. Each consists of a bony *vertebral portion*, possessing two articular processes (tubercle and capitulum), and a much shorter *sternal part* of more or less ossified cartilage.

> The tubercles of the ribs articulate with the tubercular facets, and the capitulum of each of the first 9 articulates with a capitular facet, formed in part by the anterior half-facet of its own vertebra, in part by the posterior half-facet of the preceding vertebra (the seventh cervical in the case of the first rib). The last 3 (or 4) possess no tubercles, and their capitula articulate with the corresponding facets on their own vertebræ. The sternal parts of the first 7 ribs unite distally with the sternum; those of the remaining ones do not. This is the distinction between "true" and "false" ribs.

The **lumbar vertebræ** are 7 (or 6) in number. They are large, with elongated centra, laterally flattened neural spines, directed forwards, and strong elongated transverse processes running downwards and forwards.

> They also possess large *metapophyses*, stout processes overhanging the pre-zygapophyses, and small *anapophyses*, backwardly directed processes given off below the post-zygapophyses. The first two also have *hypapophyses*, unpaired processes running downwards from the under side of the centra.

The **sacral vertebræ** are 4 in number, closely fused together. The two first possess *sacral ribs*, which form lateral wing-like expansions, articulating with the inner sides of the ilia.

The **caudal vertebræ**, about 15 in number, gradually get smaller towards the end of the tail, losing first their processes, and then their neural arches.

(c) The **sternum** is a narrow rod, made up of six laterally compressed segments or *sternebræ*, and placed in the mid-ventral line of the thorax. The first segment (*manubrium*) is much larger than the others, and possesses a prominent ventral ridge. It is composed of two segments completely fused together, and this is indicated in the adult by the attachment of the first sternal ribs to the middle of its length. The remaining six pairs of sternal ribs are connected with the junctions of the sternebræ, one pair each to the first four, and two pairs to the last. The hindmost sternebra is elongated, slender, and terminated by a rounded plate of cartilage, with which it forms the *xiphisternum*.

(2) *Appendicular Endoskeleton.*—The long bones are terminated as usual by epiphyses. As in other cases the skeleton of fore- or hind-limb is divisible into girdle and free part.

(a) **Fore-limb** (Fig. 84).—The *shoulder girdle* is mainly composed of the triangular **scapula**, placed external to the anterior ribs, where it is held in place by muscles and ligaments. Its enlarged apex presents a shallow *glenoid cavity* for the head of the humerus. The base, anterior, and posterior sides of the triangle are termed *supra-scapular, coracoid,* and *glenoid borders* respectively. With the first a strip of cartilage, the *supra-scapula*, is connected. A conspicuous ridge, the *spine*, runs along the outer surface of the scapula, and is produced below into a freely projecting process, the *acromion* (*ac*), from which a more slender *metacromion* is given off behind. The coracoid border is continued below into a hook-like **coracoid process**. This is originally distinct, and represents the *coracoid bone* of lower Vertebrates. A ligament, in the centre of which is a slender curved *clavicle*, runs from the tip of the manubrium to the acromion. To the sternal end of this ligament a minute nodule of cartilage is attached which represents a *precoracoid*.

Free Limb.—The **humerus** presents a proximal *head*, on its upper (dorsal) surface, for articulation with the glenoid cavity, and a distal pulley-like *trochlea*, which assists to form the elbow-joint. The radius and ulna, which support the antebrachium, are immovably articulated, but not fused together in the position of *pronation*.

Where, as in Man, the radius and ulna are movable, they may assume two main positions:—(*a*) *Supination*, when they are parallel, the palm of the hand is upwards, and the thumb outside (like the radius). (*b*) *Pronation*, when the radius crosses over the ulna towards the inside, the palm of the hand is downwards, and the thumb inside (like the distal end of the radius).

The preaxial **radius** possesses a proximal head for articulation with the trochlea, and two distal concavities for the carpal bones. The postaxial **ulna** has a proximal *sigmoid cavity* for articulation with the trochlea, and a convex distal end. Proximally the ulna is produced into the *olecranon*, a process which, in the extended limb, fits into the *olecranon fossa*, a pit situated above the trochlea.

The *carpus* consists of a proximal and a distal row of small bones, and a diminutive *centrale*. The *proximal* row is made up of four bones, which are (beginning on the inner (preaxial) side) —(*a*) *radiale* (scaphoid), and (*β*) *intermedium* (lunar), articulating with the concavities on the radius, (*γ*) *ulnare* (cuneiform), and (δ) *pisiform* (not shown), articulating with the convexity on the ulna. The *distal* row is made up of four carpalia, which are, beginning as before, *carpale* 1 (trapezium), *carpale* 2 (trapezoid), *carpale* 3 (magnum), *carpale* 4 + *carpale* 5 (unciform). Five *digits* are present supported by five **metacarpals**, articulating with the corresponding carpalia, and completed by **phalanges**, of which the short *pollex* (1st digit) possesses 2, the others 3 each. The terminal phalanges are conical, and grooved to support the claws.

(*b*) **Hind-limb.**—The *hip-girdle* is formed by an *innominate bone* on each side, made up of four fused bones, *ilium, ischium, pubis*, and *cotyloid*. Upon the outer side of each innominate is a deep cup, the *acetabulum*, the floor of which is complete, and in the young rabbit, marked by the three-rayed junction of the ilium, ischium, and cotyloid.

The **ilium** is placed above and in front. It is laterally flattened, and the sacrum articulates with its inner surface. The **ischium** is placed above and behind. It is separated from the pubis by the oval *obturator foramen*, and presents posteriorly a rough and thickened tuberosity. A plate-like expansion runs from its hinder part towards the mid-ventral line, and unites there with its fellow to form the posterior portion of the *ischiopubic symphysis*.

The **cotyloid** is a very small bone helping to form the ventral side of the acetabulum, from which it shuts out the pubis.

The small **pubis** runs from the cotyloid bone towards the mid-ventral line, and unites with its fellow to form the anterior part of the ischio-pubic symphysis.

Free Limb.—The elongated **femur** presents a proximal *head* upon its inner (preaxial) side, which articulates with the acetabulum, and two distal *condyles* which articulate with the tibia to form the knee-joint. The actual proximal end of the femur is a large projection known as the *great trochanter*. There is a *lesser trochanter* below the head, and opposite this on the outer (postaxial) side a *third trochanter*.

A large **patella** is present on the front of the knee-joint, at the back of which are two other sesamoid bones, the **fabellæ**.

The crus is supported by the large preaxial **tibia**, which possesses two proximal concavities for the femur, and two distal surfaces for the tarsus. With the tibia a small, postaxial, rod-like **fibula**, is completely fused distally.

The ankle-joint in the rabbit is between the distal ends of the above bones and the tarsus.

The **tarsus** is made up of proximal and distal rows partly separated by a large *centrale* (navicular).

The *proximal* row contains two bones, the preaxial (*a*) *tibialo-intermedium* (astragalus), and postaxial (β) *fibulare* (calcaneum) produced into a large projecting *heel*. The *distal* row is composed of three bones, which are, beginning preaxially, (*a*) *tarsale* 2 (mesocuneiform), (β) *tarsale* 3 (ectocuneiform), and (γ) *tarsale* 4 + *tarsale* 5 (cuboid).

There are four *digits*, the first or hallux being absent. Each possesses a **metatarsal**, the first of which is produced proximally into a process, originally separate, which may represent the missing hallux. Each digit is completed by three **phalanges**, similar to those of the manus.

Small nodular *sesamoid bones* are developed in the tendons, opposite the joints on the under (ventral) side of manus and pes.

4. The **digestive organs** (Fig. 85) consist of an alimentary

canal (mouth-cavity, pharynx, gullet, stomach, intestine) with appended glands (salivary glands, liver, pancreas, and rectal glands). They begin in the head and thorax, but are contained,

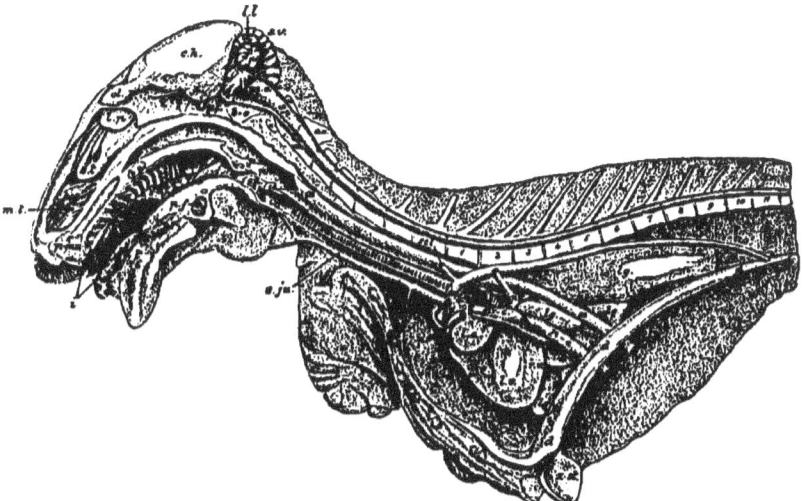

Fig. 85.—GENERAL DISSECTION OF HEAD AND THORAX OF RABBIT (reduced).—*a*, Atlas ; *ax*, axis ; 1 *to* 11, placed on centra of thoracic vertebræ ; *b-o*, basi-occipital ; *pt*, is placed on basi-sphenoid, and 2, on pre-sphenoid ; *l.p*, lamina perpendicularis ; *e-t*, ethmo-turbinal, and *m-t*, maxillo-turbinal, of right nasal cavity, the turbinal bones of left cavity having been removed and the septum nasi cut through ; *p-mx*, pre-maxilla ; *b.p*, bony palate ; *x-st*, xiphisternum ; *d,d,d*, diaphragm ; *i*, incisors ; *t*, tongue ; *p.f*, papilla foliata, above which is seen a circumvallate papilla ; * placed just below tonsil ; *ph*, pharynx ; *p.n*, internal nares, the opening of *p.n.c*, the posterior nasal chamber, which is mostly floored by soft palate — the letter *n* is placed just under Eustachian opening ; *œ*, œsophagus ; *l.a*, left auricle ; *l.v* and *r.v*, left and right ventricles ; *pr-c*, left pre-caval ; *e.ju*, right external jugular, formed in front by union of facial veins, and running back to help to form right pre-caval ; *p-c*, post caval ; *ao*, arch of aorta—left common carotid is seen running forwards along trachea (*tr* is placed just above it) ; *s-cl*, left subclavian artery ; *ao′*, dorsal aorta ; *pa* and *pv*, pulmonary artery and veins—left lung has been removed ; *ep*, epiglottis ; ×, placed on vocal chord ; *tr*, trachea ; a little way behind *ao*, opening of right bronchus, end of trachea, and left bronchus cut open, are seen ; *lg, lg, lg, lg*, lobes of right lung ; *ol*, olfactory lobe, below which is cribriform plate ; *c.h*, cerebral hemisphere ; *s.v, l.l*, and *f*, superior vermis, lateral lobe, and flocculus of cerebellum ; *m.o*, medulla oblongata —in front of these letters are the roots of the eighth, seventh, and fifth cranial nerves ; *op*, optic nerve ; *sp*, spinal cord.

for the most part, in the abdominal cavity, which is separated from the thoracic cavity by a firm partition, the *diaphragm*, and lined by the thin *peritoneum*, which forms suspensory mesenteric folds for the viscera.

The small *mouth*, guarded by flexible *lips*, leads into a large **mouth-cavity**, the roof of which is elongated, and presents a central part, the *palate*, bounded by the margin of the upper jaw. The anterior half of the palate, marked by firm transverse ridges, and the soft flexible posterior half, are called respectively *hard* and *soft palates*. The latter ends in a free notched edge, on each side of which is a small thickening, the *tonsil*, indented by a pit. The *floor* of the mouth, bounded by the margins of the mandible, has the long muscular *tongue* attached to the greater part of its extent. This exhibits in front a free tip, in the neighbourhood of which are numerous small *papillæ* that also extend back some distance on the upper surface. There is a hard elongated area above and behind, and, on each side of the posterior end of this, a *circumvallate papilla*, consisting of a small projection encircled by a groove. An oval elevation, the *papilla foliata*, across which numerous oblique ridges run, is present on each side of the tongue. The hair covering the body is continued into the sides of the mouth-cavity for a short distance.

The **teeth**, definite in number, are imbedded in sockets, or *alveoli*, and are of three kinds. There are two sets of them, the first, *milk teeth*, very transitory, and succeeded by *permanent teeth*, which continue to grow throughout life as they are worn away at their ends. This is called growth from "permanent pulps." The front of the jaws is occupied by cutting teeth (incisors); the remainder are grinding teeth (premolars and molars) placed a good deal further back.

In the front of the upper jaw two strongly curved incisors are found, grooved in front, and with chisel-shaped edges. Two very small incisors of similar shape are placed behind them. All the upper incisors are imbedded in the pre-maxillæ. Next follows a gap destitute of teeth (*diastema*) on each side, posterior to which are six prismatic grinding teeth, imbedded in the maxilla. The first three, *premolars*, are preceded by milk-grinders, the last three, *molars*, have no predecessors. The grinding teeth are much flattened, and, with the exception of the first and last, which are small, each possesses a deep groove on its outside, from the end of which a transverse ridge runs nearly across the flattish grinding surface or *crown*.

Two *incisors* similar to the large upper ones, but less curved and not grooved, are imbedded in the front of the mandible. Then comes a diastema on either side, succeeded by five grinding teeth, of which two are *premolars*, and three *molars*. They are similar to the upper grinders, but, except the last, more strongly ridged, and grooved internally as well as externally.

In many animals four pointed *canine* teeth are present, two above, immediately behind the pre-maxillo-maxillary suture, and two below. There is no trace of these in the Rabbit.

It is convenient to express the number and kind of teeth by a dental formula. Each kind is indicated by a fraction (preceded by its initial letter), in which the numerator and denominator signify upper and lower teeth, those of opposite sides being separated by dashes. Thus the Rabbit's dental formula is:—

$$i = \frac{2-2}{1-1} \quad c = \frac{0-0}{0-0} \quad p.m. = \frac{3-3}{2-2} \quad m. = \frac{3-3}{3-3} = 28.$$

Those before the vertical line have predecessors in the milk-dentition. As, however, the teeth of one side only need be expressed, those of the other exactly corresponding, and the kinds are easily remembered, the formula may be simplified to:—

$$\frac{2033}{1023}.$$

Two rudimentary incisors have been found in the front of each jaw, which are probably persistent members of the milk dentition. They are not included in the above formulæ.

The dental formula of the dog is $\frac{3142}{3143}$. The canines are strong, pointed teeth, adapted for holding prey. All the premolars and the first lower molars have sharp cutting crowns, which are very large in the last upper premolars and first lower molars. These are known as *carnassial* teeth, and work against one another like the blades of a pair of scissors. The upper molars and last two lower molars have grinding crowns.

All the teeth have continuous covering of enamel in their crowns, and taper below into pointed "fangs." They do not continue to grow throughout life as in the rabbit.

The soft palate hangs down behind, and imperfectly separates the mouth-cavity from the **pharynx**, a small chamber which passes back into the gullet. An unpaired oval opening, the *internal naris*, opens into it in front, and on its floor, at the root of the tongue, is the *glottis*, guarded in front by a thin flexible flap, the *epiglottis*, supported by cartilage. The **gullet** (œsophagus) is a narrow thick-walled tube, which runs from the pharynx through the neck and thorax, just ventral to the backbone, pierces the diaphragm, and enters the **stomach**, a large transversely placed sac, with firm thick walls.

Its left (*cardiac*) portion is much dilated, while its right (*pyloric*) portion is thickened. Posteriorly it presents a convex margin, the *greater curvature*, and anteriorly a concave margin, the *lesser curvature*, to the left of which is the *cardia*, or opening of the gullet. The *pylorus* where the duodenum commences, is at the extreme right of the pyloric end. The mucous membrane lining the stomach is raised into irregular longitudinal ridges, or *rugæ*. Between stomach and duodenum there is an inwardly projecting rim, the *pyloric valve*.

The stomach is succeeded by a much convoluted intestine, some 15 or 16 times the length of the body, and divided into small and large. The first part of the thin-walled **small intestine**,* the *duodenum*, forms a longitudinally directed U-shaped loop nearly the length of the abdominal cavity. It passes insensibly into the much longer *ileum*, in the wall of which there are oval thickenings, *Peyer's patches*, here and there.

This part of the intestine finally opens into the cæcum (*see below*), and at this point its wall is much thickened to form the *sacculus rotundus*. The mucous membrane of the small intestine is raised into numerous irregular transverse ridges, the *valvulæ conniventes*, from which project an immense number of *villi*.

The **large intestine** is divided into cæcum, colon, and rectum. The *cæcum* is a very large thin-walled sac, nearly 2 feet in length, interposed between the ileum and colon. Its wall is marked externally by a spiral groove. One end, near which the ileum opens, is continuous with the colon, while the other terminates blindly in a thick-walled finger-like *vermiform appendix*.

A spiral fold, corresponding to the external groove, projects into the cavity of the cæcum, with which cavity the sacculus rotundus communicates by a small rounded aperture (ileo-colic valve), and which opens by a larger aperture into the colon. The lining mucous membrane is raised into minute elevations; in other words, is *papillose*.

The *colon* is a moderate-sized thin-walled tube with baggy walls, along which three smooth longitudinal bands run. It passes insensibly into the narrow *rectum*, which, after some convolutions, runs back within the pelvic cavity, below the backbone, and above the urinogenital organs, to the *anus*. The mucous membrane lining the colon is papillose, that lining the rectum

* In human anatomy this is divided, somewhat arbitrarily, into three parts—*duodenum, jejunum*, and *ileum*. The two last are here included in the term "ileum."

is smooth at first, but raised into prominent longitudinal ridges near the end.

The following large glands are connected with the alimentary canal—salivary glands, liver, pancreas, and rectal glands.

Four pairs of **salivary glands** open into the mouth-cavity,—parotid, submaxillary, sublingual, and infra-orbital.

(1) The *parotid glands* are irregular pinkish masses, extending, on each side, from the origin of the pinna to the higher end of the mandible. From each a delicate (Stenson's) duct runs forwards just below the skin, turns inwards, and opens into the side of the mouth-cavity opposite the last upper molar. (2) The *sub-maxillary glands* are oval and compact bodies, of reddish colour, lying near together between the rami of the mandible, in front of the larynx. A delicate (Wharton's) duct runs from the outer side of each to the floor of the mouth, in front of the tongue. (3) The small red *sub-lingual glands* are elongated and flat. They lie in front of and above the sub-maxillary glands among the muscles of the tongue. (4) The *infra-orbital glands* are irregular and lobulated. They lie, one on each side in front of and below the eye, partly within the orbit. A duct runs downwards from each and opens into the side of the mouth-cavity, opposite the last upper premolar.

The **liver** is a very large reddish-brown organ, closely applied to the concave abdominal side of the diaphragm in front, and largely overlapping the stomach behind.

Right and left halves may be distinguished in the liver. The former is divided into two lobes, *right central*,—and *caudate*, hollowed out for, and abutting against, the right kidney. The left half is divided into three lobes, *left central*, *left lateral*, and the very small *Spigelian*, fitting into the anterior concavity of the stomach. A pear-shaped, thin-walled **gall-bladder** lies in a slit on the posterior surface of the right central lobe. From it a short *cystic duct* opens into the *bile-duct*, a small tube formed by the union of *hepatic ducts* from the various liver-lobes, and opening into the duodenum a little way beyond the pylorus.

The **pancreas** is not compact as in the frog and pigeon, but *diffuse*—*i.e.*, made up of numerous scattered lobules. These resemble small masses of fat, and are found in the mesentery between the limbs of the U formed by the duodenum. Delicate ducts proceed from the various lobules, and by their successive unions the *pancreatic duct* is formed, which opens into the distal limb of the duodenum, about 3 inches from the bend.

The **rectal glands** are two compact elongated bodies lying alongside and above the rectum near its termination, and opening into it.

The alimentary canal is lined as usual by *mucous membrane,* external to which are submucous and muscular coats, and, in the abdominal organs, there is a thin serous layer outside all.

The **mucous membrane** consists of *epithelium* internally, which, as far as the stomach, is *stratified squamous*, but in that organ and the intestines, *simple columnar*. Goblet *cells* are everywhere common, *peptic glands* are found in the stomach, and *glands of Lieberkühn* throughout the intestines. Each peptic gland opens on the surface of the stomach by a narrow *neck*, lined with columnar epithelium. It branches at its deep end, and the branches are lined by granular cuboid *chief cells*, among which are scattered (in the glands of the *cardiac* end) a much smaller number of *ovoid cells*. Beneath the epithelium is a layer of connective tissue surrounding the glands, and the external limit of the mucous membrane is marked (in the stomach and intestines) by a very thin sheet of unstriated muscle, the *muscularis mucosæ*.

The **submucous coat** is made up of loose connective tissue, traversed by vessels and nerves.

The **muscular coat** varies very much in thickness, attaining its maximum in the stomach. It consists of an inner circular and an outer longitudinal layer in the gullet and succeeding parts. The fibres are striated in the gullet and pharynx, and the tongue is mainly composed of such fibres; but the muscle in the walls of the stomach and intestines is made up of unstriated fibres.

The **salivary glands and pancreas** consist of a number of *lobules* united together by connective tissue. Each lobule is composed of numerous *acini*, blindly ending tubes lined by glandular epithelium, from each of which a duct, lined by simple columnar epithelium, proceeds. These ducts unite again and again to form the main duct of the gland. The preceding type of structure is termed *racemose*.

The liver, as in the frog and pigeon, is mainly made up of polyhedral, granular, *hepatic cells* (Fig. 86), placed in the meshes of the network formed by the *bile-capillaries*, from which the hepatic ducts arise. The hepatic cells are aggregated into small *lobules*, around each of which is an *interlobular* capillary network formed by the ultimate branches of the portal vein. From this network veinlets pass into the centre of the lobule, and there unite into a small *intralobular* vein which carries away its blood. By the union of these small vessels *sublobular* veins are formed, which are the factors of the *hepatic veins*.

The **food** consists of vegetable substances, and the great length of the alimentary canal is correlated with this, an immense absorptive surface, largely augmented by the cæcum with its spiral valve, being given.

The food is divided by the incisor teeth, and ground up by

the molars and premolars. The hard tongue, working against the ridged palate, assists in this process of mastication. *Saliva* is poured by the salivary glands into the mouth-cavity, and not only lubricates the food, but also, in virtue of a ferment, *ptyalin*, which it contains, converts more or less of the starch into soluble grape-sugar. The *gastric juice, bile,* and *pancreatic juice* act as in preceding cases (p. 194).

5. Circulatory Organs.—Well developed *blood* and *lymph systems* are present.

(1) **Blood System.**—The **blood** consists of liquid *plasma*, in which are suspended *colourless* corpuscles of the usual type, and

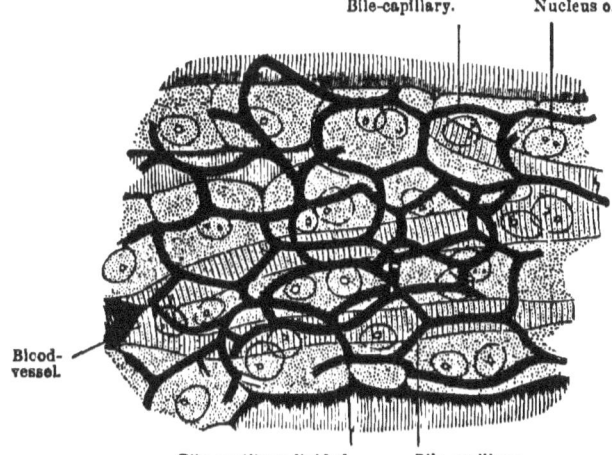

Fig. 86.—HISTOLOGY OF RABBIT'S LIVER (after *Hering*).—Relation of liver-cells, bile-capillaries, and blood-capillaries.

a much larger number of smaller *red corpuscles*, which are biconcave, *non-nucleated* discs. The blood-containing tubes comprise heart, arteries, veins, and capillaries.

The conical **heart** (Fig. 85) enclosed in its *pericardium*, lies between the lungs on the ventral side of the thoracic cavity, with its apex directed backwards. It contains four chambers, two thin-walled auricles, and two thick-walled ventricles.

The broad base of the heart is chiefly made up of the dark, thin-walled *right* and *left auricles*, to each of which is attached a plaited fold, the *auricular appendix*, while the rest of the heart is made up by the firm *right* and *left ventricles*. The internal structure of the heart is in most particulars like that of the pigeon (p. 245), but the right auriculoventricular valve is not muscular but made up of three membranous flaps (giving the name *tricuspid valve*), connected by chordæ tendineæ with papillary muscles in the right ventricle. These muscles are also found in the left ventricle, and are more numerous than is the case in the pigeon. The opening of the postcaval is guarded by a fold, the *Eustachian valve*, and that of the left precaval by a similar *Thebesian valve*.

Arteries.—The aorta takes origin from the left ventricle, and its arch curves round to the *left* to form the dorsal aorta, giving off vessels which supply head, neck, and fore-limbs. The dorsal aorta runs back ventral to the backbone, supplying the trunk with its contained viscera, and finally forks into two iliac arteries, which supply adjacent parts and continuations of which run into the hind-limbs.

The aorta dilates at its origin into the 3 *sinuses of Valsalva*, situated behind the pouches of the semilunar valves. Two *coronary arteries*, for the supply of the heart-walls, are given off here. Near the beginning of the arch a short *innominate artery* runs off, which at once gives off the *left carotid*, and soon after divides into *right subclavian* and *right carotid* arteries. The *left subclavian* arises from the left side of the arch. Each carotid runs forwards on one side of the trachea, giving off branches to the neck as it does so, and divides in front into *internal carotid* for the brain, and *external carotid* for the outside of the head. The subclavian arteries supply the fore-limbs, on entering which they become the *brachial* arteries. A *vertebral* artery runs dorsalwards from each subclavian near its commencement, and enters the vertebrarterial canal to supply the brain and spinal cord.

The dorsal aorta gives off in the thorax small paired *intercostal* arteries to the thoracic walls. In the abdomen a large unpaired *cœliac* artery takes origin a little way behind the diaphragm, and quickly divides into the *hepatic* artery for the liver, and *lieno-gastric* artery for the spleen and stomach. Another unpaired artery, the *anterior mesenteric*, for the small intestine, pancreas, cæcum, and colon, arises a short distance posterior to the cœliac. Paired *renal* and *spermatic* (or *ovarian*) arteries are next given off to the kidneys and spermaries (or ovaries), and still further back a small unpaired *posterior mesenteric* artery to the rectum. The dorsal aorta bifurcates at the posterior end of the abdominal cavity into the *iliac* arteries, each of which, after giving off an *ilio-lumbar* artery to the bodywall, divides into an *internal iliac* artery for the pelvic region and an *external iliac* artery, which, after giving off a branch that supplies the bladder (and uterus in the female), enters the thigh as the *femoral* artery for the hind-limb.

Shortly before the aorta bifurcates, it gives off from its upper surface a small *median sacral* artery, which runs back into the tail.

The **pulmonary artery** arises from a forward prolongation of the right ventricle on the ventral side, and, curving round the front of the heart, divides into right and left branches for the corresponding lungs.

Veins.—Caval, hepatic portal, and pulmonary systems can be distinguished.

(a) *Caval System.*—Right and left precavals (anterior venæ cavæ) and a postcaval (posterior vena cava) carry the impure blood from the general system into the right auricle. The precavals drain the head, neck, and fore-limbs, the postcaval the rest of the body.

Each **precaval** is formed by the union of an *external jugular*, with a *subclavian* vein. The former runs back along the side of the neck from the angle of the jaw, where it is formed by the union of *facial* veins returning blood from the outside of the head. Just before uniting with the subclavian it receives the small *internal jugular* vein from the brain. The *subclavian* vein brings back blood from the fore-limb. The *right* precaval is joined by an unpaired *azygos* vein, which lies just beneath the vertebral column in the thorax, and carries off blood from a large part of its walls. The left precaval receives a *coronary vein* from the heart-walls, just before entering the right auricle.

The **postcaval** is a very large vein formed at the posterior end of the abdomen by the union of the *internal iliac* veins from the backs of the thighs. It is joined almost at once by the *external iliac* veins, which return blood from the hind-limbs and bladder (also uterus in the female). They are direct continuations of the *femoral* veins of the thighs. The postcaval runs forwards in the abdomen close to and on the right of the aorta, receiving successively *ilio-lumbar*, *spermatic* (or *ovarian*), and *renal* veins, from the body-walls, spermaries (or ovaries), and kidneys respectively. Before reaching the diaphragm it turns ventralwards, runs through the dorsal part of the liver (from which it receives four chief *hepatic veins*), and pierces the diaphragm ventral to the gullet. In the thorax it runs forwards between the lungs and finally enters the right auricle.

(b) *Hepatic Portal System.*—The **portal vein** is formed by the union of *lieno-gastric*, *duodenal*, *anterior mesenteric*, and *posterior mesenteric* veins, which return blood from the stomach and spleen —duodenum and pancreas—ileum and most of large intestines— and last part of rectum, respectively. The portal vein soon divides into branches, which break up in the various liver-lobes.

(c) *Pulmonary System.*—There are two **pulmonary veins** from each lung. All four open into the dorsal side of the left auricle.

The muscle of the *heart* (Fig. 87) is composed of transversely-striated fibres, devoid of sarcolemma, which are united into a close network. This is marked into short lengths corresponding to the constituent cells, and in each of these a muscle-corpuscle, not placed superficially, is imbedded. The arteries and veins are constructed as described on p. 202. The *veins* contain numerous valves, which are flap-like projections of their inner coat.

The **capillaries** are networks of fine tubes (*cf.* p. 202), connecting the ultimate branches of arteries and veins.

Course of the Circulation (*cf.* p. 247).—The impure blood of the body is returned by the caval veins to the right auricle, thence passing into the right ventricle and into the lungs through the pulmonary artery. The oxygenated blood is returned to the left auricle by the pulmonary veins, then enters the left ventricle to reach the aorta, the branches which take it to all parts of the body.

(2) **Lymph System.**—The **lymph**, as in other cases, resembles blood, were it not for the absence of red blood-corpuscles. It is contained in the minute lymph-spaces found in most of the tissues, larger lymph-cavities, and lymph-vessels opening into a thoracic duct. The lymphatic vessels resemble small veins in structure, and are of two kinds—(a) *lymphatics*, belonging to the body generally; (β) *lacteals*, belonging to the gut, and commencing in blind branched tubules, one of which is found in

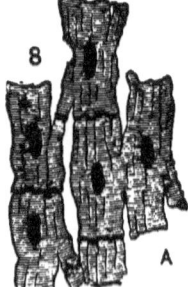 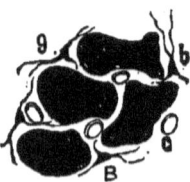

[Fig. 87.—Muscle-Fibres from Mammalian Heart (from *Landois* and *Stirling*), much enlarged.—8, Side view; 9, cross-section.

each intestinal villus. The *thoracic duct*, to which the lacteals and most of the larger lymphatic trunks run, is a slender tube lying above the dorsal aorta in the thorax, and opening anteriorly into the junction between the left subclavian and external jugular veins. Some of the lymphatics unite to form a much smaller trunk, opening similarly on the right side. The most important large lymph-space is the **cœlom**, here divided up into abdominal, pleural, and pericardial cavities.

Numerous small **lymphatic glands**, which resemble the cervical glands of the pigeon, are present in various situations. Those belonging to the lacteals are known, from their position, as *mesenteric glands*.

An elongated flattened **spleen**, of a dark-red colour, is attached by a flap of mesentery to the cardiac end of the stomach.

The **thymus gland** is a fat-like mass, largest in young animals, closely connected with the base of the heart. The **thyroid gland** consists of two small elongated bodies of reddish colour, closely applied to the sides of the front end of the trachea, and connected together by a ventral bridge of tissue.

New *colourless corpuscles* are developed in the lymphatic glands, thymus, and spleen, new *red corpuscles* in the red marrow of bones, and from pre-existing colourless corpuscles. The worn-out red corpuscles are probably broken down in the spleen.

6. The **respiratory organs** (Fig. 85) are lungs contained in the thorax, and communicating with the exterior by a trachea, the front end of which is modified into an organ of voice (larynx).

The slit-like *glottis*, situated at the base of the tongue, and guarded in front by an elastic flap (epiglottis), leads into the **larynx**. This is supported by a broad *thyroid cartilage* (which is a bent plate, incomplete dorsally), and, posterior to this, by a ring-like *cricoid cartilage*, the dorsal side of which is thickened. Two elastic folds, the *vocal chords*, project from the sides of the larynx. They are attached below to the thyroid cartilage, above to two small *arytenoid cartilages*, which articulate dorsally with the front edge of the cricoid. The larynx may be regarded as the modified front end of the **windpipe** or **trachea**, which runs along the neck ventral to the œsophagus, and is supported by cartilaginous hoops. The trachea bifurcates within the thorax, just anterior to the base of the heart, into two *bronchi*.

The **lungs** are two spongy bodies, pink in colour, which mainly fill the thoracic cavity, as long as its walls remain intact. The left lung is subdivided into two, the right lung into four *lobes* of very unequal size. Each lung is enveloped by a delicate membranous *pleura*, which consists (like the peritoneum) of a *visceral* layer, closely applied to the surface of the lung, and a *parietal* layer, lining one half of the thorax. The two layers are continuous at the root of the lung.

The parietal layers of the two pleuræ meet together in the middle line to form, in the posterior part of the thorax, a longitudinal vertical partition, the *mediastinum*. The two sheets of which this is composed diverge further forwards, and leave between them the *mediastinal space*, occupied by the heart in its pericardium. The bronchi enter the anterior end of the lungs, and, like the trachea, are supported by cartilage. Each of them forms a *bronchial tree* in its lung by giving off alternating right and left branches.

The **diaphragm** is a thin partition which forms the posterior boundary of the thorax, and separates it from the abdomen, towards which it is strongly concave when at rest. The centre of the diaphragm is transparent and tendinous (*i.e.*, composed of white connective-tissue fibres), while its margins are muscular, especially on the dorsal side, where they are continued into two muscular *pillars*, which take origin on the under sides of the lumbar vertebræ.

Histology.—The branches of the bronchial tree divide repeatedly (the supporting cartilages being at the same time gradually lost), to form delicate *bronchial tubes*. These end in clusters of dilations, the *infundibula*, with sacculated walls. They are lined by simple squamous epithelium, beneath which is a very rich network of capillaries. The trachea and larger air-passages are lined by stratified columnar epithelium, ciliated in part.

The lungs contain a large quantity of elastic connective tissue and unstriated muscle.

Respiration.—Inspiration and expiration are partly effected by movements of the *ribs*, and partly by movements of the *diaphragm*, and respiration is hence said to be both *costal* and *diaphragmatic*.

It must be remembered that each lung is enclosed in an airtight bag, the *pleura*, and is obliged to follow all the movements of the thoracic walls, or else a vacuum would be formed between the two layers of this. The elastic tissue in the lungs causes them to be always *trying* to contract, but this is prevented by the pressure of the air in the lung-passages. If, however, the pleura is perforated, this pressure is counterbalanced and the lung at once contracts.

Inspiration.—The ribs and sternum are moved downwards and forwards, and thus increase the dorso-ventral capacity of the thorax. At the same time the muscular parts (especially the pillars) of the diaphragm contract, thus flattening it and increasing the antero-posterior capacity of the thorax.

Air consequently rushes into the larger air-passages from the exterior.

Expiration is the exact converse of this, but is largely passive, the various muscles ceasing to contract. It is aided by the elasticity of the lungs and walls of the thorax. As a result, the air rushes out again.

The air passes *directly* in and out of the *larger* air-passages *only*. The rest is effected by diffusion, and the *essential* part of respiration (exchange of gases) is carried on in the infundibula.

Respiration is less active than in the pigeon, but here also the body is maintained at a temperature (about 100° F.) much above that of the surrounding medium.

7. The **urino-genital organs** consist of excretory and reproductive organs.

Excretory Organs (Fig. 89).—In both sexes two oval, compact, reddish-brown **kidneys** are closely applied to the dorsal wall of the abdominal cavity, the right rather further forwards than the left. They are covered ventrally by peritoneum.

Each kidney presents a notch, the *hilus*, on its inner margin, from which the *ureter*, a narrow tube with muscular walls, proceeds. The ureters open obliquely into the dorsal side of the pear-shaped *urinary bladder*, towards its narrow portion or *neck*. The walls of the bladder are translucent and muscular. It projects into the body-cavity just in front of the symphysis pubis.

A median horizontal section of either kidney shows that it is differentiated into a marginal *cortex*, dark red in colour, with numerous small dots (Malpighian bodies), and a central paler *medulla*, which presents a striated appearance. The striæ converge internally towards a conical eminence, the *urinary pyramid*, which projects into the *pelvis*, or dilated end of the ureter.

The kidney is essentially made up of **uriniferous tubules** (Fig. 88), each commencing with a *Bowman's capsule*, which with its *glomerulus* constitutes a *Malpighian body*, and is situated in the cortex. The *neck* of the capsule passes into a thickened *glandular part*, also in the cortex, from which a narrow part runs into the medulla, and then loops back into the cortex to dilate into a second *glandular part*, which passes into a collecting part. The collecting parts run towards the urinary pyramid, often uniting in their course, and finally open upon it.

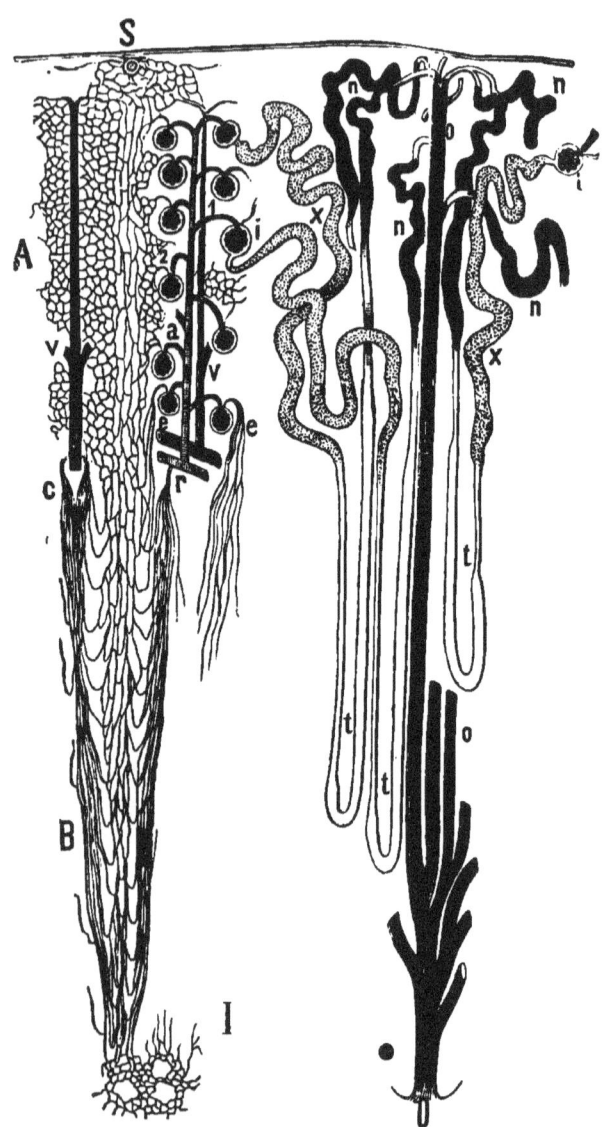

Fig. 88.—DIAGRAM OF BLOOD-VESSELS AND URINIFEROUS TUBULES OF KIDNEY (from *Landois* and *Stirling*).—A and B, Capillaries of cortex and medulla. On the left numerous glomeruli are shown with (1) afferent artery; (2), efferent vein. The glomeruli are enclosed in Bowman's capsules. $x, x, x,$ and $n, n, n,$ glandular parts, and $o, o, o,$ collecting parts of tubules; O, opening on urinary pyramid.

The *renal artery* divides within the kidney, and from its branches small *afferent arteries* run to Bowman's capsules and break up into *glomeruli*. A small *efferent vein* arises from each glomerulus. The efferent veins now break up into a network of capillaries surrounding the glandular parts of the tubules, and the factors of the *renal vein* take origin in this network.

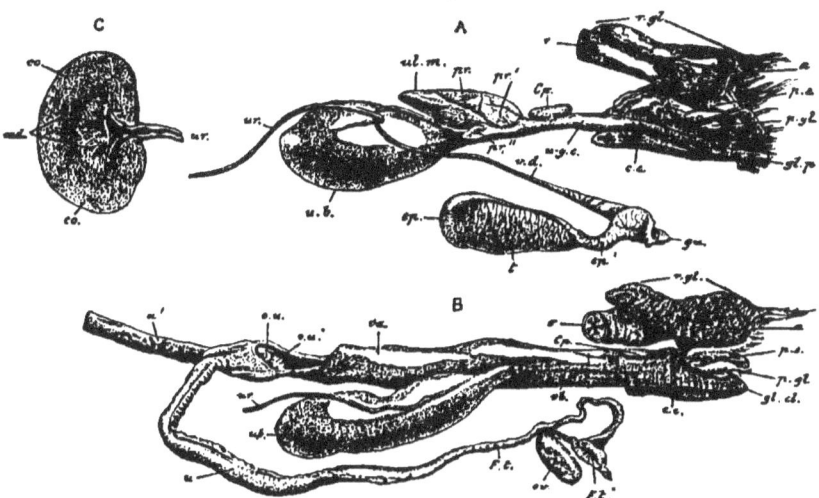

Fig. 89.—URINOGENITAL ORGANS OF RABBIT.—A, male; B, female; C, horizontal section of right kidney; *co* and *md*, cortex and medulla of kidney (in C); ×, placed on urinary pyramid; *ur*, ureter; *u.b*, urinary bladder; *r*, rectum; *r.gl*, rectal gland; *a*, anus; *p.s.*, perineal space; *p.gl*, perineal gland (in A, the aperture is shown by a dot); *t*, spermaries (testes); *ep* and *ep'*, caput and cauda epididymis; *gu*, gubernaculum; *v.d*, spermiduct (vas deferens); *ut.m.*, uterus masculinus; *pr*, *pr'*, and *pr''*, lobes of prostate; *u.-g.c*, urinogenital canal; *c.c*, corpus cavernosum; *gl.p*, glans penis; *ov*, left ovary; *F.t*, left Fallopian tube; *F.t'*, funnel of ditto; *u,u'*, left and right uteri, the latter cut short; *o.u* and *o.u'*, the mouths of the left and right uteri, to show which the vagina, *va*, has been cut open; *vb*, vestibule; *gl.cl*, glans clitoridis.

Excretion (*cf.* p. 206).—The nitrogenous waste-products are urea (CH_4N_2O) and *hippuric acid* ($C_9H_9NO_3$).

The **male reproductive organs** (Fig. 89) consist of spermaries (testes), from which spermiducts (vasa deferentia) proceed. These communicate with a urinogenital canal traversing a cylindrical penis. Other accessory parts are present.

The oval **spermaries** (testes) are contained in the *scrotal sacs*, which, in the breeding season, project at the sides of the urinogenital aperture. Each spermary is attached to its sac by a short fibrous cord (*gubernaculum*), and is also suspended by a mesenteric fold (*mesorchium*). Attached to the spermary is a compact mass of coiled tubules (*epididymis*), divided into the *caput epididymis* in front, and the *cauda epididymis* behind. These cap the anterior and posterior ends of the testis, and are connected by a tubule running along its surface.

The *spermatic artery* and *vein* are connected with the anterior end of the testis. They form, with nerves, connective tissue, &c., the *spermatic cord*.

The **spermiduct** (vas deferens) proceeds from the cauda epididymis, runs forwards, and then curves over the ureter to enter the *uterus masculinus*, a large, slightly bilobed sac, which opens into the dorsal side of a **urinogenital canal** (urethra), a backward continuation of the neck of the bladder. After receiving the uterus masculinus, the ducts of a five-lobed *prostate gland*, and, further back, the openings of two small *Cowper's glands*, the urinogenital canal traverses the *penis*, an elongated backwardly-directed structure, the ventral side of which is strengthened by two fibrous rods, the *corpora cavernosa*, attached in front to the ischia. The dorsal side of the penis is made up of a mass of vascular tissue, the *corpus spongiosum*, through which the urinogenital canal runs, and which projects beyond the corpora cavernosa as the *glans penis*, a soft conical body, upon which is the slit-like *urinogenital aperture*.

The *spermary* is essentially composed of *seminiferous tubules* lined by germinal epithelium, the cells of which divide frequently to form sperms, each producing several, but not being entirely used up in the process. The *sperms* (spermatozoa) have oval flattened *heads* and vibratile *tails*. The tubules of the spermary unite to form *vasa efferentia*, which make up part of the epididymis.

The **female reproductive organs** (Fig. 89) consist of ovaries, oviducts, vagina, and urinogenital canal, together with accessory parts.

The **ovaries** are whitish, oval bodies, placed one on either side of the dorsal abdominal wall, posterior to the kidneys. In a mature female, *Graafian* (ovarian) *follicles* appear as clear, rounded projections on their surfaces. Near each ovary is the fringed funnel-shaped opening of the corresponding **oviduct**. This consists of a narrow convoluted *Fallopian tube*, which merges into a cylindrical *uterus*, with thick muscular walls, which may be of very large size in the pregnant female, and then exhibits a series of oval swellings, in each of which an embryo is contained. The two uteri open by separate apertures (*ora uterorum*) with rounded projecting margins, into the **vagina**, a large tube which runs back dorsal to the bladder, and becomes continuous with the **urinogenital canal** (vestibule), a somewhat smaller but still wide tube, which runs back from the bladder in the same relative position as the narrow urinogenital canal of the male. The walls of this tube are very vascular, and it ends in an elongated *urinogenital aperture* (vulva). Imbedded in the ventral wall of the vestibule is a small elongated body, the *clitoris*, supported by *corpora cavernosa*, similar to, but very much smaller than, those of the penis, and similarly situated. The clitoris ends in a soft, flattened *glans clitoridis*. Small *Cowper's glands* open into the dorsal wall of the vestibule.

The **ovary** consists of Graafian follicles in various states of development imbedded in a connective-tissue basis or *stroma*. The mature follicles project from the surface of the ovary, and each, when mature, is a fluid-containing vesicle, the wall of which consists of a fibrous coat, internal to which are two or more layers of columnar epithelial cells (membrana granulosa). The ovum is placed on the outer side, and is surrounded by several layers of these cells, with which it forms a mass, the *discus proligerus*, that projects into the cavity of the follicle. The ripe *ovum* is invested by a radiately striated membrane, the *zona radiata*, but no distinct traces remain of the *vitelline membrane* present in the young ovum. The *vitellus* contains hardly any food-yolk, and within it is a large *germinal vesicle* with *germinal spot*.

The ova burst from their follicles into the body-cavity, and are taken up by the funnels of the Fallopian tubes. *Fertilization* is effected in the upper part of these tubes, by fusion with sperms previously ejected by the male into the female passages, the penis being used as a copulatory organ.

8. Muscular System (*cf.* p. 208).—The muscles are very numerous and arranged in a very complicated manner. The voluntary muscles are from their appearance divided into *red* and *white*, the darker colour of the former being due to their richer blood-supply.

300 AN ELEMENTARY TEXT-BOOK OF BIOLOGY.

The striated muscle-fibres are smaller than those of the frog (cf. p. 208), and the *muscle-corpuscles* are situated superficially, just beneath the sarcolemma. The fibres of the *red* muscles are larger than the rest.

9. The Nervous System (Fig. 85) consists of cerebro-spinal axis, cranio-spinal nerves, and sympathetic system.

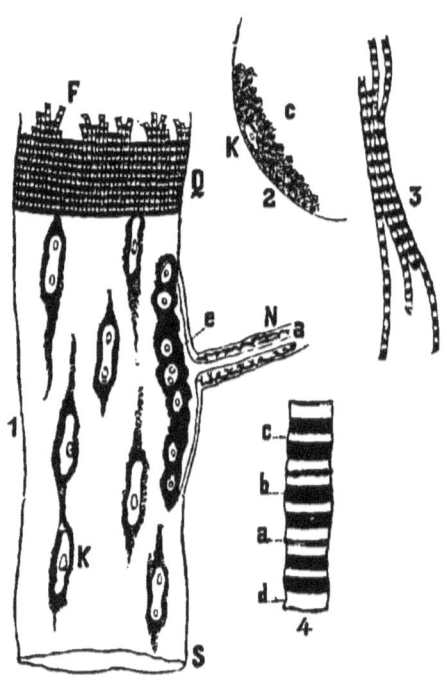

Fig. 90.—HISTOLOGY OF STRIATED MUSCLE (from *Landois* and *Stirling*), all much enlarged.—1. Part of a fibre—Q, transverse striations; F, primitive fibrillæ; K, muscle-corpuscles; S, sarcolemma; N, an entering nerve-fibre, with *a*, axis cylinder, and *e*, end-plate. 2. Part of a fibre in cross-section—K, muscle-corpuscle. 3. Primitive fibrillæ.

(1) The cerebro-spinal axis comprises brain and spinal cord contained in the neural canal and invested by three membranes, —*pia mater* carrying blood-vessels for supply of the nervous tissue, *arachnoid*, a delicate mesh-work traversed by lymph-spaces, and firm *dura mater* lining the neural canal.

The **brain** is elongated, but not particularly well developed. It presents the usual division into fore-, mid-, and hind-brain.

Fore-Brain.—The **thalamencephalon** possesses two parts not present in preceding types. One is the *middle commissure*, a broad band of grey matter connecting the side-walls of the third ventricle between the anterior and posterior commissures. The other is a rounded eminence, the *corpus mammillare*, seen on the under surface immediately behind the *pituitary body*, which is attached as previously to the *infundibulum*. With the hinder part of the thin roof of the third ventricle the *pineal gland* is connected by a hollow bifurcated stalk.

The **cerebral hemispheres** are broad behind and taper in front to blunt points. Externally they are almost smooth, but each is divided by grooves into three lobes, *frontal, parietal*, and *temporal*, placed respectively in front, postero-dorsally and postero-ventrally. As in the frog and pigeon the hemispheres are closely applied in the middle line, but here, in addition, they are firmly united by an elongated band of transverse fibres, the *corpus callosum*. Below this and continuous with it behind is another fibrous band, the body of the *fornix*, which divides in front into four smaller bands, the *pillars* of the fornix.

Two of these, the *anterior pillars*, pass downwards in the lamina terminalis and substance of the optic thalami to reach the corpus mammillare, while the other two, *posterior pillars*, curve outwards into the hemispheres. Each *lateral ventricle* is divisible into three parts—a narrow *anterior cornu* in front, and towards the inner side—a *posterior cornu*, similarly placed behind, and a *descending cornu*, which passes down into the temporal lobe. A prominent oval mass, the *hippocampus major*, projects into the last-named cornu, and along the anterior edge of this the posterior pillar of the fornix runs. In front of this pillar there is a vascular fold, the *choroid plexus*, which extends into the ventricle from the pia mater. The *corpus striatum* forms the outer side and floor of the anterior cornu. The cerebral hemispheres fuse where they meet in front of the lamina terminalis, and form a party-wall, the *septum lucidum*, to the hinder parts of the anterior cornua, immediately in front of the foramen of Monro on each side. This partition is not solid, but contains a small slit-like *fifth ventricle*.

This is not, therefore, like the other ventricles, a part of the cavity of the original neural tube of the embryo. These were originally named 1st and 2nd (*i.e.*, the two lateral), 3rd and 4th ventricles, and when another brain-cavity was discovered it naturally received the name 5th ventricle.

The **olfactory lobes** are elongated and dilated at their ends. They are attached to the under sides of the frontal lobes, and run forwards in front of them to rest on the cribriform plate.

Each contains a small *olfactory ventricle* continuous with the corresponding anterior cornu.

The *mid-brain* is traversed by the *Sylvian aqueduct*. Above it presents the **optic lobes** (corpora quadrigemina) which are overlapped by the cerebral hemispheres. Each of them is subdivided into two, so that there are *four* elevations, two anterior and larger *nates*, and two posterior *testes*. There are no optic ventricles. Two very distinct longitudinal masses of fibres, the **crura cerebri**, form the floor of the mid-brain.

Hind-brain. — The **bulb** (medulla oblongata) presents well-marked *dorsal* and *ventral fissures* bounded by narrow bands, the *dorsal* and *ventral pyramids*. The dorsal pyramids diverge in front and sweep round the sides of the thin roof of the fourth ventricle, on each side of which is a projection, the *corpus restiforme*. On the under surface the ventral fissure and pyramids are interrupted by a broad band of transverse fibres, the *pons Varolii*, just behind the crura cerebri. Immediately behind this there is a rectangular area, the *corpus trapezoideum*, on each side of the ventral pyramid, occupied by transverse fibres.

The **cerebellum** is very large and made up of a median *vermis* marked by deep transverse furrows, and a pair of much convoluted *lateral lobes*, to each of which a small rounded *flocculus* similarly furrowed, is attached. The cerebellum is united to the rest of the brain by three pairs of commissures or *peduncles*. The anterior peduncles connect it with the testes, the middle peduncles with the pons Varolii, and the posterior peduncles with the corpora restiformia.

The **spinal cord** is cylindrical, and possesses dorsal and ventral fissures. There is no sinus rhomboidalis.

(2) **Cranio-Spinal Nerves.**—There are twelve pairs of **cranial nerves**, as in the pigeon, which take origin and are distributed in a similar manner. The *olfactory* lobes send numerous fibres through the pores in the cribriform plate to the olfactory mucous membrane. The *optic nerves* form a chiasma, but the optic tracts are relatively narrow. The *trigeminal* nerves arise from the sides of the pons, and the *facial* and *auditory* from the sides of the corpora trapezoidea. The *abducent* nerves come off from the extreme front of the ventral pyramids. The exits of the cranial nerves have already been described (p. 274). Outside the skull it may be noted that the *mandibular* ramus of the fifth

sends a *lingual* branch to the tongue. The *vagi* run back along the neck by the sides of the common carotids. Each gives off to the larynx a *superior laryngeal* nerve, from which a *depressor* branch runs, closely united with the sympathetic, to the heart. At the posterior end of the neck a *recurrent laryngeal* nerve is given off, which curves round the aorta on the left, or the subclavian on the right, and runs forwards to the larynx. Behind this the vagi run through the thorax and pierce the diaphragm, finally breaking up into branches for the abdominal viscera.

The **spinal** nerves arise by two roots as usual, and are named *cervical, thoracic,* &c., according to the region of the vertebral column to which they correspond.

The last four cervical and the first thoracic nerves form the *brachial plexus*, from which the fore-limb is supplied, and from which also arises the *phrenic nerve*, which runs back to supply the diaphragm.

The *lumbo-sacral plexus* for the hind-limb is constituted by the last three lumbar and the first three sacral nerves.

(3) The **sympathetic system** is formed by a ganglionated cord on each side (*cf.* p. 215) of the body, beginning in the head, traversing the neck, and then running ventral to the vertebral column through the thorax and abdomen into the tail. These are connected by *rami communicantes* with some of the cranial nerves and with the spinal nerves. They are also connected by commissures with each other. In the neck there are only three ganglia, the *posterior* and *middle cervical* at the root of the neck, and the *anterior cervical* placed near and connected with the *vagus ganglion.* In the thorax two large *splanchnic nerves* run off from the cords, and, piercing the diaphragm, fuse with a large *cœliac ganglion,* placed just in front of the origin of the anterior mesenteric artery, and united with an *anterior mesenteric ganglion* just behind it.* This latter ganglion is again connected with a *posterior mesenteric ganglion* situated near the origin of the posterior mesenteric artery. From these ganglia numerous branches run off to the abdominal viscera. The sympathetic also supplies the vascular system.

10. **Sense Organs**—(1) **Tactile Organs.**—Numerous *touch-corpuscles* are present in the skin (p. 271), especially at the bases of the vibrissæ.

* *Cœliac ganglion + anterior mesenteric ganglion = semilunar ganglion* of human anatomy.

(2) **Gustatory Organs.**—Imbedded in the sides of the circumvallate papillæ and papillæ foliatæ are numerous *taste-buds*, containing elongated *taste-cells*, connected with glossopharyngeal fibres.

(3) **Olfactory Organs.**—The nasal cavities (Fig. 85) are largely lined by *olfactory mucous membrane*, which extends over the ethmoturbinals, and contains elongated *olfactory cells* (*cf.* Fig. 62).

(4) **Auditory Organs.**—(*a*) The *external ear* consists of a *pinna* and an *external auditory meatus*, closed below by the *tympanic membrane*. This last also forms the upper and outer boundary of (*b*) the *middle ear* or *tympanic cavity*, which has already been described. (*c*) The *inner ear* (Fig. 91, C) is, broadly speaking,

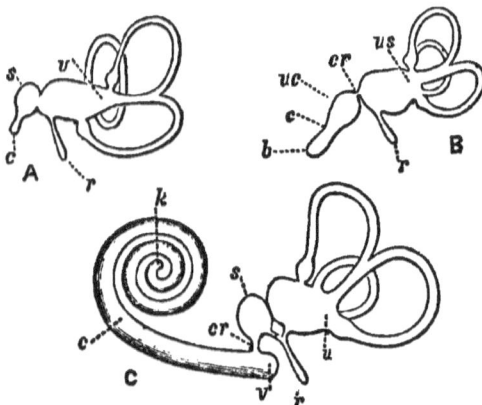

Fig. 91.—Diagrams of the Membranous Labyrinth (from *Haddon*). Internal side of left labyrinth.—A, Fish, B, Bird. C. Mammal; *us*, utriculus and sacculus; *u*, utriculus; *s*, sacculus; *c*, cochlea.

similar to that of the pigeon, but the *semicircular canals* are arranged rather differently (the anterior vertical not crossing over the horizontal), the *sacculus* and *utriculus* do not communicate directly, and the *cochlea* is produced into a delicate tube, spirally coiled, and contained within the promontory. A special series of special sense cells (hair-cells) are formed in the cochlea. This also contains a large number of pairs of elastic fibres, of different lengths, constituting the problematical *organ of Corti* and forming a series of ∧-shaped arches projecting into the endolymph.

(5) **Visual Organs** see Figs. 63 and 64).—The *eye* and the

eye-muscles do not differ widely from the condition found in the frog (p. 221). But the sclerotic is not supported by cartilage, and the lens is much flatter (as in Fig. 63). The accessory structures, besides muscles, are the *conjunctiva*, three *eyelids*, and *lachrymal* and *Harderian glands*.

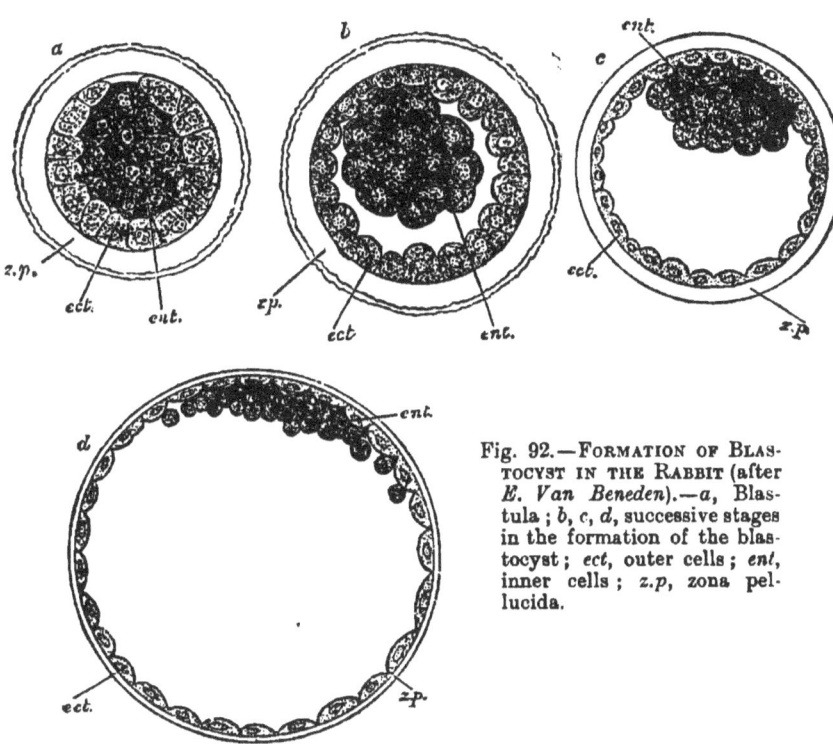

Fig. 92.—FORMATION OF BLASTOCYST IN THE RABBIT (after E. Van Beneden).—*a*, Blastula ; *b, c, d*, successive stages in the formation of the blastocyst; *ect*, outer cells ; *ent*, inner cells ; *z.p*, zona pellucida.

DEVELOPMENT.

1. Early Stages.—Cleavage (segmentation) (Fig. 92) is complete and almost regular, there being only a very small quantity of food-yolk in the minute oösperm to impede it. By successive bipartitions eight cells are produced, four of which are larger than the others, and are called *outer cells*, from their subsequent fate. The other four are termed *inner cells*, for a similar reason. Both inner and outer cells continue to divide, but the latter increase more rapidly and grow over the others. At the close of segmentation, which takes place in the Fallopian tube, the **blastula** consists (Fig. 92) of a layer of clear *outer cells*, which cover a central mass of granular *inner cells*, except at one point. The embryo is still invested by the zona radiata, external to which is an albuminous coating, the *zona pellucida*, received from the wall of the Fallopian tube. The changes that follow take place in the uterus, and lead to the formation of the embryonic layers.

The *outer cells* first of all grow completely over the inner cells, and then increase and become flattened to form the **blastocyst** (blastodermic vesicle), an internal cavity forming meanwhile, which gets larger and larger. The blastocyst is a delicate vesicle, the walls of which are mainly formed by outer cells, to which the inner cells adhere over a small internal area. It continues to increase in size, and the inner cells gradually extend at the edges. In the centre, however, of the patch of inner cells a circular thickening appears (Fig. 93), the *embryonic area*, corresponding to the area pellucida of the Chick. The inner cells here become divided into an upper layer of rounded cells and a lower layer of flattened cells. The former unite with the outer cells to form the **ectoderm** (epiblast) of the embryonic area, while the latter constitute the **endoderm** (hypoblast) of the same region. *Outside* this the epiblast and hypoblast are formed by the outer and inner cells respectively.

	Embryonic Area.	*Rest of Vesicle.*
Ectoderm, .	Outer cells.	Outer cells.
Endoderm, .	Rounded inner cell layer, Flattened inner cell layer,	Inner cells.

The **mesoderm** (mesoblast) is first formed in the embryonic area, and has a double origin similar to that described for the

Chick (p. 259). A side view of the blastocyst on the seventh day is shown in Fig. 93.*

2. **General Growth.** — This is essentially the same as in the case of the Chick (p. 260). The embryo is formed in the

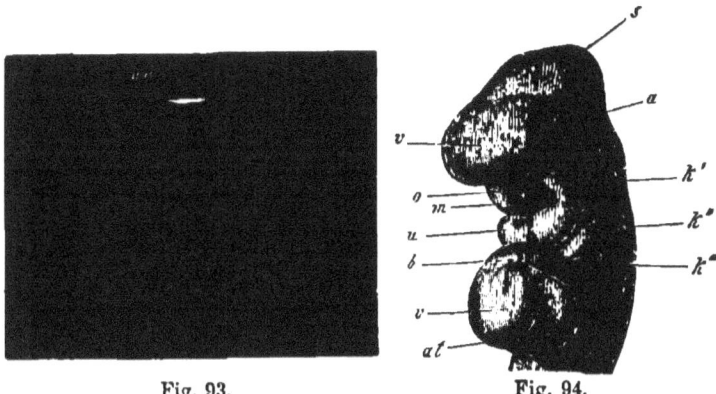

Fig. 93. Fig. 94.

Fig. 93.—BLASTOCYST OF RABBIT SEVEN DAYS AFTER FERTILIZATION. Side view. Magnified about 10 diameters (after *Kölliker*).—*ag*, Embryonic area; *ge*, lower limit of endoderm; below this line the vesicle consists only of a single layer of ectoderm.

Fig. 94.—HEAD OF TEN-DAY RABBIT (after *Kölliker*).—*a*, Eye; *at*, *v*, and *b*, atrium, ventricle, and truncus arteriosus of heart; *v* (upper one) and *s*, fore- and mid-brains; *k'*, *k"*, *k'''*, mandibular, hyoid, and first branchial arches; *o*, superior maxillary process of *k'*; *m*, mouth.

embryonic area (area pellucida), and is at first flattened out on the surface of the blastodermic vesicle, just as the Chick is flattened out on the yolk. But the vesicle only contains *fluid*,

* The above account of the origin of the germinal layers is the one usually given, but comparison with other forms renders it probable that the wall of the blastocyst outside the embryonic area consists at first of **endoderm** only, which becomes continuous with the endoderm of that area. If so, the cavity of the blastocyst = an extended archenteron. The lowest mammals are developed outside the body of the mother, from large eggs resembling those of birds, and there can be no doubt that the rabbit is descended from similar forms, the transition to an intra-uterine mode of development, resulting in loss of the food-yolk. If in a fowl's egg ready for incubation (*cf.* Fig. 73), we suppose the shell and white removed, and the food-yolk replaced by a much smaller amount of albuminous fluid—if, further, we suppose that the lower layer of cells extends right round, and encloses this mass of fluid, then we shall get something resembling the rabbit's blastocyst on the view given in this note.

there being hardly any food-yolk, and its wall, from the first, is partly formed of ectoderm, the other layers subsequently extending into it. In the Chick *all* the layers grow round to invest the yolk, which is at first covered only by the vitelline membrane. The embryo is folded off as described previously for the Chick, and it is only at a comparatively late date that special characteristics are developed. It must, however, be remembered that though four visceral clefts (Fig. 94) are present, the last one is not bounded behind by an arch.

3. **Fate of the Germinal Layers.**—The three layers give rise to the same parts as in the Chick (p. 263), but a few remarks are necessary concerning the **mesodermic** structures.

Cartilaginous bars are developed in the first three **visceral arches**, which develop as follows:—

I. *Mandibular.*—Incus, Malleus, and Meckel's cartilage of either side, Eustachian cartilage, Pterygoid, and Palatine.

II. *Hyoid*, { fuse ventrally to form } anterior
{ body of hyoid, } cornua; stapes.

III. *Branchial* 1, ... { posterior
cornua.

The **aortic arches** develop in a similar way to those of the Chick (p. 265), but the *left 4th* arch becomes the *aorta*, and the *right 4th* the *right subclavian*, while the *right 5th* arch disappears, and the *left 5th* forms the *pulmonary artery*.

The *anterior cardinals* become the *external jugulars*, and the *Cuvierian ducts* the *precavals*.

The history of the **excretory organs** is much as in the Chick (p. 265), but there is no trace of a pronephros.

4. **Embryonic Appendages** (Fig. 95).—These are modified in accordance with the important fact, that while in the Chick development mainly goes on outside the parent's body, at the expense of the bulky food-yolk, the embryo Rabbit develops mainly within the body of the mother, and is born "alive." In other words, the Bird is *oviparous*, the Mammal *viviparous*.

It must be borne in mind that the blastocyst is closely invested by the remains of the *zona radiata* of the ovum.

The **yolk-sac** (umbilical vesicle) is formed by that part of the blastocyst which is outside the embryonic area. From the first its wall is formed within the zona radiata by ectoderm, beneath which the other layers grow round. As the embryo is folded off, the yolk-sac comes to form a ventral appendage to its body. A yolk-sac blood-circulation is established, and this appendage becomes much flattened and shaped like an umbrella, with the handle attached to the embryo (Fig. 95).

The **amnion** is formed as in the Chick (Fig. 82), but there the outer limbs of the fold contain ectoderm and mesoderm, and, with the ectoderm and somatic mesoderm of the yolk-sac, with which they are continuous, fuse with the vitelline membrane.

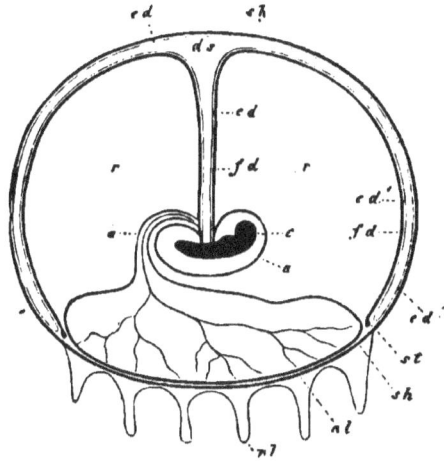

Fig. 95.—EMBRYONIC APPENDAGES OF THE RABBIT—Diagrammatic longitudinal section (from *Kölliker*, after *Bischoff*).—*c*, Embryo; *a*, amnion, and stalk of allantois (*al*); *sh*, subzonal membrane; *pl*, placental villi; *fd* and *ed*, vascular and hypoblast layers of yolk-sac; *ds*, placed in narrow cavity of yolk-sac; *r*, space filled with fluid.

Here the outer limbs of the fold contain *ectoderm* alone (probably), and are continuous with the *ectoderm* of the yolk-sac, which splits off. This continuous layer of ectoderm forms a *subzonal membrane*, lining the zona radiata. To this membrane the flattened yolk-sac attaches itself closely over a large area.

The **allantois** (*cf.* Fig. 82) grows out from the hinder end of the alimentary canal, and becomes closely united with the subzonal membrane for a small circular area on the dorsal side of the embryo, to which it remains attached by a narrow stalk. From this area numerous slender processes, *villi*, grow out, which fit into depressions in a corresponding area of the uterine mucous membrane, which becomes thickened and vascular. A blood-circulation is developed in the allantois, and blood-vessels extend into the villi. Ultimately, close union takes place between the villi and the wall of the uterus. A circular flattened structure, the *placenta*, is thus formed, consisting of an embryonic part, mainly formed by the allantois, and a *maternal* part, constituted by the thickened area of the uterine mucous membrane. The placenta is connected with the embryo by the stalk of the allantois, and certain structures which surround and form with it the *umbilical cord*. The impure blood of the embryo passes into the embryonic part of the placenta by two *allantoic* arteries which break up into branches, and, in the villi, come into close relation with the blood-vessels of the mother. There is not, however, an actual union. Waste-products thus pass from the blood of the embryo into the blood of the mother, which, on the other hand, supplies nutritive material and oxygen. The lack of food-yolk is thus compensated for. (For details of the *embryonic circulation*, see p. 267.)

Several embryos develop in each uterus at the same time, occupying separate swellings. At birth the embryonic membranes are cast off, including the embryonic part of the placenta.

CHAPTER XIII.—COMPARATIVE ANIMAL MORPHOLOGY AND PHYSIOLOGY.

THE comparison of animals is perhaps best conducted on a physiological basis, taking the headings of nutrition, katabolism, reproduction, contractility, irritability, and spontaneity (see Amœba) and adding to these the subordinate one of protection and support.

Amœba, Vorticella, and Gregarina, *unicellular* animals, perform all the functions of life, the first being but little differentiated, the second a great deal, the third modified as a result of parasitism. From this point onwards the *multicellular* condition leads to increasing specialization, different cells performing different functions. Tissues and cellular organs result.

Amœba is usually irregular in form, but, when at rest, tends to become spherical. This is simply a physical result of its semi-fluid nature. The firm cuticle possessed by Vorticella and Gregarina necessitates a definite form.

Bilateral symmetry (exhibited by most of the forms) may be conceived of as derived from radial symmetry (seen in Hydra), by elongation transverse to the long axis, which, at the same time, shortens. Meanwhile, ventral and dorsal surfaces, and anterior and posterior ends, become differentiated, as the result of a creeping mode of progression. The anterior end ultimately becomes the head, in which some of the most important organs are located.

The higher forms mostly exhibit **metameric segmentation**, that is, a division from before backwards into segments or metameres. These are similar in the Earthworm and Leech, but in the Crayfish are more or less dissimilar, and largely fused. It is thus that distinct body-regions arise. The Fluke and Ascaris are unsegmented, and the Mussel and Snail have lost all traces of segmentation. The last has also become twisted and asymmetrical owing to the development of a spiral shell. Amphioxus is clearly segmented but very unsymmetrical. Segmentation is obvious in the Dogfish, but in the Frog, Pigeon, and Rabbit, is much obscured in the adult. It is best seen in the spinal nerves. The jointing of the backbone appears to be a case of "secondary" segmentation—*i.e.*, the segments do not correspond to the primitive ones indicated by the mesodermic somites (protovertebræ) of the embryo.

Homology and **Analogy** are important kinds of agreement. Homologous organs agree in relative position and mode of development, analogous organs correspond merely in function. For example, the *quadrate* bone of the Pigeon is probably *homologous* with the *incus* of the Rabbit, though their functions are different, while the breathing-organs of the Snail, Crayfish, and Rabbit are *analogous*, agreeing in function. Homology is, in fact, *morphological*, analogy *physiological* equivalence. It must not be forgotten that organs are often both homologous *and* analogous—*e.g.*, the *eyes* of Dogfish, Frog, Pigeon, and Rabbit.

Serial homology is agreement in plan of structure and mode of development between series of structures—*e.g.*, the segments of the Earthworm, or spinal nerves and vertebræ of the Frog.

1. **Protection and Support.**—In Amœba there is, under ordinary

conditions, no special modification for this purpose, but in Vorticella and Gregarina a firm cuticle is present. All three pass through an encysted stage. A cuticle is also found in Tapeworm, Ascaris, Leech, and Earthworm, and to a more marked extent in the Fluke, where it is much thickened and provided with numerous minute spines. The maximum of cuticular protection is reached in the Crayfish, Mussel, and Snail, where a firm *exoskeleton* is produced by the impregnation of a thickened and laminated cuticle with salts of lime.

In Hydra the thread-cells serve as a very effective protection. The *skin* of the Dogfish, Frog, Pigeon, and Rabbit serves for protection, and respectively possesses scales, poison glands, feathers, and hairs. The colour of the skin is protective in Frog and Rabbit. From the Fluke upwards, support is largely afforded by what may be termed *indifferent tissues*, *i.e.*, *connective tissue*, and its modifications, *cartilage* and *bone*. Cartilage first appears in the odontophore of the Snail, and, together with bone, forms in the Dogfish, Frog, Pigeon, and Rabbit a more or less complete *endoskeleton*. The most primitive part of this endoskeleton appears to be the backbone, the foundation of which is the *notochord* which, in Amphioxus, persists without modification, except that it extends further forwards than in higher animals. In the Chordate forms, the body is divisible into an upper, smaller, *neural tube*, contained in the skull and backbone, and a lower larger *visceral tube*.

2. **Nutrition.**—The food of Amœba can be ingested, and refuse egested at any point of the shapeless body, and the *pseudopodia* are used for overwhelming any minute organisms that may serve as nutriment. *Vorticella* is usually fixed, and the place of pseudopodia is taken by *cilia*, which do not themselves seize food, but cause food-bearing currents to enter the vestibule and gullet. Within the body the food takes a definite course, and the undigested parts are cast out at a special anal area. The necessity for definite points at which food may be taken in, and innutritious residues ejected is brought about by the presence of the firm cuticle. In Gregarina, however, the ready prepared food is absorbed through the cuticle.

Hydra presents a great advance upon this state of things. The tentacles can seize small animals (which are first paralyzed by the thread-cells) and convey them to the "mouth." Within the large digestive cavity, they are partly taken in amœba-like by the endoderm cells, and partly reduced to a state of solution by digestive fluids poured out by those cells. The flagella produce currents, which assist in the circulation of food in the cavity, and the rejection of waste by the "mouth." Ascaris, Earthworm, and Leech are far more highly differentiated in this and other respects. The conversion of radial into bilateral symmetry, in the way already indicated, would cause the "mouth" of a hydra-like animal to be pulled out into a long slit. The closure of this, in the middle of its extent, would lead to the formation of two apertures, one at each end, placing the gut in communication with the exterior.

The mouth and anus of the Earthworm, &c., can be conceived of as derived in this way from the "mouth" of a form like Hydra. The digestive organs of the Earthworm present a number of special contrivances. The pharynx helps to draw in the food, and the gizzard to crush it, while the lining of the alimentary canal is *glandular* and secretes digestive fluids. The typhlosole increases the absorptive surface. The Leech is specially modified as an ectoparasitic blood-sucker.

The Fluke is a very much modified worm, adapted for a parasitic life, and the digestive organs are much specialized in accordance with this. The Tapeworm has no gut, and lives by absorption; but the equally endoparasitic Ascaris retains a well-developed alimentary canal.

The segments of the Crayfish bear jointed appendages, which, in the anterior part of the body, are adapted for seizing and crushing food, the reduction of which to a fine state of division is completed by the complex gastric mill. The alimentary canal, like that of the Earthworm, exhibits bilateral symmetry, and its absorptive surface is increased by the ridges present in the intestine. Two very large digestive glands are present, which must be regarded, like all such glands, as extremely complicated pouches or diverticula of the alimentary canal, from which, indeed, they grow out, as the study of development shows.

In the Mussel, although the mouth and anus are in the middle line, the gut is convoluted, as in the higher forms (except Amphioxus), by which absorptive surface is gained, augmented in this case by a typhlosole. The food is procured by the aid of ciliary currents, and digested by the fluid secreted in the digestive gland surrounding the stomach.

The asymmetry of the Snail affects the internal organs, including those of digestion, especially as regards the position of the anus. The odontophore is a very special modification for rasping food, and not only is a large digestive gland present, but also salivary glands, the function of which is, however, doubtful.

Amphioxus feeds on small organisms carried into the gut by ciliary currents. The nature of the so-called liver is doubtful.

The Dogfish is carnivorous and possessed of rows of sharp teeth fitted for securing prey. As in all the vertebrate forms there are two digestive glands—liver and pancreas. The gut is short, as in all cases where the food is of easily digestible animal nature, but the spiral valve offers a considerable absorptive surface.

In the Frog, insects are secured as food by means of the tongue, aided by the sticky secretion of numerous small glands. Their escape from the mouth-cavity is prevented by the minute, pointed teeth. The absorptive surface is much increased, not only by the length but by the ridged internal surface of the small intestine. The gut is relatively much longer in the herbivorous tadpole.

The Pigeon, by means of its horny beak, picks up food, which is temporarily stored up in the crop, and acted upon mechanically in the gizzard. The finger-shaped absorptive projections from the lining of the intestine are here, and in the Rabbit, called villi.

The Rabbit possesses teeth adapted for cutting and grinding vegetable food. The flexible, fleshy lips largely aid in seizing food. The alimentary canal is here immensely long, as a result of the vegetable food, and its absorptive surface is, in consequence, extremely large. Not only are liver and pancreas present, but also several pairs of salivary glands.

Circulation.—In Amœba and Vorticella this is aided by the presence of a *contractile vacuole*, and in the Tapeworm and Fluke one of the functions of a circulatory system—*i.e.*, the conveyance of waste-products to organs which get rid of them—is rendered superfluous by the large extent of those (excretory) organs themselves. The general movements of the body effect circulation in Ascaris.

The Earthworm possesses a well-defined and closed blood system as distinct from a cœlomic system. The propulsion of blood is effected by the paired hearts. The Leech exhibits a well-developed blood system in communication with a reduced cœlomic system. The Crayfish, Mussel, and Snail show no distinction between blood and lymph systems. The blood system in each case is *lacunar*—*i.e.*, the ramifications of the blood-vessels terminate in the lacunæ everywhere permeating the tissues. The heart of the Crayfish possesses no auricles; there are two in the Mussel, but only one (a result of asymmetry) in the Snail. All three hearts are *systemic*, that is to say, they receive oxygenated blood, which they pump to the body at large. Amphioxus apparently possesses freely communicating blood and lymph systems. The absence of a heart is made up for by the contractility of the vessels. In the Dogfish, Frog, Pigeon, and Rabbit, the blood system is closed, owing to the presence of the capillaries, which connect the ultimate ramifications of the arteries and veins. The lymph system is largely lacunar, including also certain large spaces, the cœlom, &c., but definite trunks are also present, best marked in the Rabbit and worst in the Dogfish. The Frog possesses propulsive lymph-hearts. In the Dogfish, the heart contains impure blood *only*, which it pumps to the gills for oxygenation and elimination of CO_2. The heart in Frog, Pigeon, and Rabbit is both systemic and pulmonary, supplying the body and also pumping impure blood to the lungs (and in the Frog to the skin also) for purification. Only one ventricle is possessed by the Frog, which involves complicated mechanical contrivances by which the complete mixing of oxygenated and impure blood in the heart is prevented. The Pigeon and Rabbit have two completely separated ventricles.

3. **Katabolism**—(1) **Respiration.**—This is aided in Amœba and Vorticella, by the contractile vacuole. In Gregarina, Hydra, Fluke, Tapeworm, Ascaris, Earthworm, and Leech, it is effected by the general surface of the body, beneath which, in the last two cases, there is a rich supply of blood-vessels. The Crayfish and Mussel possess gills (and the latter mantle-lobes as well) for breathing the oxygen dissolved in water, and in Amphioxus and Dogfish the pharynx is modified for the same purpose, while the Snail, Frog, Pigeon, and Rabbit possess lungs, by means of which the oxygen in the air is utilized. The gills of Crayfish and Mussel are specialized outgrowths of the body; the mantle lobes of Mussel and lung-roof of Snail may be similarly described; the lungs of Frog, Pigeon, and Rabbit are ventral outgrowths of the gut.

Respiratory oxygen-carrying substances are commonly present in the blood, as, for example, hæmocyanin in Crayfish and Snail; hæmoglobin in Earthworm, Dogfish, Frog, Pigeon, and Rabbit. In the Crayfish water is renewed in the gill-chamber mainly by the action of the scaphognathite, and in the Mussel the same end is effected by ciliary action. Water streams in at the mouth, and out through the branchial clefts in Amphioxus and Dogfish, partly as a result of ciliary action in the former, where, too, matters are complicated by the presence of an atrial cavity. In the four air-breathing forms inspiration and expiration may be distinguished, and special provisions are made for these—the muscular floor of the Snail's lung—the buccal respiration of the Frog—the contraction and elastic expansion of the body-walls in the Pigeon—and costal, together with diaphragmatic respiration, in the Rabbit. An increase of surface is gained in various ways both in the case of lungs and gills.

(2) **Excretion.**—The contractile vacuole assists this in Amœba and Vorticella. Ascaris possesses two lateral excretory tubes opening anteriorly by a common ventral pore. The branched excretory system of Tapeworm and Fluke is very complicated, and almost every segment in Earthworm and Leech possesses its own pair of complex glandular nephridia. A pair of glandular kidneys (? nephridia) are found in Mussel, and one (another example of asymmetry) in the Snail. The renal function is performed in the Crayfish by the paired green glands. Amphioxus appears to possess various excretory structures. The kidneys of the Dogfish, Frog, Pigeon, and Rabbit are very complicated, but essentially consist of glandular tubules, each commencing in a filtering apparatus (Malpighian body). The kidney in the Pigeon and Rabbit is a metanephros, not morphologically equivalent to the kidney of the Frog, but it is preceded in development by a transitory mesonephros, which is homologous. The kidney of the Dogfish is partly mesonephros, partly metanephros. In all these forms the mesonephros is at first made up of tubules in which a segmental arrangement can be detected. If the nephridia on each side of the Earthworm communicated with a duct which opened into the posterior part of the gut, a structure would be produced of similar nature. A urinary bladder is present in the Frog and Rabbit, which in one case is

a rudimentary allantois, and in the other is the proximal end of that structure, and receives the urinary ducts.

4. Reproduction.—This is effected **asexually** in Amœba and Vorticella by means of fission, and in Hydra by means of gemmation.

Sexual Reproduction is first hinted at in **conjugation**, which differs somewhat from the same phenomenon as exhibited by plants, in that it is merely a temporary or permanent (as in Vorticella) fusion of two similar or dissimilar individuals which leads to more vigorous asexual reproduction. In Gregarina it is followed by spore-formation.

In Hydra both male and female reproductive organs, spermaries and ovaries, producing male and female reproductive cells (sperms and ova) respectively, are present in the same individual. Fertilization is, probably, preceded in all cases by the formation of two polar cells from the ovum, and in most forms at any rate the germinal cells producing sperms are not entirely used up in the process.

The spermaries and ovaries may be termed the *essential* organs, and advance upon the conditions seen in Hydra principally consists in the development of various *accessory* parts, such as ducts, special glands, and copulatory organs. Amphioxus, however, possesses no such supplementary structures. In Dogfish and Frog the mesonephros performs, in the male, a double function. In the Pigeon and Rabbit the mesonephros aborts, but its duct remains in the male as the spermiduct. A cloaca is present in the Dogfish, Frog, and Pigeon, but, in the Rabbit, the termination of the alimentary canal is quite distinct from that of the urinogenital organs, and opens to the exterior by an anus, while they do so by an urinogenital opening.

Considerable variations are presented in **development**, which are largely dependent upon the amount of food-yolk in the egg. In the Rabbit this is present in but very small quantity, and the nutritive material and oxygen required by the embryo are supplied by the mother, the embryonic and maternal blood-vessels becoming closely related in the placenta. Waste is partly carried off in the same manner.

Hydra differs markedly from the higher forms, in that *two* germinal layers only, ectoderm and endoderm, are present. This animal is in fact little more than a permanent gastrula. The embryos of the higher forms, possess *three* germinal layers—ectoderm, endoderm, and mesoderm. Alternation of generations is exemplified in the life-history of Fluke, and possibly of Tapeworm.

5. **Contractility.**—This is exhibited in the three forms of movement, amœboid, ciliary, and muscular.

Amœboid movement is typically seen in Amœba, from which animal its name is derived. It is also observable in the endoderm cells of Hydra and the epithelial cells lining the intestine of the Fluke. From this animal upwards the power of amœboid movement is mainly retained by the colourless corpuscles found in lymph and blood (leucocytes).

Ciliary movement is first seen in Vorticella, and is also shown by the endodermal cells of Hydra, and the external surface of the ciliated embryo in the Fluke. Cilia are found in the segmental organs and oviducts of the Earthworm, and cover the gills, labial palps, and inner side of the mantle-lobes in the Mussel, in which animal they play a very important part in respiration and nutrition, as is also the case in Amphioxus. In the Frog cilia line the mouth-cavity, and are present in the lungs, oviducts, and elsewhere. In the Pigeon and Rabbit ciliary action is mainly exemplified in the trachea and oviducts. Most sperms are propelled by flagella.

Muscular movement is foreshadowed in the cortical layer of Vorticella and Gregarina, and tailed cells of the ectoderm in Hydra. The muscles are at first largely connected with the skin, as seen in the Fluke, Ascaris, Earthworm, and Leech. The exoskeleton of the Crayfish, Mussel, and Snail, serves for the attachment of muscles, as does the endoskeleton of the Dogfish, Frog, Pigeon, and Rabbit.

Muscular fibres play an important part in the working of most of the organs of the body, and a distinction arises in the vertebrate forms between unstriated and striated muscle, which belong to the internal organs, and the muscles respectively. These varieties of muscle fibre are distinguished by histological characters (*cf.* p. 208), and their involuntary and voluntary nature, the latter difference depending upon nerve-supply.

Locomotion is one of the most important outcomes of contractility. It is effected by amœboid movement in Amœba, ciliary action in free-swimming Vorticellæ, and muscular contractions in the higher forms. The Earthworm progresses in a characteristic fashion, by contractions of its dermal musculature, while Amphioxus and Dogfish can swim by means of their lateral muscles, aided in the latter by paired fins. The tadpole swims pretty much like a fish, but the Frog is adapted to terrestrial locomotion by means of transversely jointed limbs, which are also effective swimming organs. The limbs of the Rabbit are completely given up to movements on land, and in it, as in the Frog, the hind-limbs are the longer, this being connected with the leaping habit. In the Pigeon the fore-limbs are modified into organs of flight.

6. **Irritability and Spontaneity**—(1) **Nervous System.**—Undifferentiated in Amœba, Vorticella, and Gregarina, and very diffuse in Hydra. The Fluke and Ascaris possess a fairly well-developed nervous system with ganglia and nerve-cords. The two elements, ganglion-cells and nerve-fibres, are here found, the former being mainly confined to the ganglia. In the Earthworm the segmentally arranged nervous system consists of a nerve-collar and ganglionated ventral cord, but the ganglia, with the exception of the cerebral, are not very distinct, and the ganglion-cells are not confined to them. Nerve-fibres are afferent and efferent, the former being mainly sensory, the latter mainly motor. The nervous system of the Leech is similar, but the ganglia are very distinct, and this is also the case in the Crayfish, where, however, a good deal of fusion has taken place, especially anteriorly. At the same time a distinct head, containing the main ganglionic masses and with which the chief sense organs are connected, is present. The nervous system of the Mussel is made up of three pairs of widely separated ganglia connected by nerve-cords. Three pairs can also be recognized in the Snail, but here they are very much localized and a distinct head is present. Amphioxus possesses a dorsal nerve-tube contained in a neural canal. The Dogfish, Frog, Pigeon, and Rabbit possess very complicated nervous systems, consisting of brain with spinal cord, contained in the neural canal, cranio-spinal nerves, and sympathetic system. The brain becomes more complicated, mainly by the increase in size of the cerebral hemispheres and cerebellum, the former of which are connected together in the Rabbit by the corpus callosum and fornix. At the same time, histological distinction may be drawn between medullated and non-medullated

nerve-fibres. The latter are especially characteristic of the sympathetic system.

There is reason to believe that the nervous system is derived from the epidermis, with which, in lower forms, it often remains closely connected. The deeper position in higher forms has been assumed for protective purposes. The development of the brain and spinal cord (see Frog) exemplifies the steps in such a change of position. A nervous plexus underlying the epidermis appears to have preceded definite nerve-cords, which have probably arisen by condensation of parts of such a plexus, with which they may co-exist. In Amphioxus, for example, there is such a plexus.
The typical *form* of the nervous system (when differentiated nerves are present) in a radially symmetrical animal, is that of a ring. The assumption of bilateral symmetry would pull this out into a long loop, the anterior end of which might thicken into cerebral ganglia. The sides of the loop might remain separate, and even (*cf.* lateral nerves of Fluke) become disconnected behind. By the ventral fusion of such lateral nerves a circumœsophageal ring and double ventral cord (Earthworm and Crayfish) might be produced.

(2) **Sense Organs.**—These are not differentiated in Amœba, Vorticella, and Gregarina, and are absent in Tapeworm as a result of parasitism. It may be mentioned here that a sense organ essentially consists of one or more usually elongated *sense-cells* (end-organs), of epithelial nature, and connected on the one hand with the nervous system, while on the other they are adapted to receive impressions from various stimuli coming, in the large majority of cases, from the exterior.

The function of the lateral line organs and ampullæ of Dogfish are not certainly known.

(*a*) *Tactile Organs.*—Under this heading may be classified with more or less certainty cnidocils and palpocils of Hydra, head-papilla of Fluke, tactile papillæ of Ascaris, segmental papillæ of Leech, setæ of Crayfish, labial palps and tentacles round inhalent opening in Mussel. Definite tactile cells are present in the skin of Snail, Amphioxus, Dogfish, Frog, Pigeon, and Rabbit.

(*b*) *Gustatory Organs.*—These are not certainly known except in the Frog, Pigeon, and Rabbit, where, as taste-cells, they are supplied by the glosso-pharyngeal nerve. In the last case they occur aggregated into well-marked taste-buds, localized in the circumvallate papillæ and papillæ foliatæ.

(*c*) *Olfactory Organs.*—These are probably represented in the Crayfish by olfactory setæ, and in the Snail by the epithelium on the tips of the tentacles. The Dogfish, Frog, Pigeon, and

Rabbit possess nasal sacs largely lined by olfactory epithelium containing numerous olfactory cells.

(*d*) *Auditory Organs.*—These are first recognizable in the Crayfish, where they exist in the form of small, open, auditory sacs, lodged in the basal joints of the antennules, and lined by sensory epithelium. In the Mussel and Snail closed sacs are found, similar, in essential respects, to the open sacs of the Crayfish. By the growth of such simple vesicles into a complicated shape, the membranous labyrinths found in the Dogfish, Frog, Pigeon, and Rabbit are formed during development. To these may be superadded accessory parts making up middle and external ears.

The function of the "auditory organs" of Invertebrates is ill understood. In some cases they have been shown to be concerned with the perception of position in space, as is the case with the semicircular canals of Vertebrates.

(*e*) *Visual Organs.*—Two eyes in the form of pigment-spots are possessed by the ciliated embryo of the Fluke. In the Earthworm the cerebral ganglia themselves appear to be sensitive to light acting upon them through the translucent skin; and the pigmented part of the skin is sensitive to light (as also in the Snail). The Leech possesses eyes, which appear to be modified tactile organs. The Crayfish possesses compound eyes, to which the theory of "mosaic vision" is applicable, while the simple eyes of the Snail resemble in principle those of the Frog, Pigeon, and Rabbit, which have a double origin, being partly formed by involutions from the exterior, and partly by outgrowths from the fore-brain. It is doubtful whether the so-called "eye" of Amphioxus has a visual function.

CHAPTER XIV.—MAN.

THE aim of this chapter is to form a connecting link between the subjects of General Biology and Human Anatomy as successively taken in a medical course, and it mainly consists of a statement of the leading differences between Rabbit and Man, so that a student who has carefully dissected the former will be able to turn the knowledge thus gained to useful account when he comes to dissect the latter. Comparative Anatomy is also of importance in relation to therapeutics and pathology. Those characters are also given by which Man is distinguished anatomically from the forms most closely related—*i.e.*, the higher apes.

The proportions of the human body and the structure of many of its parts are influenced by two chief factors, (1) the erect attitude, and (2) the mental development.

(1) Man is the only Mammal in which the erect position is habitual, easy, and maintained without the help of the upper limbs. In accordance with this the lower limbs, which are much the longer, have become specialized as supports and means of progression, with corresponding loss of mobility in the feet. The relatively short upper limbs, on the contrary, are capable of performing exceedingly complex movements, and the hand is capable of the most delicate manipulation as a combined result of extreme flexibility, the presence of a long thumb opposable to the remaining digits, flat nails, and a delicate sense of touch. These characteristics are associated with corresponding features in the structure of the skeletal and muscular systems.

(2) Omitting psychological characteristics, although these are by far the most distinctive, the relatively great mental development of Man is associated with exceedingly large cerebral hemispheres, and this again profoundly influences the proportions of the skull.

Comparison with Rabbit.

In human anatomy the body is supposed to be in a vertical position, so that upper (superior), lower (inferior), front (anterior), back (posterior) are used as the respective equivalents of the terms front (anterior), back (posterior), lower (ventral), upper (dorsal), as used in Comparative Anatomy. The latter set of terms will be employed here.

1. Skin.—The covering of hair is very much reduced as regards strength and thickness, but the only large areas entirely devoid of hair are the palms of the hands and soles of the feet, this being associated in the former case with very great tactile sensibility.

2. Endoskeleton.

I. *Axial Skeleton.*—(1) The **skull** is accurately poised upon the vertebral column, the two occipital condyles being situated about the middle of its base and the occipital plane being horizontal. The cranium is of enormous relative size, in correlation with the dimensions of the brain, while the facial region is reduced, owing to the smaller development of the olfactory capsules, and the reduction of the jaws, which are no longer directly used as weapons of offence and defence, nor in procuring food. The facial * and basicranial axes are not, as in the Rabbit, nearly in the same straight line, but the former is bent sharply downwards, so that the craniofacial angle is not more than 120°, and in many races may be much less, 90° being the usual inferior limit. Many of the bones of the skull are very strong and thick, and fusion has taken place to a much larger extent than in the Rabbit. (a) *Cranium.*—The four elements of the occipital ring (basi-, ex-, and supra-occipitals) are united into a single **occipital** bone, and a median **sphenoid** bone is formed by the fusion of the ventral portions of the parietal and frontal rings. The sphenoid consists of a central body and two pairs of lateral expansions or "wings." The anterior part of the body = presphenoid, and the posterior part (on the dorsal side of which is the sella turcica for the pituitary body) = basisphenoid. The anterior or lesser wings = orbitosphenoids, and the posterior or greater wings, with which the small pterygoids are fused ventrally, = alisphenoids. The **parietals** are distinct, but the **frontals** are usually united together into one bone. The **ethmoid** bone, which completes the morphologically anterior end of the cranium and forms a large part of the bony framework of the olfactory capsules, consists of (a) a small horizontally placed cribriform plate, (β) a vertical bony plate (lamina perpendicularis), which with its cartilaginous continuation (septum nasi) constitutes a party-wall between the olfactory capsules, and (γ) two lateral masses of the ethmoid (= ethmoturbinals) projecting into the capsules. In the auditory region of the cranium a great deal of fusion has taken place leading to the formation of a large **temporal** bone = periotic + tympanic + squamosal, which are represented by different regions, as follows :—Petrous and mastoid regions = periotic, tympanic plate = tympanic, and squamous region = squamosal. Besides this, the temporal possesses a prominent styloid process mainly formed by the fusion of two small elements (tympano-hyal and stylo-hyal) belonging to the upper end of the hyoid arch. (b) *Face.*—The olfactory capsules are relatively short and deep. The part taken by the ethmoid in their formation has been mentioned above. Each lateral mass of the ethmoid is divided into upper and lower parts, known as the superior and middle turbinate bones. These are

* The facial axis is a line joining the front of the premaxillæ with the anterior end of the basicranial axis.

much less complex than the corresponding ethmoturbinal of the rabbit, which is also the case with the inferior turbinate bone representing the maxilloturbinal. The equivalent of the naso-

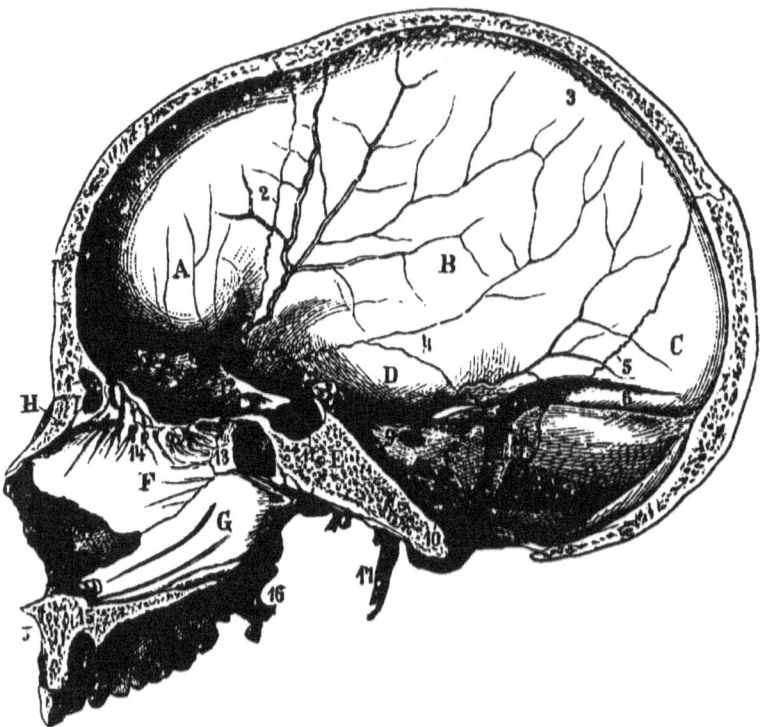

Fig. 96.—MAN. Skull in Longitudinal Vertical Section (from *Macalister*).—A, Frontal; B, parietal; C, occipital; D, squamous part of temporal; E, sphenoid body; F, lamina perpendicularis; G, vomer; H, nasal; J, superior maxilla. 2, Suture between frontal and parietal; 3, suture between parietals; 4, boundary of D; 5, suture between parietal and occipital; 9, internal auditory meatus on inner side of petrous part of temporal; 10, condylar foramen; 11, sella turcica; 12, sphenoidal fissure; 14, placed below cribriform plate; 16, external pterygoid process; 17, styloid process.

turbinal is here found in the uppermost lamella of the superior turbinate. The nasals are small, but the vomer is of considerable size and produced ventrally into a vertical plate which divides the posterior nasal passage into right and left halves.

[The orbits are forwardly directed and almost completely shut off by bone from the temporal fossæ.

The **superior maxillary** bones (= premaxillæ + maxillæ) which support the upper jaw, are of great importance in the formation of the face. Above, they partly wall in the olfactory capsules, and together with the **malars**, frontal, and **lachrymals**, form the margins of the orbits. Below, they present a horseshoe-shaped ridge excavated by the sockets of the upper teeth, and large palatine processes,* making up the greater part of the hard palate, which is relatively much larger than in the rabbit. The hard palate

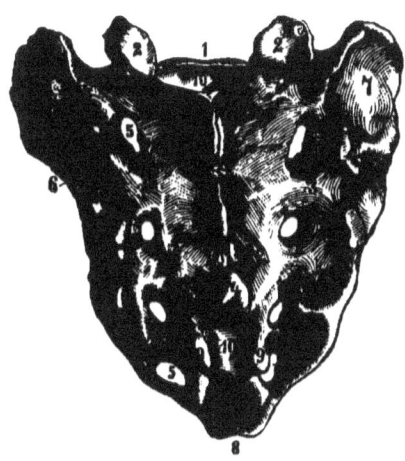

Fig. 97.—MAN. Dorsal view of Sacrum (from *Macalister*).

Fig. 98.—MAN. Dorsal view of Coccyx (from *Macalister*).

is completed by **palatal** bones, and immediately behind it on each side there is a pterygoid plate projecting downwards from the greater wing of the sphenoid (= alisphenoid) and divided into an external pterygoid plate, and a narrow internal pterygoid plate (= pterygoid bone) produced into a hook-like process.

The most noteworthy features in the **inferior maxilla** or bone of the lower jaw (= mandible) are the very complete fusion of its two rami and the presence of a chin-projection or mental prominence at the symphysis. The inferior maxilla also presents, in the upper side of its horizontal part, a horseshoe-shaped ridge excavated by the sockets of the lower teeth.

The **vertebral column** is both strong and flexible. To support the vertically-placed body great strength is necessary, and in

* If these processes fail to unite in the middle line cleft palate results.

accordance with this the column broadens out below, especially in the sacrum which transmits the weight to the innominate bones. Strength and flexibility are both given by the series of gentle curves into which the spine is thrown, and a kind of spring is thus constituted by which the transmission of shocks to the brain is prevented. The convexities of these curves are ventral in the neck and loins. The vertebræ are usually thirty-three in number, as follows:— Cervical 7, dorsal (= thoracic) 12, lumbar 5, sacral 5, coccygeal (= caudal) 4. The sacral vertebræ are fused into a large, strong **sacrum**. The caudal (coccygeal) vertebræ are very much reduced, and, in the adult, generally become anchylosed together into a small triangular **coccyx**.

The **sternum** is relatively broad and flat, nor is it divided into distinct sternebræ, although manubrium and xiphisternum commonly remain distinct. The former does not possess a ventral keel.

There are twelve pairs of **ribs**, of which the first seven are "true," their costal cartilages (= sternal ribs) uniting directly with the sternum. Of the remaining five pairs of "false" ribs, the first three pairs have costal cartilages connected with the preceding ones, while the last two pairs ("floating" ribs) possess free ventral ends.

II. *Appendicular Skeleton*—(1) **Fore-Limb**.—The most noteworthy point regarding the shoulder girdle is the presence of a strong *f*-shaped **clavicle** or collar-bone stretching from sternum to **scapula**. The presence of clavicles is associated with the power of free lateral movement possessed by the arms, these bones acting as props to keep the shoulders well apart, and giving points of attachment to several muscles. *Free Limb.*—The articulation of the **humerus** with the very shallow glenoid cavity forms an exceedingly free ball-and-socket joint, contrasting with the hinge-joint at the elbow.

The **radius** has a very large distal end which plays the chief part in the support of the hand, and it is capable of rotation upon the ulna at both ends. When the palm is placed upwards (supine position) radius and ulna are parallel, but if it is then turned downwards (prone position) the radius rotates on the ulna, carrying the hand round with it. In the rabbit the arm is immovably fixed in the prone position. The **ulna** is larger than the radius, and its large proximal end takes a larger share in forming the elbow joint. The large olecranon process prevents over-extension.

The **carpus** differs from that of the rabbit in the absence of a distinct centrale, it being fused with the radiale. The proportionate size of the different elements is not the same, carpale 3, for example, being relatively very large. The following will give some idea of the method of arrangement, the theoretical names being placed in parentheses:— [R = radius; U = ulna.]

R			U
scaphoid	semilunar	cuneiform	pisiform
(radiale + centrale);	(intermedium);	(ulnare);	
trapezium	trapezoid	magnum	unciform
(carpale 1);	(carp. 2);	(carp. 3);	(carp. 4 + carp. 5)

The skeleton of the hand is completed by five **metacarpals**, and fourteen **phalanges**, of which two belong to the thumb, and three to each of the fingers. The first metacarpal articulates with a saddle-shaped surface on the trapezium, which allows of very free movement, so that the thumb can be readily opposed to the other digits.

(2) **Hind-Limb.**—The ossa innominata which constitute the hip-girdles are very broad, and together with the sacrum and coccyx make up the basin-shaped **pelvis**, by which the weight of the trunk is supported and transmitted to the legs. Each **os innominatum** is made up of the same four elements as in the rabbit—*i.e.*, ilium, pubis, ischium, and acetabular (= cotyloid), but the last is very small and fused with the pubis, of which for all practical purposes it is considered a part. The **ilia** are very broad and expanded; the **pubes** but not the **ischia** meet in a median ventral symphysis. The large size

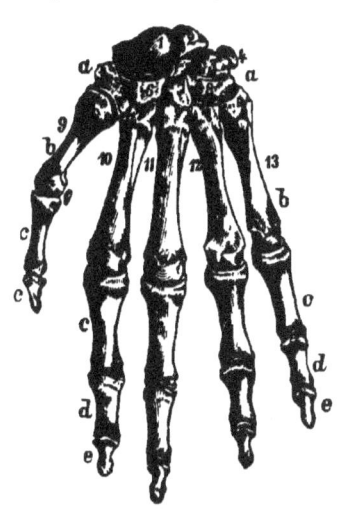

Fig. 99.—MAN. Skeleton of Left Hand. Dorsal view (from *Macalister*).—1, Scaphoid; 2, semilunar; 3, cuneiform; 4, pisiform; 5, trapezium; 6, trapezoid; 7, magnum; 8, unciform; 9-13, metacarpals; *b,c,d,e*, phalanges.

and strength of the bones supporting the free limb are worthy of notice. The **femur** is very long, **tibia** and **fibula** are well developed and separate, and there is a large **patella**. The bones of the foot are arranged so as to constitute strong arched supports, the parts which touch the ground being the outer sides, heels, and front part of the sole. The **tarsal bones** are as follows:— [T = tibia ; F = fibula.]

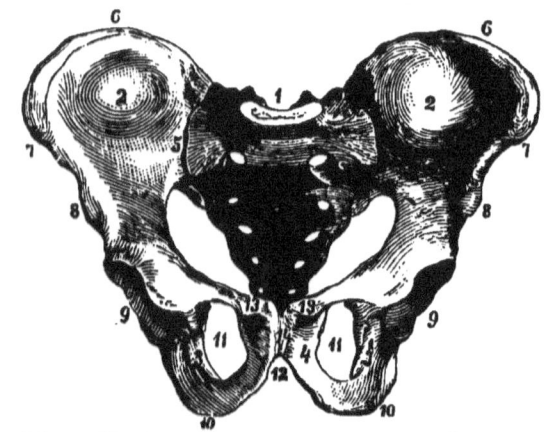

Fig. 100.—MAN. Ventral view of Pelvis (from *Macalister*).—1, Sacrum; 2, 5, 6, 7, 8, ilium; 4, 13, pubis; 14, pubic symphysis; 3, 10, ischium; 9, acetabulum; 11, obturator foramen.

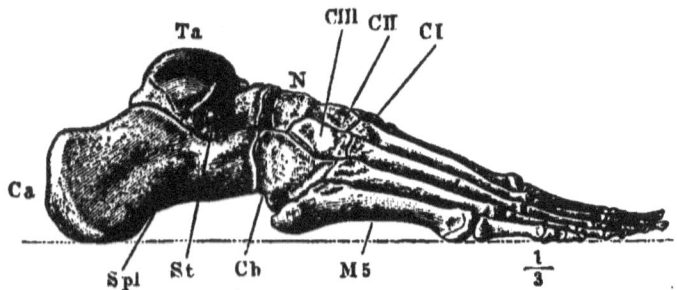

Fig. 101.—MAN. Outer Side of Right Foot (from *Macalister*).—Ta, Astralagus; Ca, calcaneum; N, navicular; CI, CII, CIII, internal, middle, and external cuneiforms; Cb, cuboid; M5, fifth metatarsal.

T			F
astragalus (tibiale + intermedium);			calcaneum (fibulare)
	navicular (centrale)		
internal cuneiform (tarsale 1);	middle cuneiform (tars. 2);	external cuneiform (tars. 3);	cuboid (tars. 4 + tars. 5)

The foot is completed by five **metatarsals**, and fourteen **phalanges**, of which two belong to the great toe and three to each of the other digits. In the rabbit the great toe (hallux) is only represented by a process of the 2nd metatarsal, there being no distinct equivalents of the internal cuneiform (tarsale 1) and 1st metatarsal with the two corresponding phalanges, of the human foot.

3. **Digestive Organs.**—The only parts of the alimentary canal which call for special notice are mouth-cavity, pharynx, and cæcum.

Mouth and **Mouth-cavity.**—The pathological condition known as hare-lip must not be confounded with the cleft upper lip of rabbit and hare. The upper boundary of the mouth is formed to begin with by a median fronto-nasal process and a maxillary process growing forward on each side (*cf.* Fig. 94) from the mandibular arch. This is the permanent condition in a dogfish (Fig. 46), but in chick, rabbit, and man, the fronto-nasal and maxillary processes fuse together and constitute a continuous ridge within which premaxillæ and maxillæ develop. Failure of this union on one side is the cause of single hare-lip, while failure on both sides results in double hare-lip.

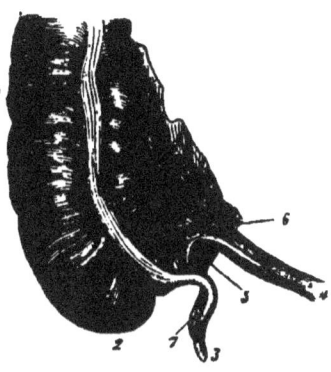

Fig. 102.—MAN. Junction of Small and Large Intestines (from *Macalister.*—1, Colon; 2, cæcum; 3, vermiform appendix; 4, ileum.

The cleft in a rabbit's lip is median, and does not affect the hard parts.

There are no papillæ foliatæ on the **tongue**, the circumvallate papillæ are from 7 to 12 in number, and arranged on the posterior part of the back of the tongue in the form of a V with backwardly directed apex.

The upper and lower **teeth** are arranged in two horseshoe-shaped curves. There are no gaps, and the canines are not disproportionately large. The dental formula is—

$$i = \frac{2-2}{2-2}, \quad c = \frac{1-1}{1-1}, \quad p.m. = \frac{2-2}{2-2}, \quad m = \frac{3-3}{3-3} = 32$$

The premolars are termed bicuspids.

The cæcum and appendix vermiformis are extremely small as compared with the same structures in rabbit.

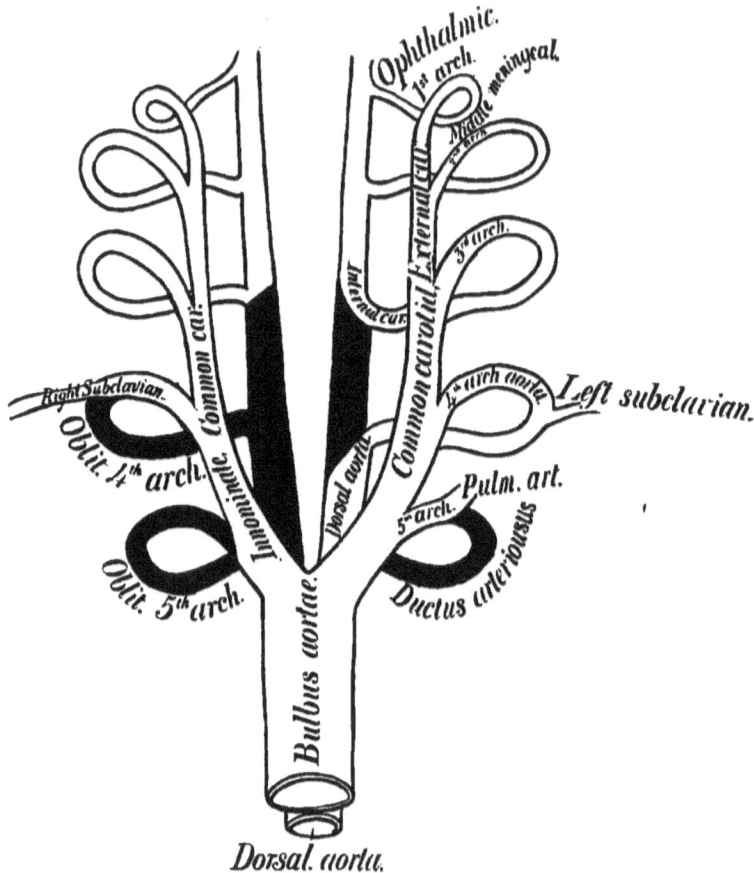

Fig. 103.—MAN. Diagram to explain fate of Aortic Arches (from *Macalister*).—The parts in black are obliterated.

The large **glands** which open into the alimentary canal are salivary glands, liver, and pancreas. Infra-orbital salivary glands are not represented, the liver is not so deeply cleft into lobes, and the duct of the compact pancreas joins the bile duct in the wall of the duodenum.

4. Circulatory System.—Three arteries come off from the aortic arch, (*a*) innominate, from which right subclavian and right common carotid arise, (*b*) left common carotid, (*c*) left subclavian. The fate of the embryonic **aortic arches** is shown in Fig. 103.

1st (= mandibular) arches become ophthalmic branches of external carotids.

2nd (= hyoid) ,, ,, { middle meningeal } ,, ,, ,,

3rd (= 1st branchial) ,, ,, internal carotids.

4th (= 2nd branchial)
left }
right } ,, ,, { aortic arch represented by a very variable arteria aberrans.

5th (= 3rd branchial)
left }
right } ,, ,, { pulmonary artery. obliterated.

Veins.—The development of the **caval system** is illustrated by Fig. 104. There are to begin with anterior and posterior car-

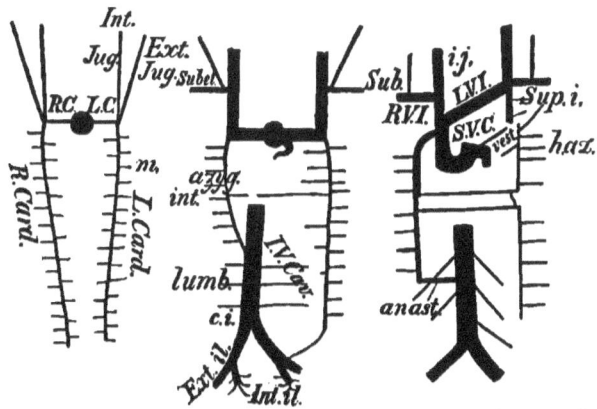

Fig. 104.—MAN. Diagram to explain Development of Systemic Veins (from *Macalister*).—Three successive stages are represented, R.C. and L.C, Cuvierian veins ; R.V.I., L.V.I., vena innominata ; S.V.C., superior vena cava ; *sup. i*, superior intercostal; *h.az*, hemi-azygos.

dinals, which, as in the adult Dogfish (*cf*. p. 165), unite together on each side to constitute a Cuvierian vein which opens into the heart. The **anterior cardinals** persist as the *external jugulars;* internal jugulars and subclavians are developed later on. So far there is agreement with the Rabbit, but in Man an oblique cross-connection is established between the junction of the left jugular and subclavian and the middle of the right Cuvierian vein. This cross-piece becomes the *left innominate vein*, the part of the **right Cuvierian vein** distal to it becomes the *right innominate*, and the part proximal to it becomes the *superior vena cava*. The **left Cuvierian vein** partly aborts, but its proximal end persists as a *coronary sinus*, which receives the coronary veins and opens into the right auricle, while its distal end becomes the *superior intercostal vein*.

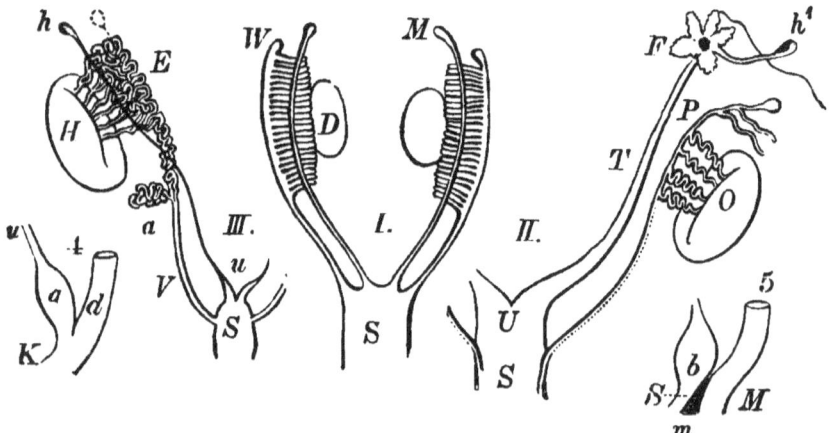

Fig. 105.—MAN. Development of Urinogenital Organs (from *Landois* and *Stirling*).—I, Undifferentiated condition—D, Gonad, resting on tubules of mesonephros, W; M, Müllerian duct; S, urinogenital sinus. II, Transformations in the female—F, Funnel, with hydatid (h'), of Fallopian tube, T; U, utero-vaginal region formed by fusion; O, ovary, with parovarium, P, &c. III, Tranformations in male—h, Hydatid of Morgagni; u, uterus masculinus; H, spermary (testis), with vasa efferentia running to epididymis, E; a, vas aberrans; V, spermiduct (vas deferens). 4, Shows bladder (a) and rectum (d) opening into cloaca. 5, Later stage, where bladder (b) and urinogenital sinus (S) are separated from end of rectum (M) by a perineum (m).

The **posterior cardinals** atrophy in the middle of their course; their hinder ends become the *internal iliac veins*. The anterior part of the right posterior cardinal is converted into the *azygos* *vein*, with which is connected a *hemi-azygos vein* developed from part of the left posterior cardinal.

The **inferior vena cava** (posterior v.c.) only becomes important when the hind-limbs have been developed to some extent. It is formed by the union of iliac veins.

5. **Urinogenital System.**—The **excretory organs** of the embryo consist, as in the rabbit, of a **mesonephros** (Wolffian body) on each side, from which a **mesonephric duct** (Wolffian duct) runs back to open into an urinogenital sinus. The adult kidney is a **metanephros**, and the **ureter** (metanephric duct) is an outgrowth from the posterior end of the mesonephric duct. As before, the **bladder** is the proximal end of the allantois. The reproductive organs at first consist of a pair of **gonads** and two **Müllerian ducts**, and later on the excretory organs of the embryo in part become genital ducts, and in part are reduced to rudiments, which are of considerable importance from a medical point of view, since they may enlarge and become the seat of disease. The following table exhibits the most important details regarding the fate of the embryonic excretory organs and Müllerian ducts. Rudiments are italicized.

As compared with the Rabbit it should be noticed that (1) in the male there are vesiculæ seminales, while the uterus masculinus is relatively small. (2) In the female the most important part is the single uterus, though in abnormal cases the Müllerian ducts may fuse less completely.

6. **Nervous System.**—The **brain** (Figs. 106 and 107) is exceedingly large, one result of which is to profoundly influence the shape and proportions of the skull, as already explained. This increased size is mainly due to the enormously developed **cerebral hemispheres**, which overlap and largely conceal the other regions, extending in all directions to such an extent that nothing else is visible in a brain viewed from above. In the rabbit the hemispheres leave the cerebellum quite uncovered, while they are nearly smooth and a division into lobes is only indicated. In each hemisphere of the human brain, on the contrary, six chief lobes can be distinguished, and its surface is marked by elaborately arranged convolutions (gyri) separated from one another by furrows (sulci). In this way the surface is $5\frac{1}{2}$ times as great as if it were smooth, and the extent of the cortex is correspondingly increased. It is an established fact that the intelligence of an

ADULT MALE.	EMBRYO.	ADULT FEMALE.
	I. Mesonephros.	
Vasa efferentia (coni vasculosi),	1. Anterior tubules, .	*Parovarium* (organ of Rosenmüller, epoöphoron). A series of tubules (some of which may be converted into hydatids) connected with front end of ovary.
Vasa aberrantia and parepididymis (organ of Giraldès). Blindly ending tubules connected with the epididymis,	2. Posterior tubules, .	*Paroöphoron;* a much smaller rudiment lying near the hind end of the ovary.
	II. Mesonephric Duct.	
Epididymis, . . .	1. Anterior part, . .	*Parovarian duct*, terminating anteriorly in a *hydatid*, and sometimes continued back as *Gærtner's duct*.
Spermiduct (vas deferens) and branched vesicula seminalis,	2. Middle part, .	
	3. Posterior ,, .	
	III. Müllerian Duct.	
Hydatids of Morgagni (one more constant and of larger size than rest). Situated between spermary (testis) and caput epididymis,	1. Anterior part, . .	Fallopian tube, to the fimbriated funnel of which a *hydatid* is attached.
Aborts, ...	2. Middle part, . . .	Unites with its fellow to form a pyriform uterus.
?Unites with its fellow to form a small uterus masculinus (sinus prostaticus),	3. Posterior part, . .	Unites with its fellow to form vagina.

* Hydatids are small stalked pear-shaped bodies.

animal is proportional to (1) the size of its hemispheres, and to
(2) the complexity of their convolutions. It must be remembered,
however, that the absolute size of the animal has also an influence
on these factors, as might be expected, since the hemispheres are

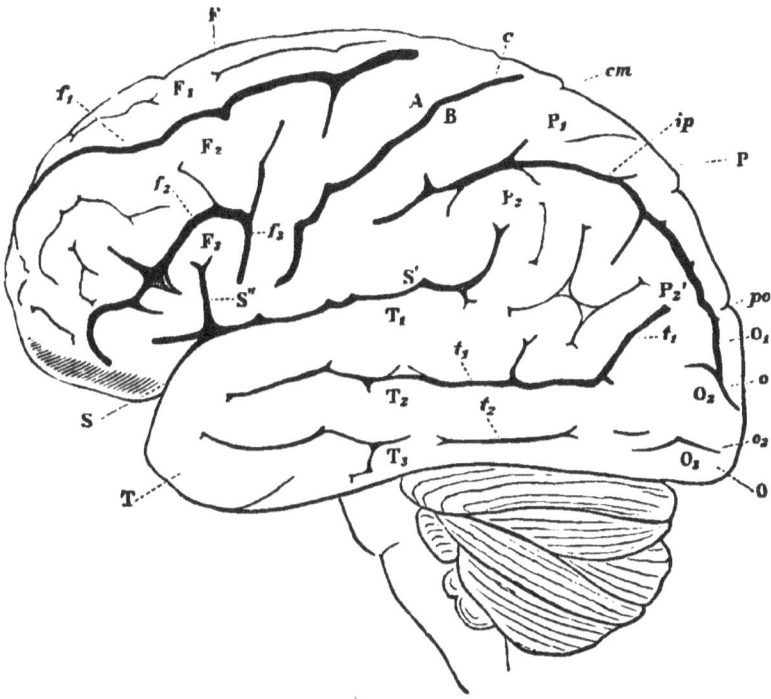

Fig. 106.—MAN. Left Side of Brain, showing convolutions on outer side of left hemisphere (from *Macalister*).—S, Sylvian fissure, running nearly to P_2; c, Rolandian fissure; A, B, S' S'', convolutions of opercular lobe; F, frontal lobe, including convolutions F_1, F_2, F_3; P, parietal lobe, including convolutions P_1, P_2, $P_{2'}$; T, temporal lobe, including convolutions T_1, T_2, T_3; O, occipital lobe, including convolutions O_1, O_2, O_3. N.B.—These lobes also include convolutions on the inner flattened side of the hemisphere (see Fig. 107). The bulb and cerebellum are also shown.

not concerned with sensation and intelligence alone, but also
contain the highest motor centres. In fact, the cortex has been
partly mapped out into motor centres, regulating definite groups
of muscles, and sensory centres, concerned with specific sensations.

The most important sulcus is the *Sylvian fissure* (just indicated in the rabbit), which marks off the *temporal lobe* below. Above this another important sulcus, the *central* or *Rolandian fissure*, the

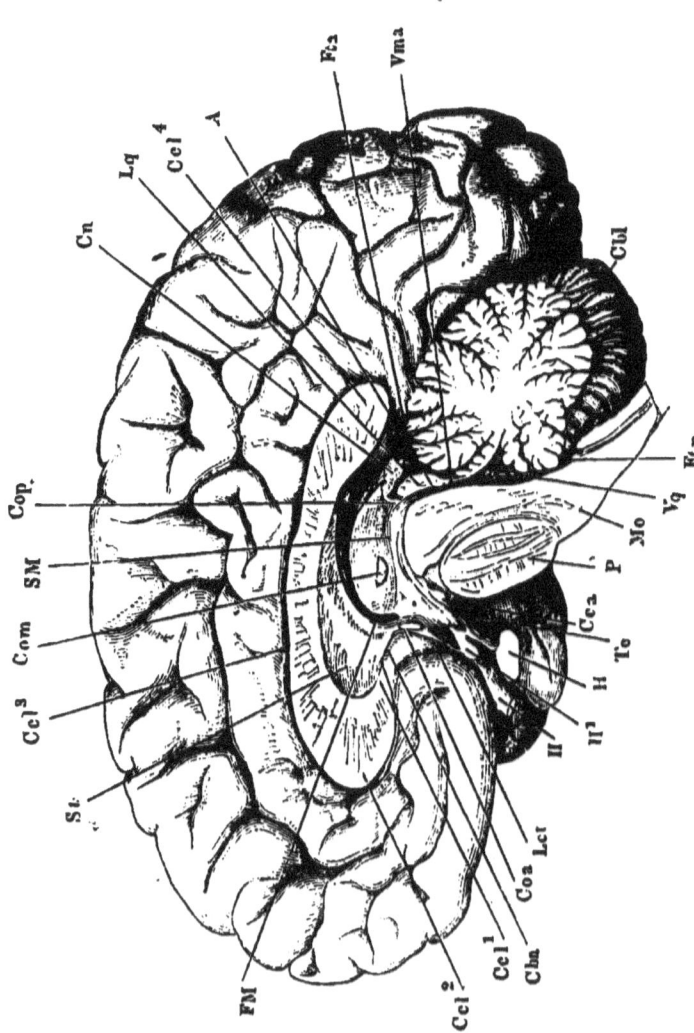

Fig. 107.—Man. Median Longitudinal Section of Brain (from *Macalister*).—FM, Foramen of Monro; Ccl, corpus callosum; Coa, anterior commissure; Lct, lamina terminalis; II, optic nerve; III, optic chiasma, cut through; H, pituitary body; Cca, corpus albicans; P, pons Varolii; Mo, medulla oblongata; Vq, fourth ventricle; Cbl, cerebellum; Vma, valve of Vieussens; A, Sylvian aqueduct; Lq, optic lobe; Cn, pineal body; Cop, posterior commissure; Com, 3rd commissure; Sl, septum lucidum.

convolutions bounding which constitute the *opercular lobe*.* The *frontal lobe* is anterior to this, the *parietal lobe* posterior to it, and there is an *occipital lobe* at the extreme back of the hemisphere. Besides this, there is a small *central lobe* (island of Reil) which can only be seen by separating the lips of the Sylvian fissure. The opercular lobe has been mapped out into lower, middle, and upper motor areas for the muscles of the face, leg, and arm respectively. The following sensory centres have also been

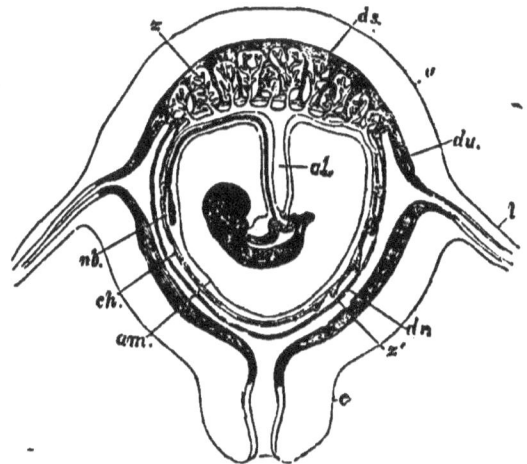

Fig. 108.—MAN. Embryo in Uterus, showing Fœtal Membranes (from *Macalister*, after *Longet*).—*al*, Stalk of allantois; *am*, amnion; *c*, neck of uterus; *ch*, chorion; *dr*, decidua reflexa; *ds*, decidua serotina; *du*, decidua vera; *nb*, yolk-sac; *z*, placental villi; z^1, transient villi.

recognized (with less certainty) in other lobes, (a) *Sight*, occipital lobe; (b) *Hearing*, uppermost convolution of temporal lobe (T); (c) *Smell*, inner side of temporal lobe, just below Sylvian fissure. The **corpus callosum** is relatively very large, and it is much more strongly curved than in the rabbit. The **anterior commissure** is smaller, and so are the **olfactory lobes**. There are two **corpora mammillaria** (c. albicantia).

The **cerebellum** is well-developed, but, in comparison with the

* Many anatomists do not recognize the opercular lobe as a distinct subdivision, and take the Rolandian fissure as the line of demarcation between frontal and parietal lobes.

hemispheres, smaller than in the rabbit; its lateral lobes are far larger than the median one, and the flocculi are insignificant. There are no **corpora trapezoidea.**

7. **Placenta.**—This is of the type known as *metadiscoidal*, for, although it is circular like that of the Rabbit, yet, to begin with, villi are developed over the entire surface of the chorion, though they only persist over a limited area. These villi, too, are much branched, and their relation to the maternal tissue is complex. It will also be observed that the placenta is opposite the ventral surface of the embryo, the reverse being the case in the Rabbit (Fig. 95). The **yolk-sac** (umbilical vesicle), again, is of less importance. The thickened mucous membrane lining the uterus is known as the **decidua**, because it is shed at birth. That part of it which forms the maternal portion of the placenta is the *decidua serotina*, that lining the rest of the uterus is the *decidua vera*, and a *decidua reflexa* surrounding the embryo is formed by the outgrowth and fusion of folds.

Comparison of Man with the Highest Apes.

The apes which come nearest to Man in structure are the Gibbons (*Hylobates*), Orang-outan (*Simia satyrus*), Chimpanzee (*Troglodytes niger*), and Gorilla (*T. gorilla*). In none of these is the erect attitude maintained without the assistance of the forelimbs, the easiest and most habitual position. Consequently there are not the same perfect adaptations to this attitude which are formed in Man, and of which the most striking are the following:—(1) Double S-shaped curve of the vertebral column, (2) hind-limbs longer than fore-limbs, (3) feet completely specialized for terrestrial locomotion, (4) pelvis broad and short. Both pollex and hallux are proportionately much larger in Man, but the latter is not opposable as in apes.

The dental formula of Man and the higher apes is the same, but in Man the teeth form a continuous curve above and below (there being no diastema), and the canines are not so well developed. The canines of apes are disproportionately large; when the jaws are closed their points fit into gaps between the incisors and canines above, between the canines and premolars below.

The most distinctive anatomical features of Man, besides those mentioned, are found in the relatively large cranium and brain. In the Gorilla, for example, which is better developed in these respects than other apes, the brain volume and brain weight are 30-35 cubic inches and about 24 oz., as compared with 55-115 cubic inches and over 40 oz., in Man. The cerebellum of apes is larger in comparison with the cerebral hemispheres. The cranial capacity of Man is, of course, proportionately large, while the facial part of the skull is relatively small (*cf.* p. 322), and the cranio-facial angle is less than in apes. The nasal bones of Man, however, project more, and there is a mental prominence (chin) at the mandibular symphysis.

CHAPTER XV.—CLASSIFICATION AND DISTRIBUTION OF ANIMALS.

THE following is a brief outline of the classification of animals (*cf.* p. 3) :—

DIVISION A.—PROTOZOA.

Unicellular animals (or if multicellular, as *Collozoum*, the cells are all similar and independent), not reproducing by means of sperms and ova.

GROUP 1. RHIZOPODA (Amœba).—Protozoa devoid of cuticle, and possessing pseudopodia.

GROUP 2. INFUSORIA (Vorticella).—Protozoa covered by a cuticle and possessing flagella or cilia.

GROUP 3. SPOROZOA (Gregarinida) (*Gregarina*).— Endoparasitic spore-producing Protozoa, covered by a cuticle but devoid of cilia or flagella.

DIVISION B.—METAZOA.

Multicellular animals, made up of tissues, and reproducing by sperms and ova.

A. DIPLOBLASTICA.—*Embryo* two-layered.

PHYLUM I. CŒLENTERATA (*Hydra*).—Radially symmetrical Metazoa with body-wall composed of ectoderm, endoderm, and mesoglœa. There is a digestive cavity opening by a mouth, but no body-cavity.

B. TRIPLOBLASTICA (Cœlomata). — Bilaterally symmetrical Metazoa, usually with a body-cavity, and possessing mesoderm in addition to ectoderm and endoderm.

PHYLUM II. PLATYHELMIA (*Distoma, Tænia*).—Unsegmented usually flattened worms, without lateral appendages or distinct body-cavity. Nervous system of cerebral ganglia, nerve-cord (*not* as double ventral cord), and nerve-plexus. Life-history often complex.

PHYLUM III. NEMATHELMIA (*Ascaris*).—Elongated cylindrical unsegmented worms, usually with complete alimentary canal, but no blood-system. Cuticle well developed; ciliated epithelium absent. Excretory organs as two unbranched lateral canals opening by a common anterior ventral pore. Sexes distinct. Gullet surrounded by a nerve-ring from which two principal nerves (lateral) run forwards and two (a dorsal and a ventral) backwards.

PHYLUM IV. ANNELIDA.—Elongated segmented worms, with paired nephridia, a nerve-ring, and a ventral ganglionated nerve-cord of double nature. Possess both blood-system and a cœlom.

    Class 1. Chætopoda (*Lumbricus*).—Annelids possessing setæ, and with a well-developed cœlom.

    Class 2. Hirudinea (*Hirudo*).—Annelids devoid of setæ, with segmentation obscured by secondary annulation, and with a reduced cœlom communicating with the blood-system.

PHYLUM V. ARTHROPODA (*Astacus*).—Segmented animals possessing well-marked body-regions as a result of the differentiation of segments. Lateral jointed appendages, and a nervous system of the same type as in Phylum II., but with better-developed ganglia, especially anteriorly. Blood-system largely lacunar, replacing the cœlom.

CLASSIFICATION AND DISTRIBUTION OF ANIMALS. 341

PHYLUM VI. MOLLUSCA.—Unsegmented animals possessing a mantle and (usually) a shell. The body is ventrally produced into a muscular locomotor appendage, the foot. The blood-system is largely lacunar, and the cœlom is reduced to a pericardial cavity communicating with the exterior by two tubular kidneys (? nephridia), one of which is often aborted. Nervous system more or less concentrated, and mainly formed of ganglionated œsophageal ring.

GROUP 1. LAMELLIBRANCHIATA (Anodonta, Unio).—Molluscs without a distinct head or an odontophore. Shell bivalve and lateral. Nerve-ring long.

GROUP 2. ODONTOPHORA (Helix).—Molluscs with odontophore, and a distinct head bearing sense-organs and containing a short nerve-ring. Mantle never bilobed. A univalve shell commonly present.

PHYLUM VII. CHORDATA.—Segmented animals possessing a notochord and visceral clefts, transitorily or permanently. Central nervous system dorsal and (usually) tubular. (See p. 135).

SUB-PHYLUM I. HEMICHORDA (Balanoglossus, Cephalodiscus, Rhabdopleura).—A small group including three genera of worm-like Chordates possessing a proboscis supported by a small notochord. [Rhabdopleura has no visceral clefts.]

SUB-PHYLUM II. UROCHORDA (Ascidians).—A group of degenerate Chordates in which the larva possesses a tail supported by a notochord, both, however, disappearing as a rule before the adult stage is reached.

SUB-PHYLUM III. CEPHALOCHORDA (Amphioxus).—A Chordate group containing a single genus, and characterized by asymmetry, very complete segmentation, a notochord extending from end to end, and the presence of an atrial cavity.

SUB-PHYLUM IV. VERTEBRATA.—Chordates with never more than two pairs of limbs, and usually possessing jaws that move up and down and are not constituted by modified appendages. Skin consists of a several-layered epidermis and a connective tissue dermis traversed by vessels, nerves, &c. The notochord does not extend further forward than the middle of the ventral side of the brain. It is supplemented, or more or less replaced by a well-developed endoskeleton consisting of at least a vertebral column, braincase, and sense-capsules; also supports for limbs when these are present. The blood contains red as well as colourless corpuscles, and it is enclosed in a system of tubes comprising a ventral heart, arteries, veins, and capillaries. There is a lymph-system, communicating with the blood-vessels, and consisting of the large cœlomic body-cavity, smaller lymphatic spaces, lymphatic vessels, and "ductless" glands, of which the most conspicuous is the spleen. The body may be regarded as made up of two tubes, (a) a smaller dorsal tube enclosing brain and spinal cord, (β) a larger ventral tube (cœlom) containing most of the other organs. Two compact kidneys are present, each provided with a special longitudinal duct. Ova in most cases shed into cœlom and are carried off by two oviducts. When the mesonephros persists it not only performs excretory functions, but may also serve as a channel for carrying off the sperms. When entirely replaced by a metanephros it is converted into a conducting-apparatus for sperms. The outlets of the intestine, excretory and reproductive organs, are situated close together and ventrally. There is a distinct brain, outgrowths from which develop into the retinæ of two eyes, each of which also possesses a lens formed as an ectodermic involution, and a muscular iris of mesodermic origin. Two olfactory sacs and two membranous labyrinths are present, all of which are developed from ectodermic involutions.

A. A n a m n i o t a.—Embryo without amnion. Allantois, when present, is not an embryonic appendage.

GROUP I. ICHTHYOPSIDA.—Vertebrates with epidermic exoskeleton slight or absent. Lateral line sense-organs and a median fin present at some period of life. The notochord often persists, and the vertebræ, when present, are without epiphyses. The cranium always possesses extensive tracts of cartilage, there is never a well-developed basisphenoid, and the cranial membrane bones, when present, include a parasphenoid. The gut either ends in a cloaca or opens by an anus in front of the urinogenital apertures. Blood at about same temperature as surrounding medium, and containing large oval nucleated red corpuscles. Heart two- or three-chambered, and aortic arches never less than two. Gills, and at least four pairs of branchial arches, always present at some period of life. The cœlom is not divided into thoracic and abdominal sections by a respiratory diaphragm. The mesonephros persists. The cerebral hemispheres are never united by a conspicuous corpus callosum, the hypoglossal nerve does not perforate the brain-case, and the sclerotic coat of eye, when present, is cartilaginous or bony.

Class 1. Pisces (*Scyllium*).—Ichthyopsida with fin-like limbs, the margins of which are supported by fin-rays, as also are the unpaired fins. A well-developed exoskeleton mainly of dermal origin. No postcaval vein and no allantois (see p. 153).

Class 2. Amphibia (*Rana*).—Ichthyopsida in which the unpaired fins always present in the larva and sometimes in the adult are not supported by fin-rays. The limbs are transversely jointed and their extremities divided into digits. There is no epidermic exoskeleton, and the skin is very glandular. A cloaca and an allantoid bladder.

B. **Amniota.**—Embryo with amnion and respiratory allantois. Gills never present.

GROUP 2. SAUROPSIDA.—Air-breathing Amniota, with epidermic exoskeleton of scales or feathers. The vertebræ are without epiphyses, the cranium is well ossified and articulates with the vertebral column by a single occipital condyle, partly belonging to the basi- and partly to the ex-occipital region. There is an interorbital septum, the otic bones do not fuse into a periotic before uniting with other elements, there is no separate parasphenoid, and the complex mandible articulates with a quadrate bone. The ankle-joint is between the proximal and distal tarsal bones. The oval nucleated red corpuscles are smaller than those of the Ichthyopsida, the heart possesses two auricles and one (physiologically double) or two ventricles, and the aortic arches are seldom more than two. There is never a complete respiratory diaphragm. The kidneys are metanephric, and their ducts open into a cloaca, together with the intestine and sexual ducts. The cerebral hemispheres are not united by a well-marked corpus callosum, and the hypoglossal nerve perforates the brain-case. The sclerotic coat of the eye is supported by cartilage or bone, the cochlea is straight or slightly curved, and there is a columella. The large telolecithal ova are invested in firm calcareous shells, the cleavage is meroblastic, and most of the development takes place, with few exceptions, outside the body of the parent.

Class 1. Reptilia (*Reptiles*).

Class 2. Aves (*Columba*).—Warm-blooded Sauropsida with fore-limbs modified into (usually) functional wings, and a reduced tail-region (uropygium). There is an epidermic exoskeleton, in the form of feathers. The cervical vertebræ possess saddle-shaped articular surfaces, the true sacral vertebræ are without expanded ribs, and fuse with numerous other vertebræ to form a pseudosacrum, and the posterior caudal vertebræ (usually) unite together into a ploughshare-bone. The manus never possesses more than two free carpals nor more than three digits; there is a carpo-metacarpus. The ilia extend far behind and in front of the perforated acetabulum, and the pubes and ischia are directed backwards, the former rarely, the latter never meeting in a ventral symphysis. The tarsal bones are never separate, but fuse with tibia proximally and metatarsus distally to form a tibio-tarsus and tarso-metatarsus; digits of pes, never more than four. The jaws are covered by a horny sheath, functional teeth never being present in recent forms; a bursa Fabricii opens into the cloaca. The heart is four-chambered, and the right auriculo-ventricular valve muscular. There is a single aortic arch, curving to the right. The lungs are immobile, and the bronchia communicate with air sacs. The right ovary and oviduct are rudimentary. The optic lobes are displaced to the sides of brain.

CLASSIFICATION AND DISTRIBUTION OF ANIMALS. 345

GROUP 3. MAMMALIA.

Class 1. Mammalia.—Warm-blooded Amniota in which an epidermic exoskeleton, in the form of hair, is always present at some period of life. The vertebræ possess epiphyses, and the cervical vertebræ are usually seven. The occipital region is well ossified, and there are two occipital condyles. The otic bones early fuse together into a periotic. The mandible is of two pieces only, and articulates directly with the squamosal. The lips are usually definite and muscular, and the teeth, when present, are only replaceable to a limited extent by new ones. The large intestine is very long, and usually opens independently, behind a urinogenital aperture. The red corpuscles are non-nucleated, the heart four-chambered, the right auriculo-ventricular valve membranous, and there is a single aortic arch, curving to the left. There is an epiglottis, the lungs are mobile, there are no air-sacs, the bronchial tubes end in infundibula, and there is a complete respiratory diaphragm. The cerebral hemispheres are united by a corpus callosum, a third commissure and corpus mammillare are present, and the optic lobes are double. The sclerotic coat of eye is fibrous, there are three auditory ossicles, and the cochlea is (almost always) spirally coiled. Usually viviparous, and the young always nourished by milk for some time after birth.

Sub-Class 1. Prototheria (Ornithodelphia).—Mammals devoid of functional teeth, and possessing distinct coracoids, epicoracoids, interclavicle, and epipubic (marsupial) bones. There is a cloaca, the young are hatched from large eggs with a calcareous shell and much food-yolk, the cleavage being consequently meroblastic. The mammary glands have no teats.

Order 1. Monotremata.–Ornithorhynchus, Echidna, Proechidna.

The remaining sub-classes of Mammals include viviparous animals which usually have functional teeth, but do not possess distinct interclavicle, epicoracoids and coracoids, the last being represented by a process of the scapula. There is no cloaca. The mammary glands possess teats.

Sub-Class 2. Metatheria (Didelphia).—Mammals with inflected angle to the mandible and epipubic (marsupial) bones. The oviducts are distinct in typical cases, and there is not an allantoic placenta. The female usually possesses a pouch.

Order 1. Marsupialia.—Pouched animals:—Kangaroo, opossum, wombat, &c.

Sub-Class 3. Eutheria (Monodelphia).—Mammals in which the angle of the jaw is not inflected, and there are no epipubic bones. The oviducts are more or less fused, and there is an allantoic placenta.

Orders of Eutheria.—(1) *Edentata*—Armadillo, sloth, &c. (2) *Ungulata*—(a) Artiodactyla, none of the digits bilaterally symmetrical—Swine, hippopotamus, ruminants; (b) Perissodactyla, with a bilaterally symmetrical 3rd digit in each manus and pes— Tapir, rhinoceros, horse, &c. (3) *Sirenia*—Dugong, manatee. (4) *Rodentia*—Rabbit, rat, porcupine, squirrel, guinea pig, &c. (5) *Proboscidea*—Elephant. (6) *Hyracoidea*—Hyrax. (7) *Insectivora*— Hedgehog, mole, shrew, &c. (8) *Cheiroptera*—Bats. (9) *Carnivora* —Dog, cat, bear, seal, &c. (10) *Prosimiæ*—Lemurs. (11) *Primates* —(a) Arctopitheci, Marmozets; (b) Platyrrhini, lower monkeys; (c) Catarrhini, higher monkeys (Cynomorpha) and man-like apes (Anthropomorpha), see p. 338; (d) Anthropidæ—Man.

PRINCIPLES OF DISTRIBUTION.

Explanatory Theories.

I. *Theory of Special Creations*, now abandoned. According to it:—1. Organisms were created where now found. 2. The fauna and flora of any particular region must be better adapted to it than any other fauna and flora. 3. Climate, soil, and position explain all the phenomena of distribution.

II. *Theory of Evolution*, generally accepted. Regards modern distribution as the result of innumerable changes that have affected—(1) organisms; (2) the surface of the globe.

1. Geological history shows that there has been a succession of faunas and floras passing gradually into one another, old species becoming extinct, and new ones being evolved by the combined influence of variation and heredity, which respectively originate and accumulate new characters. Upon the whole a gradual advance in complexity has taken place, but the geological record is extremely imperfect, especially as regards land organisms. The process of change is still going on, and examples of modern extinction (*e.g.*, the gigantic wingless New Zealand bird **Dinornis** and Sirenian form **Rhytina**) are well known, but the detection of newly evolved species involves greater difficulties.

Species once established have extended themselves over smaller or larger areas, according to their powers of migration and surrounding conditions—*i.e.*, their environment. Physical barriers, such as oceans, mountains, climate, and soil, have played an important part in limiting such extension, but the competition of other forms has had a still greater influence. Introduced forms often increase prodigiously, and even supersede the indigenous ones, whence it follows that these last are not necessarily the best adapted. *Exs.* Rabbits in Australia; the brown rat in England, which has almost ousted the indigenous black rat.

2. Owing to the wearing away or erosion of the land by various agencies (chiefly the different forms of water), and the action of subterranean forces by which upward and downward movements of the earth's crust are produced, the distribution of land and sea has constantly varied. Europe and North America, for example, have most likely been connected at various times by land occupying part of what is now the North Atlantic, and Australia appears to have been once united with Asia. On the other hand, evidence is found on every continental land surface of the former presence of the sea. In spite of what has been said, the theory of "permanence of oceanic and continental areas" finds much support. According to it the great oceans are of extreme antiquity, and, *on the whole*, more or less land has

existed from very remote times within the present continental areas. It is perhaps best to accept this theory only for the deeper parts of the great oceans. An accurate knowledge of the contours of the ocean floor is important in this connection, and serves as an important check upon speculations regarding former land-unions. On this basis islands have been divided into *oceanic* and *continental*, which are believed respectively to have been always isolated, and to have been connected with an adjoining continent.

Oceanic Islands are:—(a) generally remote from continents; (b) separated from them by very deep (usually over 1,000 fathoms) water; (c) of volcanic or coral nature; (d) inhabited by forms which possess powers of migration capable of carrying them, actively or passively, over more or less broad ocean tracks; (e) characterised by numerous peculiar species. *Exs.* The Azores, St. Helena, Ascension, coral islands of Pacific.

Continental Islands are:—(a) comparatively near a continent; (b) separated from it by comparatively shallow (under 1,000 fathoms) water; (c) of similar geological structure, and not entirely volcanic or coral; (d) inhabited by similar organisms, irrespective of powers of migration. Such islands are:—(1) *Ancient continental*, separated from the nearest continent by fairly deep (over 100 fathoms) water, and presenting only a general resemblance in the fauna and flora; many peculiar species. *Exs.* Madagascar, Celebes. (2) *Recent continental*, separated from the adjacent continent by shallow (not more than 100 fathoms) water, and with closely similar fauna and flora; very few peculiar species. *Exs.* British Islands, Japan. Both (1) and (2) have presumably been united with the adjoining continents, the latter at a recent date, geologically speaking.

The surface of the globe has also undergone numerous mutations as regards *climate*. The temperate parts of N. America and Europe, for example, were at a geologically recent period passing through a glacial epoch (the "great ice age"), as proved by ice-worn and scratched rock-surfaces and rocks, boulder-clay, &c. On the other hand, fossil plants evidencing subtropical conditions have been found in the Arctic regions. Many theories have been advanced to account for secular changes of climate. The most satisfactory is one by Wallace which attributes them to geographical revolutions (previously suggested by Lyell), influenced by astronomical changes (variations in excentricity of earth's orbit and movements of precession, as advanced by Croll).

All the preceding changes must have exerted a profound influence upon organisms, and throw light upon many problems of distribution.

Areas of Distribution.—May be mapped out for species, genera, families, and orders. In all cases : (1) Size and nature of boundaries very variable. (2) Need not be continuous.

(a) *Exs. of Limited Areas.*—The marmot only found in the Alps. A species of humming-bird confined to the crater of the extinct volcano Chiriqui in Veragua.

Six genera of Lemurs are peculiar to Madagascar. The family *Galeopithecidæ* (including the single genus **Galeopithecus**) is limited to Malacca, Sumatra, Borneo, and the Philippines. The order *Monotremata* only occurs in Australia, Tasmania, and New Guinea.

(b) *Exs. of Extensive Areas.*—The Leopard is distributed through the whole of Africa and S. Asia to Borneo and E. China. The genus **Felis** (cat, lion, leopard, &c.) ranges over most of the globe except Australia, the Pacific Islands, W. Indies, Madagascar, and the more northerly parts of North America and Asia. The family *Vespertilionidæ*, including 200 species of small insect-eating bats, occurs everywhere within the tropical and temperate zones; while the family *Muridæ* (rats, mice, &c.) is only absent from Polynesia and New Zealand. 2. Discontinuity is generally a sign of antiquity, the two or more parts being remains of a once continuous distributional area, in part of which extinction has occurred. Changes in the distribution of land and sea have broken up many once continuous areas. *Examples.*—The variable hare (**Lepus variabilis**), Europe and Asia N. of 55°; Alps, Pyrenees, and Caucasus. The genus **Tapirus**, S. America, S. E. Asia. *Centetidæ* (a family of the Insectivora), Madagascar, Cuba. Hayti.

Ganoid fishes are now represented by genera with the following distribution :—**Acipenser**, N. temperate and Arctic regions. Most species marine, others are found in the Caspian Sea, Black Sea, and N. American lakes, with their rivers, also in the Danube, Mississippi, and Columbia. **Scaphirhynchus**, Mississippi and tributaries. **Polyodon**, Mississippi and Yang-tse-Kiang. **Polypterus**, Nile and W. African rivers. **Calamoichthys**, rivers of Old Calabar. **Amia,** fresh-water, United States. **Lepidosteus,** fresh-water, N. America to Mexico and Cuba. Ganoid fishes are of great geological antiquity, and were formerly a widely spread *marine* group. Most of the forms now surviving have gradually accommodated themselves to a life in rivers, lakes, &c., where the struggle for existence is less severe.

The peculiar distribution of the *Dipnoi* (p. 153) can be explained similarly.

The *Marsupialia* and *Edentata* (p. 354) are also good examples of interrupted areas of distribution.

Zoological Regions, characterized by the presence of peculiar families and genera, and by the absence of other families and genera, have been formed for sea and land. The most useful division of the latter is chiefly based on the Mammalia, but applies very well to birds and reptiles and fairly to other groups. The regions thus established are six in number.

I. *Palæarctic Region.*—Temperate Europe and Asia and N. temperate Africa. Extends W. to Iceland, the Azores, and Cape Verde Islands, and E. to Behring Straits and Japan. Southern boundary somewhat indefinite,—tropic of Cancer in Africa and Arabia, river Indus, Himalayas, Nanling mountains.

II. *Ethiopian Region.*—Africa and Arabia south of the tropic of Cancer, and including Madagascar.

III. *Oriental Region.*—Asia, S. of Region I., and the western part of the E. Indies. The eastern boundary of this region (*Wallace's line*) passes between Bali, Borneo, and the Philippines on the one hand, Lombok and Celebes on the other. The former islands are therefore in the Oriental Region, the latter in the Australian Region.

IV. *Australian Region.*—Australia, New Guinea, New Zealand, with the smaller islands from Wallace's line to the Marquesas and Low Archipelago, and the tropic of Cancer to the Macquarie Islands.

V. *Neotropical Region.*—S. America, the W. Indies, and tropical N. America, with the exception of the central part of the Mexican table-land.

VI. *Nearctic Region.*—Arctic and temperate N. America, with the central part of the Mexican table-land.

DISTRIBUTION OF MAMMALS.

I. *Palæarctic Region.*—Thirty-five families represented. *Peculiar Genera.* The camel, six deer, the yak, six antelopes (including the chamois), and all wild sheep and goats, except two species. Six p.g. of *Muridæ* (rats and mice), two of mole-rats, and one other; dormice and pikas (calling hares) are almost confined to this region. Six p.g. of moles; the remaining two genera of which (**Talpa, Urotrichus**) extend, respectively, into N. India and N. W. America. Five p.g. of Carnivora, including the racoon-dog, a seal, and the badger; the last just enters the Oriental region in China.

II. *Ethiopian Region.*—Fifty families represented, of which nine are peculiar—*i.e., Orycteropodidæ. Hippopotamidæ; Camelopardidæ. Potamogalidæ* (including **Potamogale**, an otter-like Insectivore); *Chrysochloridæ* (golden moles). *Cryptoproctidæ* (**Cryptoprocta** is a small civet-cat-like form peculiar to Madagascar); *Protelidæ* (**Proteles**, the aard-wolf, is allied to the hyænas and weasels); *Cheiromyidæ* (contains **Cheiromys**, the aye-aye, peculiar to Madagascar).

Peculiar Genera (besides those in the above families).—**Potamochærus** (river hog), and **Phacochærus** (wart hog); **Hyomoschus** (a small deer-like form), twelve p.g. of antelopes. Thirteen p.g. of *Muridæ;* **Pedetes** (a jerboa or jumping mouse); **Anomalurus** (a flying squirrel); three other p.g. of Rodents. Three p.g. of elephant-shrews, and the Insectivorous family *Centetidæ*, except one genus from Cuba and Hayti. Three p.g. of bats. Seventeen p.g. of *Viverridæ* (civets and ichneumons), two p.g. of dogs, and two p.g. of *Mustelidæ* (weasels, otters, &c.) Nine p.g. of Lemurs. Eight p.g. of apes and monkeys, the most important being **Troglodytes** (gorilla and chimpanzee).

Peculiar Species.—Among these are several species of Manis, the two-horned African rhinoceroses, the zebras, African elephant, and lion.

Absent Palæarctic Forms.—The genera **Bos** (wild ox), and **Sus** (wild boar), camels, deer, goats and sheep, moles, bears.

III. *Oriental Region.*—Forty-two families represented, of which two are peculiar, and one other almost so—*i.e., Galeopithecidæ* (including **Galeopithecus**, the flying lemur), and *Tupaiidæ* (treeshrews), among Insectivora. *Tarsiidæ*, a family of lemurs (including only **Tarsius spectrum**, found in Sumatra, Banca, and Borneo; also outside the Oriental region in Celebes).

Peculiar Genera (besides those in the above families).—**Tragulus** (chevrotain); **Cervulus** (a deer); **Bibos** (wild cattle); three p.g. of antelopes. **Platanista** (a dolphin found in Ganges and Indus). Three p.g. of *Muridæ*; **Pteromys** (a flying squirrel); **Acanthion** (a porcupine); **Gymnura** (a hedgehog). Eleven p.g. of bats. Twelve p.g. of *Viverridæ*; **Cuon** (a dog); five p.g. of *Mustelidæ*; two p.g. of bears. **Loris** and **Nycticebus** (lemurs). Four p.g. of apes, including **Simia** (orang-utan), and **Hylobates** (gibbon).

Peculiar Species.—Among these are the Indian tapir, several species of rhinoceros, and the Indian elephant.

IV. *Australian Region.*—Twenty-eight families represented, of which eight are peculiar—*i.e.*, seven out of the eight families of Marsupials, and the two families of Monotremes. All of these, however, are absent from Polynesia and New Zealand.

Peculiar Genera (besides those in the above families).—**Babirusa** (a hog), and **Anoa** (a small kind of cow) in Celebes. Five p.g. of *Muridæ* in Australia, and one of these in Tasmania also. Three p.g. of bats.

Absent Forms.—Australia and New Guinea possess no non-aquatic Mammals higher than Marsupials, except some bats, mice, and rats. This points to extremely long-continued isolation, which has afforded time for the Marsupials to become modified in the most diverse directions, thus enabling them to fill places elsewhere occupied by other orders. New Zealand is remarkable for the absence of all indigenous Mammals, so far as certainly known, with the exception of two bats.

V. *Neotropical Region.*—Thirty-seven families represented, of which seven are peculiar—*i.e.*, *Bradypodidæ* (sloths), *Dasypodidæ* (armadilloes), and *Myrmecophagidæ* (true ant-eaters). *Chinchillidæ* (chinchillas) and *Caviidæ* (cavies) among Rodents. *Cebidæ* (New World monkeys); *Hapalidæ* (marmozets). The *Phyllostomidæ* (leaf-nosed bats) are peculiar, with the exception of a Californian species.

Peculiar Genera (besides those in the above families).—**Chironectes** and **Hyracodon** (opossums). **Dicoctyles** (peccary, also in Texas); **Auchenia** (llama); **Elasmognathus** (a tapir). **Inia** (a dolphin, upper part of Amazon basin). Six p.g. of *Muridæ*; six p.g. of *Octodontidæ* (rat-like forms), two of them peculiar to W.

Indies; eight p.g. of *Echimyidæ* (spiny rats); two p.g. of *Cercolabidæ* (tree porcupines). **Solenodon** (a hedgehog-like form from Cuba and Hayti). Twenty-six p.g. of bats, including the Vampires. Five p.g. of *Canidæ;* three p.g. of *Mustelidæ;* **Nasua** (coati) and **Cercoleptes** (Kinkajou); **Tremarctos** (spectacled bear); **Otaria** (an eared seal).

Peculiar Species.—Among these are twenty species of **Didelphys** (to which genus most opossums belong). The American tapir. A species of racoon (genus **Procyon**).

Absent Forms.—*Ungulata* are scarce, deer and llamas being the only ruminants, tapirs and peccaries the only non-ruminants. The only insectivores are **Solenodon** and a species of shrew (**Sorex**). The *Viverridæ* are absent.

VI. *Nearctic Region.*—Thirty-two families are represented, of which one is peculiar, while one other is almost so—*i.e.*, *Haploödontidæ* (rat-like forms allied to beavers and marmots), and *Saccomyidæ* (the pouched rats, of which one genus ranges into the N. of the Neotropical region).

Peculiar Genera (besides those in the above families).—**Antilocapra** (prong-horned antelope), **Aplocerus** (a goat-like antelope), and **Ovibos** (the musk-sheep). Three p.g. of *Muridæ;* **Jaculus** (a jerboa), **Cynomys** (the so-called prairie dog), and **Erethizon** (the tree porcupine). Three p.g. of moles; two p.g. of bats; two p.g. of *Mustelidæ;* **Eumetopias** (an eared seal); **Halicyon** (a seal).

Peculiar Species.—Among these are two of **Didelphys**, a peccary, several deer, the American bison, racoons, and the grizzly bear (**Ursus ferox**).

Absent Forms.—Ungulates are ill represented, deer, the American bison, two antelopes, a sheep, and the musk-sheep being the only ruminants, while a peccary (Texas to Red River) is the only non-ruminant. Hedgehogs, *Viverridæ*, and monkeys are all unrepresented.

Distribution of Orders.

1. *Monotremata.*—Consists of only three genera, limited to part of the Australian region. **Ornithorhynchus**, Australia and Tasmania; **Echidna**, Australia, Tasmania, and N. New Guinea; **Proechidna**, S. New Guinea.

No fossil forms are found elsewhere, so that the place of origin of this order is unknown.

2. *Marsupialia.*—Consists of eight families (comprising thirty-seven genera), of which only one, the Opossum family (including three genera), occurs outside the Australian region. The Opossums are Neotropical and Nearctic.

Fossil opossums occur in the Pleistocene of America, and in much older European deposits (Eocene to Miocene). The secondary rocks of Europe contain a number of small forms, which probably resemble the ancestors of the Australian Marsupials. We may therefore suppose that this order originated in the Palæarctic region, and then extended into what is now Australia (at that time united by land with Asia), isolation occurring soon after, followed by specialization in various directions. The Opossums seem first to have existed in Europe, from whence they spread into America by former northerly land-connections.

3. *Edentata.*—This order is now chiefly limited to S. America, but **Orycteropus** is peculiar to the Ethiopian region, while **Manis** is found both in that and the Oriental region.

The geological evidence is in favour of considerable development in Africa, whence the order would spread north to the Oriental and Palæarctic regions, and thence on to America. The competition with higher forms has caused its extinction in most areas, and Edentates appear to be most abundant in S. America, because the competition with other animals is there comparatively small. The peculiar burrowing or climbing habits of most of the genera also tend to preserve them, and these habits no doubt represent attempts to escape from the severe competition with higher forms. The size of existing Edentates is insignificant compared with that of Pleistocene S. American and European genera.

4. *Ungulata.*—(*a*) Artiodactyla. *Non-ruminantia.* Swine are only represented in America by peccaries (**Dicotyles**); true swine are found in all the other regions, but only extend into the Australian as far as New Guinea. These animals are first known in the European Eocene, and during Miocene and Pliocene times were as common in N. America as Europe, but since then have almost entirely disappeared from the former area.

The hippopotamus is now limited to the Ethiopian region, but fossil forms occur in Europe (Pliocene and Pleistocene) and India (Miocene).

CLASSIFICATION AND DISTRIBUTION OF ANIMALS. 355

Ruminantia.—Recent *Camelidæ* are only found in the Neotropical and Palæarctic regions, but numerous forms occur in the Miocene and later deposits of N. America, where the group originated.

Tragulidæ or mouse-deer have also a discontinuous area of distribution—W. Africa (**Hyomoschus**) and Oriental region (**Tragulus**). This is accounted for by the presence of Miocene forms in Europe, whence the family extended south.

Deer occur in all the regions except the Ethiopian, but do not extend far into the Australian region. They appear to have taken origin in the Old World, from whence they reached N. America in Miocene times, and afterwards passed to S. America.

Giraffes are at the present time confined to the Ethiopian region, but fossil forms are known from S. Europe and India, and a northern temperate origin is probable.

Bovidæ (oxen, sheep, antelopes, &c.) are present in all the regions except the Neotropical, though they only just pass into the Australian, and are scarce in the Nearctic. The family appears to have originated in the Palæarctic and Oriental regions during Miocene times.

(*b*) Perissodactyla. Tapirs present a striking example of discontinuous distribution, being found, on the one hand, in the Malay peninsula, Sumatra, and Borneo, and on the other, in S. and Central America. True tapirs occur in the W. of Europe as far back as Miocene times, but in America are not found further back than the Pleistocene. Migration from the Palæarctic region is thus indicated.

Rhinoceroses are now only Ethiopian and Oriental, but they appear to have originated in the Palæarctic region, where they extend back to the Miocene period. In Pliocene times they also ranged into N. America.

The genus **Equus** (horse, ass, zebra) is now limited to the Ethiopian and Palæarctic regions. It appears to have originated in the latter area during Miocene times, and then migrated not only into the Ethiopian, but also into the Oriental, Nearctic, and Neotropical regions, as proved by fossil forms.

5. *Sirenia.*—**Manatus** (manatee), E. coast of S. America, and W. coast of Africa. **Halicore** (dugong), shores of Indian Ocean and Red Sea.

6. *Rodentia.*—Very widely distributed; occurring in all the

regions, but in Madagascar and Australia only represented by *Muridæ*. They attain their largest development in S. America.

The order is of great geological antiquity, for some living genera extend back to the Eocene. Rodents probably originated in the Palæarctic region, whence migrations took place at an early date to S. America and S. Africa, allowing time for great specialization.

7. *Proboscidea.*—Elephants are now limited to the Ethiopian and Oriental regions, but formerly had a much wider extension.

Palæarctic forms occur from Miocene to Pleistocene times, and elephants have lived in India since the Miocene period. Numerous fossil examples occur in the Pliocene and Pleistocene deposits of N. and S. America.

8. *Hyracoidea.*—Almost entirely limited to the Ethiopian region, but range northwards as far as Syria.

9. *Insectivora.*—Very widely distributed, and represented by numerous specialized forms. Absent from S. America and Australia.

This order is a very ancient one, as shown by the fact that Miocene forms mostly belong to existing families. Extinction appears to be slowly taking place, and has led to many cases of discontinuous distribution—*e.g.*, *Centetidæ*, represented by **Solenodon** in the W. Indies, and **Centetes**, with four other genera, in Madagascar.

10. *Cheiroptera.*—Bats, as might be expected, are found in all the great areas, but the Frugivora are absent from the Nearctic and Neotropical regions, as are the Horse-shoe bats (*Rhinolophidæ*) among insectivorous forms. On the other hand, the Leaf-nosed bats (*Phyllostomidæ*) are almost exclusively Neotropical.

Fossil bats, very like recent species, date from Eocene times, and the order is undoubtedly one of extreme antiquity.

11. *Carnivora.*—(*a*) Fissipedia. Occur in all the regions, except, perhaps, the Australian (the "native dog" of Australia is only doubtfully indigenous), but are especially characteristic of the Ethiopian and Oriental, which possess almost all the *Viverridæ* and *Hyænidæ*, with a great many of the *Felidæ* and *Mustelidæ*. Two genera, **Cryptoprocta** and **Proteles**, constituting distinct families, are limited respectively to Madagascar and S. Africa. Bears, however, are absent from the Ethiopian region, and are only represented by one species in the Neotropical region, which

is also very poor in other Carnivora. The *Procyonidæ* are small bear-like forms, found in the Nearctic and Neotropical regions, and include the racoons (**Procyon**), coatis (**Nasua**), and kinkajous (**Cercoleptes**).

Fossil Carnivores date back to the Lower Eocene, but the recent families were not then differentiated. The order appears to have originated in the northern half of the Old World.

(*b*) Pinnipedia.—Seals are limited to cold and temperate seas, and are also found in the Caspian, Sea of Aral, and Lake Baikal, all of which, at no very distant epoch, were connected with the Arctic Ocean.

Walruses characterize the North Polar regions.

12. *Prosimiæ.*—(*a*) Cheiromyini. Only one form, the aye-aye (**Cheiromys**), which is restricted to Madagascar.

(*b*) Lemurini.—Practically limited to the Ethiopian and Oriental regions. **Indris, Lemur,** and four other genera are only found in Madagascar. **Tarsius** constitutes a distinct family, and is limited to Sumatra, Banca, Borneo, and Celebes.

Lemurs date back to the Eocene in Europe.

13. *Primates.*—(*a*) Arctopitheci and (*b*) Platyrrhini are confined to the Neotropical region.

(*c*) Catarrhini are found only in the Old World.

C y n o m o r p h a are especially Ethiopian and Oriental, but also extend into the Palæarctic region, and into the Australian region as far as Timor.

A n t h r o p o m o r p h a present a marked example of discontinuous distribution. Gibbon—S. E. Asia and Malay Archipelago. Orang—Borneo and Sumatra. Chimpanzee and Gorilla—W. Africa.

The order dates back to the Eocene.

Origin and Migrations of the Mammalia.—The Class, and most likely all the Orders, originated in the Northern Hemisphere. Australia was isolated at a very early date, and therefore has preserved a very ancient Mammalian fauna. S. America and S. Africa were severed somewhat later, to be afterwards reunited, and they also have preserved some very ancient forms. The northerly connection between the Eastern and Western Hemispheres was then broken, not only by submergence of land, but also by a lowering of temperature. The Oriental and Ethiopian regions were also marked off by the formation of the Himalayas and the desert zone stretching from the Sahara to Central Asia.

INDEX-GLOSSARY TO PART II.

A

ABDOMEN, Astacus, 83, 84; Lepus, 269.

Abduction (*ab*, from; *duco, ductum*, draw, lead) of appendages—Movement away from the middle line, 91.

Abiogenesis (ἀ-, not; βίος, life, γένεσις, birth)—Derivation of living from non-living matter, 7.

Aboral (*ab*, away from; *os*, mouth)—Situated at the end further away from the mouth, 19.

Absorption, 5, 194.

Accretion (*ad*, to; *cresco, cretum*, grow)—Growth by addition of layers to the outside, 7.

Acetabulum (*acetabulum*, properly a vessel for vinegar; the socket of the hip bone)—The socket into which the head of the femur fits, 186, 239, 281, 327.

Achromatin (ἀ-, not; χρῶμα, colouring matter)—The part of the nucleus which does not readily take up stain, 8, 73.

Acinus (ἄκινος, *acinus*, a berry, a grape) of a gland—One of the ultimate (spherical or tubular) subdivisions, lined by secreting cells, 33, 288.

Acipenser, 153, 349.

Acrania, 135.

Acromion (ἀκρώμιον, used with the modern meaning) in Mammalia—A spine-like process of the scapula, 280.

Action, 102.

Adduction (*ad*, to; *duco, ductum*, draw) of appendages—Movement towards the middle line, 91.

Adrenal, 205, 250.

Aetiology (αἴτιον, a cause; λόγος) = Phylogeny, *q.v.*

Afferent (*ad*, to; *fero*, I carry)—Of nerve-fibres in conducting impulses to a ganglion cell or to the central nervous system, 102, 146, 217; of vessels carrying impure blood to gills, 164.

Africa, 349.

After-shaft, in the Fowl—A minute vane attached to the superior umbilicus of an ordinary feather, 231.

Air-sac, 249, 252.
Alimentary Canal. *See* Digestive Organs.
Allantois (ἄλλᾱς, ἀλλαντ-, a sausage), in the Bird and Mammal—An embryonic appendage which grows out from the posterior part of the intestine, 309.
Alps, 349.
Alternation of Generations, 39 ; Hydra, 27 ; Distoma, 35, 39.
Alveolus (*alveolus*, a little hollow)—A socket for a tooth, 284.
America—North A., 349, 350, 353 ; South A., 349, 350, 352.
Amia, 349.
Amnion (ἄμνιον, used with the modern meaning), in the Bird and Mammal —A sac-like appendage by which the embryo is surrounded, 265, 309.
Amniota (ἄμνιον), 342.
Amœba (ἀμοιβός, changing), 7, 11, 311.
Amœboid—Resembling an amoeba, especially as regards movement by slow protrusion of pseudopodia, 11, 60, 318.
Amphibia (ἄμφω, both ; βίος, life), 343 ; 3, 4, Chap. x., 174.
Amphicœlous (ἄμφω, both ; κοῖλος, hollow) of the centrum of a vertebra— Biconcave, 158, 184.
Amphioxus (ἀμφί, on both sides ; ὀξύς, sharp), 341, 134-152.
Ampulla (*ampulla*, a flask, a bottle), in the Vertebrate ear—A dilatation at one end of each semicircular canal, 173, 219, 256.
Anabolism (ἀναβολή, a heaping up, an ascent) = Assimilation, *q.v.*
Analogy—Physiological equivalence, 311.
Anamniota (ἀν-, negative ; ἄμνιον), 343.
Anapophysis; in a Vertebra—A process situated below the postzygapophysis, 279.
Angle ; of jaw, 276.
Annelida (*annellus*, a little ring)—Segmented worms, 340 ; Chap. v., 65.
Annulus, 75.
Animals—Distinction from plants, 5.
Ankle. *See* Tarsus.
Antebrachium (*ante*, before ; brachium)—The fore-arm. *See* Arm.
Antenna (*antenna*, a yard-arm), in Crustacea—One of the second pair of slender sensory appendages of the head, 89, 103.
Antennule (dim. of *antenna*, a yard-arm), in Crustacea—One of the first pair of slender sensory appendages of the head, 89, 103.
Anti-trochanter (ἀντί, against ; trochanter), in the Bird—A facet above the acetabulum, against which the great trochanter plays, 239.
Anodonta (ἀν-, negative ; ὀδούς, ὀδοντ-, a tooth)—Fresh-water Mussel, 341 ; 109-122.
Anthropidæ, 346.

Anura, 4.

Anus (*anus*, in modern sense)—The opening by which the undigested products are ejected from the food-canal, 14, 29, 48, 86, 94, 113, 123, 269, 286, 313.

Aorta, 128, 164, 197, 227, 246, 265, 308; Anterior, 115; Cardiac, 142; Dorsal, 142; Posterior, 115.

Apertures, pores, &c.—Abdominal p., 154; Auditory a., 89, 269; Auriculo-ventricular, 162, 196, 245; Cloacal, 48, 111, 154, 175, 191, 230, 242; Eustachian, 241; Excretory, 28, 31, 48, 55, 61, 76, 80, 89, 112; Exhalent, 111; Genital, 28, 32, 62, 63, 76, 80, 87, 99, 112; Renal, 89, 112; Urinogenital, 269, 298, 299.

Apodeme (ἀπόδημος, absent from home), in the Crayfish—One of the elements of the endophragmal system, 90.

Aponeurosis (ἀπονεύρωσις, used with the modern meaning)—The connective tissue sheath of a muscle, 208.

Appendages, 83, 84, 87, 89, 90, 108.

Appendicular, of skeleton—Belonging to limbs, 178, 239, 280, 326.

Appendix, Vermiform, 286, 330.

Apteria (ἀ-, negative; πτερόν, a feather), in the Bird—Featherless patches, 229.

Aqueduct, Sylvian, 169.

Aqueductus vestibuli, 173.

Arachnoid (ἀράχνη, a spider's web; ἰδ-εῖδος, ἰδ-, appearance, 210.

Archoplasm (ἀρχή, a beginning; πλάσμα, anything formed)—The protoplasm composing the centrosoma, 73.

Archenteron (ἀρχή, the beginning; ἔντερον, an intestine)—The digestive cavity of the gastrula, 68, 106, 148, 224, 260, 307.

Arch—Gill a., 141; Hæmal a., 158; Neural a., 158, 184.

Arches—Arterial or Aortic, 197, 246, 265, 290, 308; Carotid, Pulmocutaneous, Systemic, 197.

Arches, Visceral, in Chordata—Thickenings in the wall of the pharynx between which are the visceral clefts, 141, 157, 225, 227, 262, 264, 308; Mandibular a.: The first bar of the visceral skeleton, 157, 225, 227, 264, 331; Hyoid a.: The second bar of the visceral skeleton, 157, 225, 227, 236, 264, 276, 331; Branchial a.: The third and succeeding bars of the visceral skeleton, 157, 225, 227, 236, 276, 308, 331.

Area—a. opaca, in the Bird and Mammal—The part of the blastoderm outside the area pellucida, 257, 260; a. pellucida (embryonic area), in the Bird and Mammal—That part of the blastoderm from which the body of the embryo is mainly formed, 257, 260, 307.

Areas of distribution, 349.

Arm, 175, 230, 269.

INDEX-GLOSSARY. 361

Artery—A blood-vessel carrying blood away from the heart—Abdominal, 96; Allantoic, 267, 310; Antennary, 96; Aorta, 164, 197, 227, 246, 265, 308; Brachial, 246, 290; Branchial, 164, 227; Carotid, 164, 197, 246, 265, 290, 331; Caudal, 164, 246; Cœliac, 164, 197, 246, 290; Cœliaco-mesenteric, 197; Coronary, 290; Cutaneous, 198; Dorsal, 142; Femoral, 246, 290; Hepatic, 96, 290; Hyoidean, 165; Iliac, 197, 246, 290; Ilio-lumbar, 290; Innominate, 246, 290, 331; Intercostal, 290; Lateral, 142; Lieno-gastric, 164, 290; Lingual, 197; Mesenteric, 164, 197, 246, 290; Ophthalmic, 196; Parietal, 164; Pectoral, 246; Pulmonary, 198, 246, 265, 290, 308, 331; Renal, 164, 246, 290, 297; Sacral, 296; Sciatic, 198, 246; Sternal, 96; Subclavian, 164, 197, 246, 265, 290, 331; Thoracic, 96; Vertebral, 246, 290; Vitelline, 267.

Arthrobranchia (ἄρθρον, a joint; branchia), in the Crayfish)—One of the eleven brush-like gills attached on either side to the membranous junctions of certain appendages and the body, 97.

Arthropoda (ἄρθρον, a joint; πούς, ποδ-, a foot), 340; Chap. vi., 83.

Artiodactyla (ἄρτιος, even; δακτυλός, toe), 346, 355.

Ascaris, 340; 47–53, 74.

Ascension Is., 348.

Ascidia, 135, 341.

Asia, 349, 350, 357, 352.

Assimilation (assimilo, I make like to)—The building up of food into protoplasm, 9.

Astacus (ἀστακός, a species of crayfish), (crayfish), 340; 83–109, 135, 138.

Asterias, 73, 74.

Atlas—The ring-like first vertebra, 184, 238, 278.

Atrial cavity, in Amphioxus—A chamber surrounding the pharynx through which water passes out, 136, 134, 152.

Atriopore, 136, 139, 142.

Atrium, genital (atrium, an entrance hall), 42, 131.

Auditory Organs, 369; Astacus, 103; Anodonta, Unio, 121; Helix, 134; Scyllium, 172; Rana, 218; Columba, 255; Lepus, 304; Comparison of, 321.

Auricle (auricula, dim. of auris, the ear)—In a two or more chambered heart, the thin walled receptive part, 114, 162, 195, 245, 289.

Australian region, 350, 352.

Aves, 344; Chap. xi., 229.

Axial, of skeleton—Belonging to the head and trunk, 156, 178.

Axis, in Bird and Mammal—The second vertebra, 238, 279; Basi-cranial a., 271; Facial a., 323.

Axis-cylinder, of nerve-fibre, 215.

Azores, 348.

B

Backbone. *See* Vertebral Column.
Bacteria, 58.
Balanoglossus, 341.
Bali, 350.
Banca, 352.
Barb, in the *feather*—One of the lateral appendages of the shaft, 231.
Barbule, in the *feather*—One of the lateral appendages of a barb, 231.
Basipterygium (*basis*, the base ; πτερύγιον, a fin), 160.
Bats, 346, 356.
Bell-animalcule = Vorticella, *q.v.*
Bile—The secretion of the liver, 160, 161, 191, 194.
Bile-duct, 94, 126, 160, 191.
Bilharzia, 30.
Biogenesis (βίος, life ; γένεσις, birth)—The production of life from life, 7.
Bivalve, 79.
Black Sea, 349.
Bladder, Gall b., 160, 191 ; Urinary, 99, 205, 267, 295, 316, 333.
Bladder-worm, 44.
Blastocœle (βλαστός, a germ ; κοῖλος, hollow)—The cavity of the blastosphere (segmentation cavity), 27, 67, 147, 224, 257.
Blastoderm (βλαστός ; δέρμα, the skin), in Meroblastic Ova—The cellular patch resulting from segmentation, 105.
Blastomeres (βλαστός ; μέρος, a part)—Cells resulting from the cleavage of the oösperm, 27.
Blastopore (βλαστός ; πόρος, a passage)—The orifice by which the archenteron of a gastrula communicates with the exterior, 68, 70, 106, 148, 224, 260.
Blastosphere (βλαστός ; σφαῖρα, a sphere)—A hollow blastula, 27.
Blastula (dim. of βλαστός, a germ)—The embryo at the conclusion of cleavage, 27, 67, 105, 147, 224, 306.
Blood, 95.
Blood-corpuscles, 59, 78, 142, 166, 195, 245, 289.
Blood-system, Lumbricus. *See* Circulatory System.
Body (of vertebra). *See* Centrum.
Body-cavity—A space or system of spaces between the viscera and the body wall, 31, 49 ; Lumbricus, 56, 71 ; Hirudo, 79. *See also* Cœlom.
Body-wall—Hydra, 20 ; Distoma, 28 ; Ascaris, 48.
Bojanus, organs of, in Mollusca—The kidneys, 118.
Bone, 312. I. Cartilage b.—One replacing pre-existing cartilage, 179, 188, 232, 271 ; Acetabular (cotyloid), 327 ; Alisphenoid, 272, 323, 325 ;

Articular, 236; Astragalus, 187, 282, 328; Basioccipital, 232, 272; Basisphenoid, 233, 272, 323; Calcaneum, 187, 282, 328; Carpale, 186, 281, 327; Centrale, 186, 187, 281, 282, 327, 328; Coccyx, 326; Columella, 183, 236, 255, 264; Coracoid, 185, 239, 280; Cotyloid, 241, 327; Cuboid, 282, 328; Cuneiform, 281, 327; Epicoracoid, 185; Epiotic, 232; Ethmoid, 323; Ethmoturbinal, 275, 323; Exoccipital, 179, 180, 232, 272; Femur, 187, 241, 327; Fibula, 187, 241, 282, 327; Fibulare, 282, 328; Humerus, 185, 239, 280, 326; Ilium, 187, 239, 281, 327; Incus, 273, 311; Innominate, 239, 281, 327; Intermedium, 186, 187, 281, 282; Ischium, 187, 239, 281, 327; Lunar, 281; Magnum, 281, 327; Malleus, 273; Maxilla (man), 325; Maxillo-turbinal, 275; Mento-meckelian, 183; Mesethmoid, 233, 272; Metacarpal, 186, 239, 281, 327; Metatarsal, 188, 241, 282, 328; Nasal, 182; Naso-turbinal, 275; Navicular, 282, 328; Occipital, 323; Omosternum, 184; Opisthotic, 232; Orbiculare, 274; Orbitosphenoid, 233, 272, 323; Palatal, 325; Palatine (rabbit), 275; Periotic, 273, 323; Phalanx, 188, 239, 241, 281, 282, 327, 329; Ploughshare b., 237; Precoracoid, 280; Presphenoid, 233, 272, 323; Pro-otic, 180, 232; Pterygoid (rabbit), 275; Pubis, 187, 239, 281, 327; Quadrate, 234, 264, 311; Radiale, 186, 239, 281; Radio-ulna, 185; Radius, 186, 239, 281, 326; Sacrum, 237, 326; Scaphoid, 281, 327; Scapula, 185, 239, 280, 326; Stapes, 274, 276, 308; Sternum, 185, 238, 280, 326; Sphenethmoid, 179; Sphenoid, 323; Supraoccipital, 232, 272; Tarsale, 187, 282, 328; Tarso-metatarsal, 230, 241; Temporal, 323; Tibia, 187, 284, 327; Tibiale, 187; Tibio-fibula, 187: Tibio-tarsus, 241; Trapezium, 281, 327; Trapezoid, 281, 327; Turbinal, 234; Turbinate, 323; Tympano-hyal, 276; Ulna, 186, 237, 281, 326; Ulnare, 186, 239, 281; Uncinate, 281, 327. II.—Membrane bone—One replacing pre-existing connective tissue, 179, 188, 232, 271; Angular, 237; Angulo-splenial, 183; Clavicle, 185, 239, 326; Dentary, 236; Frontal, 232, 272, 323; Furcula, 239; Interclavicle, 239; Jugal (malar), 234, 325; Lachrymal, 233, 275, 325; Maxilla, 182, 189, 234, 275; Maxillary, 325; Nasal, 182, 234, 275, 324; Palatine, 182, 236; Parasphenoid, 180, 233, 236; Parietal, 232, 272, 323; Parieto-frontal, 179; Premaxilla, 182, 189, 234, 275; Pterygoid, 182, 236, 264; Quadrato-jugal, 182, 234; Splenial, 236; Squamosal, 183, 232, 272, 323; Tympanic, 273, 323; Vomer, 182, 234, 273, 324. III.—Sesamoid b.—One developed in a tendon, 232; Fabellæ, 282; Patella, 282, 327; Pisiform, 281.
Borneo, 349, 350.
Bothriocephalus, 46.
Botryoidal (βότρυς, a bunch of grapes; β-ιδος, β-, appearance), 79.
Bovidæ, 355.

364 INDEX-GLOSSARY.

Bowman's Capsule—The dilated commencement of a uriniferous tubule, into which a glomerulus projects, 205, 251, 295.
Brachium (*brachium*, the arm)—The upper arm. *See* Arm.
Brain—Scyllium, 168: Rana, 210; Columba, 252; Lepus, 301; Homo, 333, 339; Gorilla, 339.
Braincase = Cranium, *q.v*,
Branchiæ (βράγχια, *branchiæ*, the gills of a fish). *See* Gills.
Branchio-cardiac-groove, 86.
Branchiostegite, in the Crayfish—The fused thoracic pleura, which form the outer wall of the branchial chamber, 86.
British Isles, 348.
Bronchus—One of the main subdivisions of the trachea, 248, 293.
Brow-spot, in the Frog, 2.
Buccal cavity = Mouth cavity, *q.v.*
Buccal mass, in the Snail—The anterior part of the alimentary canal, which contains the mouth-cavity and odontophore, 125.
Buccal pouch, 57.
Buddha, 113.
Budding, 39.
Bulb (Medulla Oblongata)—The axial part of the hind brain, 169, 211, 253, 302.
Bulla (*bulla*, anything swelling up and so becoming round)—The dilated part of the tympanic bone, in which the tympanic cavity is contained, 273.
Bursa; B. Entiana, 160; B. Fabricii, 242.
Butterfly, 4.
Byssus, in some Lamellibranchs—Threads developed by a gland near the foot, and subserving attachment to foreign bodies, 121.

C

Cæcum (*cæcus*, blind)—A pouch of the intestine which ends blindly, 94, 242, 286, 330.
Calamoichthys, 349.
Calcar (*calcar*, a spur), 176, 188.
Canal—Branchio-cardiac, 96; Central c., 253; Cœlomic, 143; Neural, 168, 210, 252, 300; Neurenteric c., 151, 224; Pigmented c., 144; Pore c., 90; Semicircular c., 173, 219, 255, 304; Urinogenital, 298, 299; Vertebrarterial, 238, 278.
Canaliculi (*canaliculus*, dim. of *canalis*, a channel)—Minute tubules in bone, traversed by processes of the bone-cells, 188.
Capillaries, 60, 79, 96, 142, 166, 200, 245, 292.

Capitulum (dim. of *caput*, the head)—The head of a rib, 237, 279.
Capsule—Auditory, 157, 180, 232, 273; Nasal c., 234; Olfactory c., 157; 179, 182, 275, 323.
Carapace, 83.
Carbon dioxide, 10. *See* Respiration.
Cardia (καρδία, used with the modern meaning), 286.
Cardiac (καρδιακός, belonging to the cardia)—Belonging to or near the heart. In the stomach : the part nearest the opening of the œsophagus, 92.
Carnivora, 346, 356.
Carpus—The wrist, 186, 239, 281, 327.
Cartilage—A connective-tissue substance consisting of cells embedded in a clear matrix, 179.
Cartilages—Arytenoid, 204, 293; Basi-hyal, 157, 158, 236; Basi-branchial, 236; Cerato-hyal, 157; Cricoid, 204, 293; Epicoracoid, 185; Episternum, 184; Extra-branchial, 158; Glosso-hyal, 236; Hyomandibular, 157; Intercalary, 158; Labial, 158; Meckel's c., 183, 236, 264; Pessulus, 248; Quadrate c., 182; Septum nasi, 275, 234, 323; Suprascapular, 185, 280; Thyroid, 273; Tympanic, 183; Xiphi-sternum, 185.
Caspian, 349.
Cat, 349.
Caucasus, 349.
Cauda equina—The filum terminale together with the last few nerve-roots, 213.
Caudal—Belonging to the tail, 136.
Caval system—Rana, 198; Columba, 246; Lepus, 291; Man, 331.
Cavity—Abdominal c., 161; Atrial c., 136, 139, 152; Buccal or mouth c., 29, 57, 76, 136, 139, 151, 160, 189, 204, 241, 284, 329; Glenoid c., 185, 239, 280, 326; Pericardial, 195; Peritoneal, 162; Pleuro-peritoneal, 191; Sigmoid c., 281; Tympanic c., 189, 219, 232, 255, 273, 304.
Celebes, 348.
Cell—The morphological and physiological unit—Auditory, 221, 256; Bone c., 188; Calcareous, 127; Chief c., 288; Chloragogen c., 57; Cystogenous, 38; Ferment c., 127; Flame c., 31; Ganglion c , 205, 215; Germinal, 20, 33, 36, 62; Goblet c., 55, 193, 288; Gustatory, 254, 304; Hair c., 172, 173, 304; Hepatic, 288; Inner, 306; Liver c., 127; Lower layer c., 257; Muscle c., 81, 209; Nerve c., 35, 145, 172, 205; Olfactory, 255, 304; Outer c., 305; Ovoid c., 305; Pigment c. (Chromatophore), 178; Polar c., 26, 73, 143, 207; Sense c., 216, 320; Sensory, 138; Sperm-mother c., 33; Taste c., 254, 304; Thread c. (Nematocyst), 24, 312; Yolk c., 226.
Centetidæ, 349.
Centrolecithal, of ovum—Containing a central mass of yolk, 105.

Centrosoma—A specialized part of the cell protoplasm which probably determines the division of the nucleus, 73.
Centrum—The body of a vertebra, 158, 184.
Cephalochorda, 341, 135.
Cephalothorax, 83, 86.
Ceratodus, 153.
Cercaria, in the Fluke—The free tadpole-like stage which, after losing its tail, becomes the adult, 36, 38.
Cercaria, 338.
Cere (*cera*, wax), in the Pigeon—A bare patch near each external naris, 229.
Cerebellum—A dorsal outgrowth of the hind-brain, 169, 211, 253, 254, 302, 337, 339.
Cerebral hemispheres—Lateral hollow outgrowths of fore-brain, 169, 210, 216, 253, 254, 301, 333.
Cerebro-spinal axis—Scyllium, 168; Rana, 209; Columba, 252; Lepus, 300; Homo, 333.
Cervical groove, 86.
Chætopoda (χαίτη, hair ; πούς, ποδός, a foot), 340.
Chalaza (χάλαζα, hail), in the Bird's egg—A cord-like structure traversing the white at either end, 256.
Chamber—Branchial, 111 ; Cloacal, 112 ; Supra-branchial, 111, 117.
Cheiroptera, 346, 356.
Chelate (χηλή, a cloven hoof, a claw)—Provided with pincers (chelæ), 87.
Chiasma, optic (χίασμα, the mark of the letter χ)—The X-shaped structure formed by the crossing of the optic nerves on the ventral surface of the thalamencephalon, 169, 211, 253, 302.
Chick, 256-268.
Chimpanzee, 338.
China, 349.
Chiriqui, 349.
Chloragogen, 57, 61.
Chlorophyll (χλωρός, green ; φύλλον, a leaf)—The characteristic green colouring matter of plants, 14, 22, 44.
Chondrocranium (χόνδρος, a cartilage; cranium)—The cartilaginous groundwork of the skull ; the primordial cranium, 179, 232.
Chordinæ tendinæ—Firm bands of connective tissue uniting the flaps of an auriculo-ventricular valve with the wall of the ventricle, 245, 288.
Chordata (χορδή, a string), 341 ; 3 ; Chap. vii., 134, *et seq.*
Chorion, 338.
Choroid (χόριον, skin, membrane ; είδος, ίδ-, appearance), 174, 221 ; c. plexus, 210.
Chromatin (χρωμάτινος, coloured)—The more deeply staining part of a nucleus, 8, 73.

Chromatophore (χρῶμα, χρωματ-, colouring matter; φορός, bearing; φέρω, I bear)—A pigment-bearing cell, 178.
Cilium (cilium, an eyelash)—A short vibratile thread of protoplasm on the free surface of a cell, 13, 61, 312.
Circulatory system—Lumbricus, 59; Hirudo, 78; Astacus, 97; Anodonta, Unio, 114; Helix, 127; Amphioxus, 142; Scyllium, 161; Rana, 194; 202; Tadpole, 227; Columba, 245; Chick, 267; Lepus, 289; Homo, 331; Comparison of, 314.
Cirrus, in Distoma and Tænia—The penis, 32, 42; buccal c., 136.
Cirrus-sac, 33.
Clasper, 154, 167.
Classification of Animals, 1, 3; Chap. xv., 339.
Claw, 176, 230, 241.
Cleavage (Segmentation)—The early stages of cell division occurring in a developing oösperm, 26, 67, 105, 147, 222, 256, 306.
Clefts, visceral, in Chordata—Lateral apertures by which the pharynx communicates with the exterior, 3, 157, 225, 226, 262, 308.
Clepsidrina, 18.
Climate, 348.
Clitellum (clitellæ, a pack-saddle), 55, 64, 76.
Clitoris, 299.
Cloaca (cloaca, a sewer)—The chamber into which the intestine, genital, and, usually, urinary ducts open, 48, 49, 160, 191, 242.
Cnemial crest, 290.
Cnidoblast, 22.
Cnidocil, 22.
Coagulation, 165.
Coats, of Alimentary canal, 192, 244, 288; of Artery, 202.
Coccyx, 326.
Cockroach, 18.
Cocoon, 82.
Cochlea (cochlea, a snail-shell)—A posterior outgrowth from the membranous labyrinth, coiled in higher forms, 173, 219, 256, 304.
Cod, 153.
Cœlenterata, 340; Chap. ii., 19.
Cœlom (κοῖλος, hollow)—A body cavity which typically (1) does not contain blood, (2) communicates at some period with the exterior by excretory tubes, (3) has parts of its lining thickened into gonads, (4) arises early in development, 56, 60, 79, 95, 143, 150, 191, 248, 263.
Cœlomata, 340.
Cœnurus, 46.
Collozoum, 339.

Colon, 286.
Colpidium, 72.
Columba, 344; 229-268.
Columbia, 349.
Columella, 123, 219.
Columnæ carneæ, in the heart—Internal projections of the ventricular wall, 245.
Commissure (*commissura*, a joining together)—(1) In Invertebrates: a nerve-band uniting transversely two ganglia of the same name, 119; (2) in the Vertebrate brain: bands of connecting fibres; Anterior c., 253, 301, 337; Posterior c., 210, 252, 301; Middle c., 301.
Comparative Animal Morphology and Physiology, Chap. xiii., 310.
Condyle—A rounded projection, 276; Mandibular, 183; Occipital c., 179, 232, 238, 272, 322.
Cones—Crystal c., 104; of Retina, 174, 221.
Conjugation, in Protozoa—The temporary or permanent fusion of two individuals, 15, 18, 72, 317.
Conjunctiva (*conjungo, conjunct-*, join together), in Vertebrates—A transparent membrane, covering the front of the eyeball, and connecting the eyelids, 222, 256, 305.
Connective—A nerve cord joining ganglia of different name to each other or to the main cord, 66, 81, 119.
Continental Islands, 348.
Contractility, 318; Amœba, 10; Vorticella, 16; Hydra, 25.
Conus Arteriosus, in many Fishes—A contractile tubular part of the heart succeeding the ventricle, 162.
Copulatory Organ, 86, 154.
Coral Is., 348.
Cornea (*corneus*, horny), in the eye—The transparent area forming the front part of the sclerotic, 103, 134, 174, 221.
Cornua—(1) in Brain, of lateral ventricle, 301; (2) of hyoid apparatus, *see* Hyoid apparatus.
Corpora quadrigemina, in Mammalia—The optic lobes, each being divided into natis and testis, 302.
Corpus albicans = c. mammillare, *q.v.*
Corpora cavernosa, in Mammalia—The two firm cylindrical structures supporting the clitoris and penis, 298, 299.
Corpus adiposum, 205.
Corpus callosum, in Mammalia—A band of nerve-fibres uniting the cerebral hemispheres, 301, 337.
Corpus mammillare, in the Rabbit—A small rounded projection situated on the base of the brain, just behind the infundibulum, 301.

Corpus restiforme, in the Rabbit—A projection situated on either side of the roof of the fourth ventricle, 302.

Corpus spongiosum, in Mammalia—A mass of vascular tissue making up the dorsal part of the penis, and traversed by the urinogenital canal, 298.

Corpus striatum, in the fore-brain—A mass of grey matter projecting into the front part of either lateral ventricle, 210, 253, 301.

Corpus trapezoideum, in the Rabbit—A rectangular area situated ventrally and anteriorly on either side of the medulla oblongata, 302, 338.

Corpuscles—Blood c., 59, 78, 161, 195, 245, 289, 293; Central c. (centrosoma), 73; Colourless c., 161; Muscle c., 208, 291, 300; Red c., 161; Touch c., 178, 218, 254, 271, 303.

Cortex, of kidney, 295; of brain, 333.

Corti, Organ of, 304.

Coverts—Wing c., 229; Tail c., 229.

Cranio-facial angle, 272, 339.

Craniota, 135.

Cranium (κρανίον, the skull)—The brain-case, 179, 232, 271, 323.

Cranium—Primordial. *See* Chondrocranium.

Crayfish = Astacus, *q.v.*

Crista acustica—A projection in the ampulla of a semicircular canal, upon which is a patch of auditory epithelium, 221.

Croll, 349.

Crop—Part of the mid-gut serving for the storage of food, 57, 58, 242.

Cross, 4.

Crura cerebri—Two longitudinal masses of nerve-fibres on the floor of the mid-brain, 211, 253, 302.

Crus (*crus*, a leg), in Vertebrates—The part of the leg between thigh and ankle. *See* Hind limb.

Crustacea, 86.

Crystal cone, 104.

Crystalline style, 114.

Ctenidium (dim. of κτείς, κτενίς, a comb)—The gill of Molluscs, typically plume-like, 112.

Cuba, 349.

Cuticle (*cuticula*, dim. from *cutis*, skin)—A structureless membrane covering the epidermis, 13, 28, 29, 48, 55, 76, 312.

Cyclas, 121.

Cyclostomata, 153.

Cyst—A protective case in which some animals are enclosed during a dormant stage, 9, 14.

Cysticercus, 44, 46.

D

Danger signal, 268.
Danube, 349.
Dart-sac, 131.
Darwin, 58.
Death, 10, 11.
De Candolle, 4.
Decidua, 338.
Deer, 355.
Delamination—Formation of epiblast and hypoblast, by division of the segmented ovum parallel to the surface.
Dentine—The part of a tooth or scale developed from the dermis, 155.
Dermal—Belonging to the dermis, or, in general, to the skin, 34.
Dermis—The deeper part of the skin derived from mesoderm, 76, 90, 138, 178, 227, 231.
Deutomerite, 17, 18.
Development—Direct d., 107; Indirect d. = Metamorphosis, *q.v.*; of nervous system, 150; of feather, 231; of hair, 270; Hydra, 26; Distoma, 35; Tænia, 43; Ascaris, 53; Lumbricus, 67; Hirudo, 82; Astacus, 107; Anodonta, Unio, 121; Amphioxus, 147; Scyllium, 167; Rana, 222; Columba, 256; Lepus, 305; Homo, 338; Comparison of, 317.
Diaphragm, (1) in the Frog—The muscular anterior boundary of the body-cavity, 191; (2) in Mammalia—A muscular and tendinous partition separating the thoracic and abdominal cavities, 294.
Diastema (διάστημα, an interval)—A gap between the incisor and premolar teeth, 284, 329, 338.
Diastole, in the heart—The relaxation which follows each contraction, 97.
Didelphia (δι-, two; δελφύς, the womb), 346.
Differentiation—Morphological d., 20.
Digestion. *See* Nutrition.
Digestive System—Vorticella, 13; Hydra, 27; Distoma, 29, 36 (embryo), 38 (cercaria); Tænia, 41; Ascaris, 48; Lumbricus, 55; Hirudo, 76; Astacus, 192; Anodonta, Unio, 113; Helix, 125; Amphioxus, 139; Scyllium, 160; Rana, 189; Columba, 241; Rana, 282; Homo, 329; Comparison of, 313.
Digit, 176, 186, 239, 241, 281, 282, 327, 329.
Dimorphism (δίς, doubly; μορφή, a shape)—The occurrence of bi-sexual and uni-sexual forms in the life cycle of one animal, 73.
Dinornis, 347.
Diploblastica (διπλοῦς, double; βλαστικός, growing), 340.
Dipnoi, 153, 350.

Disc, of Vorticella, 13.
Discontinuity, 349.
Discus proligerus, in the Rabbit—The projection into the cavity of a Graafian follicle, formed by the ovum and several surrounding layers of cells, 299.
Dissepiment. *See* Septum.
Distal—Furthest removed from the main body, 11.
Distoma (the Liverfluke), 340; 28-39.
Distribution, 1, 5, 346; of Mammals, 351; of Orders, 353.
Division of labour, 20.
Dog, Skull of, 278.
Dogfish = Scyllium, *q.v.*
Dorsal (*dorsum*, the back)—Upper, 28, 322.
Duct—Archinephric, 228, 265; Bile (hepato-pancreatic), 160, 191, 244, 287; Cystic, 191, 287; Hepatic, 191, 287; Lachrymal, 275; Mesonephric, 167, 265, 333; Metanephric, 166, 265; Müllerian, 167, 228, 333; Pancreatic d., 161, 191, 244, 287; Pronephric, 228; Segmental, 228, 265; Stenson's d., 287; Sub-lingual, 287; Thoracic d., 248, 292; Urinary, 205, 206; Urinogenital, 166, 205, 206; Wharton's d., 287; Wolffian d., 166, 265, 333.
Ductus—D. Botalli, 268; in embryos: D. arteriosus—A cross branch from the pulmonary artery to the dorsal aorta, 267; D. venosus—A trunk traversing the liver, formed by the union of a vein from the gut with the vitelline and allantoic veins, 267.
Duodenum—The first part of the small intestine, into which the liver and pancreas open, 191, 242, 286.
Dura mater—A firm fibrous membrane lining the neural canal in Vertebrates, 168, 210.

E

Ear, 172, 218, 255, 304.
Earthworm = Lumbricus, *q.v.*
Ecdysis (ἔκδυσις, the act of putting off, or unclothing)—The complete casting-off of an exoskeleton at one time, 166.
Echidna, 345, 353.
Echinococcus, 45.
Ectoderm (ἐκτός, without; δέρμα, the skin)—The outer cell-layer of the body, 20, 27, 36, 70, 106, 148, 226, 263, 308.
Ectoparasite (ἐκτός, parasite)—A parasite living on the exterior of its host, 75.
Ectosarc (ἐκτός; σάρξ, σαρκός, flesh), in Protozoa—The firmer external contractile layer of the cell, 13, 16.

Edges—Pre-axial, 225; Post-axial, 225.
Edentata, 346, 349, 354.
Eel, 153.
Efferent (*efferens*, carrying out)—Of nerve fibres—carrying impulses away from a ganglion cell or the central nervous system, 102, 146, 217.
Egg—The fertilized ovum often with nutritive and protective layers, 36, 43, 52, 83, 100, 132, 167, 256, 305.
Egg-capsule, 64, 67.
Elasmobranchii, 153.
Elephant, 5, 356.
Emboly—Formation of the gastrula by inpushing of endoderm cells, 68.
Embryo (*see* Development), free e., 35, 36.
Embryology—The history of the early stages of development. *See* Development.
Embryonic appendages. *See* Amnion and Allantois.
Embryonic circulation, 267.
Emulsion, 58.
Enamel—The part of a tooth or scale developed from the epidermis, 155.
Encystment—The formation of a cyst, 14, 16, 18.
End-body, 216.
End-brush, 216.
Endoderm (ἔνδον, within; δέρμα, the skin)—The inner cell-layer of the body, 22, 27, 36, 71, 107, 148, 226.
Endolymph (ἔνδον; lymph)—The fluid contained in the membranous labyrinth, 219, 256, 304.
Endoparasite (ἔνδον; parasite)—An internal parasite, 17, 28, 39, 47.
Endophragmal system, in the Crayfish—A series of hard parts, which incompletely roof over the sternal sinus in the thorax, 90.
Endoplasm (ἔνδον; πλάσμα, anything formed), in Amœba—The internal granular part of the body, 8.
Endopodite (ἔνδον; ποῦς, ποδός, a foot), in the Crayfish appendages—The internal part of the forked end, 84.
Endosarc (ἔνδον; σάρξ, σαρκός, flesh), in Protozoa—The granular, more fluid, internal part of the protoplasm, 13, 16.
Endoskeleton—An internal skeleton. *See* Skeleton.
Endothelium—Simple squamous epithelium, lining the heart, vessel, and lymphatic sinuses, 202.
End-organs, 103, 174.
End-plate, 92, 209, 216.
Endostyle—A medium ventral thickening in the wall of the pharynx of lower chordates, 139.
Ends—Anterior and posterior, 28, 311; Oral and aboral, 19.

Energy, 9.
Enterocœle (ἔντερον, an intestine; κοῖλος, hollow)—A body cavity (cœlom) formed by the outgrowth of pouches from an archenteron, 150.
Environment, 347.
Epiblast (ἐπί, upon; βλαστός, a growth)—The ectoderm of a developing embryo, 27.
Epiboly (ἐπί, upon; βολή, a throwing, casting)—Formation of a gastrula by overgrowth of ectoderm cells, 148.
Epidermis (ἐπί, upon; dermis)—Simple columnar or stratified squamous epithelium covering the external surface. *See* Skin.
Epididymis (ἐπί, upon; δίδυμοι, the testicles)—A convoluted region of the spermiduct, 80, 298, 334.
Epiglottis (ἐπί, upon; glottis), 285.
Epimerite, 17, 18.
Epimeron (ἐπί, upon; μέρος, a part), in the Crayfish—The part of a segment situated between the pleuron and the attachment of the appendage, 84.
Epiphragm (ἐπί, upon; φράγμα, a fence, a partition), 123.
Epiphysis (ἐπί, upon; φύσις, growth, a growing), in bones—A separately ossified end, 185, 239, 280.
Epipodite (ἐπί, upon; ποῦς, ποδός, a foot), in the Crayfish appendages—A process which (usually) bears gill-filaments, 178, 185.
Episkeletal, of muscles—Superficial to the endoskeleton, 208.
Episternum, 185.
Epithelium—Cell-layers covering external and lining internal surfaces, 29, 177; Ciliated, 113, 117, 193; Germinal, 34, 202; Glandular, 94, 118, 244; Olfactory, 218; Simple columnar e., 31, 57, 121, 244; Stratified, squamous e., 244.
Ethiopian region, 350, 351.
Ethmo-palatine bar, 180, 182.
Europe, 349.
Eustachian tube, 189, 219, 233, 255, 274; e. valve, 245, 267, 290.
Eutheria (εὖ, well; θηρίον, a beast), 346.
Eversion (*everto*, I turn out)—The process of turning a part inside out, 19.
Evolution, 347.
Excretion—A waste product cast out by the body, 10, 14, 31.
Excretory System—Amœba, 10; Vorticella, 14; Distoma, 30, 31; Tænia, 41; Ascaris, 49; Lumbricus, 60; Hirudo, 80; Astacus, 98; Anodonta, Unio, 118; Helix, 129; Amphioxus, 143; Scyllium, 166; Rana, 205; Columba, 250; Chick, 265; Lepus, 295; Homo, 333; Comparison of, 316.
Exits, of nerves, 233, 274.

Exoplasm (ἔξω, without; πλάσμα, a thing formed), in Amœba—The outer clearer part of the body, 8.
Exopodite (ἔξω, without; τοῦς, τοδὶς, a foot), in the Crayfish appendages—The outer part of the forked end, 84.
Exoskeleton—An external skeleton, 84, 109, 122.
Expiration—The outbreathing of air, 129, 204, 249, 295, 316.
External characters—Amœba, 7; Vorticella, 11; Hydra, 19; Distoma, 28; Tænia, 40; Ascaris, 47; Lumbricus, 54; Hirudo, 75; Astacus, 83; Anodonta, Unio, 109; Helix, 122; Amphioxus, 136; Scyllium, 153; Rana, 175; Columba, 229; Lepus, 268.
Extrinsic, of muscle of any part—Originating outside that part, 91, 208.
Eye—Compound e., 103; Scyllium, 173; Rana, 221; Columba, 221; Lepus, 304. *See also* Visual Organs.
Eyelashes, 268.
Eyelids, 155, 174, 222, 256, 268, 305.
Eye-spot, 36.

F

Facet—Capitular, 237, 279; Tubercular, 237, 279.
Falciform (*falx*, a sickle; *forma*, shape), sickle-shaped, 18.
Fallopian tube, 299, 334.
Fang, 285.
Fat-body (corpus adiposum), 205.
Feathers, 229, 231.
Feeler, 103.
Felis, 349.
Femur—(1) The thigh, or (2) the thigh bone, 176, 230, 241, 282.
Fenestra (*fenestra*, a window)—F. ovalis, 130, 219, 233, 255, 273; f. rotunda, 180, 233, 255, 273.
Ferment—Bodies which excite chemical change without themselves entering into the reaction, 58, 94.
Fertilization, 74; Hydra, 25, 26; Tænia, 43; Ascaris, 52; Lumbricus, 64; Astacus, 100; Anodonta, Unio, 119; Helix, 131; Amphioxus, 147; Scyllium, 167; Rana, 207; Columba, 252; Lepus, 299.
Fibrilla—One of the fine threads into which muscle fibres may be split, 91.
Filoplume—A rudimentary feather, 229, 231.
Filum terminale—The filament in which the spinal cord ends, 211.
Fins, 136, 154.
Fin-ray, 138, 159.
Fish—Chap. ix.
Fission (*findo*, *fissum*, split)—Reproduction by splitting into two or more equal parts, 10, 15.

Fissure—Central or Rolandian f., 336; Dorsal f., 145, 169, 211, 253, 302; Obturator f., 241; Sphenoidal f., 274; Ventral f., 169, 211, 253, 302; Sylvian f., 336.

Flagellum (*flagellum*, a whip, a lash)—(1) In Protozoa, an elongated motile thread of protoplasm, 22; in Snail, 131.

Flexure, 262; Cephalic f., 89.

Flocculus—A lateral projection of the cerebellum, 253, 302, 338.

Fluke = Distoma, *q.v.*

Fold—Lateral f., 261; Head f., 261; Neural (medullary f.), 224, 261; Tail f., 261.

Follicle, Ovarian (Graafian)—An ovum-containing capsule, 207, 251, 299.

Fontanelles, in the endoskeleton—Spaces filled with membrane, 179, 180; Coracoid f., 185.

Food, 5, 6, 9. *See* Nutrition.

Foot—Hydra, 19; Mollusca, 111, 123; Vertebrata, = pes, *q.v.*

Foot-jaws = Maxillipedes, 87.

Foramen—Condylar, 234, 274; Intervertebral f., 184, 237, 254, 278; f. lacerum anterius (Sphenoidal fissure), 274; f. lacerum medium, 274; f. lacerum posterius, 274; f. magnum, 157, 179, 180, 232; f. of Munro, 210, 253, 301; Obturator f., 281; Olfactory f., 179, 233, 274; Optic, 179, 180, 233, 274; f. ovale, 267; Stylomastoid f., 274; Trigeminal f., 180; f. of urostyle, 184; Vagus f., 180.

Forceps, 87.

Fore-arm (antebrachium), 175, 230, 269.

Fore-brain—The part of the brain developed from the first cerebral vesicle, 168, 210, 252, 301.

Fore-gut (Stomodæum)—The anterior part of the alimentary canal, which arises as an endodermic pit, 48, 57, 77, 92.

Fore-limb, 175, 230, 239, 280, 326.

Formula—Dental, 285; Rabbit, 285; Dog, 285; Man, 329.

Fornix, in the Mammalian brain—A transverse band of fibres connecting the hemispheres and optic thalami, 301.

Fossa, of brain, 273, 274; glenoid f., 276; f. ovalis, 245, 267.

Fowl, 229.

Fresh-water Polype = Hydra, *q.v.*

Frog = Rana, *q.v.*

Funnel—Atrio-cœlomic f., 144; Excretory f., 61, 144; Seminal f., 62.

G

Galeopithecus, 349.

Gall-bladder—A swelling on the bile-duct acting as a store for bile, 160, 244, 287.

Gallus bankiva = Fowl, 229.
Ganglion (γάγγλιον, a small tumour)—An aggregation of nerve-cells, 35, 66, 81; Abdominal, 101, 102; Anal, 52; Buccal, 133; Cerebral, 35, 36, 66, 81, 100, 132; Cerebro-pleural, 119; Cervical, 303; Cœliac, 303; Gasserian, 212, 215, 253; Infra-œsophageal, 81; Lateral, 52; Mesenteric, 303; Optic, 103; Parieto-splanchnic, 119; Pedal, 119, 133; Pleuro-visceral, 133; Post-œsophageal, 101; Semilunar, 303; Spinal, 215; Sympathetic, 171; Thoracic, 101; Vagus, 213, 303; Ventral, 35, 52; Visceral, 119.
Ganoidei, 153, 349.
Gastric juice, 194.
Gastric mill, 92.
Gastrolith (γαστὴρ, the stomach; λίθος, a stone), 93.
Gastrula (dim. from *gaster*, γαστὴρ)—An embryonic stage, the typical form of which is a two-layered sac with digestive cavity opening to the exterior, 38, 68, 105, 148, 260.
Gastrulation—The formation of the gastrula, 68, 148, 224.
Gemmation (*gemma*, a bud)—Asexual reproduction by budding, 24.
Generative or Genital Organs. *See* Reproductive Organs.
Genus, 3, 4.
Geology, 347.
Germinal band, 70.
Germinal disc, in the Crayfish—A thickening of the blastoderm, which indicates the ventral surface of the embryo, 105; in the Bird's ovum— The lenticular mass which contains most of the protoplasm, 252, 256.
Germinal layers, in the Embryo—Cell-aggregates, which precede the tissues in development (*see* Layers, Germinal, fate of).
Germinal spot—The nucleolus of the nucleus (germinal vesicle) of an ovum, 23, 34, 73.
Germinal vesicle—The nucleus of an ovum, 23, 34.
Giant-fibres, 66.
Gibbon, 338.
Gill—A respiratory organ adapted for breathing the oxygen dissolved in water, 97, 115, 117, 166, 225, 315.
Gill-chamber, 86, 117.
Gill-filament, 117.
Gill slit, 141, 151, 155, 166.
Giraffe, 355.
Girdle, in the endoskeleton of limbs—The part between the body and free limb. Shoulder (pectoral),159, 185, 239, 280, 326; Hip (pelvic), 160, 186, 239, 281, 327.
Gizzard—A muscular part of the mid-gut, 57, 58, 242.

Glacial epoch, 348.
Gland—An organ essentially formed by one or more epithelial cells, which elaborate a secretion or excretion, 33 ; Accessory g., 32 ; Albumen g., 131 ; Byssus g., 122; Calciferous, 57, 58 ; Capsulogenous, 55; Cement g., 90 ; Clasper g., 155 ; Cowper's g., 298, 299 ; Digestive, 94, 127 ; Ductless, 165, 201 ; Gastric, 193 ; Green g., 98, 316 ; Harderian, 222, 256, 305 ; Hermaphrodite g., 129 ; Intermaxillary, 189 ; Infra-orbital, 287 ; Lachrymal, 256, 305 ; g. of Lieberkühn, 193, 244, 288 ; Lymphatic, 248, 292 ; Mesenteric, 292 ; Milk (mammary) g., 269, 271 ; Mucus, 76, 131, 177 ; Œsophageal, 57, 193; Oil g., 230, 231 ; Oviducal, 167 ; Parotid, 287 ; Peptic, 193, 242, 244, 288 ; Perineal, 269, 271 ; Pineal, 2, 168, 210, 216, 253, 301 ; Prostate, 130, 298 ; Rectal, 161, 287 ; Salivary, 78, 125, 127, 287, 288 ; Sebaceous, 270 ; Serous, 177 ; Shell g., 33, 43 ; Sub-lingual, 287 ; Sub-maxillary, 287 ; Suprapedal, 133 ; Thymus, 201, 293 ; Thyroid, 201, 293 ; Yolk g., 33, 43.
Glans penis, 298 ; g. clitoridis, 299.
Glochidium (dim. from γλῶχες, the prickles on ears of corn)—The free-swimming embryo of Anodonta or Unio, 122.
Glomerulus (dim. of *glomus*, -*eris*, a skein), in the kidney—A tuft of capillaries projecting into Bowman's capsule, 206, 295.
Glottis (γλωττίς, used with modern sense), in air-breathing Chordata—The chink-like opening into the respiratory organs, 189, 204, 242, 285.
Glucose—Grape-sugar, 58.
Glycogen, 194.
Goblet-cell, 55, 66.
Gonad (γονή, production)—A reproductive gland, typically formed by a thickening of the epithelium (and underlying tissue) lining the body-cavity. *See* Spermary, ovary.
Gorilla, 338.
Grape-sugar = Glucose, *q.v.*
Great toe = Hallux, *q.v.*
Green gland, 98.
Gregarina, 17-19, 64, 311, 339.
Gregarinida, 339.
Grey matter, in the brain and spinal cord—The part mainly composed of ganglion-cells, 172, 215.
Groove—Epibranchial g., 139 ; Neural (medullary) g., 224, 261.
Guanin, 32, 41, 99, 118.
Gubernaculum (*gubernaculum*, a rudder), in the Rabbit—A fibrous cord connecting the spermaries with the scrotal sac, 298.
Gullet—The part of the alimentary canal which next succeeds the pharynx ; the œsophagus, 29, 49, 57, 92, 242, 284.

Gustatory Organs, 369; Anodonta, Unio, 121; Amphioxus, 146; Rana, 218; Columba, 255; Lepus, 304; comparison of, 320.

Gut—The alimentary canal, composed of fore-gut, mid-gut, and hind-gut. See Digestive System—In Human Anatomy: the small intestine.

H

Hæmal arch, 158.
Hæmal spine, 158.
Hæmocyanin (αἷμα, blood; κυάνιος, blue), 95, 98.
Hæmoglobin (αἷμα)—The red colouring-matter of blood, 60, 78, 95, 166.
Hag, 153.
Hair, 270.
Hair-follicle—The capsule in which the base of a hair is ensheathed, 270.
Hair-germ—The first rudiment of a hair in the embryo, consisting of a thickening of the Malpighian layer of the epidermis, 270.
Hair-papilla—The projection at the bottom of a hair-follicle, upon which the hair is moulded, 270.
Half-facets, Capitular, 279.
Hallux—The great toe, 176, 230, 329.
Hand = Manus, *q.v.*
Hare-lip, 329.
Haversian canals—The cavities in bone tissue in which the small blood-vessels run, 188.
Haversian system—An Haversian canal with concentric lamellæ of bone surrounding it, 291.
Hayti, 349.
Head—Astacus, 83, 87; Helix, 123; of Ribs = Capitulum, *q.v.*; Scyllium, 154; Rana, 175; Columba, 229; Lepus, 268.
Head kidney, in Vertebrata = Pronephros, *q.v.*
Heart—Astacus, 96; Anodonta, Unio, 115; Helix, 127; Scyllium, 161; Rana, 195; Columba, 245; Chick, 265; Lepus, 289.
Hearts, of Lumbricus, 60.
Heel, 282, 328.
Hemichorda, 135, 341.
Hepatic—Belonging to the liver.
Herring, 153, 161.
Heterocercal, of tail-fin—Asymmetrical both externally and internally, 154.
Heterogamy (ἕτερος, another; γάμος, a marriage)—An alternation of normal sexual reproduction with parthenogenesis, 39.
Helix (Snail), 4, 122-134, 341.

Hepato-pancreas, 94.
Hepatic portal system—Rana, 143, 165, 199, 268.
Hermaphrodite—Bi-sexual, 23, 55, 80, 129.
Hilus, in Mammalia—The point at which the ureter leaves the kidney, 295.
Hind-brain—The part of the brain developed from the third cerebral vesicle, 169, 211, 253, 302.
Hind-gut (Proctodæum)—The posterior part of the alimentary canal which arises as an ectodermic pit, 48, 57, 78, 94.
Hind-limb, 176, 230, 241, 269, 281, 327.
Hinge-joints, 91, 326.
Hippocampus major, 301.
Hippopotamus, 5, 354.
Hippuric acid, 297.
Hirudo (Leech), 75-82, 135, 340.
Hirudinea, 340.
Histology (ἱστός, a thing woven; λόγος, speech, discourse) — The morphology of minute structure, 2.
Homarus = Lobster, *q.v.*
Homo, Chap. xiv., 321.
Homology—Morphological equivalence, 27, 90, 311.
Holoblastic (ὅλος, whole; βλαστός, a germ) of cleavage—Affecting the whole oösperm, 26, 44, 67.
Host—An organism preyed on by a parasite, 34.
Humming bird, 349.
Humour—Aqueous, 174, 222; Vitreous, 174, 222.
Hyæna, 356.
Hybernaculum (*hibernaculum*, a winter abode), 123.
Hybernation (*hiberno*, I winter) — A state of torpidity during winter; Helix, 123.
Hybrid, 4.
Hydra (ὕδρα, a water-serpent), 19-27, 340.
Hydroid—Resembling Hydra, 27.
Hydrozoa, 27.
Hylobates, 338.
Hyoid—H. apparatus: (1) Strictly speaking, structures derived from the hyoid arch or in connection with such ; (2) The above + elements from the branchial arches, 183, 236, 276. *See* Arches.
Hyomandibular, 157.
Hyostylic, of the Skull—With the lower jaw attached with the aid of hyoid elements, 158.
Hypapophysis (ὑπό, under; ἀπόφυσις, a process)—A downward process from the centrum of a vertebra, 238.

Hypoblast (ὑπό; βλαστός, a growth)—The endoderm of a developing embryo, 27.
Hyposkeletal, of muscles—Taking origin within the endoskeleton, 208.
Hypostome (ὑπό; στόμα, the mouth), in Hydra—The conical projection at the end of which the mouth is placed, 19.
Hyracoidea, 346.

I

Ice age, 348.
Ichthyopsida (ἰχθύς, a fish; ὄψις, an appearance), 342.
Ileum (ἴλλω, I twist; ἰλεός, εἰλεός, colic), in the small intestine—The part succeeding the duodenum, 286.
Impregnation. *See* Fertilization.
Incubation, 257.
Infundibulum (*infundibulum*, a funnel)—(1) A ventral projection of the thalamencephalon, to the end of which the pituitary body is united, 168, 210, 253, 301; (2) in Mammalia—One of the vesicles in which the bronchial tubes end, 294.
Infusoria, 339.
Ingestion (*ingero, ingestum*, carry in)—The taking-in of solid food-particles, 9.
Insectivora, 346, 349.
Insertion, of muscle—The end attached to a relatively movable part, 91.
Inspiration—The in-breathing of air, 129, 204, 249, 294, 316.
Intelligence, 333.
Intercellular—Situated between cells, 61.
Interstitial—In the interstices between cells, 21.
Intestine—The part of the alimentary canal in which digestion is completed, 29, 49, 57, 78, 94, 113, 125, 141, 160, 190, 242, 286.
Intracellular—Within the cell, 24, 31, 61, 80, 209.
Intrinsic, of muscle of any part—With origin and insertion within that part, 91, 208.
Intussusception (*intus*, inside; *suscipio, susceptum*, take up)—The intercalation of new particles between pre-existing ones, 6.
Invagination—The process by which a part originally external becomes internal, 38, 148, 224.
Invertebrata, 135.
Iris, 174, 221.
Irritability, 319—Amœba, 11; Vorticella, 16; Hydra, 25; 319.
Islands, 348.
Island of Reil, 338.
Iter—The ventricle of the mid-brain; the Sylvian aqueduct, 169.

J

Japan, 348.
Jaws, 125; Upper j., 182, 236, 275, 325; Lower j. = mandible, *q.v.*
Jejunum, in human anatomy—The part of the small intestine which immediately succeeds the duodenum, and is followed by the ileum, 286.
Jelly-fish, 27.

K

Karyokinesis (κάρνον, a kernel; κίνησις, movement)—A method of cell-division in which the nucleus undergoes complex changes, 72.
Katabolism (καταβολή, a casting down)—The breaking down of complex substances into simple ones, 9, 315.
Keber, organs of, 118.
Keel, 239.
Kidney, 98, 118, 129, 166, 228, 250, 265, 295, 316, 333.
Knee, 176, 230, 269.

L

Labial palps, 112.
Labrum (*labrum*, a lip), in Astacus—The upper lip, 89.
Labyrinth, Membranous, in Vertebrates—The sac which forms the essential part of the auditory organ, 172, 219, 255, 273; Bony l.—The compact layer of bone which, in higher vertebrates, surrounds the membranous l., 255.
Lacteal—A lymph-vessel belonging to the gut, 248.
Lacuna (*lacuna*, a cavity)—An interstice between the constituents of a tissue, 188.
Lamella (dim. of *lamina*, a plate), 117.
Lamellibranchiata (lamella; βράγχια, branchiæ, gills of fish), 341.
Lamina inferior, 328; l. perpendicularis, 275, 323; l. terminalis, 225.
Lamprey, 153.
Lancelet = Amphioxus, *q.v.*
Larva—A free-living embryo differing from the adult in form, 107, 122, 151, 222.
Larynx (λάρυγξ, used with the modern meaning), in Vertebrate—The modified top of the trachea, usually serving as an organ of voice, 204, 248, 293.
Lateral line, 155.
Laurer's canal, in the Fluke—A tube by which the oviduct opens on the dorsal surface, 28, 33.

Leech = Hirudo, q.v.
Lemurs, 349, 357.
Lens, in the Eye—A firm refracting structure, 104, 134, 174, 222, 256, 305.
Leopard, 349.
Lepidosiren, 153.
Lepidosteus, 349.
Lepus (Rabbit), 135, 268-310, 347 ; L. variabilis, 349.
Life History, 7. *See* Development.
Ligament, 109, 178.
Light. *See* Eye.
Limb-girdle. *See* Girdle.
Limbs—Rana, 185 ; Columba, 239 ; Lepus, 280.
Lime, 90, 113, 121, 188.
Limnæa (λιμναῖος, living in a λίμνη or marsh), 36.
Line, pallial, 110.
Lion, 349.
Liver—A gland, usually digestive, derived from and opening into the midgut, 57, 94, 126, 141, 160, 191, 242, 287, 288.
Liver-rot, 28.
Lobes—Of Liver, 160, 191, 244, 287, 330 ; of cerebral hemispheres, 301, 337 ; Olfactory l., 168, 210, 252, 301 ; Optic, 169, 211, 253, 302.
Lobi inferiores, in Scyllium—Two oval swellings on the ventral surface of the thalamencephalon, 168.
Lobster, 107.
Lobules, 288.
Locomotion—Amœba, 10 ; Hydra, 25 ; Distoma, 34 ; Lumbricus, 65 ; Hirudo, 81 ; Astacus, 91 ; Anodonta, Unio, 119 ; Comparison of, 318.
Lombok, 350.
Lumbricus (Earthworm), 17, 54-71, 135, 340.
Lung—A respiratory organ adapted for breathing ordinary air, 129, 204, 248, 293, 315.
Lyell, 348.
Lymph, 165.
Lymphatics—A system of vessels by which (1) Part of the digested food, (2) Plasma and white corpuscles exuded in excess through the capillaries—are carried into the blood-system, 200, 248, 292.
Lymph-heart, 200, 204.
Lymph-sinus, 192, 200.
Lymph-space, 150, 168, 172, 210, 248, 292.
Lymph-system ; Amphioxus, 142 ; Scyllium, 165, Rana, 200 ; Columba, 248 ; Lepus, 292 ; Comparison of, 314.

M

Macronucleus (μακρὸς, large ; nucleus), 72.
Macrozooid (μακρὸς; ζῶον, a living creature), 15, 72.
Macula (*macula*, a spot), in the membranous labyrinth—One of several sensory patches situated in the utriculus, sacculus, and (when rudimentary) the cochlea, 221.
Madagascar, 348, 349.
Malacca, 349.
Malpighian Body, in the Kidney—A Bowman's capsule with its glomerulus, 206, 295.
Malpighian layer, 177.
Mammalia, 345; Chap. xii., 268, *et seq.*; Distribution of, 357; Origin of, 357. The lower jaw, 183, 236, 276, 325.
Man = Homo, *q.v.*
Mandible—In Crustacea : the third pair of head appendages, which act as jaws, 89.
Mantle, in Mollusca—A flap-like outgrowth of the body-wall which usually shelters the gills, 111.
Manubrium—(1) A process of the sternum, 238, 280, 326; (2) a process of the malleus, 273.
Manus—The hand, 176, 186, 230, 239, 281.
Marmot, 349.
Marrow—The central soft tissue contained within bone, 188, 293.
Marsupialia, 346, 349, 352, 354.
Mastoid—A term applied to the spongy part of the periotic, 32, 273.
Matrix—Intercellular substance in which tissue elements are imbedded, 126, 188.
Maxilla (*maxilla*, a jaw)—(1) In Crustacea: one of the second or third jaw-like appendages belonging to the head, 89; (2) In Vertebrata: one of the membrane bones of the upper jaw, 182, 234, 275, 325.
Maxillipede—A flattened jaw-like appendage belonging to the head, 89.
Meatus (*meatus*, a passage)—External auditory m., 230, 255.
Mediastinum—A median partition in the thorax, formed by the apposition of the parietal layers of the two pleuræ, 294.
Medulla—In nerve fibre: the fatty layer surrounding the axis cylinder, 81 ; m. of hair, 270 ; m. of kidney, 295.
Medulla oblongata—The axial part of the hind-brain ; the bulb, 169, 211, 216, 253, 302.
Medusa, 27.
Membrane—Arachnoid, 210 ; m. granulosa, 299 ; Mucous, 189, 192, 244,

289; Nictitating, 229; m. semilunaris, 248; Shell m., 256; Subzonal, 309; Tympanic, 219, 255, 273, 304; m. tympaniformis interna, 248; Vitelline, 73, 252, 299.

Mesenteron (μέσος, middle; ἔντερον, a bowel)—That part of the alimentary tube which is lined by epithelinm of hypoblastic origin; the mid-gut, 48.

Mesentery (μέσος; ἔντερον, a bowel)—(1) In Earthworm (see Septum; (2) in Vertebrates, a fold of peritoneum by which the alimentary canal is suspended, 161, 192.

Mesoblast (μέσος; βλαστός, properly anything grown; βλαστάνω, I grow) —The mesoderm of a developing embryo, 36.

Mesoblastic bands, 71.

Mesoblastic somites, 227, 263.

Mesoblasts—Cells in the embryo which gives rise to the mesoderm, 67.

Mesoderm (μέσος; δέρμα, the skin)—The middle germinal layer and parts derived from it, 36, 68, 71, 106, 107, 149, 224, 226, 259, 263, 306, 308.

Mesoglœa, 20.

Mesonephros (μέσος; νεφρός, a kidney), in Vertebrates—The second-formed excretory organ, which, as in the Frog, may persist throughout life, 166, 216, 265, 333, 334.

Mesopterygium, 159.

Mesorchium (μέσος; ὄρχις, a testicle)—A fold of peritoneum by which the testis is suspended, 206, 251, 298.

Mesosoma (μέσος; σῶμα, a body), in Molluscs—The main body-mass, 111.

Mesotarsal, applied to an ankle-joint situated in the middle of the tarsus, 241.

Mesovarium (μέσος; ovarius, adjective from ovum, an egg)—A fold of peritoneum by which the ovary is suspended, 207, 251.

Metabolism (μεταβολή, change)—The sum total of the chemical changes which constantly occur in protoplasm, 6.

Metacromion, 280.

Metadiscoidal, 338.

Metagenesis—Simple alternation of sexual and asexual stages, 39.

Metamere = Segment, q.v.

Metamorphosis—The series of changes by which a larval form becomes adult, 107, 122.

Metanephros (μετά, behind; νεφρός, a kidney)—The third-formed excretory organ—typically developed in the Pigeon and Rabbit, 265, 316, 333, 334.

Metapleural, 136.

Metapophysis (μετά, in addition; ἀπόφυσις, a projecting part), in Vertebræ— A process on the anterior end of the arch, above the præzygapophysis, 273.

Metapterygium (μετά, after; πτερύγιον, a fin), 159, 160.

Metastoma (μετὰ, behind; στόμα, a mouth), in the Crayfish—The lower lip, 89.
Metatarsus, 188.
Metatheria, 346.
Metazoa (μετὰ, after, later; ζῶα, living creatures)—Animals consisting of more than one cell, 339, Chap. ii. to end, 71.
Micronucleus, 72.
Micropyle (μικρὸς, little; πύλη, door), 119.
Microzooid, 15, 72.
Mid-brain—The part of the brain developed from the second cerebral vesicle, 169, 211, 253, 302.
Mid-gut (Mesenteron)—That part of the alimentary canal which is lined by epithelium of endodermic origin, 48, 57, 78, 94.
Mitosis (μίτος, a thread) = Karyokinesis, *q.v.*
Mississippi, 349.
Mollusca (*mollis*, soft), 340; Chap. vii., 109, 161.
Monocystis, 17, 18, 64.
Monodelphia, 346.
Monotremata, 345, 349, 352, 353.
Morphology (μορφή, a shape; λόγος, speech, discourse), 1, 2; Amœba, 7; Vorticella, 11; Hydra, 19; Distomum, 28; Tænia, 40; Ascaris, 47; Lumbricus, 54; Hirudo, 75; Astacus, 83; Anodonta, Unio, 109; Helix, 122; Amphioxus, 136; Scyllium, 153; Rana, 175; Columba, 229; Lepus, 268; Homo, 322.
Morula (dim. of *morum*, a mulberry)—A solid blastula, 38.
Mosaic vision, 104.
Motor, of Nerves—Supplying muscular tissue and conveying impulse leading to its contraction, 102, 146, 209, 216.
Moults, 106.
Mouth, 19, 29, 38, 47, 54, 57, 70, 76, 92, 113, 125, 157, 241, 268, 284, 313, 329.
Mouth-cavity = Cavity, buccal, *q.v.*
Mud-fish, 153.
Multipolar, 145, 214.
Muridæ, 349.
Muscle, 209; Involuntary m., 208; Striped or striated m., 91, 145, 202, 208, 209; Unstriped m., 132, 192: Voluntary m., 208; Abductor, 91; Adductor, 91, 109; Circular, 65, 81; of Eye, 174, 222, 256, 305; Extensor, 91; Flexor, 91; Gastric, 93; of Heart, 291; Longitudinal, 65, 81; Oblique, 81, 174, 222; Papillary, 245; Protractor, 29, 65, 119; Rectus, 174, 222; Retractor, 29, 81, 119; Spindle m., 123, 132.
Muscle-cell, 81.

Muscle-corpuscle, 91.
Muscle-fibre, 91, 192, 201, 209.
Muscularis mucosæ, in Vertebrates—A thin layer of unstriated muscle traversing the mucous membrane of the stomach and intestines, 192, 288.
Muscular system—Distoma, 34; Tænia, 43; Ascaris, 48; Lumbricus, 58, 64; Hirudo, 81; Astacus, 91; Anodonta, Unio, 119; Helix, 122; Amphioxus, 144; Scyllium, 168; Rana, 208; Columba, 252; Lepus, 299.
Musculature—The muscle system, 34.
Mussel (salt-water), 122.
Mussel (fresh-water), 100-122.
Mustelus, 167.
Myocœle (μῦς, a muscle; κοῖλος, hollow)—The cavities of the mesodermic somites, 149.
Myocœlomic, 149.
Myophan striations (μῦς; φαίνομαι, φαν-, appear), 13, 16.
Myomere = Segment, muscle, *q.v.*
Myotome = Segment, muscle, *q.v.*
Mysis, 109.

N

Nail, 176.
Nares (*nares*, the nostrils)—External (anterior), 175, 218, 229, 268, 275; Internal (posterior), 189, 218, 241, 285.
Nates, 302.
Nearctic, 350, 353.
Nebalia, 86.
Nemathelmia (νῆμα, a thread; ἕλμινς, intestinal worm)—Thread worms, 340; Chap. iv., 47.
Nematocyst (Thread-cell), 22, 24, 25, 312.
Neotropical, 350, 352.
Nephridiopore—The external opening of a nephridium, 61.
Nephridium (dim. of νεφρός, a kidney)—An excretory tube by which waste products are carried from the cœlom, or a section of it, to the exterior, 60, 70, 80, 118, 129, 316.
Nephrops, 109.
Nephrostome (νεφρός; στόμα, a mouth)—The internal opening of a nephridium, 61.
·Nerve—Nerve-exits, 233, 274; Cranio-spinal nerves, 319; Scyllium, 169; Rana, 211; Columba, 253; Lepus, 302; Abducent (vi.), 170, 212; Auditory (viii.), 133, 171, 213; Brachial, 213; Branchial, 171; Buccal,

133; Coccygeal, 184, 213; Cranial, 169, 211, 217; Facial (vii.), 171, 212; Glossopharyngeal (ix.), 171, 213; Hyoid, *see* Facial; Hyomandibular, *see* Facial; Hypoglossal (xii.), 254; Labial, 133; Lateral, 35; Lateral line n., 213; Mandibular, *see* Trigeminal and Facial; Maxillary, *see* Trigeminal; Oculomotor (iii.), 169, 212; Olfactory (i.), 169, 211; Ophthalmic, *see* Trigeminal and Facial; Optic (ii.), 103, 133, 169, 174, 211, 221; Palatine, *see* Facial; Pathetic (iv.), 170, 212; Phrenic, 303; Pneumogastric (x.) = Vagus, *q.v.*; Post-spiracular = Hyoid, *q.v.*; Sciatic, 213; Spinal, 171, 213, 217; Spinal accessory (xi.), 254; Splanchnic, 303; Tentacular, 133; Trigeminal (v.), 171, 212, 253; Trochlear (iv.) = Pathetic, *q.v.*; Unpaired n. (impar), 254; Vagus (x., Pneumogastric), 171, 213, 254; Visceral, 101, 103.

Nerve-cell, 35, 145, 172, 205.
Nerve-cord, 145; ventral n.-c., 66, 81.
Nerve-corpuscle, 101.
Nerve-fibre, 35, 92, 101, 145, 209, 215.
Nerve-ring, 35, 65, 81, 119.
Nervous system—Development of, 150, 320; Distoma, 34; Tænia, 43; Ascaris, 52; Lumbricus, 65; Hirudo, 81; Astacus, 100; Anodonta, Unio, 119; Helix, 132; Amphioxus, 145; Scyllium, 169; Rana, 209; Columba, 252; Lepus, 300; Homo, 333; Comparison of, 319.
Neural arch, 158.
Neural fold, 151, 224.
Neural groove, 151, 224.
Neural plate, 151, 224.
Neuroglia (νεῦρον, a nerve; γλία, glue) — The delicate connective-tissue framework of the nervous system, 66, 216.
New Zealand, 347, 349.
Nile, 349.
Nodes—Of Ranvier, 215; of Schmidt, 215.
Non-chordata, 135.
Norway Lobster (Nephrops), 109.
Nostrils (*see* Nares, external).
Notochord (νῶτον, the back; χορδή, a string), in the Chordata—A cellular rod which prefigures the vertebral column, 3, 135, 138, 150, 156, 158, 189, 226, 259, 263, 312.
Nucleolus (dim. of *nucleus*, a nut, a kernel)—A dense particle within a nucleus, 8.
Nucleus (*nucleus*, a nut, a kernel)—A more highly differentiated part of the protoplasm occurring in most cells, 8; Segmentation n.—The nucleus of the oösperm, 26, 74.
Nutrition—Amœba, 9; Vorticella, 14; Hydra, 23; Distoma, 31; Astacus,

83; Anodonta, Unio, 114; Helix, 127; Amphioxus, 141; Scyllium, 161; Rana, 194; Columba, 244; Lepus, 288; Comparative, 312.

O

Oceanic Is., 348.
Occipital (*occiput*, the back of the head)—Belonging to the hinder region of the skull, 179.
Odontophora, 341.
Odontophore (ὀδούς, ὀδοντ-, a tooth; φέρω, I bear; φορός, bearing), in the Snail—A rasping structure which projects into the floor of the mouth cavity, 125.
Odontophoral cartilages, 215.
Odours, 369.
Œsophagus. *See* Gullet.
Old Calabar, 349.
Olfactory Organs, 66; Astacus, 103; Anodonta, Unio, 121; Helix, 123, 133; Scyllium, 172; Rana, 218; Columba, 255; Lepus, 304; Comparison of, 320.
Ommatidium (dim. of ὀμμάτιον, a little eye)—One of the elements of a compound eye, 103, 104.
Omo-sternum (ὦμος, the shoulder; στέρνον, the breast), 184.
Ontogeny (ὄντα, beings; γεννάω, I produce)—The development of individuals, 2.
Operculum (*operculum*, a lid, a cover), 36, 225.
Oögenesis (ᾠόν, an egg; γίνεσις, birth)—The development of the ovum, 52.
Oösperm (ᾠόν; σπέρμα, seed)—The sexual cell produced by fertilization of the ovum, 25, 74.
Oötheca (ᾠόν; θήκη, a cover, a case)—A female organ for the storage of ova (receptaculum ovorum), 64.
Opossum, 354.
Optic thalami — The thickened side-walls of the thalamencephalon, 210, 252.
Orang-outan, 338.
Orbit, 182, 232, 271, 325.
Organ, 1.
Organic acids, 58.
Oriental Region, 350, 351.
Origin of species, 5; of muscle—The end attached to a relatively fixed part, 91.
Ornithodelphia, 345.
Ornithorhynchus, 345, 353.

Osphradium (ὀσφράδιον, anything sharp-smelling; ὀσφραίνομαι, I smell)—
The olfactory organ of Molluscs, probably serving to test the respiratory quality of water, 121.
Ossicle—Auditory o. = malleus, incus, stapes, and os orbiculare, 273.
Ossification, 188.
Osteoblast (ὀστέον, a bone; βλαστάνω, I grow; βλαστὸς, something that grows)—A bone-forming cell, which becomes modified into a bone-corpuscle, 188.
Ostium (ostium, a door, an entrance), in the Crayfish—One of six valvular slits in the wall of the heart, 97.
Otocyst (οὖς, ὠτὸς, the ear; κύστις, a bag, a bladder)—The auditory sac, 121, 134.
Otoliths (οὖς, ὠτὸς, the ear; λίθος, a stone)—Concretions (or foreign particles) found within an auditory sac or membranous labyrinth, 121, 134, 219.
Ovary (ovum, an egg)—The organ producing ova; Hydra, 23; Distoma, 33; Tænia, 42; Ascaris, 52; Lumbricus, 64; Hirudo, 80; Astacus, 100; Anodonta, Unio, 118; Amphioxus, 144; Scyllium, 167; Rana, 207; Columba, 251; Lepus, 298.
Oviduct—The duct conducting the ova to the exterior or into the urinogenital canal, 33, 64, 80, 100, 119, 131, 167, 207, 251, 299, 333.
Oviparous (ovum, an egg; pario, I produce)—Laying eggs, 168, 308.
Ovogenetic, 74.
Ovum—The female sexual cell, 23, 34, 72, 73, 80, 100, 131, 144, 167, 207, 251, 299.
Oxidation, 10. *See* Respiration.
Oxygen, 10. ,, ,,
Oxyuris, 53.

P

Pacific islands, 349.
Palæarctic, 350, 351.
Palate, 275, 284, 325; cleft p., 325.
Pallial—Belonging to the mantle, 110.
Palp, labial, 112, 113, 120.
Pancreas (πάγκρεας, the sweet bread)—A digestive gland opening into the intestine, and secreting a fluid which converts starch into sugar, proteid into peptone, &c., 160, 191, 244, 287, 288.
Pancreatic juice, 161, 194, 245, 289.
Panniculus carnosus, 270.
Papilla—A small projection; Head p., 28, 36; Circumvallate p., 284, 329; Filiform p., 189; p. Foliata, 284, 329; Fungiform p., 189; Genital p., 41, 207, 251; Tactile p., 53; Urinary p., 167.

Papillose—Covered with small papillæ, 286.

Paranucleus (micronucleus)—A small reproductive nuclear structure by the side of the main nucleus, 13,'16, 72.

Parasite—An organism living in, or upon, another organism and deriving nourishment from it, 17, 313; endoparasite, 17, 28, 39, 47, 313; intracellular p., 18.

Parenchyma (παρένχυμα, the spongy[substance of the lungs)—Tissue composed of cells which are fairly equal in their different dimensions, 41.

Parthenogenesis (παρθένος, a maiden; γεννάω, I produce)—Development of an ovum without fertilisation, 39, 74.

Pearls, 113.

Pearly layer, 110, 125.

Pecten (*pecten*, a comb), in the Pigeon—A plaited vascular fold projecting into the vitreous humour of the eye, 256.

Peduncle (*pedunculus*, a little stalk)—One of the bands of fibres uniting the cerebellum with the rest of the brain, 253, 302.

Pelvis = (1) The hip-girdle + the sacrum. See Hip-girdle. (2) The dilated end of the ureter, within the kidney, 295.

Penis—A copulatory organ, 32, 42, 80, 131.

Pentadactyle (πέντε, five; δάκτυλος, a finger)—With five digits, 176.

Pepsin—A proteid-digesting ferment found in the gastric juice, 194.

Peptone—A soluble form of proteid matter, 58.

Perch, 153.

Pericardial cavity, 195.

Pericardium (περί, around; καρδία, the heart)—The space surrounding the heart, 245, 289.

Perichondrium (περί; χόνδρος, cartilage)—The fibrous membrane which invests cartilage, 188.

Perilymph (περί; *lympha*, water)—A clear fluid surrounding the membranous labyrinth, 219, 255, 304.

Perineum, in the Rabbit—The area between the anus and urinogenital aperture, 269.

Periosteum (περί; ὀστέον, a bone)—The fibrous membrane investing bone, 188.

Periostracum (περί; ὄστρακον, a shell), in the Mussel and Snail—The pigmented horny membrane which covers the shell, 110, 125.

Perissodactyla (περισσός, odd; δάκτυλος, toe), 346, 355.

Peristaltic action—The wave-like contraction of muscle or of the muscular wall of a tube, 209.

Peristome (περί; στόμα, the mouth), 13, 123.

Peristomium (περί; στόμα, the mouth), 55, 75.

Peritoneum (περιτόνειον, used with modern meaning; περί, around; τείνω,

τεν-, τον-, stretch), in Vertebrates—The membrane lining the abdominal cavity, 161, 244, 293.
Perivisceral—Surrounding the viscera, 60.
Permanence, of areas, 347.
Pes, in Vertebrata—The foot, 176, 230, 241, 269, 282.
Pessulus (*pessulus*, a bolt), in the syrinx—A slender bar of cartilage running dorso-ventrally, 248.
Petrous part (of periotic), 273.
Peyer's patches, in the Rabbit—Lymphatic thickenings in the wall of the small intestine, 286.
Phalanx—A joint of a digit, 188, 239, 241, 281, 282, 327, 329.
Pharynx (φάρυγξ, used in modern sense)—The part of the alimentary canal next succeeding the mouth-cavity, 13, 28, 38, 57, 76, 139, 160, 189, 242, 285.
Phillipines, 349.
Phylogeny (φῦλον, a tribe; γιννάω, I produce)—The development of groups (phyla), 1.
Phylum (φῦλον)—One of the main groups of organisms, 1.
Physiology (φύσις, nature; λόγος, a discourse), 1, 2; Amœba, 9; Vorticella, 14; Hydra, 23; Astacus, 83; Anodonta, Unio, 109; Helix, 122; Amphioxus, 136; Scyllium, 153; Rana, 175; Columba, 229; Lepus, 268; Homo, Chap. xiv.
Pia mater—A delicate vascular membrane investing the cerebro-spinal axis, 168, 210.
Pigeon = Columba, *q.v.*
Pillar, of Fornix, 301.
Pineal body, 2, 168, 210, 216, 253, 301.
Pinna (*pinna*, a feather), in the Rabbit—The external flap of the ear, 319.
Pisces (*pisces*, fishes)—Fishes, 343; Chap. ix., 152.
Pituitary body, in the brain—A rounded non-nervous structure connected with the infundibulum, 169, 210, 253, 273, 301.
Placenta (*placenta*, a cake) in Mammals—A vascular structure, bringing the embryo into relation with the mother, 168, 310, 338.
Placoid (πλάξ, anything flat and broad; ἰδος, ἰδ-, appearance), of Scale—Developed both from epidermal and dermal elements, 155.
Plants, 5.
Plasma (πλάσμα, anything which has been formed)—The liquid part of the blood or lymph, 78; germ Plasma—The part of the nuclear protoplasm of a sexual cell, which enables it to develop into an embryo, 74.
Plate—Cribriform p., 275, 323; Neural (Medullary) p., 151, 224, 261; Orbital p. 233; Sclerotic p., 256.

Platyhelmia (πλατύς, flat; ἕλμινς, ἕλμινθος, a worm)—Flat worms, 340; Chap. iii., 28.

Pleura (πλευρά, a rib), in Vertebrates—The membrane lining the thorax, 293.

Pleuro-branchia (πλευρὸν, a rib; branchia), in the Crayfish—A gill attached to an epimeron, 97.

Pleuron (πλευρὸν), a rib, 84.

Pleuro-peritoneum, in the Frog—The membrane lining the body-cavity, 191.

Pleuro-peritoneal cavity, in the Frog—The body-cavity, 191.

Plexus—A network; Brachial p., 254, 303; Choroid p., 210, 301; Lumbar p., 254; Lumbo-sacral, 303; Nerve-p, 146, 171, 215, 320; Sciatic p., 213, 254.

Podobranchia (πούς, ποδός, a foot; branchia), in the Crayfish—One of the gills attached to the bases of the appendages, 97.

Polar-cell—The cell, or one of the cells cut off from the ovum before it develops into the embryo, 26, 73, 143, 207.

Pollex—The thumb, 176, 230, 281, 338.

Polynesia, 349.

Polyodon, 349.

Polype, fresh water = Hydra, *q.v.*

Polypterus, 349.

Pons Varolii—A band of fibres running transversely across the ventral surface of the medulla, at its front end, 302.

Pore—Excretory p., 31; Genital pore, *see* Aperture; Dorsal p., 65; Ventral p., 80; Abdominal, 155.

Pore-Canal, 90.

Portal System—A meshwork of capillaries interposed in the course of a vein, 143, 165, 199, 247, 291.

Position of body, 176, 230, 269, 322; Erect p., 338; p. of fore-limb, 281, 326; Primitive p., 175.

Post-axial—Lying posterior to the axis, 154, 176.

Postzygapophysis—An articular process projecting from the posterior end of a vertebra's neural arch, 184, 237.

Præzygapophysis—An articular process projecting from the anterior side of a vertebra's neural arch, 184, 237.

Pre-axial—Lying anterior to the axis, 154, 176.

Prehallux, 176.

Premolars, in the Rabbit—Grinding teeth which succeed the milk molars, 284.

Primates, 346, 357.

Primitive fibrillæ—Longitudinal elements into which a muscle-fibre or nerve-fibre may break up, 214, 300.

Primitive sheath—The delicate membranous investment of a nerve-fibre, 215.
Prismatic layer, 110, 125.
Proboscidea, 346, 356.
Process—Acromion p., 250; Articular, 184; Clinoid, 273; Coracoid, Coronoid, 183, 276; Costal, 238; Fronto-nasal, 329; Hamular, 275; Iliac, 160; Mastoid p., 273; Maxillary p., 329; Maxillo-palatine, 234; Odontoid, 238, 279; Olecranon p., 186, 239, 281, 326; Orbital p., 233, 272; Palatine, 275; Par-occipital p., 272; Pterygoid p., 273; Styloid p., 323; Supra-orbital p., 272; Transverse, 158, 184, 237, 278; Uncinate p., 237; Xiphoid, 238; Zygomatic, 273, 275.
Procœlous (πρό, before; κοῖλος, hollow), of the centra of vertebræ—Concave only in front, 184.
Pro-echidna, 345, 353.
Proctodæum (πρωκτός, the rectum, the anus)—The posterior part of the alimentary canal which arises as an epiblastic pit; the hind-gut, 48.
Proglottis, 40-44.
Promontory, 273.
Pronation, of the fore-limb—The position in which the palm is downwards and the radius crosses the ulna, so that its distal end is inwards, 281, 326.
Pronephros (πρό, before; νεφρός, a kidney), in Vertebrates—The first-formed excretory organ, which is usually very transitory; the head kidney, 166, 228, 265, 308.
Pronucleus—Male p., The name given to the nucleus of the sperm after it enters the ovum, 74; female p.—The nucleus of the ovum after the formation of polar cells, 26, 74.
Propterygium (πρό, before; πτερύγιον, a fin), 159.
Prosencephalon (πρός, in addition; ἐγκέφαλος, the brain)—An outgrowth or two lateral outgrowths of the fore-brain. *See* Fore-brain.
Pro-scolex, 44.
Prosimiæ, 346, 357.
Prostate—An accessory gland on the male organs, 130.
Prostomium (πρό; στόμα, the mouth), 54, 75.
Protection, 25, 178, 268, 312.
Proteids (Πρωτεύς, a sea-god with remarkable power of changing his form) —Complex compounds of C, O, N and H with small amounts of S and P, 58.
Proteus animalcule = Amœba, *q.v.*
Protomerite, 17, 18.
Protoplasm (πρῶτος, first; πλάσμα, that which has been formed)—An extremely complex substance upon which life-manifestations depend, 6.
Protopodite, in the Crayfish—The basal, undivided part of an appendage, 84.

Protopterus, 153.
Prototheria (πρῶτος, first; θηρίον, a beast), 345.
Protovertebræ = Somites, mesodermic, q.v.
Protozoa (πρῶτος, first; ζῶον, a living creature)—The simplest animals, consisting each of one cell, 339; Chap. i. 7, 71.
Proventriculus (pro, in front; ventriculus, dim. of venter, the belly), in the Pigeon—The part of the stomach which secretes the gastric juice. It precedes the gizzard, 242.
Proximal (proximus, nearest, next)—Nearest to the main body, 11.
Pseudobranch (ψιυδής, false; βράγχια, the gills of a fish), 166.
Pseudonavicellæ (ψιυδής; navicella, dim. of navis, a ship), 18.
Pseudopodium (ψιυδής; πόδιον, dim. of πούς, a foot)—A temporarily projecting, blunt lobe of protoplasm, 7, 10, 312.
Pterygoid bar, 182.
Pterylæ, 229.
Ptyalin (πτύαλον, spittle)—A ferment, found in saliva, which converts starch into grape-sugar, 289.
Pulmonary—Belonging to the lungs. See Respiration.
Pupil, 174, 221.
Pylangium (πύλη, a gate; ἀγγεῖον, a vessel), in the Frog—That part of the truncus arteriosus which immediately succeeds the ventricle, 196.
Pyloric—Stomach : next the intestine, 91.
Pylorus—The valve between stomach and intestine, 190, 286.
Pyramid—Urinary, 295; of Brain, 302.
Pyrenees, 349.

Q

Quill, 229, 230.

R

Rabbit = Lepus, q.v.
Racemose (racemus, a bunch or cluster of berries), 288.
Radula (radula, a scraper)—A horny, tooth-bearing ribbon, which forms part of the odontophore, 125.
Ramus (ramus, a branch)—Of Nerves, 213, 215; of Mandible, 236, 276, 325; r. communicans, 213, 215, 254, 303.
Rana, 343; 1-5, 135, 175-228.
Ranidæ, 4.
Rat, 347, 349.
Receptaculum—R. ovorum = Oötheca, q.v.; r. seminis = Spermotheca, q.v.
Rectrices (rectrix, fem. of rector, a ruler, a steersman)—Quill-feathers of the tail, 229.

Rectum (*rectus*, straight)—The last part of the intestine, 49, 57, 78, 112, 160, 286.
Redia, 36, 37.
Reflex, of action—Not directly under the control of the will, 62, 216.
Regions—Zoological, 350.
Regular, of segmentation, 27.
Remiges (*remex, remigis*, an oarsman)—Quill-feathers of the wing, 229, 230.
Renal Portal System, 199.
Reproduction—
    (1) Asexual—Amœba, 10; Vorticella, 15; Hydra, 24; Distoma, 37; Comparison of, 317.
    (2) Sexual, 71; Protozoa, 72; Metazoa, 73; Vorticella, 15, 72; Hydra, 25. *See* Reproductive Organs.
Reproductive Organs—Hydra, 22; Distoma, 32; Tœnia, 42; Ascaris, 49; Lumbricus, 62; Hirudo, 80; Astacus, 199; Anodonta, Unio, 118; Helix, 129; Amphioxus, 144; Scyllium, 166; Rana, 206; Columba, 251; Lepus, 297; Homo, 333; Comparison of, 317.
Reptiles, 344.
Resemblance, 178.
Respiration, 10, 141; Amœba, 10; Vorticella, 14; Lumbricus, 60; Hirudo, 76; Astacus, 98; Anodonta, Unio, 115; Helix, 129; Amphioxus, 139; Scyllium, 166; Rana, 204; Columba, 248; Lepus, 293; Comparison of, 315.
Respiration—Anal, 98.
Restiform bodies—Lateral lobes of the medulla oblongata, 169.
Retina (*rete*, a net), in the eye—The layer which contains sense-cells, 134, 174, 221.
Retinula, 104.
Rhabdom, 104.
Rhabdopleura, 341.
Rhinoceros, 5, 355.
Rhizopoda, 339.
Rhytina, 347.
Ribs—Scyllium, 158; Columba, 237; Lepus, 278; Homo, 326.
Ridge—Condylar r., 185; Deltoid r., 185.
Rodentia, 346, 355.
Rods—Certain of the retinal sense-cells, 174, 221.
Rostrum (*rostrum*, a beak), 86, 87; parisphenoidal r. = parasphenoid, 233.
Round-worm = Ascaris, *q.v.*
Rugæ (*ruga*, a wrinkle), in the Rabbit—Longitudinal folds into which the mucous membrane of the stomach is thrown, 286.
Ruminantia, 355.

S

Sac—Air, 249, 252; Dart, 131; Scrotal, 298; Vocal s., 175, 189; Yolk s., 167, 265, 308, 338.

Sacculus (dim. of *saccus*, a bag), in Vertebrates—The lower part of the membranous labyrinth, with which the cochlea is connected, 173, 219, 255, 304.

Sacculus rotundus, 285.

Sacrum—That part of the vertebral column with which the hip-girdles are connected, 184, 238, 325.

Saliva, 289.

Salts, 206.

Sarcolemma (σάρξ, σαρκός, flesh; λῆμμα, used here in the sense of that which receives, a sheath)—The delicate membranous sheath of a striated muscle-fibre, 91.

Sauropsida (σαῦρος, a lizard; ὄψις, sight, appearance), 343.

Scales, 155, 230.

Scaphirhynchus, 349.

Scaphognathite (σκάφος, a boat; γνάθος, a jaw), in the Crayfish—The large epipodite of the second maxilla, 89, 98.

Sclerite (σκληρός, hard)—A definite cuticular plate or band, 92.

Sclerotic coat (σκληρός)—The firm outer coat of the eyeball, 173, 221, 305; sclerotic plate, 256.

Scolex, 40, 44.

Scyllium (Dogfish) (Σκύλλα, a monster barking like a dog), 343; 135, 152-174.

Secretion—A katastate which performs some definite functions before it is excreted, 9.

Segment—One of a series of transverse divisions into which all or part of the body may be divided, 54, 75, 83, 90, 136, 311; muscle s. (myotome, myomere), 136, 145, 168, 208.

Segmentation, of embryo = Cleavage, *q.v.*

Segmentation, of body—Division into originally similar parts (segments) from before backwards, 311; secondary s., 311.

Segmentation-cavity = Blastocœle, *q.v.*

Segmental Organ = Nephridium, *q.v.*

Self-fertilization, 43.

Sella turcica, 273, 323.

Sensation. *See* Sense Organs.

Sense organs—Ascaris, 53; Lumbricus, 66; Hirudo, 82; Astacus, 103; Anodonta, Unio, 121; Helix, 133; Amphioxus, 146; Scyllium, 172; Rana, 218, 226; Columba, 254; Lepus, 303; Comparison of, 320.

INDEX-GLOSSARY. 397

Sensory, of nerves—Conveying impulses from sense-cells, 102, 146, 209, 216.
Sensory tubes, 155.
Septum (*saeptum, septum*, a partition)—A partition, 18, 56 ; Auricular s., 245; Interorbital s., 233 ; S. lucidum, 301 ; Nasal s., 218 ; S. nasi, 271, 323 ; Ventricular s., 245.
Serous coat, 192.
Seta (*seta*, a bristle)—A bristle or hair-like cuticular structure, 55, 84, 90 ; Auditory, 103 ; Coxopoditic, 97 ; Olfactory, 103 ; Tactile, 103.
Sex, or Sexual Organs. *See* Reproductive Organs.
Shaft, of bone, 185; of feather, 230.
Shark, 153.
Shell, 109, 113, 121, 122, 256.
Shoulder—s. girdle, 159, 185, 239, 280, 326.
Sight. *See* Eye.
Sigmoidal cavity, 281.
Simia, 338.
Simple—s. epithelium, 177.
Sinus (*sinus*, a curve, a bay)—(1) An irregular space, or (2) a much dilated blood-vessel, 95, 96, 128; Cardinal s., 165; Coronary s., 332; Cuvierian s., 165 ; Genital s., 324 ; Lymph s., 192 ; Pericardial s., 95 ; s. rhomboidalis, 253 ; Urinary, 167 ; Urinogenital s., 166 ; s. of Valsalva, 290, 333.
Sinus Venosus—The receptacle for impure blood entering the heart of a lower vertebrate, 162, 195.
Sirenia, 346, 355.
Skate, 153.
Skeleton—Amphioxus, 138; Scyllium, 156 ; Rana, 178 ; Columba, 231 ; Lepus, 271 ; Homo, 322; Visceral or branchial s.—The skeleton supporting the respiratory part of the pharynx. *See* Arches.
Skin—Lumbricus, 55 ; Hirudo, 76 ; Astacus, 90 ; Anodonta, Unio, 112 ; Helix, 123; Amphioxus, 138; Scyllium, 155; Rana, 176, 204; Columba, 230 ; Lepus, 269 ; Homo, 322.
Skull, 156 ; Scyllium, 156 ; Rana, 179 ; Columba, 232 ; Lepus, 271 ; Dog, 278; Homo, 322.
Smell. *See* Olfactory Organs.
Snail = Helix, *q.v.*
Somatic (σῶμα, σώματος, a body)—Relating to the body-wall, 149, 226.
Somatopleure (σῶμα, σώματος; πλευρον, a side)—The body-wall of certain embryos, formed by epiblast, together with somatic mesoderm, 227, 263.
Somites, mesodermic, in certain embryos—Transverse segments into which the mesoderm (or part of it) is divided, 107, 149, 227, 263.

Sound-waves, 271.
Space. (*See* Sinus), perineal, 268.
Special creations, 346.
Species, 3, 4.
Sperm (σπέρμα, seed)—A male sex-cell (= spermatozoon), 23, 51, 63, 72, 73, 99, 119, 206, 251, 298.
Sperm-blastophor, 51, 63, 74, 207.
Spermary (σπέρμα) [= Testis]—The organ producing sperms, 23; Hydra, 23; Distoma, 32; Tænia, 42; Ascaris, 51; Lumbricus, 62; Hirudo, 80; Astacus, 99; Anodonta, Unio, 119; Amphioxus, 144; Scyllium, 167; Rana, 206; Columba, 251; Lepus, 298.
Spermatocyte (σπέρμα; κύτος, anything that contains)—A cell which produces sperms directly or by division, 51, 63, 99.
Spermatogenesis (σπέρμα; γένεσις, birth)—Sperm development, 51, 74.
Spermatophore (σπέρμα; φορός, bearing)—An aggregation of sperms, 64, 80, 102, 132.
Spermatospore (σπέρμα, σπέρματος, seed; σπορά, seed)—A cell from which several spermatozoa are produced, 18.
Spermatozoon (σπέρμα, σπέρματος; ζῶον, a living creature) = Sperm, *q.v.*
Spermiduct—The duct by which the sperms are carried to the exterior, or into the urinogenital canal [= vas deferens], 32, 42, 62, 80, 119, 130, 199, 251, 265, 298, 334.
Sperm-morula, 63.
Spermotheca (σπέρμα; θήκη, a cover, a case)—In certain female organs—The sac in which sperms are stored; the receptaculum seminis, 64, 131.
Sperm-sac, 167.
Sphenoidal fissure, 274.
Spiculum amoris, 131.
Spinal cord, in Vertebrates—The posterior part of the cerebro-spinal axis, 145, 169, 211, 253, 302.
Spindle; Nuclear s.—A form assumed by threads of cell protoplasm in process of karyokinesis, 73.
Spine, of scapula, 280; in a Vertebra—A median dorsal process of the arch, 158, 184, 237, 278.
Spiracle (*spiraculum*, an air-hole), (1) in the Tadpole: the opening of the gill-chambers to the exterior, 155; (2) in many fishes: the opening of the hyomandibular cleft to the exterior, 153.
Splanchnic (σπλάγχνα, the viscera)—Relating to the gut or viscera, 149, 226.
Splanchnocœle (σπλάγχνα, the viscera; κοῖλον, a hollow), 149.
Splanchnopleure—The wall of the alimentary canal of certain embryos, formed by hypoblast and splanchnic mesoderm, 227, 264.

Spleen, 248, 292.
Spontaneity—Amœba, 11; Vorticella, 16; Hydra, 26; Rana, 216, 319.
Spontaneous, of action—Not directly dependent on external stimuli, 102.
Spore (σπορά, seed), in Protozoa—One of many minute individuals into which the encysted individual may break up,
Sporocyst (σπορά, seed ; κύστις, a bag, a bladder), in the Fluke—A sac-like asexual stage, 35, 37.
Sporoduct, 19.
Sporozoa, 339.
Spot, germinal—The nucleolus of the ovum, *q.v.*
Squama (*squama*, a scale), 89.
Squamous—Scale-like, 177.
Starch, 58.
Sternebræ (στέρνον, the breast), in Mammalia—The segments of the sternum, 280.
Sternum—In Crayfish, 84; in Vertebrata, the breastbone. *See* Bones.
Stomach, 42, 141, 160, 190, 286.
Stomodæum (στόμα, the mouth)—The anterior part of the alimentary canal originating as an epiblastic pit; the fore-gut, 48.
Strainer, 92, 93.
Stratified, 177.
Streak, Primitive—A mesodermic thickening in the posterior part of the embryonic area, which represents the blastopore of other forms, 256.
Stroma (στρῶμα, anything which is spread out)—The connective-tissue framework of an ovary, 251, 299.
Structure—Amœba, 8; Vorticella, 13.
Sturgeon, 153.
Subcutaneous—Lying beneath the skin, 269.
Sucker, 28, 34, 75.
Sumatra, 349.
Supination (*supinus*, bent backwards, lying on its back), of the fore-arm— The position in which the palm is upwards and the radius and ulna parallel, 281, 326.
Support, 311.
Supra-scapula—A cartilage united to the upper edge of the scapula, 185, 280.
Supra-scapular border—The upper edge of the scapula, 280.
Suspensorium (*suspendo, suspensum*, hang up)—The arrangement by which the lower jaw is suspended to the skull, 182.
Suture, in the Skull—A jagged union between two bones, 271; Coronal s., 272; Frontal s., 272; Lambdoidal, 272; Sagittal, 272.
Swimbladder, 153.
Swimmeret, 84.
Swimming, 318, 319.

Swine, 354.
Sylvian aqueduct, 169, 211, 253, 302.
Sylvian fissure, 336.
Symmetry—Bilateral s., 28, 54, 75, 154, 311 ; Radial s., 20, 311.
Sympathetic system—A special set of nerves supplying the internal organs, 82, 101, 104, 171, 215 ; Columba, 254 ; Lepus, 303.
Symphysis (σὺν, together ; φύσις, a growth)—A median fusion of two bones ; Ischio-pubic, 18, 282 ; Mandibular, 183, 236, 276, 325 ; Pubic s., 327.
Synangium (σὺν, together ; ἀγγεῖον, a vessel), in the Frog—That part of the truncus arteriosus from which the aortic arches arise, 196.
Syrinx (σύριγξ, a reed-pipe)—The vocal organ of birds, situated at the bifurcation of the trachea, 248.
Systemic, of heart—Distributing oxygenated blood to the body in general, 97, 115, 128, 315.
Systole (συστολὴ, used with the modern meaning)—The contraction of a heart, or the chamber of a heart, 97.

T

Tactile Organs—Hirudo, 82 ; Astacus, 103 ; Anodonta, Unio, 120 ; Helix, 133 ; Amphioxus, 146 ; Scyllium, 172 ; Rana, 218 ; Columba, 254 ; Lepus, 303 ; Comparison of, 320.
Tænia, 340 ; 39-45.
Tail, 229, 269.
Tapeworm = Tænia, *q.v.*
Tapir, 349, 355.
Tarsometatarsus, in the Pigeon—A bone formed by the union of the distal part of the tarsus with three of the metatarsals, 241.
Tarsus (ταρσός, the sole of the foot)—The ankle or the ankle bone, 176, 187, 241, 282, 328.
Taste. *See* Gustatory Organs.
Teats, 269.
Teloblast (τέλος, the end ; βλαστός, a germ)—An embryonic cell which by division gives rise to a cellular band, 70, 148.
Teleostei, 153.
Telson (τέλσον, the termination, the end), in the Crayfish—The last segment of the body, 86.
Tendon—A fibrous cord by which muscle is attached, 91, 178, 208.
Tentacles—A small projecting finger-like part of the body, usually sensory, 19, 312.
Tergum (*tergum*, the back), in the Crayfish—The dorsal part of the exoskeleton in each segment, 84.

INDEX-GLOSSARY. 401

Testis—(1) The Spermary, *q.v.*; (2) the posterior half of a divided optic lobe, 302.

Tetradactyle (τίσσαρις, τιτρα-, four; δάκτυλος, a finger)—With four digits, 176.

Thalamencephalon—The axial part of the fore-brain, containing the 3rd ventricle, the side-walls of which are formed by the optic thalami. *See* Fore-brain.

Thalamus, Optic—A mass of grey matter in the thalamencephalon, forming either side-wall of the 3rd ventricle, 210, 252.

Thigh (Femur), 176.

Thorax (θώραξ, the chest), the chest, 83, 86, 269.

Thumb (Pollex), 176, 230, 281, 338.

Thyrohyals—The posterior cornua of the hyoid, 183.

Tissue—An aggregate of similar cells adapted for the performance of some special function, 20; Connective t., 29, 188; Botryoidal, 29.

Tongue, 160, 189, 194, 242, 284, 329.

Tonsil—A thickening of the free edge of the soft palate, 284.

Tooth, of Crayfish, 93; Helix, 126; Scyllium, 160; Rana, 182, 189, 193; Lepus, 284; Homo, 329.

Touch—Lumbricus, 66. *See* Tactile Organs.

Touch corpuscles, 178.

Trachea (τραχὺς, τραχεῖα, rough)—The windpipe, 248, 293.

Transverse process, 158.

Tragulidæ, 355.

Tracts, Optic—The posterior limbs of the X-shaped optic chiasma, 169, 211, 253, 302.

Tree, Bronchial, 343.

Trichina, 54.

Tridactyle (τρεῖς, τρι-, three; δάκτυλος, a finger)—With three digits, 230.

Triploblastica (τριπλοῦς, triple; βλαστός, a germ), 340.

Trochanter, 241, 282.

Trochlea (*trochlea*, a set of blocks and pulleys; τροχὸς, a wheel)—The pulley-like articular surface on the distal end of the humerus, 239, 280.

Troglodytes, 338.

Truncus arteriosus, in the Frog—The tubular part of the heart which succeeds the ventricle, and from which the arterial arches take origin, 195, 203.

Trunk,

Tube—Eustachian, 189, 219, 233, 255, 274; Fallopian, 299, 334; Neural, 151, 224, 261.

Tubercle, of rib, 237, 279.

Tubules—Uriniferous, 167, 205, 250, 295.
Tympanic area, 175.
Tympanic cavity, 189.
Tympanum (τύμπανον, *tympanum*, a drum)—The cavity of the middle ear, 219, 255, 304.
Typhlosole (τυφλίς, blind; σωλήν, a channel, a fold), in some Invertebrates—A longitudinal fold of the intestinal wall which projects inwards, 57, 133, 313.

U

Umbilical cord, 310.
Umbilicus, 231.
Umbo (*umbo*, the boss in the centre of a shield)—A projection constituted by the oldest part of the shell, 109.
Uncinus (ὄγκινος), *uncinus*, a small hook), in the Snail—One of the symmetrical median teeth of the radula, 126.
Ungulata, 346, 354.
Unicellular, 20, 310.
Unio (Fresh-water Mussel), 109-122, 341.
Unipolar, 145.
United States, 349.
Univalve, 123.
Urea, 118, 206, 297.
Ureter—The kidney-duct, 98, 118, 129, 166, 250, 265, 295, 333, 334.
Urethra—The urinogenital canal of the male Rabbit, 298.
Uric acid, 99, 118, 251.
Urine, 62, 251.
Urinogenital system—Rana, 205; Scyllium, 166; Columba, 250; Lepus, 293; Homo, 233.
Urochorda (οὐρὰ, the tail; χορδή, a string), 341, 135.
Uropygium (ὑροπύγιον, ὀρροπύγιον, the tail of a bird or fish)—The stumpy tail of the Pigeon, 229.
Urostyle (οὐρα, a tail; στῦλος, a pillar), in the Frog—A bony unsegmented rod, forming the posterior part of the backbone, 183, 184.
Uterus—The dilated part of an oviduct, 33, 43, 52, 207, 299, 334.
Uterus masculinus, in the male Rabbit—A pouch on the dorsal side of the bladder, into which the vasa deferentia open, 298.
Utriculus (dim. of *uter*, a bag or vessel of skin)—A sac-like central part of the membranous labyrinth, from which the semicircular canals take origin, 173, 219, 255, 304.

V.

Vacuole (dim. of *vacuum*, an empty space)—A fluid-containing space in the protoplasm of a cell, 8, 31; contractile v., 8, 10, 13, 314, 315, 316; food v., 9, 14.

Vagina—The lower part of the oviduct leading from the uterus to the exterior, or into the urinogenital canal or the cloaca, 43, 52, 80, 131, 299.

Valve; of Shell, 109; of Heart, 162, 164, 196; Auriculo-ventricular, 115, 162, 196, 245; Bicuspid (mitral), 245; Eustachian, 245, 290; Semilunar, 196, 245; Spiral v., 160, 161; Thebesian v., 290.

Valvulæ conniventes, 285.

Vane—The expanded part of a feather, 230.

Variety, 4.

Vas deferens = Spermiduct, *q.v.*

Vas efferens—A duct leading from spermary (testis) to spermiduct, 62, 80, 298, 334.

Vein—A vessel carrying blood to or towards the heart, 115; Allantoic, 267; Anterior abdominal, 199, 297; Azygos, 291, 332; Brachial, 246; Branchial (Afferent and Efferent), 115; Cardinal, 265, 331; Caudal, 247; Caval, 198; Coronary, 291; Cuvierian, 265, 331; Facial, 291; Femoral, 197, 247, 291; Gastric, 247; Hepatic, 142, 193, 198, 247, 291; Hepatic portal, 165; Hypogastric, 247; Iliac, 247, 291, 332; Innominate, 198, 331; Intercostal, 332; Jugular, 198, 246, 265, 291, 332; Lieno-gastric, 292; Mesenteric, 247, 291; Pectoral, 246; Pelvic, 199; Portal, 142, 198, 199, 247, 291; Postcaval (Posterior vena cava), 198, 246, 291, 333; Precaval (Superior vena cava), 198, 246, 265, 291; Pulmonary, 200, 247, 291; Renal, 128, 165, 198, 247, 297; Renal portal, 165, 291; Sciatic, 199, 247; Subclavian, 198, 291, 331; Subscapular, 198; Venæ cavæ (Caval veins), 198, 246, 291.

Velum, 139.

Vena cava, 198, 246, 291.

Ventral (*venter*, the stomach)—Lower, 28, 322.

Ventricle (*ventriculus*, used with the modern meaning), in the heart—In the two or more chambered heart a muscular propulsive chamber, 162, 195, 245, 290, 315.

Ventricle, of nerve-cord, 145, 168; of brain, 160, 252, 301.

Veragua, 349.

Vertebra (*vertebra*, a joint, especially of the backbone)—A joint of the backbone, 158, 183; Atlas, 184.

Vertebral Column—Scyllium, 158; Rana, 183; Columba, 257; Lepus, 278; Homo, 325.

Vertebrata (*vertebratus*, fitted with *vertebræ* or joints), 3, 135, 152, 341.
Vesicle, Auditory (*see* Sacs, Auditory)—Brain (Cerebral), 168, 210, 260; Germinal, 23, 24; Optic, 226; Umbilical v. = yolk sac.
Vesicula seminalis—A part of the spermiduct (vas deferens) which serves for the storage of sperms, 17, 33, 43, 51, 62, 166, 206, 251, 333.
Vespertilionidæ, 349.
Vessel, 59; Dorsal v., 59; Lateral, 59, 78; Lateral-neural, 59; Parietal (commissural), 59; Subneural, 59.
Vestibule, in Vorticella—A ciliated depression on one side of the disc, which leads to the gullet, 13; in the vertebrate ear—The central part of the membranous labyrinth, 173, 219.
Vibrissæ (*vibro*, I vibrate), in the Rabbit—The whiskers, 268, 271.
Villus (*villus*, shaggy hair)—A minute, finger-like projection, 242, 286, 309.
Visceral-mass, 111.
Vision, mosaic, 104.
Visual Organs—Hirudo, 76; Astacus, 103; Helix, 134; Scyllium, 173; Rana, 221; Columba, 256; Lepus, 304; Comparison of, 321.
Visual-pyramids, 103.
Vitelline membrane—The cell-wall of an ovum, 73.
Vitellus (*vitellus*, the yolk of an egg)—The protoplasm of an ovum, 23, 34, 73.
Vitreous body, 104.
Viviparous—Bringing forth well-developed young, as opposed to eggs, 68, 308.
Vocal Cords, in the Larynx—Two elastic folds, the edges of which can be brought parallel and thrown into vibrations, 204.
Vocal Sac, 175, 189.
Voice, 205.
Volition, 102, 216.
Voluntary, 102.
Vorticella (dim. of *vortex*, a whirlpool), 339; 8, 11-16, 311.
Vulva, in the Rabbit—The female urinogenital aperture, 299.

W

Walking, 83, 229, 268.
Walking-legs, 87.
Wallace, 348.
Wallace's line, 350.
Waste products, 10, 14, 61, 80.
Water Flea, 23.
Water Snail, 83.
Weismann, 74.

Whiskers, 268, 271.
White matter, of Brain and Spinal Cord—The part mainly composed of nerve fibres, 172, 215.
Will, 102.
Wind-pipe (Trachea), 248, 293.
Wing, 230—Bastard w., 230.
Wolffian body = Mesonephros, *q.v.*
Wolffian duct = Duct, mesonephric, *q.v.*
Worm-castings, 59.
Wrist = Carpus, *q.v.*

X

Xanthin, 41.
Xiphisternum — The posterior, more or less cartilaginous part of the sternum, 185.

Y

Yang-tse-Kiang, 34.
Yeast, 58.
Yolk, 23, 252.
Yolk-gland, 33, 43.
Yolk-reservoir, 33.
Yolk-sac, 167, 265, 308, 338.

Z

Zona radiata, in Mammalia—A radiately striated membrane which invests the ovum, 299, 306, 308.
Zooid (ζῶον, a living creature); in Protozoa and colonies of other groups— The individual, 15.
Zoology (ζῶον; λόγος, a speech, discourse), 1.
Zoological Regions, 350.
Zoophyte (ζῶον; φυτόν, a plant), 27.
Zonitic constrictions, in the Earthworm—Shallow grooves, of which one or more encircle each segment, 54.
Zygapophysis (ζυγὸν, a yoke; ἀπόφυσις, a process)—A process projecting from the front or back of a neural arch, and assisting in the articulation of adjacent vertebræ, 184, 237, 278.
Zygoma (ζύγωμα, a bar or bolt), in the Mammalian skull—A bony bar below the orbit and temporal fossa, 272.

BELL AND BAIN, LIMITED, PRINTERS, GLASGOW

OTHER WORKS

By J. R. AINSWORTH DAVIS, B.A.,
Professor of Biology and Geology, University College, Aberystwyth.

THE FLOWERING PLANT,
AS ILLUSTRATING THE FIRST PRINCIPLES OF BOTANY.

With Numerous Illustrations. SECOND EDITION, 3s. 6d.

"It would be hard to find a Text-book which would better guide the student to an accurate knowledge of modern discoveries in Botany. . . . The SCIENTIFIC ACCURACY of statement, and the concise exposition of FIRST PRINCIPLES make it valuable for educational purposes. In the chapter on the Physiology of Flowers, an *admirable résumé* is given, drawn from Darwin, Hermann Müller, Kerner, and Lubbock, of what is known of the Fertilization of Flowers."—*Journal of the Linnean Society.*

A ZOOLOGICAL POCKET-BOOK:
OR, SYNOPSIS OF ANIMAL CLASSIFICATION.

Comprising Definitions of the Phyla, Classes, and Orders, with explanatory Remarks and Tables.

BY DR. EMIL SELENKA,
Professor in the University of Erlangen.

Authorised English translation from the Third German Edition.

In Small Post 8vo, Interleaved for the use of Students. 4s.

"Dr. Selenka's Manual will be found useful by all Students of Zoology. It is a COMPREHENSIVE and SUCCESSFUL attempt to present us with a scheme of the natural arrangement of the animal world."—*Edin. Med. Journal.*

"Will prove very serviceable to those who are attending Biology Lectures."—*The Lancet.*

LONDON: CHARLES GRIFFIN & CO., LTD., EXETER STREET, STRAND.

SECOND EDITION, Revised.

PRACTICAL GEOLOGY
(AIDS IN):
WITH A SECTION ON PALÆONTOLOGY.

BY

GRENVILLE A. J. COLE, F.G.S.,
Professor of Geology in the Royal College of Science for Ireland.

With Numerous Illustrations and Tables. Large Crown 8vo. Cloth, 10s. 6d.

GENERAL CONTENTS.

PART I.—SAMPLING OF THE EARTH'S CRUST.

Observations in the field. | Collection and packing of specimens.

PART II.—EXAMINATION OF MINERALS.

Some physical characters of minerals. | Blowpipe-tests.
Simple tests with wet reagents. | Quantitative flame reactions of the felspars
Examination of minerals with the blowpipe. | and their allies.
Simple and characteristic reactions. | Examination of the optical properties of minerals.

PART III.—EXAMINATION OF ROCKS.

Introductory. | The more prominent characters to be ob-
Rock-structures easily distinguished. | served in minerals in rock-sections.
Some physical characters of rocks. | Characters of the chief rock-forming minerals
Chemical examination of rocks. | in the rock-mass and in thin sections.
Isolation of the constituents of rocks. | Sedimentary rocks.
The petrological microscope and microscopic | Igneous rocks.
preparations. | Metamorphic rocks.

PART IV.—EXAMINATION OF FOSSILS.

Introductory. | Scaphopoda; Gastropoda; Pteropoda
Fossil generic types.—Rhizopoda; Spongiæ; | Cephalopoda.
Hydrozoa; Actinozoa. | Echinodermata; Vermes.
Polyzoa; Brachiopoda. | Arthropoda.
Lamellibranchiata. | Suggested list of characteristic invertebrate fossils.

"Prof. Cole treats of the examination of minerals and rocks in a way that has never been attempted before . . . DESERVING OF THE HIGHEST PRAISE. Here indeed are 'aids' INNUMERABLE and INVALUABLE. All the directions are given with the utmost clearness and precision. Prof. Cole is not only an accomplished Petrologist, he is evidently also a thoroughly sympathetic teacher, and seems to know intuitively what are stumbling-blocks to learners—a rare and priceless quality."—*Athenæum*.

"To the younger workers in Geology, Prof. Cole's book will be as INDISPENSABLE as a dictionary to the learners of a language."—*Saturday Review*.

"That the work deserves its title, that it is full of 'AIDS,' and in the highest degree 'PRACTICAL,' will be the verdict of all who use it."—*Nature*.

"A MOST VALUABLE and welcome book . . . the subject is treated on lines wholly different from those in any other Manual, and is therefore very ORIGINAL."—*Science Gossip*.

"A more useful work for the practical geologist has not appeared in handy form."—*Scottish Geographical Magazine*.

"This EXCELLENT MANUAL . . . will be A VERY GREAT HELP. . . . The section on the Examination of Fossils is probably the BEST of its kind yet published. . . . FULL of well-digested information from the newest sources and from personal research."—*Annals of Nat. History.*

LONDON: CHARLES GRIFFIN & CO., LTD., EXETER STREET, STRAND.

In Large 8vo. Cloth, 12s. 6d.

THE PHYSIOLOGIST'S NOTE-BOOK:

A Summary of the Present State of Physiological Science, for the use of Students.

BY

ALEX HILL, M.A., M.D.,
Master of Downing College, Cambridge.

With Numerous Illustrations and Blank Pages for MS. Notes.

GENERAL CONTENTS.

The Blood—The Vascular System—The Nerves—Muscle—Digestion—The Skin—The Kidney—Respiration—The Senses—Voice and Speech—Central Nervous System—Reproduction—Chemistry of the Body.

** The object of this work is not to supersede the larger Text-Books, still less to take the place of Lectures and Laboratory work, but to assist the Student in CODIFYING HIS KNOWLEDGE.

"The 'Note-book' deals with the ARGUMENTS OF PHYSIOLOGY, and it is as well that the Student should from the outset recognise that the subject, although it has made rapid strides during the last twenty years, is still in a transitional state, and that many of its most important issues can only be summed up as leaving a balance of evidence on the one side or the other—a balance which subsequent investigation may possibly disturb. I have made an attempt, which might, perhaps, be carried further with advantage in other scientific text-books, to show the logical sequence of each portion of the argument by its position on the page.

"If a Student could rely on remembering every word which he had ever heard or read, such a book as this would be unnecessary; but experience teaches that he constantly needs to recall the form of an argument and to make sure of the proper CLASSIFICATION OF HIS FACTS, although he does not need a second time to follow the author up all the short steps by which the ascent was first made. With a view to rendering the book useful for rapid recapitulation, I have endeavoured to strike out every word which was not essential to clearness, and thus, without I hope falling into 'telegram' English, to give the text the form which it may be supposed to take in a well kept Note-book; at the same time, space has been left for the INTRODUCTION IN MS. of such additional facts and arguments as seem to the reader to bear upon the subject-matter. For the same reason the drawings are reduced to DIAGRAMS. All details which are not necessary to the comprehension of the principles of construction of the apparatus or organ, as the case may be, are omitted, and it is hoped that the drawings will, therefore, be easy to grasp, remember, and reproduce.

"As it is intended that the 'Note-book' should be essentially a Student's book, no references are given to foreign literature or to recondite papers in English; but, on the other hand, references are given to a number of CLASSICAL ENGLISH MEMOIRS, as well as to descriptions in text-books which appear to me to be particularly lucid, and the Student is strongly recommended to study the passages and papers referred to."—*Extract from Author's Preface.*

LONDON: CHARLES GRIFFIN & CO., LTD., EXETER STREET, STRAND.

A

CATALOGUE

OF

STANDARD WORKS

PUBLISHED BY

CHARLES GRIFFIN & COMPANY,

LIMITED.

	PAGE
I.—Religious Works,	4
II.—Works on Medicine and the Allied Sciences,	9
III.—General Scientific and Technical Works,	23
IV.—Educational Works,	54
V.—Works in General Literature,	62

LONDON:
12 EXETER STREET, STRAND.

INDEX TO AUTHORS.

	PAGE
AITKEN (Sir W., M.D.), Science and Practice of Medicine,	9
—— Outlines,	9
ANDERSON (M'Call) on Skin Diseases,	14
ANGLIN (S.), Design of Structures,	23
BELL (R.), Golden Leaves,	
BERINGER (J. J. and C.), Assaying,	62
BLYTH (A. W.), Hygiène and Public Health,	24
—— Foods and Poisons,	10
BROTHERS (A.), Photography,	10
BROUGH (B. H.), Mine-Surveying,	25
BROWNE (W. R.), Works,	26
BRYCE (A. H.), Works of Virgil,	26
BUNYAN'S Pilgrim's Progress (Mason),	55
BURNET (Dr.), Foods and Dietaries,	4
CAIRD and CATHCART, Surgical Handbook,	11
CHEEVER'S (Dr.), Religious Anecdotes,	12
CLARK (Sir A.), Fibroid Disease of Lung,	4
COBBETT (Wm.), Works,	16
COBBIN'S Mangnall's Questions,	55, 63
COLE (Prof.), Practical Geology,	55
COLERIDGE on Method,	27
CRAIK (G.), History of English Literature,	55
—— Manual of do.,	56
CRIMP (W. S.), Sewage Disposal Works,	56
CROOM (Halliday), Gynæcology,	28
CRUDEN'S CONCORDANCE, by Eadie,	16
CRUTTWELL'S History of Roman Literature,	6
—— Specimens of do.,	57
—— Early Christian Literature,	57
CURRIE (J.), Works of Horace,	4
DAVIS (J. R. A.), An Introduction to Biology,	55
—— The Flowering Plant,	29
—— Zoological Pocket-Book,	29
DICK (Dr.), Celestial Scenery,	30
DOERING and GRÆME'S Hellas,	4
D'ORSEY (A. J.), Spelling by Dictation,	58
DUCKWORTH (Sir D., M.D.), Gout,	58
DUPRÉ & HAKE, Manual of Chemistry,	14
EADIE (Rev. Dr.), Biblical Cyclopædia,	30
—— Cruden's Concordance,	6
—— Classified Bible,	6
—— Ecclesiastical Cyclopædia,	6
—— Dictionary of the Bible,	6
ELBORNE (W.), Pharmacy and Materia Medica,	12
EMERALD SERIES OF POETS,	64
ETHERIDGE (R.), Stratigraphical Geology,	42
EWART (Prof.), The Preservation of Fish,	30
FIDLER (T. Claxton), Bridge-Construction,	31
FLEMING (Prof.), Vocabulary of Philosophy,	58
FOSTER (Chas.), Story of the Bible,	5
FOSTER (C. Le N.), Ore and Stone Mining,	32
GARROD (Dr. A. E.), Rheumatism,	14
GILMER (R.), Interest Tables,	63
GRÆME (Elliott), Beethoven,	63
—— Novel with Two Heroes,	63
GRIFFIN'S ELECTRICAL PRICE-BOOK,	32
GRIFFIN (J. J.), Chemical Recreations,	32
GURDEN (R.), Traverse Tables,	32
GUTTMANN (O.), Rock Blasting,	35
HADDON (Prof.), Embryology,	13
HORSLEY (Victor), Brain and Spinal Cord,	15
HUGHES (H. W.), Coal-Mining,	36
HUMPHRY, Manual of Nursing,	18
HURST (G. H.), Colours and Varnishes,	35
JAKSCH (M.) and CAGNEY,Clinical Diagnosis,	14
JAMES (W. P.), From Source to Sea,	58
JAMIESON (A.), Manual of the Steam Engine	37
——Steam and the Steam Engine—Elementary,	37
——Applied Mechanics,	38
——Magnetism and Electricity,	38

	PAGE
JEVONS (F. B.), A History of Greek Literature,	59
—— Athenian Democracy,	59
—— Journal of Anatomy and Physiology,	67
—— Journal of State Medicine,	67
KEBLE'S Christian Year,	4
KNECHT and RAWSON, Dyeing,	35
LANDIS (Dr.), Management of Labour,	18
LANDOIS and STIRLING'S Physiology,	13
LEWIS (W. B.), Mental Diseases,	15
LINN (Dr.), On the Teeth,	18
LONGMORE (Prof.), Sanitary Contrasts,	19
MACALISTER (Prof.), Human Anatomy,	13
MACREADY (J. F. C.), on Ruptures,	16
MANN (Prof.), Forensic Medicine,	16
MACKEY (A. G.), Lexicon of Freemasonry,	65
MAYHEW (H.), London Labour,	65
M'BURNEY (Dr.), Ovid's Metamorphoses,	60
MEYER AND FERGUS' Ophthalmology,	14
MILLER (W. G.), Philosophy of Law,	60
M'MILLAN (W. G.), Electro-Metallurgy,	39
MUNRO (R. D.), Steam Boilers,	38
MUNRO and JAMIESON'S Electrical Pocket-Book,	40
NYSTROM'S Pocket-Book for Engineers,	40
OBERSTEINER and HILL, Central Nervous Organs,	15
PAGE (H.W.), Railway Injuries,	16
PARKER (Prof.), Mammalian Descent,	19
PHILLIPS and BAUERMAN, Metallurgy,	43
POE (Edgar), Poetical Works of,	64
PORTER (Surg.-Maj.), Surgeon's Pocket-Book,	19
RAMSAY (Prof.), Roman Antiquities,	60
—— Do. Elementary,	60
—— Latin Prosody,	60
—— Do. Elementary,	60
RANKINE'S ENGINEERING WORKS,	44-45
REED (Sir E. J.), Stability of Ships,	46
REID (Geo.), Practical Sanitation,	20
ROBERTS-AUSTEN (Prof.), Metallurgy,	47
ROBINSON (Prof.), Hydraulics,	48
RANSOM (A. E.), Diseases of the Heart,	15
SCHRADER and JEVONS, The Prehistoric Antiquities of the Aryan Peoples,	61
SCHWACKHÖFER and BROWNE, Fuel and Water,	49
SEATON (A. E.), Marine Engineering,	50
SEELEY (Prof.), Physical Geology,	41
SENIOR (Prof.), Political Economy,	62
SEXTON (Prof.), Quantitative Analysis,	49
—— Qualitative Analysis,	49
SHELTON-BEY (W. V.), Mechanic's Guide,	49
SOUTHGATE (H.), Many Thoughts of Many Minds,	66
—— Suggestive Thoughts,	7
—— (Mrs.), Christian Life,	7
STIRLING (William), Human Physiology,	13
—— Outlines of Practical Physiology,	21
—— Outlines of Practical Histology,	21
TAIT (Rev. J.), Mind in Matter,	8
THE MASSES: How shall we reach them?,	8
THOMSON (Dr. Spencer), Domestic Medicine,	68
THOMSON'S Seasons,	62
THORBURN (Wm), Surgery of the Spine,	15
THORNTON (J. K.), Surgery of the Kidneys,	16
TRAILL (W.), Boilers, Land and Marine,	51
WESTLAND (A.), Wife and Mother,	22
WHATELY (Archbishop), Logic, and Rhetoric,	62
WORDS AND WORKS OF OUR LORD,	8
WRIGHT (Alder), The Threshold of Science,	52
YEAR-BOOK OF SCIENTIFIC SOCIETIES,	53

INDEX TO SUBJECTS.

	PAGE
ANATOMY, Human,	13
——Journal of,	67
ANECDOTES, Cyclopædia of,	4
ANTIQUITIES, Prehistoric,	61
——Greek,	59
——Roman,	60
ARYAN PEOPLES,	61
ASSAYING,	24
BIBLE (The Holy), Classified,	6
——Concordance to,	6
——Cyclopædia of,	6
——Dictionary of,	6
——Story of,	5
BIOLOGY,	29
BOILERS, Marine and Land,	51
——Management of,	38
BOTANY,	29
BRAIN, The,	15
BRIDGE-CONSTRUCTION,	23, 31
CHEMISTRY, Inorganic,	30
——Experiments in,	32, 52
——Qualitative Analysis,	49
——Quantitative ,,	49
——Recreations in,	32
CLINICAL Diagnosis,	14
DAILY Readings,	7, 8
DICTIONARY of Anecdotes,	4
——of the Holy Bible,	6
——Ecclesiastical,	6
——of Medicine (Domestic),	68
——of Philosophical Terms,	58
——of Quotations,	66
——of ,, (Religious),	7
DIETARIES for the Sick,	11
EDUCATIONAL Works,	54
ELECTRICAL Price-Book,	32
ELECTRICITY,	37, 38, 40
ELECTRO-METALLURGY,	39-40
EMBRYOLOGY,	13, 19
ENGINEERING, Civil,	44
——Marine,	50
——Useful Rules in,	40, 45
EYE, Diseases of the,	14
FISH, Preservation of,	30
FOODS, Analysis of,	10
FOODS and Dietaries,	11
FORENSIC MEDICINE,	16
FREEMASONRY,	65
FUEL and Water,	49
GEOLOGY, Practical,	27
——Physical	41
——Stratigraphical,	42
GOUT,	14
GRAMMARS,	55
GYNÆCOLOGY,	16
HEART, Diseases of the,	15
HISTOLOGY,	21
HORACE, Works of,	55
HYDRAULIC Power,	48
HYGIENE and Public Health,	10, 20
INTEREST Tables,	63
LABORATORY Handbooks—	
Histology,	21
Pharmacy,	12
Physiology,	21
LATIN Prosody,	60
LAW, Philosophy of,	60
LITERATURE, General,	62
——Early Christian,	4
——English,	56
LITERATURE, Greek,	59
——Roman,	57
LOGIC,	62
LUNG, Fibroid Disease of,	16
MACHINERY, Hydraulic,	48
——and Millwork,	44
MAGNETISM,	37, 38
MAMMALIAN Descent,	19
MARINE Engineering,	50
MECHANICS,	26, 38, 40, 44, 45, 49
MEDICAL Series,	13-16
——Works,	9
MEDICINE, Science and Practice of,	9
——Domestic,	68
——Journal of State,	67
MENTAL Diseases,	15
——Science,	60
METALLURGY,	43, 47
MINE-SURVEYING,	26
MINING, Coal,	36
——Ore and Stone,	32
MYTHOLOGY, Greek,	58
NAVAL Construction,	46
NERVOUS ORGANS, Central,	15
——System,	15
NURSING, Medical and Surgical,	18, 22, 68
OBSTETRICS,	18, 22
PALÆONTOLOGY,	27, 41, 42
PHARMACY,	12
PHILOSOPHY, Vocabulary of,	58
PHOTOGRAPHY,	25
PHYSICS, Experiments in,	52
PHYSIOLOGY, Human,	13
——Practical,	21
POCKET-BOOK, Electrical,	40
——Engineering,	40
——Surgical,	12, 19
——Zoological,	30
POETS, Emerald Series of,	64
POISONS, Detection of,	10
POLITICAL ECONOMY,	62
POOR, Condition of the,	8, 65
RAILWAY Injuries,	16
RELIGIOUS Works,	4
RHETORIC,	62
RHEUMATISM,	14
ROCK-BLASTING,	35
ROOFS, Design of,	23
RUPTURES,	16
SCIENCE, Popular Introduction to,	52
SCIENTIFIC Societies, Papers read before,	53
SEWAGE Disposal Works,	28
SHIPS, Stability of,	46
——Wave-forms, Propulsion, &c. (Rankine),	45
SKIN, Diseases of the,	14
SPINAL Cord,	15
STEAM-ENGINE,	37, 45
STRUCTURES, Design of,	23
STUDENTS' Text-Books,	17, 34, 54
SURGERY, Civil,	12
——Military,	19
——of Kidneys,	16
——of Spinal Cord,	15
SURVEYING,	26, 32
TEETH, Care of the,	18
THERMODYNAMICS, (Rankine),	45
TRAVERSE Tables,	32
VIRGIL, Works of,	55
ZOOLOGY,	30

CHARLES GRIFFIN & COMPANY'S LIST OF PUBLICATIONS.

RELIGIOUS WORKS.

ANECDOTES (CYCLOPÆDIA OF RELIGIOUS AND MORAL). With an Introductory Essay by the Rev. GEORGE CHEEVER, D.D. *Thirty-sixth Thousand.* Crown 8vo. Cloth, 3/6.

⁎ These Anecdotes relate to no trifling subjects; and they have been selected, not for amusement, but for instruction. By those engaged in the tuition of the young, they will be found highly useful.

THE LARGE-TYPE BUNYAN.

BUNYAN'S PILGRIM'S PROGRESS. With Life and Notes, Experimental and Practical, by WILLIAM MASON. Printed in large type, and Illustrated with full-page Woodcuts. *Twelfth Thousand.* Crown 8vo. Bevelled boards, gilt, and gilt edges, 3/6.

CHRISTIAN YEAR (The): With Memoir of the Rev. JOHN KEBLE, by W. TEMPLE, Portrait, and Eight Engravings on Steel, after eminent Masters. *New Edition.* Small 8vo, toned paper. Cloth gilt, 5/.

⁎ The above is the only issue of the "Christian Year" with Memoir and Portrait of the Author. In ordering, Griffin's Edition should be specified.

CRUTTWELL (REV. CHARLES T., M.A.) A HISTORY OF EARLY CHRISTIAN LITERATURE. In large 8vo, handsome cloth. [*In preparation.*

⁎ This work is intended not only for Theological Students, but for General Readers, and will be welcomed by all acquainted with the Author's admirable "*History of Roman Literature*," a work which ha- now reached its Fifth Edition.

DICK (Thos., LL.D.): CELESTIAL SCENERY; or, The Wonders of the Planetary System Displayed. This Work is intended for general readers, presenting to their view, in an attractive manner, sublime objects of contemplation. Illustrated. *New Edition.* Crown 8vo, toned paper. Handsomely bound, gilt edges, 5/.

LONDON: EXETER STREET, STRAND.

RELIGIOUS WORKS. 5

Now Ready. FIFTH AND GREATLY IMPROVED EDITION. In LARGE 8vo.
Cloth Elegant, 6/. Gilt and Gilt Edges, 7/6.

THE STORY OF THE BIBLE,
From GENESIS to REVELATION.

Including the Historical Connection between the Old and New Testaments.
Told in Simple Language.

By CHARLES FOSTER.

With Maps and over 250 Engravings
(Many of them Full-page, after the Drawings of Professor CARL SCHÖNHERR and others),

Illustrative of the Bible Narrative, and of Eastern Manners and Customs.

OPINIONS OF THE PRESS.

"A book which, once taken up, is not easily laid down. When the volume is opened, we are fairly caught. Not to speak of the well-executed wood engravings, which will each tell its story, we find a simple version of the main portions of the Bible, all that may most profitably be included in a work intended at once to instruct and charm the young —a version couched in the simplest, purest, most idiomatic English, and executed throughout with good taste, and in the most reverential spirit. *The work needs only to be known to make its way into families*, and it will (at any rate, it *ought* to) become a favourite Manual in Sunday Schools."—*Scotsman.*

"A HOUSEHOLD TREASURE."—*Western Morning News.*

"This attractive and handsome volume . . . written in a simple and transparent style. . . . Mr. Foster's explanations and comments are MODELS OF TEACHING."—*Freeman.*

"This large and handsome volume, abounding in Illustrations, is just what is wanted. . . . The STORY is very beautifully and reverently told."—*Glasgow News.*

"There could be few better Presentation Books than this handsome volume."—*Daily Review.*

"WILL ACCOMPLISH A GOOD WORK."—*Sunday School Chronicle.*

"In this beautiful volume no more of comment is indulged in than is necessary to the elucidation of the text. Everything approaching Sectarian narrowness is carefully eschewed."—*Methodist Magazine.*

"This simple and impressive Narrative . . . succeeds thoroughly in riveting the attention of children; . . . admirably adapted for reading in the Home Circle."—*Daily Chronicle.*

"The HISTORICAL SKETCH connecting the Old and New Testaments is a very good idea; it is a common fault to look on these as distinct histories, instead of as parts of *one grand whole.*"—*Christian.*

"Sunday School Teachers and Heads of Families will best know how to value this handsome volume."—*Northern Whig.*

⁎ The above is the original English Edition. In ordering, Griffin's Edition, by Charles Foster, should be distinctly specified.

LONDON: EXETER STREET, STRAND.

STANDARD BIBLICAL WORKS

BY

THE REV. JOHN EADIE, D.D., LL.D.,

Late a Member of the New Testament Revision Company.

This SERIES has been prepared to afford sound and necessary aid to the Reader of Holy Scripture. The VOLUMES comprised in it form in themselves a COMPLETE LIBRARY OF REFERENCE. The number of Copies already issued greatly exceeds A QUARTER OF A MILLION.

I. EADIE (Rev. Prof.): BIBLICAL CYCLO- PÆDIA (A); or, Dictionary of Eastern Antiquities, Geography, and Natural History, illustrative of the Old and New Testaments. With Maps, many Engravings, and Lithographed Fac-simile of the Moabite Stone. Large post 8vo, 700 pages. Handsome cloth, 7/6. *Twenty-sixth Edition.*

II. EADIE (Rev. Prof.): CRUDEN'S CON- CORDANCE TO THE HOLY SCRIPTURES. With Portrait on Steel, and Introduction by the Rev. Dr. KING. Post 8vo. *Fifty-third Edition.* Handsome cloth, 3/6.

** Dr. EADIE'S has long and deservedly borne the reputation of being the COM-PLETEST and BEST CONCORDANCE extant.

III. EADIE (Rev. Prof.): CLASSIFIED BIBLE (The). An Analytical Concordance. Illustrated by Maps. Large Post 8vo. *Sixth Edition.* Handsome cloth, 8/6.

"We have only to add our unqualified commendation of a work of real excellence to every Biblical student."—*Christian Times.*

IV. EADIE (Rev. Prof.): ECCLESIASTICAL CYCLOPÆDIA (The). A Dictionary of Christian Antiquities, and of the History of the Christian Church. By the Rev. Professor EADIE, assisted by numerous Contributors. Large Post 8vo. *Sixth Edition.*

Handsome cloth, ... 8/6.

"The ECCLESIASTICAL CYCLOPÆDIA will prove acceptable both to the clergy and laity of Great Britain. A great body of useful information will be found in it."—*Athenæum.*

V. EADIE (Rev. Prof.): DICTIONARY OF THE HOLY BIBLE (A); for the use of Young People. With Map and Illustrations. Small 8vo. *Thirty-eighth Thousand.*

Cloth, elegant, ... 2/6.

LONDON: EXETER STREET, STRAND.

MR. SOUTHGATE'S WORKS.
(*See also p. 66.*)

"No one who is in the habit of writing and speaking much on a variety of subjects can afford to dispense with Mr. SOUTHGATE'S WORKS."—*Glasgow News.*

THIRD EDITION.

SUGGESTIVE THOUGHTS ON RELIGIOUS SUBJECTS:
A Dictionary of Quotations and Selected Passages from nearly 1,000 of the best Writers, Ancient and Modern.

Compiled and Analytically Arranged

By HENRY SOUTHGATE.

In Square 8vo, elegantly printed on toned paper.

Presentation Edition, Cloth Elegant, ... 10/6.
Library Edition, Roxburghe, ... 12/.
Ditto, Morocco Antique, ... 20/.

"The topics treated of are as wide as our Christianity itself: the writers quoted from, of every Section of the one Catholic Church of JESUS CHRIST."—*Author's Preface.*

"Mr. Southgate's work has been compiled with a great deal of judgment, and it will, I trust, be extensively useful."—*Rev. Canon Liddon, D.D., D.C.L.*

"A casket of gems."—*English Churchman.*

"This is another of Mr. Southgate's most valuable volumes. . . . The mission which the Author is so successfully prosecuting in literature is not only highly beneficial, but necessary in this age. . . . If men are to make any acquaintance at all with the great minds of the world, they can only do so with the means which our Author supplies."—*Homilist.*

"Many a busy Christian teacher will be thankful to Mr. Southgate for having unearthed so many rich gems of thought; while many outside the ministerial circle will obtain stimulus, encouragement, consolation, and counsel, within the pages of this handsome volume."—*Nonconformist.*

"Mr. SOUTHGATE is an indefatigable labourer in a field which he has made peculiarly his own. . . . The labour expended on 'Suggestive Thoughts' must have been immense, and the result is as nearly perfect as human fallibility can make it. . . . Apart from the selections it contains, the book is of value as an index to theological writings. As a model of judicious, logical, and suggestive treatment of a subject, we may refer our readers to the manner in which the subject 'JESUS CHRIST' is arranged and illustrated in 'Suggestive Thoughts.'"—*Glasgow News.*

"*Every day is a little life.*"—BISHOP HALL.

THE CHRISTIAN LIFE:
Thoughts in Prose and Verse from 500 of the Best Writers of all Ages. Selected and Arranged for Every Day in the Year.

BY MRS. H. SOUTHGATE.

Small 8vo. With Red Lines and unique Initial Letters on each page.
Cloth Elegant, 5/. *Second Edition.*

"A volume as handsome as it is intrinsically valuable."—*Scotsman.*
"The Readings are excellent."—*Record.*
"A library in itself."—*Northern Whig.*

LONDON: EXETER STREET, STRAND.

MIND IN MATTER:

A SHORT ARGUMENT ON THEISM.

BY THE
REV. JAMES TAIT.

[*Third Edition.* Demy 8vo. Handsome Cloth, 6/.

GENERAL CONTENTS.—Evolution in Nature and Mind—Mr. Darwin and Mr. Herbert Spencer—Inspiration, Natural and Supernatural—Deductions.

*** Special attention has in this, the Third Edition, been directed to Embryology and later Darwinism.

"An able and original contribution to Theistic literature. . . . The style is pointed, concise, and telling to a degree."—*Glasgow Herald.*
"Mr. TAIT advances many new and striking arguments . . . highly suggestive and fresh."—*Brit. Quarterly Review.*

THE MASSES:
HOW SHALL WE REACH THEM?

Some Hindrances in the way, set forth from the standpoint of the People, with Comments and Suggestions.

BY
AN OLD LAY HELPER.

Cloth, 2s. 6d. *Second Edition.*

*** An attempt to set forth some deficiencies in our present methods of reaching the poor, in the language of the People themselves.

"So full of suggestiveness that we should reprint a tithe of the book if we were to transcribe all the extracts we should like to make."—*Church Bells.*
"'Hindrances in the way' exactly describes the subject-matter of the Book. Any one contemplating Missionary work in a large town would be helped by studying it."—*Guardian.*
"'The Masses' is a book to be well pondered over and acted upon."—*Church Work.*
"A very useful book, well worth reading."—*Church Times.*
"A most interesting book. . . . Contains a graphic description of work among the masses."—*English Churchman.*

WORDS AND WORKS OF OUR BLESSED LORD:
AND THEIR LESSONS FOR DAILY LIFE.

Two Vols. in One. Foolscap 8vo. Cloth, gilt edges, 6/.

LONDON: EXETER STREET, STRAND.

Works in Medicine, Surgery, and the Allied Sciences.

∗ *Special Illustrated Catalogue sent Post-free on application.*

WORKS

By Sir WILLIAM AITKEN, M.D., Edin., F.R.S.,

LATE PROFESSOR OF PATHOLOGY IN THE ARMY MEDICAL SCHOOL; EXAMINER IN MEDICINE FOR THE MILITARY MEDICAL SERVICES OF THE QUEEN; FELLOW OF THE SANITARY INSTITUTE OF GREAT BRITAIN; CORRESPONDING MEMBER OF THE ROYAL IMPERIAL SOCIETY OF PHYSICIANS OF VIENNA; AND OF THE SOCIETY OF MEDICINE AND NATURAL HISTORY OF DRESDEN.

SEVENTH EDITION.

THE SCIENCE AND PRACTICE OF MEDICINE.

In Two Volumes, Royal 8vo., cloth. Illustrated by numerous Engravings on Wood, and a Map of the Geographical Distribution of Diseases. To a great extent Rewritten; Enlarged, Remodelled, and Carefully Revised throughout, 42/.

Opinions of the Press.

"The work is an admirable one, and adapted to the requirements of the Student, Professor, and Practitioner of Medicine. . . . The reader will find a large amount of information not to be met with in other books, epitomised for him in this. We know of no work that contains so much, or such full and varied information on all subjects connected with the Science and Practice of Medicine."—*Lancet.*

"The SEVENTH EDITION of this important Text-Book fully maintains its reputation. . . . Dr. Aitken is indefatigable in his efforts. . . . The section on DISEASES of the BRAIN and NERVOUS SYSTEM is completely remodelled, so as to include all the most recent researches, which in this department have been not less important than they are numerous."—*British Medical Journal.*

OUTLINES

OF THE

SCIENCE AND PRACTICE OF MEDICINE.

A TEXT-BOOK FOR STUDENTS.

Second Edition. Crown 8vo, 12/6.

"Students preparing for examinations will hail it as a perfect godsend for its conciseness." —*Athenæum.*

"Well-digested, clear, and well-written, the work of a man conversant with every detail of his subject, and a thorough master of the art of teaching."—*British Medical Journal.*

LONDON : EXETER STREET, STRAND.

10   CHARLES GRIFFIN & CO.'S PUBLICATIONS.

WORKS by A. WYNTER BLYTH, M.R.C.S., F.C.S.,
Barrister-at-Law, Public Analyst for the County of Devon, and Medical Officer
of Health for St. Marylebone.

I. FOODS: THEIR COMPOSITION AND
ANALYSIS. Price 16/. In Crown 8vo, cloth, with Elaborate Tables and Litho-Plates. *Third Edition.* Revised and partly rewritten.

. General Contents.

History of Adulteration—Legislation, Past and Present—Apparatus useful to the Food Analyst—"Ash "—Sugar—Confectionery—Honey—Treacle—Jams and Preserved Fruits—Starches—Wheaten-Flour—Bread—Oats—Barley—Rye—Rice—Maize—Millet —Potato—Peas—Chinese Peas—Lentils—Beans—MILK—Cream—Butter—Cheese—Tea —Coffee — Cocoa and Chocolate — Alcohol — Brandy—Rum—Whisky—Gin—Arrack— Liqueurs—Beer—Wine—Vinegar—Lemon and Lime Juice—Mustard—Pepper—Sweet and Bitter Almond—Annatto—Olive Oil—Water. *Appendix:* Text of English and American Adulteration Acts.

" Will be used by every Analyst."—*Lancet.*
" STANDS UNRIVALLED for completeness of information. . . . A really 'practical' work for the guidance of practical men."—*Sanitary Record.*
" An admirable digest of the most recent state of knowledge. . . . Interesting even to lay-readers."—*Chemical News.*
** The THIRD EDITION contains many Notable Additions, especially on the subject of MILK and its relation to FEVER-EPIDEMICS, the PURITY of WATER-SUPPLY, the MARGARINE ACT, &c., &c.

II. POISONS: THEIR EFFECTS AND DE-
TECTION. Price 16/.

General Contents.

Historical Introduction—Statistics—General Methods of Procedure—Life Tests— Special Apparatus—Classification : I.—ORGANIC POISONS: (*a.*) Sulphuric, Hydrochloric, and Nitric Acids, Potash, Soda, Ammonia, &c. ; (*b.*) Petroleum, Benzene, Camphor, Alcohols, Chloroform, Carbolic Acid, Prussic Acid, Phosphorus, &c. ; (*c.*) Hemlock, Nicotine, Opium, Strychnine, Aconite, Atropine, Digitalis, &c. ; (*d.*) Poisons derived from Animal Substances; (*e.*) The Oxalic Acid Group. II.—INORGANIC POISONS: Arsenic, Antimony, Lead, Copper, Bismuth, Silver, Mercury, Zinc, Nickel, Iron, Chromium, Alkaline Earths, &c. *Appendix:* A. Examination of Blood and Blood-Spots. B. *Hints for Emergencies:* Treatment—Antidotes.

"Should be in the hands of every medical practitioner."—*Lancet.*
" A sound and practical Manual of Toxicology, which cannot be too warmly re-commended. One of its chief merits is that it discusses substances which have been overlooked."—*Chemical News.*
" One of the best, most thorough, and comprehensive works on the subject."— *Saturday Review.*

HYGIÈNE AND PUBLIC HEALTH (a Dic-
tionary of) : embracing the following subjects :—

I.—SANITARY CHEMISTRY : the Composition and Dietetic Value of Foods, with the Detection of Adulterations.

II.—SANITARY ENGINEERING : Sewage, Drainage, Storage of Water, Ventilation, Warming, &c.

III.—SANITARY LEGISLATION : the whole of the PUBLIC HEALTH ACT, together with portions of other Sanitary Statutes, in a form admitting of easy and rapid Reference.

IV.—EPIDEMIC AND EPIZOOTIC DISEASES : their History and Pro-pagation, with the Measures for Disinfection.

V.—HYGIÈNE—MILITARY, NAVAL, PRIVATE, PUBLIC, SCHOOL.

Royal 8vo, 672 pp., cloth, with Map and 140 Illustrations, 28/.
" A work that must have entailed a vast amount of labour and research. . . . Will become a STANDARD WORK IN PUBLIC HEALTH."—*Medical Times and Gazette.*
" Contains a great mass of information of easy reference."—*Sanitary Record.*

LONDON: EXETER STREET, STRAND.

SECOND and CHEAPER EDITION. *Handsome Cloth, 4s.*

FOODS AND DIETARIES:
A Manual of Clinical Dietetics.

BY

R. W. BURNET, M.D.,

Member of the Royal College of Physicians of London; Physician to the Great Northern Central Hospital, &c.

In Dr. Burnet's "Foods and Dietaries," the *rationale* of the special dietary recommended is briefly stated at the beginning of each section. To give definiteness to the directions, the HOURS of taking food and the QUANTITIES to be given at each time are stated, as well as the KINDS of food most suitable. In many instances there is also added a list of foods and dishes that are UNSUITABLE to the special case.

* To the SECOND EDITION a chapter on Diet in INFLUENZA, and numerous Fresh Rec'p's for Invalid Cookery, have been added.

GENERAL CONTENTS.

DIET in Diseases of the Stomach, Intestinal Tract, Liver, Lungs and Pleuræ, Heart, Kidneys, &c.; in Diabetes, Scurvy, Anæmia, Scrofula, Gout (Chronic and Acute), Obesity, Acute and Chronic Rheumatism, Alcoholism, Nervous Disorders, Diathetic Diseases, Diseases of Children, with a Section on Prepared and Predigested Foods, and Appendix on Invalid Cookery.

"The directions given are UNIFORMLY JUDICIOUS and characterised by good-sense. . . . May be confidently taken as a RELIABLE GUIDE in the art of feeding the sick."— *Brit. Med. Journal.*

"To all who have much to do with Invalids, Dr. Burnet's book will be of great use. . . . It will be found all the more valuable in that it deals with BROAD and ACCEPTED VIEWS. There are large classes of disease which, if not caused solely by errors of diet, have a principal cause in such errors, and can only be removed by an intelligent apprehension of their relation to such. Gout, Scurvy, Rickets, and Alcoholism are instances in point, and they are all TREATED with ADMIRABLE SENSE and JUDGMENT by Dr. Burnet. He shows a desire to allow as much range and VARIETY as possible. The careful study of such books as this will very much help the Practitioner in the Treatment of cases, and powerfully aid the action of remedies."—*Lancet.*

"Dr. Burnet's work is intended to meet a want which is evident to all those who have to do with nursing the sick. . . . The plan is METHODICAL, SIMPLE, and PRACTICAL. . . . Dr. Burnet takes the important diseases *seriatim* . . . and gives a Time-table of Diet, with Bill of Fare for each meal, quantities, and beverages. . . . An appendix of Cookery for Invalids is given, which will help the nurse when at her wits' end for a change of diet, to meet the urgency of the moment or tempt the capricious appetite of the patient."— *Glasgow Herald.*

LONDON: EXETER STREET, STRAND.

12   CHARLES GRIFFIN & CO.'S PUBLICATIONS.

FOURTH EDITION, *Pocket-Size, Leather,* 8s. 6d.   *With very Numerous Illustrations.*

A SURGICAL HANDBOOK,

For Practitioners, Students, House=Surgeons, and Dressers.

BY F. M. CAIRD, M.B., F.R.C.S. (ED.),

AND

C. W. CATHCART, M.B., F.R.C.S. (ENG. & ED.),

Assistant-Surgeons, Royal Infirmary, Edinburgh.

GENERAL CONTENTS.

Case-Taking—Treatment of Patients before and after Operation—Anæsthetics: General and Local—Antiseptics and Wound-Treatment—Arrest of Hæmorrhage—Shock and Wound Fever—Emergency Cases—Tracheotomy: Minor Surgical Operations—Bandaging—Fractures—Dislocations, Sprains, and Bruises—Extemporary Appliances and Civil Ambulance Work—Massage—Surgical Applications of Electricity—Joint-Fixation and Fixed Apparatus—The Syphon and its Uses—Trusses and Artificial Limbs—Plaster-Casting—Post-Mortem Examination—Sickroom Cookery Receipts, &c., &c., &c.

"THOROUGHLY PRACTICAL AND TRUSTWORTHY. Clear, accurate, succinct."—*The Lancet.*
"ADMIRABLY ARRANGED. The best practical little work we have seen. The matter is as good as the manner."—*Edinburgh Medical Journal.*
"THIS EXCELLENT LITTLE WORK. Clear, concise, and very readable. Gives attention to important details, often omitted, but ABSOLUTELY NECESSARY TO SUCCESS."—*Athenæum.*
"A dainty volume."—*Manchester Medical Chronicle.*

In Extra Crown 8vo, with Litho-plates and Numerous Illustrations. Cloth, 8s. 6d.

PHARMACY AND MATERIA MEDICA

(A Laboratory Course of):

Including the Principles and Practice of Dispensing.

Adapted to the Study of the British Pharmacopœia and the Requirements of the Private Student.

BY W. ELBORNE, F.L.S., F.C.S.,

Late Assistant-Lecturer in Materia Medica and Pharmacy in the Owens College, Manchester.

"A work which we can very highly recommend to the perusal of all Students of Medicine. . . . ADMIRABLY ADAPTED to their requirements."—*Edinburgh Medical Journal.*
"Mr. Elborne evidently appreciates the Requirements of Medical Students, and there can be no doubt that any one who works through this Course will obtain an excellent insight into Chemical Pharmacy."—*British Medical Journal.*
"The system . . . which Mr. Elborne here sketches is thoroughly sound."—*Chemist and Druggist.*

LONDON: EXETER STREET, STRAND.

GRIFFIN'S MEDICAL SERIES.

Standard Works of Reference for Practitioners and Students,

ISSUED UNIFORMLY IN LIBRARY STYLE.

Large 8vo, Handsome Cloth, very fully Illustrated.

Full Catalogue, with Specimens of the Illustrations, sent Post-free on application.

VOLUMES ALREADY PUBLISHED.

HUMAN ANATOMY.

BY ALEXANDER MACALISTER, M.A., M.D., F.R.S., F.S.A.,
Professor of Anatomy in the University of Cambridge, and Fellow of St. John's College. 36s.

"BY FAR THE MOST IMPORTANT WORK on this subject which has appeared in recent years."
— *The Lancet.*
"Destined to be a main factor in the advancement of Scientific Anatomy. . . . The fine collection of Illustrations must be mentioned."—*Dublin Medical Journal.*
"This SPLENDID WORK."—*Saturday Review.*

HUMAN PHYSIOLOGY.

BY PROFESSOR LANDOIS OF GREIFSWALD
AND
WM. STIRLING, M.D., Sc.D.,
Brackenbury Professor of Physiology in Owens College and Victoria University, Manchester; Examiner in the University of Oxford. FOURTH EDITION. With some of the Illustrations in Colours. 2 Vols., 42s.

"The Book is the MOST COMPLETE *résumé* of all the facts in Physiology in the language. Admirably adapted for the PRACTITIONER. . . . With this Text-book at command, NO STUDENT COULD FAIL IN HIS EXAMINATION."—*The Lancet.*
"One of the MOST PRACTICAL WORKS on Physiology ever written. EXCELLENTLY CLEAR, ATTRACTIVE, and SUCCINCT."—*British Medical Journal.*

EMBRYOLOGY (An Introduction to).

BY ALFRED C. HADDON, M.A., M.R.I.A.,
Professor of Zoology in the Royal College of Science, Dublin. 18s.

"An EXCELLENT RÉSUMÉ OF RECENT RESEARCH, well adapted for self-study. . . . Gives remarkably good accounts (including all recent work) of the development of the heart and other organs. . . . The book is handsomely got up."—*The Lancet.*

LONDON: EXETER STREET, STRAND.

GRIFFIN'S MEDICAL SERIES—*Continued.*

CLINICAL DIAGNOSIS.

The Chemical, Microscopical, and Bacteriological Evidence of Disease.

BY PROF. VON JAKSCH, OF PRAGUE.

From the Third German Edition, by JAS. CAGNEY, M.A., M.D., of St. Mary's Hospital.

SECOND ENGLISH EDITION. With all the Original Illustrations, many printed in Colours. 25s.

"Prof. v. Jaksch's 'Clinical Diagnosis' stands almost alone in the width of its range, the THOROUGHNESS OF ITS EXPOSITION and the clearness of its style. . . . A STANDARD WORK, as TRUSTWORTHY as it is scientific."—*Lancet.*

GOUT (A Treatise on).

BY SIR DYCE DUCKWORTH, M.D.EDIN., F.R.C.P.,

Physician to, and Lecturer on Clinical Medicine at, St. Bartholomew's Hospital. 25s.

"At once thoroughly practical and highly philosophical. The practitioner will find in it an ENORMOUS AMOUNT OF INFORMATION."—*Practitioner*

RHEUMATISM AND RHEUMATOID ARTHRITIS.

BY ARCH. E. GARROD, M.A., M.D.OXON.,

Assistant-Physician to the West London Hospital, &c. 21s.

"We gladly welcome this Treatise. . . . The amount of information collected and the manner in which the facts are marshalled are deserving of ALL PRAISE."—*Lancet.*

DISEASES OF THE SKIN.

BY T. M'CALL ANDERSON, M.D.,

Professor of Clinical Medicine in the University of Glasgow. 25s.

"Beyond doubt, the MOST IMPORTANT WORK on Skin Diseases that has appeared in England for many years."—*British Medical Journal.*

DISEASES OF THE EYE.

BY DR. ED. MEYER, OF PARIS.

From the Third French Edition,

BY A. FREELAND FERGUS, M.B.,

Ophthalmic Surgeon, Glasgow Royal Infirmary. 25s.

"An EXCELLENT TRANSLATION of a Standard French Text-Book. . . . Essentially a PRACTICAL WORK. The publishers have done their part in the tasteful and substantial manner characteristic of their medical publications."—*Ophthalmic Review.*

LONDON: EXETER STREET, STRAND.

GRIFFIN'S MEDICAL SERIES—*Continued.*

DISEASES OF THE HEART
(THE DIAGNOSIS OF).

By A. ERNEST SANSOM, M.D., F.R.C P.,

Physician to the London Hospital; Consulting Physician, North-Eastern Hospital for Children; Examiner in Medicine, Royal College of Physicians (Conjoint Board for England), and University of Durham; Lecturer on Medical Jurisprudence and Public Health, London Hospital Medical College, &c. 28s.

THE BRAIN AND SPINAL CORD:
THEIR STRUCTURE AND FUNCTIONS.

By VICTOR HORSLEY, F.R.C.S., F.R.S.,

Professor of Pathology, University College; Assist.-Surgeon, University College Hospital; Surg. Nat. Hosp. for Paralysed and Epileptic, &c., &c. 10s. 6d.

THE CENTRAL NERVOUS ORGANS
(The Anatomy of), in Health and Disease.

By PROFESSOR OBERSTEINER, OF VIENNA.
TRANSLATED BY ALEX HILL, M.A., M.D.,
Master of Downing College, Cambridge. 25s.

"Dr. Hill has enriched the work with many Notes of his own. . . . Dr. Obersteiner's work is admirable. . . . INVALUABLE AS A TEXT-BOOK."—*British Medical Journal.*

MENTAL DISEASES:
With Special Reference to the Pathological Aspects of Insanity.

By W. BEVAN LEWIS, L.R.C.P. (LOND.), M.R.C.S. (ENG.),
Medical Director of the West Riding Asylum, Wakefield. 28s.

"Without doubt the BEST WORK in English of its kind."—*Journal of Mental Science.*
"This ADMIRABLE TEXT-BOOK places the study of Mental Diseases on a SOLID BASIS. . . . The plates are numerous and admirable. To the student the work is INDISPENSABLE."
—*Practitioner.*

The SURGERY of the SPINAL CORD.

By WILLIAM THORBURN, B.S., B.Sc., M.D., F.R.C.S.,
Assistant-Surgeon to the Manchester Royal Infirmary. 12s. 6d.

"Really the FULLEST RECORD we have of Spinal Surgery, and marks an important advance."
—*British Medical Journal.*

LONDON: EXETER STREET, STRAND.

GRIFFIN'S MEDICAL SERIES—*Continued*.

RAILWAY INJURIES:

With Special Reference to those of the Back and Nervous System, in their Medico-Legal and Clinical Aspects.

By HERBERT W. PAGE, M.A., M.C. (CANTAB), F.R.C.S. (ENG.), Surgeon to St. Mary's Hospital; Dean, St. Mary's Hospital Medical School, &c., &c. 6s.

"A work INVALUABLE to those who have many railway cases under their care pending litigation. . . . A book which every lawyer as well as doctor should have on his shelves."
—*British Medical Journal.*

THE SURGERY OF THE KIDNEYS.

Being the Harveian Lectures, 1889.

By J. KNOWSLEY THORNTON, M.B., M.C., Surgeon to the Samaritan Free Hospital, &c.

In Demy 8vo, with Illustrations, Cloth, 5s.

"The name and experience of the Author confer on the Lectures the stamp of authority."—*British Medical Journal.*

READY SHORTLY.

ON FIBROID DISEASE OF THE LUNG AND ON FIRBOID PHTHISIS. By Sir ANDREW CLARK, Bart., M.D., Consulting Physician and Lecturer on Clinical Medicine to the London Hospital, and Drs. W. J. HADLEY and CHAPLIN, Assistant Physicians to the City of London Hospital for Diseases of the Chest. With numerous full-page coloured Illustrations.

RUPTURES (A Treatise on). By J. F. C. MACREADY, F.R.C.S., Surgeon, City of London Truss Society, City of London Hospital for Diseases of the Chest, &c. With numerous Plates engraved on Stone after Photographs.

FORENSIC MEDICINE AND TOXICOLOGY. By J. DIXON MANN, M.D., F.R.C.P., Professor of Medical Jurisprudence and Toxicology, Owens College, Manchester; Examiner in Forensic Medicine, London University, and Victoria University, Manchester.

GYNÆCOLOGY (A Practical Treatise on): by JOHN HALLIDAY CROOM, M.D., F.R.C.P.E., F.R.C.S.E., Physician to the Royal Infirmary and Royal Maternity Hospital, Edinburgh; Examiner in Midwifery, R.C.P., Edinburgh; Lecturer, Edinburgh School of Medicine, &c., &c., with the Collaboration of MM. JOHNSON SYMINGTON, M.D., F.R.C.S.E., and MILNE MURRAY, M.A., M.B., F.R.C.P.E.

*** Volumes on other subjects in active preparation.

LONDON: EXETER STREET, STRAND.

MEDICINE AND THE ALLIED SCIENCES. 17

Griffin's Medical Students' Text-Books.

		PAGE
Anatomy,	Prof. Macalister,	13
Biology,	Ainsworth Davis,	29
Botany (Elementary),	Ainsworth Davis,	29
Brain and Spinal Cord,	Victor Horsley,	15
Central Nervous Organs,	Obersteiner and Hill,	15
Mental Diseases,	Bevan Lewis,	15
Chemistry, Inorganic,	Dupré and Hake,	30
Qualitative Analysis,	Prof. Sexton,	49
Quantitative ,,	Prof. Sexton,	49
Electricity,	Prof. Jamieson,	38
Embryology,	Prof. Haddon,	13
Eye, Diseases of the,	Meyer and Fergus,	14
Foods, Analysis of,	Wynter Blyth,	10
Foods and Dietaries,	R. W. Burnet,	11
Gynæcology,	Halliday Croom,	16
Histology,	Prof. Stirling,	21
Medicine,	Sir Wm. Aitken,	9
Nursing,	L. Humphry,	18
Obstetrics,	H. G. Landis,	18
Pharmacy,	W. Elborne,	12
Physiology, Human,	Landois and Stirling,	13
Practical,	Prof. Stirling,	21
Poisons, Detection of,	Wynter Blyth,	10
Sanitation, Practical,	Dr. George Reid,	20
Skin, Diseases of the,	Prof. Anderson,	14
Surgery—		
Civil,	Caird and Cathcart,	12
Military,	Porter-Godwin,	19
Zoology,	Selenka and Davis,	30

LONDON: EXETER STREET, STRAND.

18   CHARLES GRIFFIN & CO.'S PUBLICATIONS.

EIGHTH EDITION. *In Extra Crown 8vo, with Numerous Illustrations, Cloth, 3s. 6d.*

NURSING (A Manual of):
Medical and Surgical.

BY LAURENCE HUMPHRY, M.A., M.B., M.R.C.S.,

Assistant-Physician to, and Lecturer to Probationers at, Addenbrooke's Hospital, Cambridge.

GENERAL CONTENTS.

The General Management of the Sick Room in Private Houses—General Plan of the Human Body—Diseases of the Nervous System—Respiratory System—Heart and Blood-Vessels—Digestive System—Skin and Kidneys—Fevers—Diseases of Children—Wounds and Fractures—Management of Child-Bed—Sick-Room Cookery, &c., &c.

⁎⁎⁎ A Full Prospectus Post Free on Application.

"In the fullest sense Mr. Humphry's book is a DISTINCT ADVANCE on all previous Manuals. . . . Its value is greatly enhanced by copious Woodcuts and diagrams of the bones and internal organs, by many Illustrations of the art of BANDAGING, by Temperature charts indicative of the course of some of the most characteristic diseases, and by a goodly array of SICK-ROOM APPLIANCES with which EVERY NURSE should endeavour to become acquainted."—*British Medical Journal.*

"We should advise ALL NURSES to possess a copy of the work. We can confidently recommend it as an EXCELLENT GUIDE and companion."—*Hospital.*

LANDIS (Henry G., A.M., M.D., Professor of Obstetrics in Starling Medical College):
THE MANAGEMENT OF LABOUR AND OF THE LYING-IN PERIOD. In 8vo, with Illustrations. Cloth, 7/6.

"Fully accomplishes the object kept in view by its author. . . . Will be found of GREAT VALUE by the young practitioner."—*Glasgow Medical Journal.*

LINN (S.H., M.D., D.D.S., Dentist to the Imperial Medico-Chirurgical Academy of St. Petersburg):
THE TEETH: How to preserve them and prevent their Decay. A Popular Treatise on the Diseases and the Care of the Teeth. With Plates and Diagrams. Crown 8vo. Cloth, 2/6.

LONDON: EXETER STREET, STRAND.

LONGMORE (Surgeon-General, C.B., Q.H.S., F.R.C.S., &c., late Professor of Military Surgery, Army Medical School):
THE SANITARY CONTRASTS OF THE CRIMEAN WAR. Demy 8vo. Cloth limp, 1/6.

"A most valuable contribution to Military Medicine."—*British Medical Journal.*
"A most concise and interesting Review."—*Lancet.*

PARKER (Prof. W. Kitchen, F.R.S., Hunterian Professor, Royal College of Surgeons):
MAMMALIAN DESCENT: being the Hunterian Lectures for 1884. Adapted for General Readers. With Illustrations. In 8vo, cloth, 10/6.

"The smallest details òf science catch a LIVING GLOW from the ardour of the author's imagination, . . . we are led to compare it to some quickening spirit which makes all the dry bones of skulls and skeletons stand up around him as an exceeding great army."—Prof. Romanes in *Nature.*

"Get this book; read it straight ahead, . . . you will first be interested, then absorbed; before reaching the end you will comprehend what a lofty ideal of creation is that of him, who, recognising the unity and the continuity of Nature, traces the gradual development of life from age to age . . . and has thus learned to 'look through Nature up to Nature's GOD.'"—*Scotsman.*

"A very striking book . . . as readable as a book of travels. Prof. PARKER is no Materialist."—*Leicester Post.*

PORTER AND GODWIN'S POCKET-BOOK.

FOURTH EDITION: *Revised and enlarged. Foolscap 8vo, Roan, with 152 Illustrations and Folding-plate.* 8s. 6d.

THE SURGEON'S POCKET-BOOK.

Specially adapted to the Public Medical Services.

BY SURGEON-MAJOR J. H. PORTER.

Revised and in great part rewritten

BY BRIGADE-SURGEON C. H. Y. GODWIN,
Professor of Military Surgery in the Army Medical School.

"The present editor—Brigade-Surgeon Godwin—has introduced so much that is new and practical, that we can recommend this 'Surgeon's Pocket-Book' as an INVALUABLE GUIDE to all engaged, or likely to be engaged, in Field Medical Service."—*Lancet.*

"A complete *vade mecum* to guide the military surgeon in the field."—*British Medical Journal.*

LONDON: EXETER STREET, STRAND.

PRACTICAL SANITATION:
A HAND-BOOK FOR SANITARY INSPECTORS AND OTHERS INTERESTED IN SANITATION.

BY

GEORGE REID, M.D., D.P.H.,
Fellow of the Sanitary Institute of Great Britain, and Medical Officer, Staffordshire County Council.

With an Appendix on Sanitary Law

BY

HERBERT MANLEY, M.A., M.B., D.P.H.,
Medical Officer of Health for the County Borough of West Bromwich.

In Large Crown 8vo., with Illustrations. Price 6s.

Dr. Reid's PRACTICAL SANITATION is heartily recommended to all who take an interest in the great subject on which it treats, a subject which is attracting the serious attention of the general public. The necessity for no longer allowing grave sanitary defects to exist in our houses and their surroundings is now generally acknowledged, and people are beginning to enquire for themselves into matters to which, hitherto, they have closed their eyes. While specially designed for the use of Sanitary Inspectors, the book will be found to be of value to all, as giving within small compass a thorough, clearly written, and well-illustrated digest of Sanitary Science.

GENERAL CONTENTS.

Introduction — Water Supply: Drinking Water, Pollution of Water — Ventilation and Warming — Principles of Sewage Removal — Details of Drainage; Refuse Removal and Disposal — Sanitary and Insanitary Work and Appliances — Details of Plumbers' Work — House Construction — Infection and Disinfection — Food, Inspection of; Characteristics of Good Meat; Meat, Milk, Fish, &c., unfit for Human Food — Appendix: Sanitary Law; Model Bye-Laws, &c.

"A VERY USEFUL HANDBOOK, with a very useful Appendix. We recommend it not only to SANITARY INSPECTORS, but to HOUSEHOLDERS and ALL interested in Sanitary matters."—*Sanitary Record.*

LONDON: EXETER STREET, STRAND.

WORKS

By WILLIAM STIRLING, M.D., Sc.D.,

Professor in the Victoria University, Brackenbury Professor of Physiology and Histology in the Owens College, Manchester; and Examiner in the University of Oxford.

SECOND EDITION. *In Extra Crown 8vo, with 234 Illustrations, Cloth, 9s.*

PRACTICAL PHYSIOLOGY (Outlines of):

Being a Manual for the Physiological Laboratory, including Chemical and Experimental Physiology, with Reference to Practical Medicine.

PART I.—CHEMICAL PHYSIOLOGY.

PART II.—EXPERIMENTAL PHYSIOLOGY.

*** In the Second Edition, revised and enlarged, the number of Illustrations has been increased from 142 to 234.

"A VERY EXCELLENT and COMPLETE TREATISE."—*Lancet.*

"The student is enabled to perform for himself most of the experiments usually shown in a systematic course of lectures on physiology, and the practice thus obtained must prove INVALUABLE. . . . May be confidently recommended as a guide to the student of physiology, and, we doubt not, will also find its way into the hands of many of our scientific and medical practitioners."—*Glasgow Medical Journal.*

"This valuable little manual. . . . The GENERAL CONCEPTION of the book is EXCELLENT; the arrangement of the exercises is all that can be desired; the descriptions of experiments are CLEAR, CONCISE, and to the point."—*British Medical Journal.*

In Extra Crown 8vo, with 344 Illustrations, Cloth, 12s. 6d.

PRACTICAL HISTOLOGY (Outlines of):

A Manual for Students.

*** Dr. Stirling's "Outlines of Practical Histology" is a compact Handbook for students, providing a COMPLETE LABORATORY COURSE, in which almost every exercise is accompanied by a drawing. Very many of the illustrations have been prepared expressly for the work.

"The general plan of the work is ADMIRABLE . . . It is very evident that the suggestions given are the outcome of a PROLONGED EXPERIENCE in teaching Practical Histology, combined with a REMARKABLE JUDGMENT in the selection of METHODS. . . . Merits the highest praise for the ILLUSTRATIONS, which are at once clear and faithful."—*British Medical Journal.*

"We can confidently recommend this small but CONCISELY-WRITTEN and ADMIRABLY ILLUSTRATED work to students. They will find it to be a VERY USEFUL and RELIABLE GUIDE in the laboratory, or in their own room. All the principal METHODS of preparing tissues for section are given, with such precise directions that little or no difficulty can be felt in following them in their most minute details. . . . The volume proceeds from a MASTER in his craft."—*Lancet.*

LONDON: EXETER STREET, STRAND.

In large Crown 8vo. Handsome Cloth. Price 5s. Post free.

THE WIFE AND MOTHER:

A Medical Guide

TO THE CARE OF HER HEALTH AND THE MANAGEMENT OF HER CHILDREN.

BY

ALBERT WESTLAND, M.A., M.D., C.M.

*** This work is addressed to women who are desirous of fulfilling properly their duties as wives and mothers, and is designed to assist them in exercising an intelligent supervision over their own and their children's health.

GENERAL CONTENTS.

PART   I.—Early Married Life.
PART II.—Early Motherhood.
PART III.—The Child, in Health and Sickness.
PART IV.—Later Married Life.

"WELL-ARRANGED, and CLEARLY WRITTEN. The chapter on the Nutrition of the Child is very carefully written, and the Hints as to the ARTIFICIAL FEEDING of Infants are reliable."—*Lancet.*

"A REALLY EXCELLENT BOOK. . . . The author has handled the subject conscientiously and with perfect good taste. . . . The work is what it professes to be—a guide and help, giving all that is most essential to know of the life-history of womanhood and motherhood."—*Aberdeen Journal.*

"EXCELLENT AND JUDICIOUS . . . the work of an experienced obstetricist, surgeon, and physician . . . deals with an important subject in a manner that is at once PRACTICAL AND POPULAR."—*Western Daily Press.*

"The best book I can recommend is 'THE WIFE AND MOTHER,' by Dr. ALBERT WESTLAND, published by Messrs. Charles Griffin & Co. It is a MOST VALUABLE work, written with discretion and refinement."—*Hearth and Home.*

"Will be WELCOMED by every young wife . . . abounds with valuable advice"—*Glasgow Herald.*

LONDON: EXETER STREET, STRAND.

GENERAL SCIENTIFIC WORKS

RELATING TO

CHEMISTRY (THEORETICAL AND APPLIED); ELECTRICAL SCIENCE; ENGINEERING (CIVIL AND MECHANICAL); GEOLOGY, &c.

THE DESIGN OF STRUCTURES:

A Practical Treatise on the Building of Bridges, Roofs, &c.

BY S. ANGLIN, C.E.,

Master of Engineering, Royal University of Ireland, late Whitworth Scholar, &c.

With very numerous Diagrams, Examples, and Tables.

Large Crown 8vo. Cloth, 16s.

The leading features in Mr. Anglin's carefully-planned "Design of Structures" may be briefly summarised as follows:—

1. It supplies the want, long felt among Students of Engineering and Architecture, of a concise Text-book on Structures, requiring on the part of the reader a knowledge of ELEMENTARY MATHEMATICS only.

2. The subject of GRAPHIC STATICS has only of recent years been generally applied in this country to determine the Stresses on Framed Structures; and in too many cases this is done without a knowledge of the principles upon which the science is founded. In Mr. Anglin's work the system is explained from FIRST PRINCIPLES, and the Student will find in it a valuable aid in determining the stresses on all irregularly-framed structures.

3. A large number of PRACTICAL EXAMPLES, such as occur in the every-day experience of the Engineer, are given and carefully worked out, some being solved both analytically and graphically, as a guide to the Student.

4. The chapters devoted to the practical side of the subject, the Strength of Joints, Punching, Drilling, Rivetting, and other processes connected with the manufacture of Bridges, Roofs, and Structural work generally, are the result of MANY YEARS' EXPERIENCE in the bridge-yard; and the information given on this branch of the subject will be found of great value to the practical bridge-builder.

"Students of Engineering will find this Text-Book INVALUABLE."—*Architect*.

"The author has certainly succeeded in producing a THOROUGHLY PRACTICAL Text-Book."—*Builder*.

"We can unhesitatingly recommend this work not only to the Student, as the BEST TEXT-BOOK on the subject, but also to the professional engineer as an EXCEEDINGLY VALUABLE book of reference."—*Mechanical World*.

"This work can be CONFIDENTLY recommended to engineers. The author has wisely chosen to use as little of the higher mathematics as possible, and has thus made his book of REAL USE TO THE PRACTICAL ENGINEER. . . . After careful perusal, we have nothing but praise for the work."—*Nature*.

LONDON: EXETER STREET, STRAND.

With numerous Tables and Illustrations. Crown 8vo. Cloth, 10/6.
Second Edition; Revised.

ASSAYING (A Text-Book of):

For the use of Students, Mine Managers, Assayers, &c.

BY

C. BERINGER, F.I.C., F.C.S.,

Late Chief Assayer to the Rio Tinto Copper Company, London,

AND

J. J. BERINGER, F.I.C., F.C.S.,

Public Analyst for, and Lecturer to the Mining Association of, Cornwall.

General Contents.

PART I.—INTRODUCTORY; MANIPULATION: Sampling; Drying; Calculation of Results—Laboratory-books and Reports. METHODS: Dry Gravimetric; Wet Gravimetric—Volumetric Assays: Titrometric, Colorimetric, Gasometric—Weighing and Measuring—Reagents—Formulæ, Equations, &c.—Specific Gravity.

PART II.—METALS: Detection and Assay of Silver, Gold, Platinum, Mercury, Copper, Lead, Thallium, Bismuth, Antimony, Iron, Nickel, Cobalt, Zinc, Cadmium, Tin, Tungsten, Titanium, Manganese, Chromium, &c.—Earths, Alkalies.

PART III.—NON-METALS: Oxygen and Oxides; The Halogens—Sulphur and Sulphates—Arsenic, Phosphorus, Nitrogen—Silicon, Carbon, Boron.

Appendix.—Various Tables useful to the Analyst.

"A REALLY MERITORIOUS WORK, that may be safely depended upon either for systematic instruction or for reference."—*Nature.*

"Of the fitness of the authors for the task they have undertaken, there can be no question. . . . Their book ADMIRABLY FULFILS ITS PURPOSE. . . . The results given of an exhaustive series of experiments made by the authors, showing the effects of VARYING CONDITIONS on the accuracy of the method employed, are of THE UTMOST IMPORTANCE."—*Industries.*

"A very good feature of the book is that the authors give reliable information, mostly based on practical experience."—*Engineering.*

"This work is one of the BEST of its kind. . . . Essentially of a practical character. . . . Contains all the information that the Assayer will find necessary in the examination of minerals."—*Engineer.*

LONDON: EXETER STREET, STRAND.

SCIENTIFIC AND TECHNICAL WORKS. 25

In 8vo, Handsome Cloth. Price 18s.

PHOTOGRAPHY:
ITS HISTORY, PROCESSES, APPARATUS, AND MATERIALS.

COMPRISING

WORKING DETAILS OF ALL THE MORE IMPORTANT METHODS.

BY A. BROTHERS, F.R.A.S.

WITH TWENTY-FOUR FULL PAGE PLATES BY MANY OF THE PROCESSES DESCRIBED, AND ILLUSTRATIONS IN THE TEXT.

GENERAL CONTENTS.

PART. I. INTRODUCTORY—Historical Sketch; Chemistry and Optics of Photography; Artificial Light (Electric and Oxyhydrogen Light, Compressed Gas, Ethexo-Limelight, Magnesium Light, &c.)

PART II. Photographic Processes, New and Old, with special reference to their relative Practical Usefulness.

PART III. Apparatus employed in Photography.

PART IV. Materials employed in Photography.

PART V. Applications of Photography; Practical Hints.

" Mr. Brothers has had an experience in Photography so large and varied that any work by him cannot fail to be interesting and valuable. . . . A MOST COMPREHENSIVE volume, entering with full details into the various processes, and VERY FULLY illustrated. The PRACTICAL HINTS are of GREAT VALUE. . . . Admirably got up."—*Brit. Jour. of Photography.*

" For the Illustrations alone, the book is most interesting; but, apart from these, the volume is valuable, brightly and pleasantly written, and MOST ADMIRABLY ARRANGED."—*Photographic News.*

" Certainly the FINEST ILLUSTRATED HANDBOOK to Photography which has ever been published. We have three Photogravures, four Collotypes, one Chromo-Collotype, numerous Blocks, Photo-Chromo-Typograph, Chromo-Lithograph, Woodbury-Type, and Woodbury-Gravure Prints, besides many others. . . . A work which should be on the reference shelves of every Photographic Society."—*Amateur Photographer.*

"This really IMPORTANT handbook of Photography . . . the result of wide experience . . . a manual of the best class. . . . As an album of examples of photographic reproduction alone, the book is not dear at the price. . . . A handbook so far in advance of most others, that the Photographer must not fail to obtain a copy as a reference work."—*Photographic Work.*

" The COMPLETEST HANDBOOK of the art which has yet been published. There is no process or form of apparatus which is not described and explained. The beautiful plates given as examples of the different processes are a special feature."—*Scotsman.*

" Processes are described which cannot be found elsewhere, at all events in so convenient and complete a form."—*English Mechanic.*

"The chapter on PRACTICAL HINTS will prove INVALUABLE. Mr. Brothers is certainly to be congratulated on the THOROUGHNESS of his work."—*Daily Chronicle.*

LONDON : EXETER STREET, STRAND.

MINE-SURVEYING (A Text-Book of):

For the use of Managers of Mines and Collieries, Students at the Royal School of Mines, &c.

By BENNETT H. BROUGH, F.G.S.,
Instructor of Mine-Surveying, Royal School of Mines.

With Diagrams. THIRD EDITION. Crown 8vo. Cloth, 7s. 6d.

GENERAL CONTENTS.

General Explanations—Measurement of Distances—Miner's Dial—Variation of the Magnetio-Needle—Surveying with the Magnetic-Needle in presence of Iron—Surveying with the Fixed Needle—German Dial—Theodolite—Traversing Underground—Surface-Surveys with Theodolite—Plotting the Survey—Calculation of Areas—Levelling—Connection of Underground- and Surface-Surveys—Measuring Distances by Telescope—Setting-out—Mine-Surveying Problems—Mine Plans—Applications of Magnetic-Needle in Mining—*Appendices*.

"It is the kind of book which has long been wanted, and no English-speaking Mine Agent or Mining Student will consider his technical library complete without it."—*Nature*.

"Supplies a long-felt want."—*Iron*.

"A valuable accessory to Surveyors in every department of commercial enterprise."—*Colliery Guardian*.

WORKS
By WALTER R. BROWNE, M.A., M. INST. C.E.,
Late Fellow of Trinity College, Cambridge.

THE STUDENT'S MECHANICS:
An Introduction to the Study of Force and Motion.
With Diagrams. Crown 8vo. Cloth, 4s. 6d.

"Clear in style and practical in method, 'THE STUDENT'S MECHANICS' is cordially to recommended from all points of view."—*Athenæum*.

FOUNDATIONS OF MECHANICS.
Papers reprinted from the *Engineer*. In Crown 8vo, 1s.

FUEL AND WATER:
A Manual for Users of Steam and Water.
By PROF. SCHWACKHÖFER AND W. R. BROWNE, M.A. (See p. 49).

LONDON: EXETER STREET, STRAND.

PRACTICAL GEOLOGY
(AIDS IN):
WITH A SECTION ON PALÆONTOLOGY.

BY

GRENVILLE A. J. COLE, F.G.S.,
Professor of Geology in the Royal College of Science for Ireland.

With Numerous Illustrations and Tables.   Large Crown 8vo.   Cloth, 10s. 6d.

GENERAL CONTENTS.

PART I.—SAMPLING OF THE EARTH'S CRUST.

Observations in the field. | Collection and packing of specimens.

PART II.—EXAMINATION OF MINERALS.

Some physical characters of minerals.
Simple tests with wet reagents.
Examination of minerals with the blowpipe.
Simple and characteristic reactions.
| Blowpipe-tests.
Quantitative flame reactions of the felspars and their allies.
Examination of the optical properties of minerals.

PART III.—EXAMINATION OF ROCKS.

Introductory.
Rock-structures easily distinguished.
Some physical characters of rocks.
Chemical examination of rocks.
Isolation of the constituents of rocks.
The petrological microscope and microscopic preparations.
| The more prominent characters to be observed in minerals in rock-sections.
Characters of the chief rock-forming minerals in the rock-mass and in thin sections.
Sedimentary rocks.
Igneous rocks.
Metamorphic rocks.

PART IV.—EXAMINATION OF FOSSILS.

Introductory.
Fossil generic types.—Rhizopoda ; Spongiæ ; Hydrozoa ; Actinozoa.
Polyzoa ; Brachiopoda.
Lamellibranchiata.
| Scaphopoda ;   Gastropoda ;   Pteropoda
Cephalopoda.
Echinodermata ; Vermes.
Anthropoda.
Suggested list of characteristic invertebrate fossils.

"Prof. Cole treats of the examination of minerals and rocks in a way that has never been attempted before . . . DESERVING OF THE HIGHEST PRAISE. Here indeed are 'aids' INNUMERABLE and INVALUABLE. All the directions are given with the utmost clearness and precision. Prof. Cole is not only an accomplished Petrologist, he is evidently also a thoroughly sympathetic teacher, and seems to know intuitively what are stumbling-blocks to learners—a rare and priceless quality."—*Athenæum.*

"To the younger workers in Geology, Prof. Cole's book will be as INDISPENSABLE as a dictionary to the learners of a language."—*Saturday Review.*

"That the work deserves its title, that it is full of 'AIDS,' and in the highest degree 'PRACTICAL,' will be the verdict of all who use it."—*Nature.*

"A MOST VALUABLE and welcome book . . . the subject is treated on lines wholly different from those in any other Manual, and is therefore very ORIGINAL."—*Science Gossip.*

"A more useful work for the practical geologist has not appeared in handy form."—*Scottish Geographical Magazine.*

"This EXCELLENT MANUAL . . . will be A VERY GREAT HELP. . . . The section on the Examination of Fossils is probably the BEST of its kind yet published. . . . FULL of well-digested information from the newest sources and from personal research."—*Annals of Nat. History.*

LONDON: EXETER STREET, STRAND.

SEWAGE DISPOSAL WORKS:
A GUIDE TO THE CONSTRUCTION OF WORKS FOR THE PREVENTION OF THE POLLUTION BY SEWAGE OF RIVERS AND ESTUARIES.

BY

W. SANTO CRIMP, M.INST.C.E., F.G.S.,
Assistant-Engineer, London County Council.

With Tables, Illustrations in the Text, and 33 Lithographic Plates.
Medium 8vo. Handsome Cloth, 25s.

PART I.—INTRODUCTORY.

Introduction.
Details of River Pollutions and Recommendations of Various Commissions.
Hourly and Daily Flow of Sewage.
The Pail System as Affecting Sewage.
The Separation of Rain-water from the Sewage Proper.

Settling Tanks.
Chemical Processes.
The Disposal of Sewage-sludge.
The Preparation of Land for Sewage Disposal.
Table of Sewage Farm Management.

PART II.—SEWAGE DISPOSAL WORKS IN OPERATION—THEIR CONSTRUCTION, MAINTENANCE AND COST.

Illustrated by Plates showing the General Plan and Arrangement adopted in each District.

1. Doncaster Irrigation Farm.
2. Beddington Irrigation Farm, Borough of Croydon.
3. Bedford Sewage Farm Irrigation.
4. Dewsbury and Hitchin Intermittent Filtration.
5. Merton, Croydon Rural Sanitary Authority.
6. Swanwick, Derbyshire.
7. The Ealing Sewage Works.
8. Chiswick.
9. Kingston-on-Thames, A. B. C. Process.
10. Salford Sewage Works.
11. Bradford, Precipitation.
12. New Malden, Chemical Treatment and Small Filters.
13. Friern Barnet.
14. Acton, Ferozone and Polarite Process.
15. Ilford, Chadwell, and Dagenham Sewage Disposal Works.
16. Coventry.
17. Wimbledon.
18. Birmingham.
19. Newhaven.
20. Portsmouth.
21. Sewage Precipitation Works, Dortmund (Germany).
22. Treatment of Sewage by Electrolysis.

"All persons interested in Sanitary Science owe a debt of gratitude to Mr. Crimp. . . . His work will be especially useful to SANITARY AUTHORITIES and their advisers . . . EMINENTLY PRACTICAL AND USEFUL . . . gives plans and descriptions of MANY OF THE MOST IMPORTANT SEWAGE WORKS of England . . . with very valuable information as to the COST of construction and working of each. . . . The carefully-prepared drawings permit of an easy comparison between the different systems."—*Lancet.*

"Probably the MOST COMPLETE AND BEST TREATISE on the subject which has appeared in our language. . . . Will prove of the greatest use to all who have the problem of Sewage Disposal to face. . . . The general construction, drawings, and type are all excellent."—*Edinburgh Medical Journal.*

LONDON: EXETER STREET, STRAND.

WORKS
By J. R. AINSWORTH DAVIS, B.A.,
PROFESSOR OF BIOLOGY, UNIVERSITY COLLEGE, ABERYSTWYTH.

BIOLOGY (A Text-Book of):
Comprising Vegetable and Animal Morphology and Physiology. In Large Crown 8vo, with 158 Illustrations. Cloth.
[*Second Edition in preparation.*]

GENERAL CONTENTS.
PART I. VEGETABLE MORPHOLOGY AND PHYSIOLOGY.—Fungi—Algæ—The Moss—The Fern—Gymnosperms—Angiosperms.
Comparative Vegetable Morphology and Physiology—Classification of Plants.

PART II. ANIMAL MORPHOLOGY AND PHYSIOLOGY.—Protozoa—Cœlenterata—Vermes—Arthropoda—Mollusca—Amphibia—Aves—Mammalia.
Comparative Animal Morphology and Physiology—Classification of Animals.

With Bibliography, Exam.-Questions, complete Glossary, and 158 *Illustrations.*

" As a general work of reference, Mr. Davis's manual will be HIGHLY SERVICEABLE to medical men.'—*British Medical Journal.*

" Furnishes a clear and comprehensive exposition of the subject in a systematic form."—*Saturday Review.*

" Literally PACKED with information."—*Glasgow Medical Journal.*

THE FLOWERING PLANT,
AS ILLUSTRATING THE FIRST PRINCIPLES OF BOTANY.
Specially adapted for London Matriculation, S. Kensington, and University Local Examinations in Botany. Large Crown 8vo, with numerous Illustrations. 3s. 6d.
SECOND EDITION.

" It would be hard to find a Text-book which would better guide the student to an accurate knowledge of modern discoveries in Botany. . . . The SCIENTIFIC ACCURACY of statement, and the concise exposition of FIRST PRINCIPLES make it valuable for educational purposes. In the chapter on the Physiology of Flowers, an *admirable résumé* is given, drawn from Darwin, Hermann Müller, Kerner, and Lubbock, of what is known of the Fertilization of Flowers."—*Journal of the Linnean Society.*

"We are much pleased with this volume . . . the author's style is MOST CLEAR, and his treatment that of a PRACTISED INSTRUCTOR. . . . The Illustrations are very good, suitable and helpful. The Appendix on Practical Work will be INVALUABLE to the private student. . . . We heartily commend the work."—*Schoolmaster.*

⁎ Recommended by the National Home-Reading Union; and also for use in the University Correspondence Classes.

LONDON: EXETER STREET, STRAND.

PROF. DAVIS'S WORKS—Continued.

A ZOOLOGICAL POCKET-BOOK;
Or, Synopsis of Animal Classification.

Comprising Definitions of the Phyla, Classes, and Orders, with explanatory Remarks and Tables.

BY DR. EMIL SELENKA,
Professor in the University of Erlangen.

Authorised English translation from the Third German Edition. In Small Post 8vo, Interleaved for the use of Students. Limp Covers, 4s.

"Dr. Selenka's Manual will be found useful by all Students of Zoology. It is a COMPREHENSIVE and SUCCESSFUL attempt to present us with a scheme of the natural arrangement of the animal world."—*Edin. Med. Journal.*

"Will prove very serviceable to those who are attending Biology Lectures. . . . The translation is accurate and clear."—*Lancet.*

INORGANIC CHEMISTRY (A Short Manual of).

BY A. DUPRÉ, Ph.D., F.R.S., AND WILSON HAKE,
Ph.D., F.I.C., F.C.S., of the Westminster Hospital Medical School.

SECOND EDITION, Revised. Crown 8vo. Cloth, 7s. 6d.

"A well-written, clear and accurate Elementary Manual of Inorganic Chemistry. . . . We agree heartily in the system adopted by Drs. Dupré and Hake. WILL MAKE EXPERIMENTAL WORK TREBLY INTERESTING BECAUSE INTELLIGIBLE."—*Saturday Review.*

"There is no question that, given the PERFECT GROUNDING of the Student in his Science, the remainder comes afterwards to him in a manner much more simple and easily acquired. The work IS AN EXAMPLE OF THE ADVANTAGES OF THE SYSTEMATIC TREATMENT of a Science over the fragmentary style so generally followed. BY A LONG WAY THE BEST of the small Manuals for Students."—*Analyst.*

HINTS ON THE PRESERVATION OF FISH,
IN REFERENCE TO FOOD SUPPLY.

BY J. COSSAR EWART, M.D., F.R.S.E.,
Regius Professor of Natural History, University of Edinburgh.

In Crown 8vo. Wrapper, 6d.

LONDON: EXETER STREET, STRAND.

*Royal 8vo. With numerous Illustrations and 17 Lithographic Plates.
Handsome Cloth. Price 30s.*

BRIDGE-CONSTRUCTION
(A PRACTICAL TREATISE ON):

Being a Text-Book on the Construction of Bridges in
Iron and Steel.

FOR THE USE OF STUDENTS, DRAUGHTSMEN, AND ENGINEERS.

BY

T. CLAXTON FIDLER, M. INST. C.E.,
Prof. of Engineering, University College, Dundee.

"Of late years the American treatises on Practical and Applied Mechanics have taken the lead . . . since the opening up of a vast continent has given the American engineer a number of new bridge-problems to solve . . . but we look to the PRESENT TREATISE ON BRIDGE-CONSTRUCTION, and the Forth Bridge, to bring us to the front again."—*Engineer.*

"One of the VERY BEST RECENT WORKS on the Strength of Materials and its application to Bridge-Construction. . . . Well repays a careful Study."—*Engineering.*

"An INDISPENSABLE HANDBOOK for the practical Engineer."—*Nature.*

"The science is progressive, and as an exposition of its LATEST ADVANCES we are glad to welcome Mr. Fidler's well-written treatise."—*Architect.*

"An admirable account of the theory and process of bridge-design, AT ONCE SCIENTIFIC AND THOROUGHLY PRACTICAL. It is a book such as we have a right to expect from one who is himself a substantial contributor to the theory of the subject, as well as a bridge-builder of repute."—*Saturday Review.*

"This book is a model of what an engineering treatise ought to be."—*Industries.*

"A SCIENTIFIC TREATISE OF GREAT MERIT."—*Westminster Review.*

"Of recent text-books on subjects of mechanical science, there has appeared no one more ABLE, EXHAUSTIVE, or USEFUL than Mr. Claxton Fidler's work on Bridge-Construction."—*Scotsman.*

LONDON: EXETER STREET, STRAND.

FOSTER (C. Le Neve, D.Sc., Professor of Mining,
Royal College of Science; H.M. Inspector of Mines, Llandudno):

ORE AND STONE MINING (A Text-Book of). With numerous Illustrations. Large Crown 8vo. Cloth. [*Shortly*.

GRIFFIN'S ELECTRICAL PRICE-BOOK:
For the Use of Electrical, Civil, Marine, and Borough Engineers, Local Authorities, Architects, Railway Contractors, &c., &c. Edited by H. J. DOWSING, M.INST.E.E., &c. In Crown 8vo. Cloth. [*At Press*.

GRIFFIN (John Joseph, F.C.S.):

CHEMICAL RECREATIONS: A Popular Manual of Experimental Chemistry. With 540 Engravings of Apparatus. *Tenth Edition*. Crown 4to. Cloth.

Part I.—Elementary Chemistry, 2/.

Part II.—The Chemistry of the Non-Metallic Elements, including a Comprehensive Course of Class Experiments, 10/6.

Or, complete in one volume, cloth, gilt top, . . 12/6.

GURDEN (Richard Lloyd, Authorised Surveyor
for the Governments of New South Wales and Victoria):

TRAVERSE TABLES: computed to Four Places Decimals for every Minute of Angle up to 100 of Distance. For the use of Surveyors and Engineers. *Second Edition*. Folio, strongly half-bound, 21/.

* *Published with Concurrence of the Surveyors-General for New South Wales and Victoria.*

"Those who have experience in exact SURVEY-WORK will best know how to appreciate the enormous amount of labour represented by this valuable book. The computations enable the user to ascertain the sines and cosines for a distance of twelve miles to within half an inch, and this BY REFERENCE TO BUT ONE TABLE, in place of the usual Fifteen minute computations required. This alone is evidence of the assistance which the Tables ensure to every user, and as every Surveyor in active practice has felt the want of such assistance, few knowing of their publication will remain without them."—*Engineer*.

LONDON: EXETER STREET, STRAND.

SCIENTIFIC AND TECHNICAL WORKS. 33

Griffin's Standard Publications

FOR

ENGINEERS, ELECTRICIANS, ARCHITECTS, BUILDERS, NAVAL CONSTRUCTORS, AND SURVEYORS.

		PAGE
Applied Mechanics, . .	Prof. Rankine,	44, 45
,, (Student's), .	Browne, Jamieson,	26, 38
Civil Engineering, . .	Prof. Rankine,	. 44
Bridge-Construction, .	Prof. Fidler, .	. 31
Design of Structures, .	S. Anglin, . .	. 23
Sewage Disposal Works,	Santo Crimp, .	. 28
Traverse Tables, . .	R. Gurden, .	. 32
Marine Engineering, .	A. E. Seaton, .	. 50
Stability of Ships, .	Sir E. J. Reed,	. 40
The Steam-Engine, . .	Prof. Rankine,	. 45
,, (Student's),	Prof. Jamieson,	. 37
Boiler Construction, .	T. W. Traill, .	. 51
,, Management, .	R. D. Munro, .	. 38
Fuel and Water (for Steam Users), . .	Schwackhöfer and Browne, .	49
Machinery and Millwork,	Prof. Rankine,	. 44
Hydraulic Machinery, .	Prof. Robinson,	. 48
Useful Rules and Tables for Engineers, &c., .	Profs. Rankine and Jamieson, .	45
Electrical Pocket-Book,	Munro and Jamieson,	40
Nystrom's Pocket-Book,	Dennis Marks, .	. 40
Electrical Price-Book, .	H. J. Dowsing, .	. 32

☞ For a COMPLETE RECORD of the PAPERS read before the ENGINEERING, ARCHITECTURAL, and ELECTRICAL SOCIETIES throughout the United Kingdom during each year, *vide* "THE OFFICIAL YEAR-BOOK OF THE SCIENTIFIC AND LEARNED SOCIETIES OF GREAT BRITAIN AND IRELAND" (page 53).

LONDON : EXETER STREET, STRAND.

# Griffin's Standard Publications

FOR

MINE OWNERS AND MANAGERS, GEOLOGISTS,
METALLURGISTS, AND MANUFACTURERS.

		PAGE
Geology (Stratigraphical),	R. ETHERIDGE,	42
,, (Physical),	PROF. SEELEY,	41
,, (Practical),	PROF. COLE,	27
Mine-Surveying,	B. H. BROUGH,	26
Mining, Coal	H. W. HUGHES,	36
,, Ore and Stone,	PROF. LE NEVE FOSTER,	32
Blasting and Explosives,	O. GUTTMANN,	35
Metallurgy,	PHILLIPS AND BAUERMAN,	43
,, (Introduction to),	PROF. ROBERTS-AUSTEN,	47
Assaying,	C. & J. J. BERINGER,	24
Electro-Metallurgy,	W. M'MILLAN,	39

OTHER VOLUMES IN PREPARATION.

Griffin's Students' Text-Books.

		PAGE			PAGE
Biology,	Davis,	29	Magnetism and Electricity, Jamieson,		38
Botany,	Davis,	29			
Chemistry—			Mechanics,	Rankine,	45
Inorganic, Dupré & Hake,		30	Physics (Experiments), Wright,		52
Qual. Analysis, Sexton,		49			
Quant. ,, ,,		49	Physiology,	Stirling,	21
Recreations, Griffin,		32	Steam-Engine,	Jamieson,	37
Experiments, Wright,		52	Zoology,	Davis,	30

LONDON: EXETER STREET, STRAND.

*In Large 8vo, with Numerous Illustrations and Folding Plates,
Cloth, 10s. 6d.*

BLASTING:

A Handbook for the Use of Engineers and others Engaged in
Mining, Tunnelling, Quarrying, &c.

BY

OSCAR GUTTMANN, Assoc. M. Inst. C.E.

*Member of the Societies of Civil Engineers and Architects of Vienna and Budapest,
Corresponding Member of the Imp. Roy. Geological Institution of Austria, &c.*

*** Mr. GUTTMANN'S *Blasting* is the ONLY work on the subject which gives at once full information as to the NEW METHODS adopted since the introduction of Dynamite, and, at the same time, the results of MANY YEARS PRACTICAL EXPERIENCE both in Mining Work and in the Manufacture of Explosives. It therefore presents in concise form all that has been *proved* good in the various methods of procedure. The Illustrations form a special and valuable feature of the work.

GENERAL CONTENTS.—Historical Sketch—Blasting Materials—Blasting Powder—Various Powder-mixtures—Gun-cotton—Nitro-glycerine and Dynamite—Other Nitro-compounds—Sprengel's Liquid·(acid) Explosives—Other Means of Blasting—Qualities, Dangers, and Handling of Explosives—Choice of Blasting Materials—Apparatus for Measuring Force—Blasting in Fiery Mines—Means of Igniting Charges—Preparation of Blasts—Bore-holes—Machine-drilling—Chamber Mines—Charging of Bore-holes—Determination of the Charge—Blasting in Bore-holes—Firing—Straw and Fuze Firing—Electrical Firing—Substitutes for Electrical Firing—Results of Working—Various Blasting Operations—Quarrying—Blasting Masonry, Iron Structures, Wooden Objects—Blasting in earth, under water, of ice, &c., &c.

"This ADMIRABLE work."—*Colliery Guardian.*
"Should prove a *vade-mecum* to Mining Engineers and all engaged in practical work."
—*Iron and Coal Trades Review.*

HURST (GEO. H., F.C.S.):

PAINTERS' COLOURS, OILS, AND VARNISHES : A Text-book for Students and Practical Men. With Numerous Illustrations. (Griffin's Technological Series). Crown 8vo, Cloth, 12s. 6d.

KNECHT (Dr.), RAWSON (Chr., F.C.S.), AND LOEWENTHAL (Dr.) :

A MANUAL OF DYEING. With Numerous Illustrations and Specimens of Dyed Fabrics. (Griffin's Technological Series.)

[*At Press.*

LONDON: EXETER STREET, STRAND.

COAL-MINING (A Text-Book of):

FOR THE USE OF COLLIERY MANAGERS AND OTHERS ENGAGED IN COAL-MINING.

BY

HERBERT WILLIAM HUGHES, F.G.S.,
Assoc. Royal School of Mines, Certificated Colliery Manager.

In Demy 8vo, Handsome Cloth. With very Numerous Illustrations, mostly reduced from Working Drawings. 18s.

GENERAL CONTENTS.

Geology: Rocks—Faults—Order of Succession—Carboniferous System in Britain. **Coal**: Definition and Formation of Coal—Classification and Commercial Value of Coals. **Search for Coal**: Boring—various appliances used—Devices employed to meet Difficulties of deep Boring—Special methods of Boring—Mather & Platt's, American, and Diamond systems—Accidents in Boring—Cost of Boring—Use of Boreholes. **Breaking Ground**: Tools—Transmission of Power: Compressed Air, Electricity—Power Machine Drills—Coal Cutting by Machinery—Cost of Coal Cutting—Explosives—Blasting in Dry and Dusty Mines—Blasting by Electricity—Various methods to supersede Blasting. **Sinking**: Position, Form, and Size of shaft—Operation of getting down to "Stone-head"—Method of proceeding afterwards—Lining shafts—Keeping out Water by Tubbing—Cost of Tubbing—Sinking by Boring—Kind-Chaudron, and Lipmann methods—Sinking through Quicksands—Cost of Sinking. **Preliminary Operations**: Driving underground Roads—Supporting Roof: Timbering, Chocks or Cogs, Iron and Steel Supports and Masonry—Arrangement of Inset. **Methods of Working**: Shaft, Pillar, and Subsidence—Bord and Pillar System—Lancashire Method—Longwall Method—Double Stall Method—Working Steep Seams—Working Thick Seams—Working Seams lying near together—Spontaneous Combustion. **Haulage**: Rails—Tubs—Haulage by Horses—Self-acting Inclines—Direct-acting Haulage—Main and Tail Rope—Endless Chain—Endless Rope—Comparison. **Winding**: Pit Frames—Pulleys—Cages—Ropes—Guides—Engines—Drums—Brakes—Counterbalancing—Expansion—Condensation—Compound Engines—Prevention of Overwinding—Catches at pit top—Changing Tubs—Tub Controllers—Signalling. **Pumping**: Bucket and Plunger Pumps—Supporting Pipes in Shaft—Valves—Suspended lifts for Sinking—Cornish and Bull Engines—Davey Differential Engine—Worthington Pump—Calculations as to size of Pumps—Draining Deep Workings—Dams. **Ventilation**: Quantity of air required—Gases met with in Mines—Coal-dust—Laws of Friction—Production of Air-currents—Natural Ventilation—Furnace Ventilation—Mechanical Ventilators—Efficiency of Fans—Comparison of Furnaces and Fans—Distribution of the Air-current—Measurement of Air-currents. **Lighting**: Naked Lights—Safety Lamps—Modern Lamps—Conclusions—Locking and Cleaning Lamps—Electric Light Underground—Delicate Indicators. **Works at Surface**: Boilers—Mechanical Stoking—Coal Conveyors—Workshops. **Preparation of Coal for Market**: General Considerations—Tipplers—Screens—Varying the Sizes made by Screens—Belts—Revolving Tables—Loading Shoots—Typical Illustrations of the arrangement of Various Screening Establishments—Coal Washing—Dry Coal Cleaning—Briquettes.

*** A Novel and Important Feature will be found in the BIBLIOGRAPHY given at the end of each Chapter.

"A Text-book on Coal-Mining is a great desideratum, and Mr. HUGHES possesses ADMIRABLE QUALIFICATIONS for supplying it. . . . We cordially recommend the work."
—*Colliery Guardian.*

LONDON : EXETER STREET, STRAND.

ical_Engineer.

WORKS
By ANDREW JAMIESON, M.Inst.C.E., F.R.S.E.,
Professor of Engineering, Glasgow and West of Scotland Technical College.

SEVENTH EDITION, Revised and Enlarged. Crown 8vo, Cloth, 8s. 6d.

A TEXT-BOOK ON STEAM AND STEAM-ENGINES
WITH OVER 200 ILLUSTRATIONS, FOLDING-PLATES, AND EXAMINATION QUESTIONS.

" Professor Jamieson fascinates the reader by his CLEARNESS OF CONCEPTION AND SIMPLICITY OF EXPRESSION. His treatment recalls the lecturing of Faraday."—*Athenæum.*
" The BEST BOOK yet published for the use of Students."—*Engineer.*
" Undoubtedly the MOST VALUABLE AND MOST COMPLETE Hand-book on the subject that now exists."—*Marine Engineer.*

A POCKET-BOOK of ELECTRICAL RULES and TABLES.
FOR THE USE OF ELECTRICIANS AND ENGINEERS.
Pocket Size. Leather, 8s. 6d. *Ninth Edition*, revised and enlarged.
(See under *Munro and Jamieson*.)

ELECTRICITY & MAGNETISM (An Advanced Text-Book on)
For the Use of Science and Art, City and Guilds of London, and other Students. With Illustrations.
[*Shortly.*

PROF. JAMIESON'S ELEMENTARY MANUALS FOR FIRST-YEAR STUDENTS.

1. STEAM AND THE STEAM-ENGINE
(AN ELEMENTARY MANUAL ON):
Forming an Introduction to the larger Work by the same Author. With very numerous Illustrations and Examination Questions. *Third Edition.* Crown 8vo. Cloth, 3s. 6d.

" Quite the right sort of Book . . . well illustrated with good diagrams and drawings of real engines and details, all clearly and accurately lettered. . . . CANNOT FAIL TO BE A MOST SATISFACTORY GUIDE to the apprentice and Student."—*Engineer.*
" Should be in the hands of EVERY engineering apprentice."—*Practical Engineer.*

LONDON: EXETER STREET, STRAND.

PROF. JAMIESON'S ELEMENTARY MANUALS—Continued.

SECOND EDITION. Crown 8vo, with very numerous Illustrations.
2. MAGNETISM AND ELECTRICITY
(AN ELEMENTARY MANUAL ON).
With very Numerous Diagrams and Examination Questions.
Part I.—Magnetism. Part II.—Voltaic Electricity. Part III.—Electro-Statics, or Frictional Electricity.
Complete in One Volume, 3s. 6d.

"The arrangement is as good as it well can be, . . . the diagrams are EXCELLENT. . . . The subject treated as an essentially practical one, and very clear instructions given. Teachers are to be congratulated on having such a THOROUGHLY TRUSTWORTHY TEXT-BOOK at their disposal."—*Nature*.
"An excellent and very PRACTICAL elementary treatise."—*Electrical Review*.
"An ADMIRABLE Introduction to Magnetism and Electricity . . . the production of a skilled and experienced teacher. . . . Explained at every point by simple experiments, rendered easier by admirable illustrations."—*British Medical Journal*.
"A CAPITAL TEXT-BOOK. . . . The diagrams are an important feature."—*Schoolmaster*.

3. APPLIED MECHANICS (An Elementary Manual on).
With very numerous Illustrations drawn expressly for the work, and Examination Questions. Crown 8vo. 3s. 6d.

SECOND EDITION. *Enlarged, and very fully Illustrated.* Cloth, 4s. 6d.
STEAM - BOILERS:
THEIR DEFECTS, MANAGEMENT, AND CONSTRUCTION.
By R. D. MUNRO,
Engineer of the Scottish Boiler Insurance and Engine Inspection Company.

This work, written chiefly to meet the wants of Mechanics, Engine-keepers, and Boiler-attendants, also contains information of the first importance to every user of Steam-power. It is a PRACTICAL work written for PRACTICAL men, the language and rules being throughout of the simplest nature.

GENERAL CONTENTS.—Explosions caused by Overheating of Plates : (a) Shortness of Water : (b) Deposit—Explosions caused by Defective and Overloaded Safety-Valves —Area of Safety-Valves—Explosions caused by Corrosion—Explosions caused by Defective Design and Construction.
*** To the SECOND EDITION, a Section on the Management of Upright Internally-fired Boilers, and a Specification and detailed Drawing of a Lancashire Boiler for a working pressure of 200 lbs. per sq. in., have been added.

"A valuable companion for workmen and engineers engaged about Steam Boilers, ought to be carefully studied, and ALWAYS AT HAND."—*Coll. Guardian*.
"The subjects referred to are handled in a trustworthy, clear, and practical manner. . . . The book is VERY USEFUL, especially to steam users, artisans, and young engineers."—*Engineer*.

LONDON : EXETER STREET, STRAND.

SCIENTIFIC AND TECHNICAL WORKS. 39

ELECTRO-METALLURGY (A Treatise on):

Embracing the Application of Electrolysis to the Plating, Depositing, Smelting, and Refining of various Metals, and to the Reproduction of Printing Surfaces and Art-Work, &c.

BY WALTER G. M'MILLAN, F.I.C., F.C.S.,

Chemist and Metallurgist to the Cossipore Foundry and Shell-Factory; Late Demonstrator of Metallurgy in King's College, London.

With numerous Illustrations. Large Crown 8vo. Cloth, 10s. 6d.

GENERAL CONTENTS.

Introductory and Historical—Theoretical and General—Sources of Current—General Conditions to be observed in Electro-Plating — Plating Adjuncts and Disposition of Plant—Cleansing and Preparation of Work for the Depositing-Vat, and Subsequent Polishing of Plated Goods—Electro-Deposition of Copper—Electrotyping—Electro-Deposition of Silver—of Gold—of Nickel and Cobalt—of Iron—of Platinum, Zinc, Cadmium, Tin, Lead, Antimony, and Bismuth; Electro-chromy—Electro-Deposition of Alloys—Electro-Metallurgical Extraction and Refining Processes — Recovery of certain Metals from their Solutions or Waste Substances — Determination of the Proportion of Metal in certain Depositing Solutions—Glossary of Substances commonly employed in Electro-Metallurgy—Addenda: Various useful Tables—The Bronzing or Copper and Brass Surfaces—Antidotes to Poisons.

"This excellent treatise, . . . one of the BEST and MOST COMPLETE manuals hitherto published on Electro-Metallurgy."—*Electrical Review.*

"Well brought up to date, including descriptions such as that of Elmore's recent process for the manufacture of seamless copper tubes of extraordinary strength and tenacity by electro-deposition of the pure metal. . . . Illustrated by well-executed and effective engravings."—*Journal of Soc. of Chem. Industry.*

"This work will be a STANDARD."—*Jeweller.*

"Any metallurgical process which REDUCES the COST of production must of necessity prove of great commercial importance. . . . We recommend this manual to ALL who are interested in the PRACTICAL APPLICATION of electrolytic processes."—*Nature.*

LONDON: EXETER STREET, STRAND.

40   CHARLES GRIFFIN & CO.'S PUBLICATIONS.

MUNRO & JAMIESON'S ELECTRICAL POCKET-BOOK.

NINTH EDITION, Revised and Enlarged.

A POCKET-BOOK
OF
ELECTRICAL RULES & TABLES
FOR THE USE OF ELECTRICIANS AND ENGINEERS.

BY

JOHN MUNRO, C.E., & PROF. JAMIESON, M.INST.C.E., F.R.S.E.

With Numerous Diagrams. Pocket Size. Leather, 8s. 6d.

This work is fully illustrated, and forms an extremely convenient POCKET COMPANION for reference on important points essential to ELECTRICIANS AND ELECTRICAL ENGINEERS.

GENERAL CONTENTS.

UNITS OF MEASUREMENT.	ELECTRO-METALLURGY.
MEASURES.	BATTERIES.
TESTING.	DYNAMOS AND MOTORS.
CONDUCTORS.	TRANSFORMERS.
DIELECTRICS.	ELECTRIC LIGHTING
SUBMARINE CABLES.	MISCELLANEOUS.
TELEGRAPHY.	LOGARITHMS.
ELECTRO-CHEMISTRY.	APPENDICES.

"WONDERFULLY PERFECT. . . . Worthy of the highest commendation we can give it."—*Electrician.*
"The STERLING VALUE of Messrs. MUNRO and JAMIESON'S POCKET-BOOK."—*Electrical Review.*

NYSTROM'S POCKET-BOOK
OF
MECHANICS & ENGINEERING.
REVISED AND CORRECTED BY

W. DENNIS MARKS, PH.B., C.E. (YALE S.S.S.),
Whitney Professor of Dynamical Engineering, University of Pennsylvania.

Pocket Size. Leather, 15s. TWENTIETH EDITION. Revised and greatly enlarged.

LONDON: EXETER STREET, STRAND.

Demy 8vo, Handsome cloth, 18s.
PHYSICAL GEOLOGY AND PALÆONTOLOGY,
ON THE BASIS OF PHILLIPS.
BY
HARRY GOVIER SEELEY, F.R.S.,
PROFESSOR OF GEOGRAPHY IN KING'S COLLEGE, LONDON.

With Frontispiece in Chromo=Lithography, and Illustrations.

"It is impossible to praise too highly the research which PROFESSOR SEELEY'S 'PHYSICAL GEOLOGY' evidences. IT IS FAR MORE THAN A TEXT-BOOK—it is a DIRECTORY to the Student in prosecuting his researches."—*Extract from the Presidential Address to the Geological Society*, 1885, *by Rev. Professor Bonney, D.Sc., LL.D., F.R.S.*

"PROFESSOR SEELEY maintains in his 'PHYSICAL GEOLOGY' the high reputation he already deservedly bears as a Teacher. . . . It is difficult, in the space at our command, to do fitting justice to so large a work. . . . The final chapters, which are replete with interest, deal with the Biological aspect of Palæontology. Here we find discussed the origin, the extinction, succession, migration, persistence, distribution, relation, and variation of species —with other considerations, such as the Identification of Strata by Fossils, Homotaxis, Local Faunas, Natural History Provinces, and the relation of Living to Extinct forms."—*Dr. Henry Woodward, F.R.S., in the " Geological Magazine."*

"A deeply interesting volume, dealing with Physical Geology as a whole, and also presenting us with an animated summary of the leading doctrines and facts of Palæontology, as looked at from a modern standpoint."—*Scotsman.*

"PROFESSOR SEELEY'S work includes one of the most satisfactory Treatises on Lithology in the English language. . . . So much that is not accessible in other works is presented in this volume, that no Student of Geology can afford to be without it."—*American Journal of Engineering.*

"Geology from the point of view of Evolution."—*Westminster Review.*

"PROFESSOR SEELEY'S PHYSICAL GEOLOGY is full of instructive matter, whilst the philosophical spirit which it displays will charm many a reader. From early days the author gave evidence of a powerful and eminently original genius. No one has shown more convincingly than the author that, in all ways, the past contains within itself the interpretation of the existing world."— *Annals of Natural History.*

LONDON: EXETER STREET, STRAND.

Demy 8vo, Handsome cloth, 34s.

STRATIGRAPHICAL GEOLOGY AND PALÆONTOLOGY,

ON

THE BASIS OF PHILLIPS.

BY

ROBERT ETHERIDGE, F.R.S.,

OF THE NATURAL HIST. DEPARTMENT, BRITISH MUSEUM, LATE PALÆONTOLOGIST TO THE
GEOLOGICAL SURVEY OF GREAT BRITAIN, PAST PRESIDENT OF THE
GEOLOGICAL SOCIETY, ETC.

With Map, Numerous Tables, and Thirty-six Plates.

"In 1854 Prof. JOHN MORRIS published the Second Edition of his 'Catalogue of British Fossils,' then numbering 1,280 genera and 4,000 species. Since that date 3,000 genera and nearly 12,000 new species have been described, thus bringing up the muster-roll of extinct life in the British Islands alone to 3,680 genera and 16,000 known and described species.

"Numerous TABLES of ORGANIC REMAINS have been prepared and brought down to 1884, embracing the accumulated wealth of the labours of past and present investigators during the last thirty years. Eleven of these Tables contain every known British genus, zoologically or systematically placed, with the number of species in each, showing their broad distribution through time. The remaining 105 Tables are devoted to the analysis, relation, historical value, and distribution of specific life through each group of strata. These tabular deductions, as well as the Palæontological Analyses through the text, are, for the first time, fully prepared for English students."—*Extract from Author's Preface.*

*** PROSPECTUS *of the above important work—perhaps the* MOST ELABORATE *of its kind ever written, and one calculated to give a new strength to the study of Geology in Britain—may be had on application to the Publishers.*

It is not too much to say that the work will be found to occupy a place entirely its own, and will become **an indispensable guide** to every British Geologist.

"No such compendium of geological knowledge has ever been brought together before."—*Westminster Review.*

"If Prof. SEELEY's volume was remarkable for its originality and the breadth of its views, Mr. ETHERIDGE fully justifies the assertion made in his preface that his book differs in construction and detail from any known manual. . . . Must take HIGH RANK AMONG WORKS OF REFERENCE."—*Athenæum.*

LONDON: EXETER STREET, STRAND.

THIRD EDITION, *Revised by Mr. H. Bauerman, F.G.S.*

ELEMENTS OF METALLURGY:
A PRACTICAL TREATISE ON THE ART OF EXTRACTING METALS FROM THEIR ORES.

BY J. ARTHUR PHILLIPS, M.Inst.C.E., F.C.S. F.G.S., &c.,

AND

H. BAUERMAN, V.P.G.S.

With Folding Plates and many Illustrations. Med. 8vo.
Handsome Cloth, 36s.

GENERAL CONTENTS.

Refractory Materials.	Antimony.	Iron.
Fire-Clays.	Arsenic.	Cobalt.
Fuels, &c.	Zinc.	Nickel.
Aluminium.	Mercury.	Silver.
Copper.	Bismuth.	Gold.
Tin.	Lead.	Platinum.

** Many NOTABLE ADDITIONS, dealing with new Processes and Developments will be found in the Third Edition.

"Of the THIRD EDITION, we are still able to say that, as a Text-book of Metallurgy, it is THE BEST with which we are acquainted."—*Engineer.*

"The value of this work is almost *inestimable*. There can be no question that the amount of time and labour bestowed on it is enormous. . . . There is certainly no Metallurgical Treatise in the language calculated to prove of such general utility."—*Mining Journal.*

"'Elements of Metallurgy' possesses intrinsic merits of the highest degree. Such a work is precisely wanted by the great majority of students and practical workers, and its very compactness is in itself a first-rate recommendation. The author has treated with great skill the metallurgical operations relating to all the principal metals. The methods are described with surprising clearness and exactness, placing an easily intelligible picture of each process even before men of less practical experience, and illustrating the most important contrivances in an excellent and perspicuous manner. . . . In our opinion the best work ever written on the subject with a view to its practical treatment."—*Westminster Review.*

"In this most useful and handsome volume is condensed a large amount of valuable practical knowledge. A careful study of the first division of the book, on Fuels, will be found to be of great value to every one in training for the practical applications of our scientific knowledge to any of our metallurgical operations."—*Athenæum.*

"A work which is equally valuable to the Student as a Text-book, and to the practical Smelter as a Standard Work of Reference. . . . The Illustrations are admirable examples of Wood Engraving."—*Chemical News.*

LONDON: EXETER STREET, STRAND.

SCIENTIFIC MANUALS

BY

W. J. MACQUORN RANKINE, C.E., LL.D., F.R S.,
Late Regius Professor of Civil Engineering in the University of Glasgow.

THOROUGHLY REVISED BY W. J. MILLAR, C.E.,
Secretary to the Institute of Engineers and Shipbuilders in Scotland.

In Crown 8vo. Cloth.

I. RANKINE (Prof.): APPLIED MECHANICS:
comprising the Principles of Statics and Cinematics, and Theory of Structures, Mechanism, and Machines. With numerous Diagrams. *Thirteenth Edition*, 12/6.

"Cannot fail to be adopted as a text-book. . . . The whole of the information is so admirably arranged that there is every facility for reference."—*Mining Journal.*

II. RANKINE (Prof.): CIVIL ENGINEERING:
comprising Engineering Surveys, Earthwork, Foundations, Masonry, Carpentry, Metal-work, Roads, Railways, Canals, Rivers, Water-works, Harbours, &c. With numerous Tables and Illustrations. *Eighteenth Edition*, 16/.

"Far surpasses in merit every existing work of the kind. As a manual for the hands of the professional Civil Engineer it is sufficient and unrivalled, and even when we say this, we fall short of that high appreciation of Dr. Rankine's labours which we should like to express."—*The Engineer.*

III. RANKINE (Prof.): MACHINERY AND
MILLWORK: comprising the Geometry, Motions, Work, Strength, Construction, and Objects of Machines, &c. Illustrated with nearly 300 Woodcuts. *Sixth Edition*, 12/6.

"Professor Rankine's 'Manual of Machinery and Millwork' fully maintains the high reputation which he enjoys as a scientific author; higher praise it is difficult to award to any book. It cannot fail to be a lantern to the feet of every engineer.'—*The Engineer.*

LONDON: EXETER STREET, STRAND.

PROF. RANKINE'S WORKS—(*Continued*).

IV. RANKINE (Prof.): THE STEAM ENGINE and OTHER PRIME MOVERS. With Diagram of the Mechanical Properties of Steam, Folding-Plates, numerous Tables and Illustrations. *Thirteenth Edition*, 12/6.

V. RANKINE (Prof.): USEFUL RULES and TABLES for Engineers and others. With Appendix: TABLES, TESTS, and FORMULÆ for the use of ELECTRICAL ENGINEERS; comprising Submarine Electrical Engineering, Electric Lighting, and Transmission of Power. By ANDREW JAMIESON, C.E., F.R.S.E. *Seventh Edition*, 10/6.

"Undoubtedly the most useful collection of engineering data hitherto produced."—*Mining Journal.*

"Every Electrician will consult it with profit."—*Engineering.*

VI. RANKINE (Prof.): A MECHANICAL TEXT-BOOK. by Prof. MACQUORN RANKINE and E. F. BAMBER, C.E. With numerous Illustrations. *Fourth Edition*, 9/.

"The work, as a whole, is very complete, and likely to prove invaluable for furnishing a useful and reliable outline of the subjects treated of."—*Mining Journal.*

*** THE MECHANICAL TEXT-BOOK forms a simple introduction to PROFESSOR RANKINE'S SERIES of MANUALS on ENGINEERING and MECHANICS.

VII. RANKINE (Prof.): MISCELLANEOUS SCIENTIFIC PAPERS. Royal 8vo. Cloth, 31/6.

Part I. Papers relating to Temperature, Elasticity, and Expansion of Vapours, Liquids, and Solids. Part II. Papers on Energy and its Transformations. Part III. Papers on Wave-Forms, Propulsion of Vessels, &c.

With Memoir by Professor TAIT, M.A. Edited by W. J. MILLAR, C.E. With fine Portrait on Steel, Plates, and Diagrams.

"No more enduring Memorial of Professor Rankine could be devised than the publication of these papers in an accessible form. . . . The Collection is most valuable on account of the nature of his discoveries, and the beauty and completeness of his analysis. . . . The Volume exceeds in importance any work in the same department published in our time "—*Architect.*

LONDON: EXETER STREET, STRAND.

Royal 8vo, Handsome Cloth, 25s.

THE STABILITY OF SHIPS.

BY

SIR EDWARD J. REED, K.C.B., F.R.S., M.P.,

KNIGHT OF THE IMPERIAL ORDERS OF ST. STANILAUS OF RUSSIA; FRANCIS JOSEPH OF AUSTRIA; MEDJIDIE OF TURKEY; AND RISING SUN OF JAPAN; VICE-PRESIDENT OF THE INSTITUTION OF NAVAL ARCHITECTS.

With numerous Illustrations and Tables.

THIS work has been written for the purpose of placing in the hands of Naval Constructors, Shipbuilders, Officers of the Royal and Mercantile Marines, and all Students of Naval Science, a complete Treatise upon the Stability of Ships, and is the only work in the English Language dealing exhaustively with the subject.

The plan upon which it has been designed is that of deriving the fundamental principles and definitions from the most elementary forms of floating bodies, so that they may be clearly understood without the aid of mathematics; advancing thence to all the higher and more mathematical developments of the subject.

The work also embodies a very full account of the historical rise and progress of the Stability question, setting forth the results of the labours of BOUGUER, BERNOULLI, DON JUAN D'ULLOA, EULER, CHAPMAN, and ROMME, together with those of our own Countrymen, ATWOOD, MOSELEY, and a number of others.

The modern developments of the subject, both home and foreign, are likewise treated with much fulness, and brought down to the very latest date, so as to include the labours not only of DARGNIES, REECH (whose famous *Mémoire*, hitherto a sealed book to the majority of English naval architects, has been reproduced in the present work), RISBEC, FERRANTY, DUPIN, GUYOU, and DAYMARD, in France, but also those of RANKINE, WOOLLEY, ELGAR, JOHN, WHITE, GRAY, DENNY, INGLIS, and BENJAMIN, in Great Britain.

In order to render the work complete for the purposes of the Shipbuilder, whether at home or abroad, the Methods of Calculation introduced by Mr. F. K. BARNES, Mr. GRAY, M. REECH, M. DAYMARD, and Mr. BENJAMIN, are all given separately, illustrated by Tables and worked-out examples. The book contains more than 200 Diagrams, and is illustrated by a large number of actual cases, derived from ships of all descriptions, but especially from ships of the Mercantile Marine.

The work will thus be found to constitute the most comprehensive and exhaustive Treatise hitherto presented to the Profession on the Science of the STABILITY OF SHIPS.

"Sir EDWARD REED's 'STABILITY OF SHIPS' is INVALUABLE. In it the STUDENT, new to the subject, will find the path prepared for him, and all difficulties explained with the utmost care and accuracy; the SHIP-DRAUGHTSMAN will find all the methods of calculation at present in use fully explained and illustrated, and accompanied by the Tables and Forms employed; the SHIPOWNER will find the variations in the Stability of Ships due to differences in forms and dimensions fully discussed, and the devices by which the state of his ships under all conditions may be graphically represented and easily understood; the NAVAL ARCHITECT will find brought together and ready to his hand, a mass of information which he would otherwise have to seek in an almost endless variety of publications, and some of which he would possibly not be able to obtain at all elsewhere."—*Steamship.*

"This IMPORTANT AND VALUABLE WORK . . cannot be too highly recommended to all connected with shipping interests."—*Iron.*

"This VERY IMPORTANT TREATISE, . . . the MOST INTELLIGIBLE, INSTRUCTIVE, and COMPLETE that has ever appeared."—*Nature.*

"The volume is an ESSENTIAL ONE for the shipbuilding profession."—*Westminster Review.*

LONDON: EXETER STREET, STRAND.

SCIENTIFIC AND TECHNICAL WORKS. 47

In Large Crown 8vo, Handsome Cloth, with Numerous Illustrations, 7s. 6d. Second Edition.

METALLURGY

(AN INTRODUCTION TO THE STUDY OF).

BY

W. C. ROBERTS-AUSTEN, C.B., F.R.S.,

CHEMIST AND ASSAYER OF THE ROYAL MINT; PROFESSOR OF METALLURGY IN THE ROYAL COLLEGE OF SCIENCE.

GENERAL CONTENTS.

RELATION OF METALLURGY TO CHEMISTRY.
PHYSICAL PROPERTIES OF METALS.
ALLOYS.
THE THERMAL TREATMENT OF METALS.
FUEL.

MATERIALS AND PRODUCTS OF METALLURGICAL PROCESSES.
FURNACES.
MEANS OF SUPPLYING AIR TO FURNACES.
TYPICAL METALLURGICAL PROCESSES.
ECONOMIC CONSIDERATIONS.

" No English text-book at all approaches this one either in its method of treatment, its general arrangement, or in the COMPLETENESS with which the most modern views on the subject are dealt with. Professor Austen's volume will be INVALUABLE, not only to the student, but also to those whose knowledge of the art is far advanced."—*Chemical News.*

" INVALUABLE to the student. . . . Rich in matter not to be readily found elsewhere."—*Athenæum.*

" This volume amply realises the expectations formed as to the result of the labours of so eminent an authority. It is remarkable for its ORIGINALITY of conception and for the large amount of information which it contains. . . . The enormous amount of care and trouble expended upon it. . . . We recommend every one who desires information not only to consult, but to STUDY this work."—*Engineering.*

" Will at once take FRONT RANK as a text-book.—*Science and Art.*

" Prof. ROBERTS-AUSTEN'S book marks an epoch in the history of the teaching of metallurgy in this country."—*Industries.*

LONDON: EXETER STREET, STRAND.

Medium 8vo, Handsome cloth, 25s.

HYDRAULIC POWER
AND
HYDRAULIC MACHINERY.

BY

HENRY ROBINSON, M. INST. C.E., F.G.S.,

FELLOW OF KING'S COLLEGE, LONDON; PROF. OF CIVIL ENGINEERING,
KING'S COLLEGE, ETC., ETC.

With numerous Woodcuts, and 43 Litho. Plates.

GENERAL CONTENTS.

The Flow of Water under Pressure.
General Observations.
Waterwheels.
Turbines.
Centrifugal Pumps.
Water-pressure Pumps.
The Accumulator.
Hydraulic Pumping-Engine.
Three-Cylinder Engines and Capstans.
Motors with Variable Power.
Hydraulic Presses and Lifts.
Movable Jigger Hoist.
Hydraulic Waggon Drop.
The Flow of Solids.
Shop Tools.
Cranes.
Hydraulic Power applied to Bridges.
Dock-Gate Machinery.

Hydraulic Coal-discharging Machines.
Hydraulic Machinery on board Ship.
Hydraulic Pile Driver.
Hydraulic Excavator.
Hydraulic Drill.
Hydraulic Brake.
Hydraulic Gun-Carriages.
Jets.
Hydraulic Ram.
Packing.
Power Co-operation.
Cost of Hydraulic Power.
Tapping Pressure Mains.
Meters.
Waste Water Meter.
Pressure Reducing Valves.
Pressure Regulator.

" A Book of great Professional Usefulness."—*Iron.*

A full Prospectus of the above important work—giving a description of the Plates—may be had on application to the Publishers.

LONDON: EXETER STREET, STRAND.

SCHWACKHÖFER and BROWNE:

FUEL AND WATER: A Manual for Users of Steam and Water. By Prof. FRANZ SCHWACKHÖFER of Vienna, and WALTER R. BROWNE, M.A., C.E., late Fellow of Trinity College, Cambridge. Demy 8vo, with Numerous Illustrations, 9/.

GENERAL CONTENTS.—Heat and Combustion—Fuel, Varieties of—Firing Arrangements: Furnace, Flues, Chimney—The Boiler, Choice of—Varieties—Feed-water Heaters—Steam Pipes—Water: Composition, Purification—Prevention of Scale, &c., &c.

"The Section on Heat is one of the best and most lucid ever written."—*Engineer.*
"Contains a vast amount of useful knowledge. . . . Cannot fail to be valuable to thousands compelled to use steam power."—*Railway Engineer.*
"Its practical utility is beyond question."—*Mining Journal.*

SHELTON-BEY (W. Vincent, Foreman to the
Imperial Ottoman Gun Factories, Constantinople):

THE MECHANIC'S GUIDE: A Hand-Book for Engineers and Artizans. With Copious Tables and Valuable Recipes for Practical Use. Illustrated. *Second Edition.* Crown 8vo. Cloth, 7/6.

GENERAL CONTENTS.—Arithmetic—Geometry—Mensuration—Velocities in Boring and Wheel-Gearing—Wheel and Screw-Cutting—Miscellaneous Subjects and Useful Recipes—The Steam Engine—The Locomotive—Appendix: Tables for Practical Use.
"The MECHANIC'S GUIDE will answer its purpose as completely as a whole series of elaborate text-books."—*Mining Journal.*

WORKS by Prof. HUMBOLDT SEXTON, F.I.C., F.C.S., F.R.S.E.,
Glasgow and West of Scotland Technical College.

OUTLINES OF QUANTITATIVE ANALYSIS.
FOR THE USE OF STUDENTS.
With Illustrations. THIRD EDITION. Crown 8vo, Cloth, 3s.

"A practical work by a practical man . . . will further the attainment of accuracy and method."—*Journal of Education.*
"An ADMIRABLE little volume . . . well fulfils its purpose."—*Schoolmaster.*
"A COMPACT LABORATORY GUIDE for beginners was wanted, and the want has been WELL SUPPLIED. . . . A good and useful book."—*Lancet.*
"Mr. Sexton's book will be welcome to many teachers; for the processes are WELL CHOSEN, the principle which underlies each method is always CLEARLY EXPLAINED, and the directions are both SIMPLE and CLEAR."—*Brit. Med. Journal.*

BY THE SAME AUTHOR.

OUTLINES OF QUALITATIVE ANALYSIS.
FOR THE USE OF STUDENTS.
With Illustrations. THIRD EDITION. Crown 8vo, Cloth, 3s. 6d.

"The work of a thoroughly practical chemist . . . and one which may be unhesitatingly recommended."—*British Medical Journal.*
"Compiled with great care, and will supply a want."—*Journal of Education.*

LONDON: EXETER STREET, STRAND.

Now Ready, Tenth Edition, Revised and Enlarged, Price 18s.
Demy 8vo, Cloth. With Numerous Illustrations, reduced from Working Drawings.

A MANUAL
OF
MARINE ENGINEERING:
COMPRISING
THE DESIGNING, CONSTRUCTION, AND WORKING OF MARINE MACHINERY.

BY A. E. SEATON,
Lecturer on Marine Engineering to the Royal Naval College, Greenwich; Member of the Inst. of Civil Engineers; Member of Council of the Inst. of Naval Architects; Member of the Inst. of Mech. Engineers, &c.

GENERAL CONTENTS.

Part I.—Principles of Marine Propulsion.

Part II.—Principles of Steam Engineering.

Part III.—Details of Marine Engines: Design and Calculations for Cylinders,

Pistons, Valves, Expansion Valves, &c.

Part IV.—Propellers.

Part V.—Boilers.

Part VI.—Miscellaneous.

"In the three-fold capacity of enabling a Student to learn how to design, construct, and work a modern Marine Steam-Engine, Mr. Seaton's Manual has NO RIVAL as regards comprehensiveness of purpose and lucidity of treatment."—*Times.*

"The important subject of Marine Engineering is here treated with the THOROUGH-NESS that it requires. No department has escaped attention. . . . Gives the results of much close study and practical work."—*Engineering.*

"By far the BEST MANUAL in existence. . . . Gives a complete account of the methods of solving, with the utmost possible economy, the problems before the Marine Engineer."—*Athenæum.*

"The Student, Draughtsman, and Engineer will find this work the MOST VALUABLE HANDBOOK of Reference on the Marine Engine now in existence."—*Marine Engineer.*

LONDON: EXETER STREET, STRAND.

SCIENTIFIC AND TECHNICAL WORKS.

SECOND EDITION, *Revised and Enlarged. Pocket-Size, Leather, also for Office Use, Cloth,* 12s.

BOILERS, MARINE AND LAND;
THEIR CONSTRUCTION AND STRENGTH.

A HANDBOOK OF RULES, FORMULÆ, TABLES, &C., RELATIVE TO MATERIAL, SCANTLINGS, AND PRESSURES, SAFETY VALVES, SPRINGS, FITTINGS AND MOUNTINGS, &C.

For the Use of all Steam=Users.

BY T. W. TRAILL, M. INST. C. E., F. E. R. N.,

Engineer Surveyor-in-Chief to the Board of Trade.

*** In the New Issue the subject-matter has been considerably extended; Tables have been added for Pressures up to 200 lbs. per square inch, and some of the Tables have been altered, besides which new ones and other matter have been introduced, which have been specially prepared and computed for the SECOND EDITION.

"Very unlike any of the numerous treatises on Boilers which have preceded it. . . . Really useful. . . . Contains an ENORMOUS QUANTITY OF INFORMATION arranged in a very convenient form. . . . Those who have to design boilers will find that they can settle the dimensions for any given pressure with almost no calculation with its aid. . . . A MOST USEFUL VOLUME . . . supplying information to be had nowhere else."—*The Engineer.*

"As a handbook of rules, formulæ, &c., relating to materials, scantlings, and pressures, this work will prove MOST USEFUL. The name of the Author is a sufficient guarantee for its accuracy. It will save engineers, inspectors, and draughtsmen a vast amount of calculation."—*Nature.*

"Mr. Traill has done a very useful and unpretentious piece of work. Rules and tables are given in a way simple enough to be intelligible to the most unscientific engineer."—*Saturday Review.*

"By such an authority cannot but prove a welcome addition to the literature of the subject. In the hands of the practical engineer or boilermaker, its value as a ready, reliable, and widely comprehensive book of reference, must prove almost inestimable. . . . Will rank high as a standard work on the subject. We can strongly recommend it as being the MOST COMPLETE, eminently practical work on the subject."—*Marine Engineer.*

"To the engineer and practical boiler-maker it will prove INVALUABLE. Copious and carefully worked-out tables will save much of the calculating drudgery. . . . Many exceedingly useful and practical hints are given with regard to the treatment of iron and steel, which are exceedingly valuable, and the outcome of a wide experience. The tables in all probability are the most exhaustive yet published. . . . Certainly deserves a place on the shelf in the drawing office of every boiler shop."—*Practical Engineer.*

"We give it a hearty welcome. . . . A handy pocket-book. . . . Our readers cannot do better than purchase a copy. . . . Cheap at five times the price. The intelligent engineer can make a safe investment that will yield him a rich and satisfactory return."—*Engineers' Gazette.*

"From the author's well-known character for thoroughness and exactness, there is every reason to believe that the results given in the tables may be relied on. . . . The great experience of the author in all that relates to boiler construction constitutes him an authority that no one need be ashamed of quoting, and a guide as safe as any man in Britain."—*Shipping World.*

LONDON: EXETER STREET, STRAND.

SECOND EDITION. With very Numerous Illustrations. Handsome Cloth, 6s.
Also Presentation Edition, Gilt and Gilt Edges, 7s. 6d.

THE THRESHOLD OF SCIENCE:

A VARIETY OF EXPERIMENTS (Over 400)

ILLUSTRATING

SOME OF THE CHIEF PHYSICAL AND CHEMICAL PROPERTIES OF SURROUNDING OBJECTS,
AND THE EFFECTS UPON THEM OF LIGHT AND HEAT.

BY

C. R. ALDER WRIGHT, D.Sc., F.R.S.,

Lecturer on Chemistry and Physics in St. Mary's Hospital Medical School, London.

*** To the NEW EDITION has been added an excellent chapter on the Systematic Order in which Class Experiments should be carried out for Educational purposes.

"Its capital Index, its popular phraseology, its range of subjects, and numerous Illustrations, suggest three uses which Dr. Alder Wright's *Threshold of Science* will serve :—

"1. It is a CAPITAL LIBRARY BOOK for Teachers preparing simple object-lessons.
"2. In the HOME LIBRARY it would be a practical ENCYCLOPÆDIA of Science.
"3. Finally, the book is a WORKING BOOK; every point taken up has its accompanying experiment—often a MOST USEFUL one—of which full details are given."—*Schoolmaster.*

"Any one who may still have doubts regarding the value of Elementary Science as an organ of education will speedily have his doubts dispelled, if he takes the trouble to consider the methods recommended by Dr. Alder Wright. The Additions to the New Edition will be of great service to all who wish to use the volume, not merely as a 'play-book,' but as an instrument for the TRAINING of the MENTAL FACULTIES."—*Nature.*

"A FIRST-RATE BOOK to place in the hands of a boy."—*Educational Times.*

"An ADMIRABLE COLLECTION of Physical and Chemical Experiments . . . a large proportion of these may be performed at home without any costly apparatus."—*Journal of Education.*

"The work is quite as instructive as it is entertaining. . . . The language throughout is clear and simple."—*School Guardian.*

"Will teach the young experimentalist to USE HIS HANDS, and last, but not least, to THINK FOR HIMSELF."—*Industries.*

"Just the kind of book to add to a school library."—*Manchester Guardian.*

"Dr. Alder Wright has accomplished a task that will win for him the hearts of all intelligent youths with scientific leanings. . . . Step by step the learner is here gently guided through the paths of science, made easy by the perfect knowledge of the teacher, and made flowery by the most striking and curious experiments. Well adapted to become the TREASURED FRIEND of many a bright and promising lad."—*Manchester Examiner.*

"From the nature of gases to the making of soap-bubbles, from the freezing of water to the principles of 'pin-hole' photography, Dr. Alder Wright's book is an authority."—*Liverpool Mercury.*

LONDON: EXETER STREET, STRAND.

THE OFFICIAL YEAR-BOOK
OF THE
SCIENTIFIC AND LEARNED SOCIETIES OF GREAT BRITAIN AND IRELAND. Price 7/6.

Tenth Annual Issue.   Now Ready.

COMPILED FROM OFFICIAL SOURCES.

Comprising (together with other Official Information) LISTS of the PAPERS read during 1892 before the ROYAL SOCIETIES of LONDON and EDINBURGH, the ROYAL DUBLIN SOCIETY, the BRITISH ASSOCIATION, and all the LEADING SOCIETIES throughout the Kingdom engaged in the following Departments of Research:—

§ 1. Science Generally: *i.e.*, Societies occupying themselves with several Branches of Science, or with Science and Literature jointly.
§ 2. Mathematics and Physics.
§ 3. Chemistry and Photography.
§ 4. Geology, Geography, and Mineralogy.
§ 5. Biology, including Microscopy and Anthropology.
§ 6. Economic Science and Statistics.
§ 7. Mechanical Science and Architecture.
§ 8. Naval and Military Science.
§ 9. Agriculture and Horticulture.
§ 10. Law.
§ 11. Medicine.
§ 12. Literature.
§ 13. Psychology.
§ 14. Archæology.

"The YEAR-BOOK OF SOCIETIES is a Record which ought to be of the greatest use for the progress of Science."—*Sir Lyon Playfair, F.R.S., K.C.B., M.P., Past-President of the British Association.*

"It goes almost without saying that a Handbook of this subject will be in time one of the most generally useful works for the library or the desk."—*The Times.*

"The YEAR-BOOK OF SOCIETIES meets an obvious want, and promises to be a valuable work of reference."—*Athenæum.*

"The YEAR-BOOK OF SCIENTIFIC AND LEARNED SOCIETIES meets a want, and is therefore sure of a welcome."—*Westminster Review.*

"In the YEAR-BOOK OF SOCIETIES we have the FIRST ISSUE of what is, without doubt, a very useful work."—*Spectator.*

"The YEAR-BOOK OF SOCIETIES fills a very real want. The volume will become a Scientific Directory, chronicling the work and discoveries of the year, and enabling the worker in one branch to try his hand in all that interests him in kindred lines of research. We trust that it will meet with an encouraging reception."—*Engineering.*

Copies of the FIRST ISSUE, giving an Account of the History, Organisation, and Conditions of Membership of the various Societies [with Appendix on the Leading Scientific Societies throughout the world], and forming the groundwork of the Series, may still be had, price 7/6. *Also Copies of the following Issues.*

The YEAR-BOOK OF SOCIETIES forms a complete INDEX TO THE SCIENTIFIC WORK of the year in the various Departments. It is used as a ready HANDBOOK in all our great SCIENTIFIC CENTRES, MUSEUMS, and LIBRARIES throughout the Kingdom, and will, without doubt, become an INDISPENSABLE BOOK OF REFERENCE to every one engaged in Scientific Work.

We predict that the YEAR-BOOK OF SOCIETIES will speedily become one of those Year-Books WHICH IT WOULD BE IMPOSSIBLE TO DO WITHOUT."—*Bristol Mercury.*

LONDON: EXETER STREET, STRAND.

EDUCATIONAL WORKS.

*** *Specimen Copies of all the Educational Works published by Messrs. Charles Griffin and Company may be seen at the Libraries of the College of Preceptors, South Kensington Museum, and Crystal Palace; also at the depôts of the Chief Educational Societies.*

Griffin's Standard Classical Works.

		PAGE
Prehistoric Antiquities,	SCHRADER and JEVONS,	61
Greek Antiquities,	F. B. JEVONS,	59
Roman Antiquities,	Prof. RAMSAY,	60
„ Elementary,	Prof. RAMSAY,	60
Greek Literature,	F. B. JEVONS,	59
Roman Literature,	Rev. C. T. CRUTTWELL,	57
„ Specimens of,	CRUTTWELL and BANTON,	57
Greek Geography and Mythology,	DOERING and GRÆME,	58
Latin Prosody,	Prof. RAMSAY,	60
„ Elementary,	Prof. RAMSAY,	60
Virgil: Text and Notes (Illustrated),	Dr. BRYCE,	55
Horace: Text and Notes (Illustrated),	JOS. CURRIE,	55

LONDON: EXETER STREET, STRAND.

EDUCATIONAL WORKS.

BRYCE (Archibald Hamilton, D.C.L., LL.D.,
Senior Classical Moderator in the University of Dublin):
THE WORKS OF VIRGIL. Text from HEYNE and WAGNER. English Notes, original, and selected from the leading German and English Commentators. Illustrations from the antique. Complete in One Volume. *Fourteenth Edition.* Fcap 8vo. Cloth, 6/.
Or, in Three Parts:
Part I. BUCOLICS and GEORGICS, . . 2/6.
Part II. THE ÆNEID, Books I.-VI., . 2/6.
Part III. THE ÆNEID, Books VII.-XII.,. 2/6.

"Contains the pith of what has been written by the best scholars on the subject. . . . The notes comprise everything that the student can want."—*Athenæum.*
"The most complete, as well as elegant and correct edition of Virgil ever published in this country."—*Educational Times.*
"The best commentary on Virgil which a student can obtain."—*Scotsman.*

COBBETT (William): ENGLISH GRAMMAR,
in a Series of Letters, intended for the use of Schools and Young Persons in general. With an additional chapter on Pronunciation, by the Author's Son, JAMES PAUL COBBETT. *The only correct and authorised Edition.* Fcap 8vo. Cloth, 1/6.

COBBETT (William): FRENCH GRAMMAR.
Fifteenth Edition. Fcap 8vo. Cloth, 3/6.

"Cobbett's 'French Grammar' comes out with perennial freshness. There are few grammars equal to it for those who are learning, or desirous of learning, French without a teacher. The work is excellently arranged, and in the present edition we note certain careful and wise revisions of the text."—*School Board Chronicle.*

COBBIN'S MANGNALL: MANGNALL'S
HISTORICAL AND MISCELLANEOUS QUESTIONS, for the use of Young People. By RICHMAL MANGNALL. Greatly enlarged and corrected, and continued to the present time, by INGRAM COBBIN, M.A. *Fifty-fourth Thousand. New Illustrated Edition.* 12mo. Cloth, 4/.

COLERIDGE (Samuel Taylor): A DISSER-
TATION ON THE SCIENCE OF METHOD. (*Encyclopædia Metropolitana.*) With a Synopsis. *Ninth Edition.* Cr. 8vo. Cloth, 2/.

CURRIE (Joseph, formerly Head Classical
Master of Glasgow Academy):
THE WORKS OF HORACE: Text from ORELLIUS. English Notes, original, and selected from the best Commentators. Illustrations from the antique. Complete in One Volume. Fcap 8vo. Cloth, 5/.
Or in Two Parts:
Part I.—CARMINA, ... 3/.
Part II.—SATIRES AND EPISTLES, . . 3/.
"The notes are excellent and exhaustive."—*Quarterly Journal of Education.*

LONDON: EXETER STREET, STRAND.

CRAIK'S ENGLISH LITERATURE.

A COMPENDIOUS HISTORY OF ENGLISH LITERATURE AND OF THE ENGLISH LANGUAGE FROM THE NORMAN CONQUEST.

With numerous Specimens. By GEORGE LILLIE CRAIK, LL.D., late Professor of History and English Literature, Queen's College, Belfast. *New Edition.* In two vols. Royal 8vo. Handsomely bound in cloth, 25/.

GENERAL CONTENTS.

INTRODUCTORY.

I.—THE NORMAN PERIOD—The Conquest.
II.—SECOND ENGLISH—Commonly called Semi-Saxon.
III.—THIRD ENGLISH—Mixed, or Compound English.
IV.—MIDDLE AND LATTER PART OF THE SEVENTEENTH CENTURY.
V.—THE CENTURY BETWEEN THE ENGLISH REVOLUTION AND THE FRENCH REVOLUTION.
VI.—THE LATTER PART OF THE EIGHTEENTH CENTURY.
VII.—THE NINETEENTH CENTURY (*a*) THE LAST AGE OF THE GEORGES. (*b*) THE VICTORIAN AGE.

With numerous Excerpts and Specimens of Style.

"Anyone who will take the trouble to ascertain the fact, will find how completely even our great poets and other writers of the last generation have already faded from the view of the present, with the most numerous class of the educated and reading public. Scarcely anything is generally read except the publications of the day. YET NOTHING IS MORE CERTAIN THAN THAT NO TRUE CULTIVATION CAN BE SO ACQUIRED. This is the extreme case of that entire ignorance of history which has been affirmed, not with more point than truth, to leave a person always a child. . . . The present work combines the HISTORY OF THE LITERATURE with the HISTORY OF THE LANGUAGE. The scheme of the course and revolutions of the language which is followed here is extremely simple, and resting not upon arbitrary, but upon natural or real distinctions, gives us the only view of the subject that can claim to be regarded as of a scientific character."—*Extract from the Author's Preface.*

"Professor Craik has succeeded in making a book more than usually agreeable."—*The Times.*

Crown 8vo. Cloth, 7/6. TENTH EDITION.

A MANUAL OF ENGLISH LITERATURE,

for the use of Colleges, Schools, and Civil Service Examinations. Selected from the larger work, by Dr. CRAIK. *Tenth Edition.* With an Additional Section on Recent Literature, by HENRY CRAIK, M.A., Author of "A Life of Swift."

"A Manual of English Literature from so experienced and well-read a scholar as Professor Craik needs no other recommendation than the mention of its existence."—*Spectator.*

"This augmented effort will, we doubt not, be received with decided approbation by those who are entitled to judge, and studied with much profit by those who want to learn. . . . If our young readers will give healthy perusal to Dr. Craik's work, they will greatly benefit by the wide and sound views he has placed before them."—*Athenæum.*

"The preparation of the NEW ISSUE has been entrusted to Mr. HENRY CRAIK, Secretary to the Scotch Education Department, and well known in literary circles as the author of the latest and best Life of Swift. . . . A Series of TEST QUESTIONS is added, which must prove of great service to Students studying alone."—*Glasgow Herald.*

LONDON: EXETER STREET, STRAND.

WORKS BY REV. C. T. CRUTTWELL, M.A.,
Late Fellow of Merton College, Oxford.

A HISTORY OF ROMAN LITERATURE:
From the Earliest Period to the Times of the Antonines.

FIFTH EDITION. Crown 8vo. Cloth, 8/6.

"Mr. CRUTTWELL has done a real service to all Students of the Latin Language and Literature. . . . Full of good scholarship and good criticism."—*Athenæum*.
"A most serviceable—indeed, indispensable—guide for the Student. . . . The 'general reader' will be both charmed and instructed."—*Saturday Review*.
"The Author undertakes to make Latin Literature interesting, and he has succeeded. There is not a dull page in the volume."—*Academy*.
"The great merit of the work is its fulness and accuracy."—*Guardian*.
"This elaborate and careful work, in every respect of high merit. Nothing at all equal to it has hitherto been published in England."—*British Quarterly Review*.

Companion Volume. Second Edition.

SPECIMENS OF ROMAN LITERATURE:
From the Earliest Period to the Times of the Antonines.

Passages from the Works of Latin Authors, Prose Writers, and Poets:

Part I.—ROMAN THOUGHT: Religion, Philosophy and Science, Art and Letters, 6/.

Part II.—ROMAN STYLE: Descriptive, Rhetorical, and Humorous Passages, 5/.
Or in One Volume complete, 10/6.

Edited by C. T. CRUTTWELL, M.A., Merton College, Oxford; and PEAKE BANTON, M.A., some time Scholar of Jesus College, Oxford.

"'Specimens of Roman Literature' marks a new era in the study of Latin."—*English Churchman*.
"A work which is not only useful but necessary. . . . The plan gives it a standing-ground of its own. . . . The sound judgment exercised in plan and selection calls for hearty commendation."—*Saturday Review*.
"It is hard to conceive a completer or handier repertory of specimens of Latin thought and style."—*Contemporary Review*.

** KEY to PART II., PERIOD II. (being a complete TRANSLATION of the 85 Passages composing the Section), by THOS. JOHNSTON, M.A., may now be had (by Tutors and Schoolmasters only) on application to the Publishers. Price 2/6.

A HISTORY OF EARLY CHRISTIAN LITERATURE:
For the use of Students and General Readers.

8vo, Handsome Cloth. [*In Preparation.*

LONDON: EXETER STREET, STRAND.

HELLAS:
AN INTRODUCTION TO GREEK ANTIQUITIES,

Comprising the Geography, Religion and Myths, History, Art and Culture of old Greece.

On the Basis of the German Work by E. DOERING,
With Additions by ELLIOTT GRÆME.

In Large 8vo, with Map and Illustrations.

PART I.

The Land and the People: the Religion and Myths of Old Greece.

⁎ In the English version of Mr. Doering's work, the simple and interesting style of the original—written for young Students—has been retained; but, throughout, such additions and emendations have been made as render the work suitable for more advanced Students, and for all who desire to obtain, within moderate compass, more than a superficial acquaintance with the great People whose genius and culture have so largely influenced our own. The results of the latest researches by Dr. SCHLIEMANN, MM. FOUQUÉ, CARAPANOS, and others, are incorporated. [*Shortly.*

D'ORSEY (Rev. Alex. J. D., B.D., Corpus Christi Coll., Cambridge, Lecturer at King's College, London):

SPELLING BY DICTATION: Progressive Exercises in English Orthography, for Schools and Civil Service Examinations. *Sixteenth Thousand.* 18mo. Cloth, 1/.

FLEMING (William, D.D., late Professor of Moral Philosophy in the University of Glasgow):

THE VOCABULARY OF PHILOSOPHY: PSYCHOLOGICAL, ETHICAL, AND METAPHYSICAL. With Quotations and References for the Use of Students. Revised and Edited by HENRY CALDERWOOD, LL.D., Professor of Moral Philosophy in the University of Edinburgh. *Fourth Edition, enlarged.* Crown 8vo. Cloth, 10/6.

"The additions by the Editor bear in their clear, concise, vigorous expression, the stamp of his powerful intellect, and thorough command of our language. More than ever, the work is now likely to have a prolonged and useful existence, and to facilitate the researches of those entering upon philosophic studies."—*Weekly Review.*

JAMES (W. Powell, M.A.):

FROM SOURCE TO SEA: or, Gleanings about Rivers from many Fields. A Chapter in Physical Geography. Cloth elegant, 3/6.

"Excellent reading . . . a book of popular science which deserves an extensive circulation."—*Saturday Review.*

LONDON: EXETER STREET, STRAND.

WORKS BY F. B. JEVONS, M.A.

Now Ready. SECOND EDITION, Revised. Crown 8vo, Cloth, 8s. 6d.

A HISTORY OF GREEK LITERATURE,

From the Earliest Period to the Death of Demosthenes.

By FRANK BYRON JEVONS, M.A.,
Tutor in the University of Durham.

Part I.—Epic, Lyric, and the Drama.
Part II.—History, Oratory, and Philosophy.

SECOND EDITION. With Appendix on the *Present State of the Homeric Question* and Examination-Questions for the Use of Students.

"It is beyond all question the BEST HISTORY of Greek literature that has hitherto been published."—*Spectator.*
"An admirable text-book."—*Westminster Review.*
"Mr. Jevons' work supplies a real want."—*Contemporary Review.*
"Mr. Jevons' work is distinguished by the Author's THOROUGH ACQUAINTANCE with THE OLD WRITERS, and his DISCRIMINATING USE of the MODERN LITERATURE bearing upon the subject. . . . His great merit lies in its EXCELLENT EXPOSITION of the POLITICAL AND SOCIAL CAUSES concerned in the development of the Literature of Greece."—*Berlin Philologische Wochenschrift.*
"As a Text-Book, Mr. Jevons' work from its excellence deserves to SERVE AS A MODEL."
—*Deutsche Litteraturzeitung.*

THE DEVELOPMENT OF THE ATHENIAN DEMOCRACY.

Crown 8vo, 1s.

A MANUAL OF GREEK ANTIQUITIES.

FOR THE USE OF STUDENTS.

With Maps and Numerous Illustrations.

[*In Preparation.*

PREHISTORIC ANTIQUITIES OF THE ARYAN PEOPLES,

Translated from the German of DR. O. SCHRADER by F. B. JEVONS, M.A.
(See page 61, under SCHRADER.)

LONDON: EXETER STREET, STRAND.

McBURNEY (Isaiah, LL.D.,): EXTRACTS
FROM OVID'S METAMORPHOSES. With Notes, Vocabulary, &c. Adapted for Young Scholars. *Third Edition.* 18mo. Cloth, 1/6.

MENTAL SCIENCE: S. T. COLERIDGE'S
celebrated Essay on METHOD; Archbishop WHATELY's Treatises on LOGIC and RHETORIC. *Tenth Edition.* Crown 8vo. Cloth, 5/.

MILLER (W. Galbraith, M.A., LL.B., Lecturer
on Public Law, including Jurisprudence and International Law, in the University of Glasgow):

THE PHILOSOPHY OF LAW, LECTURES ON. Designed mainly as an Introduction to the Study of International Law. In 8vo. Handsome Cloth, 12/. *Now Ready.*

"Mr. MILLER's 'PHILOSOPHY OF LAW' bears upon it the stamp of a wide culture and of an easy acquaintanceship with what is best in modern continental speculation. . . . Interesting and valuable, because suggestive."—*Journal of Jurisprudence.*

WORKS BY WILLIAM RAMSAY, M.A.,
Trinity College, Cambridge, late Professor of Humanity in the University of Glasgow.

A MANUAL OF ROMAN ANTIQUITIES.
For the use of Advanced Students. With Map, 130 Engravings, and very copious Index. *Fourteenth Edition.* Crown 8vo. Cloth, 8/6.

"Comprises all the results of modern improved scholarship within a moderate compass."—*Athenæum.*

—— AN ELEMENTARY MANUAL OF
ROMAN ANTIQUITIES. Adapted for Junior Classes. With numerous Illustrations. *Eighth Edition.* Crown 8vo. Cloth, 4/.

—— A MANUAL OF LATIN PROSODY,
Illustrated by Copious Examples and Critical Remarks. For the use of Advanced Students. *Seventh Edition.* Crown 8vo. Cloth, 5/.

"There is no other work on the subject worthy to compete with it."—*Athenæum.*

—— AN ELEMENTARY MANUAL OF
LATIN PROSODY. Adapted for Junior Classes. Crown 8vo. Cloth, 2s.

LONDON: EXETER STREET, STRAND.

EDUCATIONAL WORKS. 61

In Large 8vo, Handsome Cloth, Gilt Top, 21s.

PREHISTORIC ANTIQUITIES OF THE ARYAN PEOPLES,

A Manual of Comparative Philology and the Earliest Culture.
Being the *Sprachvergleichung und Urgeschichte* of

DR. O. SCHRADER.

Translated from the SECOND GERMAN EDITION by

F. B. JEVONS, M.A.

In DR. SCHRADER'S great work is presented to the reader a most able and judicious summary of all recent researches into the Origin and History of those Peoples, Ancient and Modern, to whom has been mainly entrusted the civilisation and culture of the world.

Dr. Schrader's pictures of the Primeval Indo-European Period in all its most important phases—the Animal Kingdom, Cattle, The Plant-World, Agriculture, Computation of Time, Food and Drink, Clothing, Dwellings, Traffic and Trade, The Culture of the Indo-Europeans, and The Prehistoric Monuments of Europe (especially the Swiss Lake-Dwellings), Family and State, Religion, The Original Home—will be found not only of exceeding interest in themselves, but of great value to the Student of History, as throwing light upon later developments.

PART I.—HISTORY OF LINGUISTIC PALÆONTOLOGY.
PART II.—RESEARCH BY MEANS OF LANGUAGE AND HISTORY.
PART III.—THE FIRST APPEARANCE OF THE METALS.
PART IV.—THE PRIMEVAL PERIOD.

" Dr. SCHRADER'S GREAT WORK."—*Times*.
" Mr. Jevons has done his work excellently, and Dr. Schrader's book is a model of industry, erudition, patience, and, what is rarest of all in these obscure studies, of moderation and common sense."—*Saturday Review*.
"Ably translated by that well-known scholar Mr. Jevons, will be found the best COMPENDIUM of the last thirty years' research into the early history and speech of the Aryan race. . . . INTERESTING FROM BEGINNING TO END."—*Manchester Guardian*.
"In comparison with the First, the Second Edition has gained greatly—not merely in point of size, but of worth. We are convinced that the success which it deserves must attend, in its new form, a book so interesting and STIMULATING."—*Litterarisches Centralblatt*.
"When a book like this reaches a Second Edition, we have in the fact a proof that it has, in a happy way, solved the problem how to rouse the sympathy of the reader."
—*Allgemeine Zeitung*.
" A work which in every respect may be described as of CONSPICUOUS EXCELLENCE."
—*B(ru)gm(ann)—Litterarisches Centralblatt*.
"I must confess that, for long, I have read no work which has roused in me so lively an interest as Dr. SCHRADER'S. Here all is FRESH, LIVING INSIGHT, AND SOLID WELL-BALANCED REASONING."—*Wilh. Geiger—Deutsche Litteraturzeitung*.
" A MOST REMARKABLE BOOK."—*St(einthal)—Zeitschrift für Völkerpsychologie und Sprachwissenschaft*.
"One of the BEST WORKS published of late years. . . . Every one who, for any reason whatsoever, is interested in the beginnings of European Civilisation and Indo-European Antiquity, will be obliged to place Dr. SCHRADER'S book on his library shelves. The work addresses itself to the general reader as well as to the learned."—
Gustav Meyer—Philologische Wochenschrift.

LONDON: EXETER STREET, STRAND.

SENIOR (Nassau William, M.A., late Professor of Political Economy in the University of Oxford):
A TREATISE ON POLITICAL ECONOMY. *Sixth Edition.* Crown 8vo. Cloth. (*Encyclopædia Metropolitana*), 4/.

THOMSON (James): THE SEASONS. With an Introduction and Notes by ROBERT BELL, Editor of the "Annotated Series of British Poets." *Fourth Edition.* Fcap 8vo. Cloth, 1/6.

"An admirable introduction to the study of our English classics."

WHATELY (Archbishop): LOGIC—A Treatise on. With Synopsis and Index. (*Encyclopædia Metropolitana*), 3/.

—— RHETORIC—A Treatise on. With Synopsis and Index. (*Encyclopædia Metropolitana*), 3/6.

WORKS IN GENERAL LITERATURE.

BELL (Robert, Editor of the "Annotated Series of British Poets"):
GOLDEN LEAVES FROM THE WORKS OF THE POETS AND PAINTERS. Illustrated by Sixty-four superb Engravings on Steel, after Paintings by DAVID ROBERTS, STANFIELD, LESLIE, STOTHARD, HAYDON, CATTERMOLE, NASMYTH, Sir THOMAS LAWRENCE, and many others, and engraved in the first style of Art by FINDEN, GREATBACH, LIGHTFOOT, &c. *Second Edition.* 4to. Cloth gilt, 21/.

"'Golden Leaves' is by far the most important book of the season. The Illustrations are really works of art, and the volume does credit to the arts of England."—*Saturday Review.*

"The Poems are selected with taste and judgment."—*Times.*

"The engravings are from drawings by Stothard, Newton, Danby, Leslie, and Turner, and it is needless to say how charming are many of the above here given."—*Athenæum.*

LONDON: EXETER STREET, STRAND.

THE WORKS OF WILLIAM COBBETT.
THE ONLY AUTHORISED EDITIONS.

COBBETT (William): ADVICE TO YOUNG
Men and (incidentally) to Young Women. *New Edition.* With admirable Portrait on Steel. Fcap 8vo. Cloth, 2/6.

―――― **COTTAGE ECONOMY.**
Eighteenth Edition, revised by the Author's Son. Fcap 8vo. Cloth, 2/6.

―――― **EDUCATIONAL WORKS.**
(See page 25.)

―――― **A LEGACY TO LABOURERS.**
With a Preface by the Author's Son, JOHN M. COBBETT, late M.P. for Oldham. *New Edition.* Fcap 8vo. Cloth, 1/6.

―――― **A LEGACY TO PARSONS.**
New Edition. Fcap 8vo. Cloth, 1/6.

GILMER'S INTEREST TABLES; Tables for Calculation of Interest, on any sum, for any number of days, at ½, 1, 1½, 2, 2½, 3, 3½, 4, 4½, 5 and 6 per Cent. By ROBERT GILMER. Corrected and enlarged. *Eleventh Edition.* 12mo. Cloth, 5/.

GRÆME (Elliott): BEETHOVEN: a Memoir.
With Portrait, Essay, and Remarks on the Pianoforte Sonatas, with Hints to Students, by DR. FERDINAND HILLER, of Cologne. *Third Edition.* Crown 8vo. Cloth gilt, elegant, 5/.

"This elegant and interesting Memoir. . . . The newest, prettiest, and most readable sketch of the immortal Master of Music."—*Musical Standard.*
"A gracious and pleasant Memorial of the Centenary."—*Spectator.*
"This delightful little book — concise, sympathetic, judicious." — *Manchester Examiner.*
"We can, without reservation, recommend it as the most trustworthy and the pleasantest Memoir of Beethoven published in England."—*Observer.*
"A most readable volume, which ought to find a place in the library of every admirer of the great Tone-Poet."—*Edinburgh Daily Review.*

―――― **A NOVEL WITH TWO HEROES.**
Second Edition. In 2 vols. Post 8vo. Cloth, 21/.

"A decided literary success."—*Athenæum.*
"Clever and amusing . . . above the average even of good novels . . . free from sensationalism, but full of interest . . . touches the deeper chords of life . . . delineation of character remarkably good."—*Spectator.*
"Superior in all respects to the common run of novels."—*Daily News.*
"A story of deep interest. . . . The dramatic scenes are powerful almost to painfulness in their intensity."—*Scotsman.*

LONDON: EXETER STREET, STRAND.

THE EMERALD SERIES OF POETS.

Illustrated by Engravings on Steel, after STOTHARD, LESLIE, DAVID ROBERTS, STANFIELD, Sir THOMAS LAWRENCE, CATTERMOLE, &c. Fcap 8vo. Cloth, gilt.

Particular attention is requested to this very beautiful series. The delicacy of the engravings, the excellence of the typography, and the quaint antique head and tail pieces, render them the most beautiful volumes ever issued from the press of this country, and now, unquestionably, the cheapest of their class.

BYRON (Lord): CHILDE HAROLD'S PILGRIMAGE. With Memoir by Professor SPALDING. Illustrated with Portrait and Engravings on Steel, by GREATBACH, MILLER, LIGHTFOOT, &c., from Paintings by CATTERMOLE, Sir T. LAWRENCE, H. HOWARD, and STOTHARD. Beautifully printed on toned paper. *Third Thousand.* Cloth, gilt edges, 3/.

CAMPBELL (Thomas): THE PLEASURES OF HOPE. With Introductory Memoir by the Rev. CHARLES ROGERS, LL.D., and several Poems never before published. Illustrated with Portrait and Steel Engravings. *Second Thousand.* Cloth, gilt edges, 3/.

CHATTERTON'S (Thomas) POETICAL WORKS. With an Original Memoir by FREDERICK MARTIN, and Portrait. Beautifully illustrated on Steel, and elegantly printed. *Fourth Thousand.* Cloth, gilt edges, 3/.

GOLDSMITH'S (Oliver) POETICAL WORKS. With Memoir by Professor SPALDING. Exquisitely illustrated with Steel Engravings. *New Edition.* Printed on superior toned paper. *Seventh Thousand.* Cloth, gilt edges, 3/.

Eton Edition, with the Latin Poems. Sixth Thousand.

GRAY'S (Thomas) POETICAL WORKS. With Life by the Rev. JOHN MITFORD, and Essay by the EARL of CARLISLE. With Portrait and numerous Engravings on Steel and Wood. Elegantly printed on toned paper. Cloth, gilt edges, 5/.

HERBERT'S (George) POETICAL WORKS. With Memoir by J. NICHOL, B.A., Oxon, Prof. of English Literature in the University of Glasgow. Edited by CHARLES COWDEN CLARKE. Antique headings to each page. *Second Thousand.* Cloth, gilt edges, 3/.

KEBLE (Rev. John): THE CHRISTIAN YEAR. With Memoir by W. TEMPLE, Portrait, and Eight beautiful Engravings on Steel. *New Edition.*
Cloth, gilt edges, ... 5/.

POE'S (Edgar Allan) COMPLETE POETICAL WORKS. Edited, with Memoir, by JAMES HANNAY. Full-page Illustrations after WEHNERT, WEIR, &c. Toned paper. *Thirteenth Thousand.* Cloth, gilt edges, 3/.

LONDON: EXETER STREET, STRAND.

MACKEY'S FREEMASONRY:

A LEXICON OF FREEMASONRY. Containing a definition of its Communicable Terms, Notices of its History, Traditions, and Antiquities, and an Account of all the Rites and Mysteries of the Ancient World. By ALBERT G. MACKEY, M.D., Secretary-General of the Supreme Council of the U.S., &c. *Eighth Edition*, thoroughly revised with APPENDIX by Michael C. Peck, Prov. Grand Secretary for N. and E. Yorkshire. Handsomely bound in cloth, 6/.

"Of MACKEY's LEXICON it would be impossible to speak in too high terms ; suffice it to say, that, in our opinion, it ought to be in the hands of every Mason who would thoroughly understand and master our noble Science. . . . No Masonic Lodge or Library should be without a copy of this most useful work."—*Masonic News*.

HENRY MAYHEW'S CELEBRATED WORK ON THE STREET-FOLK OF LONDON.

LONDON LABOUR AND THE LONDON

POOR : A Cyclopædia of the Condition and Earnings of *those that will work and those that cannot work*. By HENRY MAYHEW. With many full-page Illustrations from Photographs. In three vols. Demy 8vo. Cloth. Each vol. 4/6.

"Every page of the work is full of valuable information, laid down in so interesting a manner that the reader can never tire."—*Illustrated London News*.

" Mr. Henry Mayhew's famous record of the habits, earnings, and sufferings of the London poor."—*Lloyd's Weekly London Newspaper*.

"This remarkable book, in which Mr. Mayhew gave the better classes their first real insight into the habits, modes of livelihood, and current of thought of the London poor."—*The Patriot*.

The Extra Volume.

LONDON LABOUR AND THE LONDON

POOR : *Those that will not work*. Comprising the Non-workers, by HENRY MAYHEW ; Prostitutes, by BRACEBRIDGE HEMYNG ; Thieves, by JOHN BINNY ; Beggars, by ANDREW HALLIDAY. With an Introductory Essay on the Agencies at Present in Operation in the Metropolis for the Suppression of Crime and Vice, by the Rev. WILLIAM TUCKNISS, B.A., Chaplain to the Society for the Rescue of Young Women and Children. With Illustrations of Scenes and Localities. In one large vol. Royal 8vo. Cloth, 10/6.

"The work is full of interesting matter for the casual reader, while the philanthropist and the philosopher will find details of the greatest import."—*City Press*.

Companion volume to the preceding.

THE CRIMINAL PRISONS OF LONDON

and Scenes of Prison Life. BY HENRY MAYHEW and JOHN BINNY. Illustrated by nearly two hundred Engravings on Wood, principally from Photographs. In one large vol. Imperial 8vo. Cloth, 10/6. .

This volume concludes Mr. Henry Mayhew's account of his researches into the crime and poverty of London. The amount of labour of one kind or other, which the whole series of his publications represents, is something almost incalculable.

⁎* This celebrated Record of Investigations into the condition of the Poor of the Metropolis, undertaken from philanthropic motives by Mr. HENRY MAYHEW, first gave the wealthier classes of England some idea of the state of Heathenism, Degradation, and Misery in which multitudes of their poorer brethren languished.

LONDON: EXETER STREET, STRAND.

FIRST SERIES—THIRTY-FIFTH EDITION.
SECOND SERIES—NINTH EDITION.

MANY THOUGHTS OF MANY MINDS:

A Treasury of Reference, consisting of Selections from the Writings of the most Celebrated Authors. FIRST AND SECOND SERIES. Compiled and Analytically Arranged

By HENRY SOUTHGATE.

Each Series is complete in itself, and sold separately.

Presentation Edition, Cloth and Gold ... 12/6 each volume.
Library Edition, Half Bound, Roxburghe... 14/-   ,,
Do.,   Morocco Antique ... ... 21/-   ,,

In Square 8vo, elegantly printed on toned paper.

"'MANY THOUGHTS,' &c., are evidently the produce of years of research."—*Examiner.*

"Many beautiful examples of thought and style are to be found among the selections."—*Leader.*

"There can be little doubt that it is destined to take a high place among books of this class."—*Notes and Queries.*

"A treasure to every reader who may be fortunate enough to possess it. Its perusal is like inhaling essences; we have the cream only of the great authors quoted. Here all are seeds or gems."—*English Journal of Education.*

"Mr. Southgate's reading will be found to extend over nearly the whole known field of literature, ancient and modern."—*Gentleman's Magazine.*

"We have no hesitation in pronouncing it one of the most important books of the season. Credit is due to the publishers for the elegance with which the work is got up, and for the extreme beauty and correctness of the typography."—*Morning Chronicle.*

"Of the numerous volumes of the kind, we do not remember having met with one in which the selection was more judicious, or the accumulation of treasures so truly wonderful."—*Morning Herald.*

"The selection of the extracts has been made with taste, judgment, and critical nicety."—*Morning Post.*

"This is a wondrous book, and contains a great many gems of thought."—*Daily News.*

"As a work of reference, it will be an acquisition to any man's library."—*Publishers' Circular.*

"This volume contains more gems of thought, refined sentiments, noble axioms, and extractable sentences, than have ever before been brought together in our language."—*The Field.*

"All that the poet has described of the beautiful in nature and art, all the axioms of experience, the collected wisdom of philosopher and sage, are garnered into one heap of useful and well-arranged instruction and amusement."—*The Era.*

"The collection will prove a mine rich and inexhaustible, to those in search of a quotation."—*Art Journal.*

"Will be found to be worth its weight in gold by literary men."—*The Builder.*

"Every page is laden with the wealth of profoundest thought, and all aglow with the loftiest inspirations of genius."—*Star.*

"The work of Mr. Southgate far outstrips all others of its kind. To the clergyman, the author, the artist, the essayist, 'Many Thoughts of Many Minds' cannot fail to render almost incalculable service."—*Edinburgh Mercury.*

"We have no hesitation whatever in describing Mr. Southgate's as the very best book of the class. There is positively nothing of the kind in the language that will bear a moment's comparison with it."—*Manchester Weekly Advertiser.*

"There is no mood in which we can take it up without deriving from it instruction, consolation, and amusement. We heartily thank Mr. Southgate for a book which we shall regard as one of our best friends and companions."—*Cambridge Chronicle.*

"This work possesses the merit of being a MAGNIFICENT GIFT-BOOK, appropriate to all times and seasons; a book calculated to be of use to the scholar, the divine, and the public man."—*Freemason's Magazine.*

"It is not so much a book as a library of quotations."—*Patriot.*

"The quotations abound in that *thought* which is the mainspring of mental exercise."—*Liverpool Courier.*

"For purposes of apposite quotation, it cannot be surpassed."—*Bristol Times.*

"It is impossible to pick out a single passage in the work which does not, upon the face of it, justify its selection by its intrinsic merit."—*Dorset Chronicle.*

"We are not surprised that a SECOND SERIES of this work should have been called for. Mr. Southgate has the catholic tastes desirable in a good Editor. Preachers and public speakers will find that it has special uses for them."—*Edinburgh Daily Review.*

"The SECOND SERIES fully sustains the deserved reputation of the FIRST."—*John Bull.*

LONDON: CHARLES GRIFFIN & CO., LIMITED, EXETER STREET, STRAND.

Published Quarterly. Price 6s.

THE JOURNAL

OF

ANATOMY AND PHYSIOLOGY:
NORMAL AND PATHOLOGICAL.

CONDUCTED BY

SIR GEORGE MURRAY HUMPHRY, M.D., LL.D., F.R S.,
PROFESSOR OF SURGERY, LATE PROFESSOR OF ANATOMY IN THE
UNIVERSITY OF CAMBRIDGE;

SIR WILLIAM TURNER, M.B., LL.D., D.C.L., F.R.S.
PROF. OF ANATOMY IN THE UNIVERSITY OF EDINBURGH;

AND

J. G. M'KENDRICK, M.D., F.R.S.,
PROF. OF THE INSTITUTES OF MEDICINE IN THE UNIVERSITY OF GLASGOW

∗ MESSRS. GRIFFIN will be glad to receive at their Office Names of SUB-SCRIBERS to the above Journal.

To be Published Early in 1893.

GRIFFIN'S
Electrical Engineers' Price-Book:

For the Use of Electrical, Civil, Marine, and Borough
Engineers, Local Authorities, Architects,
Railway Contractors, &c., &c.

EDITED BY

H. J. DOWSING, M.INST.E.E.

In Crown 8vo, Cloth.

LONDON: CHARLES GRIFFIN & CO., LTD., EXETER STREET, STRAND.

A BOOK NO FAMILY SHOULD BE WITHOUT.

New Issue of this important Work—Enlarged, in part Re-written, and thoroughly Revised to date.

TWENTY-EIGHTH EDITION. *Royal 8vo, Handsome Cloth*, 10s. 6d.

A DICTIONARY OF
DOMESTIC MEDICINE AND HOUSEHOLD SURGERY,

BY

SPENCER THOMSON, M.D., EDIN., L.R.C.S.,

REVISED, AND IN PART RE-WRITTEN, BY THE AUTHOR,

AND BY

JOHN CHARLES STEELE, M.D.,

OF GUY'S HOSPITAL.

With Appendix on the Management of the Sick-room, and many Hints for the Diet and Comfort of Invalids.

In its New Form, DR. SPENCER THOMSON'S "DICTIONARY OF DOMESTIC MEDICINE" fully sustains its reputation as the "Representative Book of the Medical Knowledge and Practice of the Day" applied to Domestic Requirements.

The most recent IMPROVEMENTS in the TREATMENT OF THE SICK—in APPLIANCES for the RELIEF OF PAIN—and in all matters connected with SANITATION, HYGIENE, and the MAINTENANCE of the GENERAL HEALTH—will be found in the New Issue in clear and full detail; the experience of the Editors in the Spheres of Private Practice and of Hospital Treatment respectively, combining to render the Dictionary perhaps the most thoroughly practical work of the kind in the English Language. Many new Engravings have been introduced—improved Diagrams of different parts of the Human Body, and Illustrations of the newest Medical, Surgical, and Sanitary Apparatus.

*** *All Directions given in such a form as to be readily and safely followed.*

FROM THE AUTHOR'S PREFATORY ADDRESS.

"Without entering upon that difficult ground which correct professional knowledge and educated judgment can alone permit to be safely trodden, there is a wide and extensive field for exertion, and for usefulness, open to the unprofessional, in the kindly offices of a *true* DOMESTIC MEDICINE, the timely help and solace of a simple HOUSEHOLD SURGERY, or, better still, in the watchful care more generally known as 'SANITARY PRECAUTION,' which tends rather to preserve health than to cure disease. 'The touch of a gentle hand' will not be less gentle because guided by knowledge, nor will the *safe* domestic remedies be less anxiously or carefully administered. Life may be saved, suffering may always be alleviated. Even to the resident in the midst of civilization, the 'KNOWLEDGE IS POWER,' to do good; to the settler and emigrant it is INVALUABLE."

"Dr. Thomson has fully succeeded in conveying to the public a vast amount of useful professional knowledge."—*Dublin Journal of Medical Science.*
"The amount of useful knowledge conveyed in this Work is surprising."—*Medical Times and Gazette.*
"WORTH ITS WEIGHT IN GOLD TO FAMILIES AND THE CLERGY."—*Oxford Herald*

LONDON: CHARLES GRIFFIN & CO., LIMITED, EXETER STREET, STRAND.

www.ingramcontent.com/pod-product-compliance
Lightning Source LLC
Chambersburg PA
CBHW020834020526
44114CB00040B/779